Theory and Applications of Viscous Fluid Flows

Springer
*Berlin
Heidelberg
New York
Hong Kong
London
Milan
Paris
Tokyo*

Physics and Astronomy ONLINE LIBRARY

http://www.springer.de/phys/

Radyadour Kh. Zeytounian

Theory and Applications of Viscous Fluid Flows

With 80 Figures

Springer

Radyadour Kh. Zeytounian
12, rue Saint-Fiacre
75002 Paris, France

Honorary professor of the Université
des Sciences et Technologies de Lille;
Villeneuve d'Ascq, France

Library of Congress Cataloging-in-Publication Data
Zeytounian, R. Kh. (Radyadour Kh.), 1928-
Theory and applications of viscous fluid flows / Radyadour K. Zeytounian.
p. cm. Includes biblographical references and index.

1. Viscous flows. 2. Fluid dynamics. I. Title.
QA929.Z49 2003   532/.0533--dc21   2002044589

ISBN 978-3-642-07889-7

This work is subject to copyright. All rights are reserved, whether the whole or part of the material is concerned, specifically the rights of translation, reprinting, reuse of illustrations, recitation, broadcasting, reproduction on microfilm or in any other way, and storage in data banks. Duplication of this publication or parts thereof is permitted only under the provisions of the German Copyright Law of September 9, 1965, in its current version, and permission for use must always be obtained from Springer-Verlag. Violations are liable for prosecution under the German Copyright Law.

Springer-Verlag Berlin Heidelberg New York
a member of Springer Science+Business Media

http://www.springer.de

© Springer-Verlag Berlin Heidelberg 2010
Printed in Germany

The use of general descriptive names, registered names, trademarks, etc. in this publication does not imply, even in the absence of a specific statement, that such names are exempt from the relevant protective laws and regulations and therefore free for general use.

Cover design: *design & production* GmbH, Heidelberg

Printed on acid-free paper

Preface

This book is the natural sequel to the study of nonviscous fluid flows presented in our recent book entitled "Theory and Applications of Nonviscous Fluid Flows" and published in 2002 by the Physics Editorial Department of Springer-Verlag (ISBN 3-540-41412-6 Springer-Verlag, Berlin, Heidelberg, New York).

The physical concept of viscosity (for so-called "real fluids") is associated with both incompressible and compressible fluids. Consequently, we have a vast field of theoretical study and applications from which any subsection could have itself provided an area for a single book. It was, however, decided to attempt a global study so that each chapter serves as an introduction to more specialized study, and the book as a whole presents a necessary broad foundation for further study in depth. Consequently, this volume contains many more pages than my preceding book devoted to nonviscous fluid flows and a large number (80) of figures.

There are *three main models* for the study of viscous fluid flows: First, the model linked with *viscous incompressible fluid flows*, the so-called (dynamic) *Navier model*, governing linearly viscous divergenceless and homogeneous fluid flows.

The second is the so-called *Navier–Stokes model* (NS) which is linked to *compressible*, linearly *viscous and isentropic* equations for a polytropic viscous gas.

The third is the so-called *Navier–Stokes–Fourier* model (NSF) that governs the motion of a *compressible*, linearly *viscous, heat-conducting* gas.

The book has been written for final year undergraduates, graduates, postgraduate research workers and for young researchers in fluid mechanics, applied mathematics, and theoretical/mathematical physics, who are interested in a rational and systematic account of theoretical aspects of viscous fluid flows phenomena and modeling them in relation to practical viscous and heat-conducting problems.

The book is divided into an Introduction and ten Chapters:

In the Introduction, the reader will find a short overview (from Chap. 2 to Chap. 10) of various subjects considered in this book.

In Chap. 1, we begin with some comments about the derivation of the basic, 'exact,' NSF model equations for real/viscous heat-conducting fluids via the stress principle of Cauchy, Fourier's law, the conservation of total energy, and Gibbs's basic postulate for a homogeneous fluid. The dimensional form of the NSF equations is presented, and the various reduced nondimensional parameters in NSF equations and boundary conditions are discussed. Chapter 2 is devoted to some features and forms of the Navier, NS, and NSF model equations, and some simple (but fundamental) examples of viscous fluid flows are considered in Chap. 3. In Chaps. 4, 5, and 6, we discuss various implications of three main singular limits, the large viscosity limit, the vanishing viscosity limit, and the incompressible limit. For each of these cases, we give a theoretical account. Chapter 7 is a miscellany of various viscous fluid problems, and in the Introduction, the reader will find a detailed account of the subjects considered. Chapter 8 is devoted to basic tools for rigorous analysis of viscous (mainly Navier) equations and also to some information about recent rigorous mathematical results related to the existence and uniqueness of solutions of viscous model problems. In Chap. 9, some aspects of (hydrodynamic) stability theory are investigated, and in Chap. 10, we give a simple phenomenological presentation of the finite-dimensional dynamical system approach and routes to chaos/turbulence via strange attractors. This is followed by a 'collection' of examples of viscous fluid flow problems.

Chapters 8 to 10 cover a small part of the modern mathematical theory of viscous incompressible/compressible fluid flows.[1] Nevertheless, it was not our objective in these three chapters to give a detailed account of this mathematical theory, but rather from a fluid dynamicist's point of view, to provide an overview of recent results. In particular, Chap. 10 contains mostly arguments about current research but is essentially discursive.

The references cited (more than thousand) are listed at the end of the book before a detailed Index.

The choice of the 10 chapters mentioned and their order are, at least from our point of view, quite natural. The presentation of the material, the relative weight of the various arguments, and the general style reflect the tastes and knowledge of the author.

Naturally, the present book constitutes (with the companion book devoted to Nonviscous Fluid Flows) an advanced topic, rather than a classical course, on fluid dynamics. As for the subjects treated in these two volumes, I have been highly selective in my choice of topics. In many cases, the choice of subjects is based on my own interest and judgment. To that extent, the present text is a personal expression of my view ("à la Zeytounian"!) of the theoretical viscous fluid dynamics.

The Contents enumerates the subjects in each chapter. This content of the book is ambitious but I hope that the reader will consider, after reading,

---

[1] In the two volumes by Lions (*Mathematical Topics in Fluid Mechanics*, Volume 1; *Incompressible Models, and* Volume 2; *Compressible Models*, Oxford Lecture Series in Mathematics and its Applications, Clarendon Press, 1996 and 1998), the reader can find a rigorous mathematical theory for Navier and NS equations.

that I have (even partially) attained my objective – a modern presentation of some key problems in viscous fluid flows in terms of theoretical analysis, modeling and some applications.

Several colleagues have made many useful suggestions and criticisms during the preparation of this book for which I am grateful. However I accept final responsibility for remaining errors and omissions.

Finally, I thank: Dr. Christian Caron, Physics Editor, and the members of the Physical Editorial Department, where camera-ready manuscript in LaTeX was produced and reread by a native speaker.

Paris/Yport
June 2003
R.Kh. Zeytounian

Contents

Introduction 1

1 **Navier–Stokes–Fourier Exact Model** 11
  1.1 The Transport Theorem................................. 11
  1.2 The Equation of Continuity 12
  1.3 The Cauchy Equation of Motion 12
  1.4 The Constitutive Equations of a Viscous Fluid 13
      1.4.1 Stokes's Four Postulates: Stokesian Fluid 14
      1.4.2 Classical Linear Viscosity Theory: Newtonian Fluid .. 15
  1.5 The Energy Equation and Fourier's Law 17
      1.5.1 The Total Energy Equation 17
      1.5.2 Heat Conduction and Fourier's Law 18
  1.6 The Navier–Stokes–Fourier Equations..................... 19
      1.6.1 The NSF Equation for an Ideal Gas
          when C_v and C_p are Constants 20
      1.6.2 Dimensionless NSF Equations 21
      1.6.3 Reduced Dimensionless Parameters 22
  1.7 Conditions for Unsteady-State NSF Equations 25
      1.7.1 The Problem of Initial Conditions 26
      1.7.2 Boundary Conditions............................. 28

2 **Some Features and Various Forms of NSF Equations** 35
  2.1 Isentropicity, Polytropic Gas, Barotropic Motion,
     and Incompressibility 35
      2.1.1 NS Equations 35
      2.1.2 Navier System................................... 36
      2.1.3 Navier System with Time-Dependent Density 37
      2.1.4 Fourier Equation 38
  2.2 Some Interesting Issues in Navier Incompressible Fluid Flow . 39
      2.2.1 The Pressure Poisson Equation 41
      2.2.2 $\Psi_N - \omega_N$ and $u_N - \omega_N$ Formulations 42
      2.2.3 The Omnipotence of the Incompressibility Constraint 43
      2.2.4 A First Statement of a Well-Posed Initial
          Boundary-Value Problem (IBVP)
          for Navier Equations 46

X   Contents

        2.2.5   Cauchy Formula for Vorticity 47
        2.2.6   The Navier Equations as an Evolutionary Equation
               for Perturbations 48
  2.3 From NSF to Hyposonic
       and Oberbeck–Boussinesq (OB) Equations 50
        2.3.1   Model Equations for Hyposonic Fluid Flows 50
        2.3.2   The Oberbeck–Boussinesq Model Equations 52

**3 Some Simple Examples
of Navier, NS and NSF Viscous Fluid Flows** 57
  3.1 Plane Poiseuille Flow and the Orr–Sommerfeld Equation 57
        3.1.1   The Orr–Sommerfeld Equation 58
        3.1.2   A Double-Scale Technique
               for Resolving the Orr–Sommerfeld Equation 60
  3.2 Steady Flow Through an Arbitrary Cylinder under Pressure . 61
        3.2.1   The Case of a Circular Cylinder 62
        3.2.2   The Case of an Annular Region
               Between Concentric Cylinders 63
        3.2.3   The Case of a Cylinder of Arbitrary Section 63
  3.3 Steady-State Couette Flow Between Cylinders
       in Relative Motion 64
        3.3.1   The Classic Taylor Problem 65
        3.3.2   The Taylor Number 66
  3.4 The Bénard Linear Problem and Thermal Instability 68
  3.5 The Bénard Linear Problem
       with a Free Surface and the Marangoni Effect 71
        3.5.1   The Case when the Neutral State is Stationary 73
        3.5.2   Free-Surface Deformation 75
  3.6 Flow due to a Rotating Disc 75
        3.6.1   Small Values of ζ 77
        3.6.2   Large Values of ζ 77
        3.6.3   Joining (Matching) 78
  3.7 One-Dimensional Unsteady-State NSF Equations
       and the Rayleigh Problem 78
        3.7.1   Small M^2 Solution – Close to the Flat Plate
               but far from the Initial Time 81
        3.7.2   Small M^2 Solution – Far from a Flat Plate 83
        3.7.3   Small M^2 Solution – Close to the Initial Time 86
  3.8 Complementary Remarks 87

4 The Limit of Very Large Reynolds Numbers 89
  4.1 Introduction 89
  4.2 Classical Hierarchical Boundary-Layer Concept
       and Regular Coupling 93

		4.2.1	A 2-D Steady-State Navier Equation for the Stream Function 93

        4.2.1 A 2-D Steady-State Navier Equation
for the Stream Function 93
        4.2.2 A Local Form of the 2-D Steady-State
Navier Equation for the Stream Function 94
        4.2.3 A Large Reynolds Number and "Principal"
and "Local" Approximations 94
        4.2.4 Matching 96
        4.2.5 The Prandtl–Blasius and Blasius BL Problems 97
   4.3 Asymptotic Structure
of Unsteady-State NSF Equations at $Re \gg 1$ 103
        4.3.1 Four Significant Degeneracies of NSF Equations 105
        4.3.2 Formulation of a Simplified Initial Boundary-Value
Problem for the NSF Full Unsteady-State Equations . 108
        4.3.3 Various Facets of Large Reynolds Number
Unsteady-State Flow 109
        4.3.4 The Two Adjustment Problems.................... 114
   4.4 The Triple-Deck Concept and Singular Interactive Coupling . 118
        4.4.1 The Triple-Deck Theory
in 2-D Steady-State Navier Flow 120
   4.5 Complementary Remarks................................. 126
        4.5.1 Three-Dimensional Boundary-Layer Equations....... 130
        4.5.2 Unsteady-State Incompressible
Boundary-Layer Formulation 137
        4.5.3 The Inviscid Limit: Some Mathematical Results 140
        4.5.4 Rigorous Results for the Boundary-Layer Theory 144

5 The Limit of Very Low Reynolds Numbers 145
   5.1 Large Viscosity Limits and Stokes and Oseen Equations 145
        5.1.1 Steady-State Stokes Equation 145
        5.1.2 Unsteady-State Oseen Equation 146
        5.1.3 Unsteady-State Stokes
and Steady-State Oseen Equations 147
        5.1.4 Unsteady-State Matched Stokes–Oseen Solution
at $Re \ll 1$ for the Flow Past a Sphere 147
   5.2 Low Reynolds Number Flow
due to an Impulsively Started Circular Cylinder............. 149
        5.2.1 Formulation of the Steady-State Problem 150
        5.2.2 The Unsteady-State Problem 152
   5.3 Compressible Flow 153
        5.3.1 The Stokes Limiting Case
and Steady-State Compressible Stokes Equations 154
        5.3.2 The Oseen Limiting Case
and Steady-State Compressible Oseen Equations..... 155
   5.4 Film Flow on a Rotating Disc:
Asymptotic Analysis for Small Re 158

XII   Contents

          5.4.1   Solution for Small Re $\ll$ 1: Long-Time Scale Analysis 159
          5.4.2   Solution for Small Re $\ll$ 1: Short-Time Scale Analysis 160
   5.5   Some Rigorous Mathematical Results 164

6   Incompressible Limit: Low Mach Number Asymptotics ... 165
   6.1   Introduction 165
   6.2   Navier–Fourier Asymptotic Model 168
          6.2.1   The Initialization Problem and Equations
                  of Acoustics 171
          6.2.2   The Fourier Model 175
          6.2.3   Influence of Weak Compressibility:
                  Second-Order Equations for $\boldsymbol{u}'$ and π' 178
          6.2.4   Concluding Remarks 179
   6.3   Compressible Low Mach Number Models 181
          6.3.1   Hyposonic Model for Flow in a Bounded Cavity 181
          6.3.2   Large Channel Aspect Ratio, Low Mach Number,
                  Compressible Flow 183
   6.4   Viscous Nonadiabatic Boussinesq Equations 184
          6.4.1   The Basic State 184
          6.4.2   Asymptotic Derivation
                  of Viscous, Nonadiabatic Boussinesq Equations 186
   6.5   Some Comments 187

7   Some Viscous Fluid Motions and Problems 191
   7.1   Oscillatory Viscous Incompressible Flow................... 191
          7.1.1   Acoustic Streaming Effect 191
          7.1.2   Study of the Steady-State Streaming Phenomenon ... 196
          7.1.3   The Role of Parameters $\alpha\,\mathrm{Re} = \mathrm{Re}_S$ and $\mathrm{Re}/\alpha = \beta^2$. 198
          7.1.4   Other Examples of Viscous Oscillatory Flow 202
   7.2   Unsteady-State Viscous, Incompressible Flow
       past a Rotating and Translating Cylinder 203
          7.2.1   Formulation of the Governing Problem 203
          7.2.2   Method of Solution 204
          7.2.3   Determination of the Initial Flow 205
          7.2.4   Results of Calculations and Comparison with the
                  Visualization of Coutanceau and Ménard (1985) 206
          7.2.5   A Short Comment 207
   7.3   Ekman and Stewartson Layers 208
          7.3.1   General Equations and Boundary Conditions 210
          7.3.2   The Ekman Layer 211
          7.3.3   The Stewartson Layer 211
          7.3.4   The Inner, Outer, and Upper Regions 213
          7.3.5   Comments 214
   7.4   Low Reynolds Number Flows: Further Investigations........ 215

		7.4.1	Unsteady-State Adjustment to the Stokes Model in a Bounded Deformable Cavity $\Omega(t)$ 215

- 7.4.1 Unsteady-State Adjustment to the Stokes Model in a Bounded Deformable Cavity $\Omega(t)$ 215
- 7.4.2 On the Wake in Low Reynolds Number Flow 218
- 7.4.3 Oscillatory Disturbances as Admissible Solutions and their Possible Relationship to the Von Karman Sheet Phenomenon 220
- 7.4.4 Some References 223
- 7.5 The Bénard–Marangoni Problem: An Alternative 224
  - 7.5.1 Dimensionless Dominant Equations 226
  - 7.5.2 Dimensionless Dominant Boundary Conditions 227
  - 7.5.3 The Rayleigh–Bénard (RB) Thermal Shallow Convection Problem......................... 229
  - 7.5.4 The Bénard–Marangoni (BM) Problem 231
- 7.6 Some Aspects of Nonadiabatic Viscous Atmospheric Flow ... 233
  - 7.6.1 The L-SSHV Equations........................... 233
  - 7.6.2 The Tangent HV (THV) Equations 238
  - 7.6.3 The Quasi-Geostrophic Model 240
- 7.7 Miscellaneous Topics..................................... 246
  - 7.7.1 The Entrainment of a Viscous Fluid in a Two-Dimensional Cavity...................... 246
  - 7.7.2 Unsteady-State Boundary Layers 253
  - 7.7.3 Various Topics Related to Boundary-Layer Equations 258
  - 7.7.4 More on the Triple-Deck Theory 260
  - 7.7.5 Some Problems Related to Navier Equations for an Incompressible Viscous Fluid 266
  - 7.7.6 Low and Large Prandtl Number Flow 272
  - 7.7.7 A final comment 275

8 Some Aspects of a Mathematically Rigorous Theory 277

- 8.1 Classical, Weak, and Strong Solutions of the Navier Equations 278
- 8.2 Galerkin Approximations and Weak Solutions of the Navier Equations 283
  - 8.2.1 Some Comments and Bibliographical Notes 287
- 8.3 Rigorous Mathematical Results for Navier Incompressible and Viscous Fluid Flows 289
  - 8.3.1 Navier Equations in an Unbounded Domain 295
  - 8.3.2 Some Recent Rigorous Results..................... 298
- 8.4 Rigorous Mathematical Results for Compressible and Viscous Fluid Flows 300
  - 8.4.1 The Incompressible Limit 305
- 8.5 Some Concluding Remarks 307

9 Linear and Nonlinear Stability of Fluid Motion 311
9.1 Some Aspects of the Theory of the Stability of Fluid Motion . 311
9.1.1 Linear, Weakly Nonlinear, Nonlinear, and Hydrodynamic Stability 312
9.1.2 Reynolds–Orr, Energy, Sufficient Stability Criterion .. 316
9.1.3 An Evolution Equation for Studying the Stability of a Basic Solution of Fluid Flow..................... 317
9.2 Fundamental Ideas on the Theory of the Stability of Fluid Motion 319
9.2.1 Linear Case 320
9.2.2 Nonlinear Case 322
9.3 The Guiraud–Zeytounian Asymptotic Approach to Nonlinear Hydrodynamic Stability 324
9.3.1 Linear Theory 326
9.3.2 Nonlinear Theory – Confined Perturbations. Landau and Stuart Equations 328
9.3.3 Nonlinear Theory – Unconfined Perturbations. General Setting................................. 331
9.3.4 Nonlinear Theory – Unconfined Perturbations. Tollmien–Schlichting Waves 332
9.3.5 Nonlinear Theory – Unconfined Perturbations. Rayleigh–Bénard Convection 335
9.4 Some Facets of the RB and BM Problem 337
9.4.1 Rayleigh–Bénard Convective Instability............ 337
9.4.2 Bénard–Marangoni (BM) Thermocapillary Instability Problem for a Thin Layer (Film) with a Deformable Free Surface.................... 356
9.5 Couette–Taylor Viscous Flow Between Two Rotating Cylinders.......................... 370
9.5.1 A Short Survey 370
9.5.2 Bifurcations.................................... 376
9.6 Concluding Comments and Remarks 380

10 A Finite-Dimensional Dynamical System Approach to Turbulence 387
10.1 A Phenomenological Approach to Turbulence 387
10.2 Bifurcations in Dissipative Dynamical Systems 392
10.2.1 Normal Form of the Pitchfork Bifurcation........... 395
10.2.2 Normal Form of the Hopf Bifurcation 396
10.2.3 Bifurcation from a Periodic Orbit to an Invariant Torus............................ 398
10.3 Transition to Turbulence: Scenarios, Routes to Chaos 398
10.3.1 The Landau–Hopf "Inadequate" Scenario 399
10.3.2 The Ruelle–Takens–Newhouse Scenario 399
10.3.3 The Feigenbaum Scenario 403

>   10.3.4 The Pomeau–Manneville Scenario.................. 406
>   10.3.5 Complementary Remarks 408
>   10.4 Strange Attractors for Various Fluid Flows 414
>   10.4.1 Viscous Isochoric Wave Motions 414
>   10.4.2 The Bénard–Marangoni Problem for a Free-Falling
>   Vertical Film: The Case of Re = $O(1)$
>   and the KS Equation............................ 417
>   10.4.3 The Bénard–Marangoni Problem for a Free-Falling
>   Vertical Film: The Case of Re/$\varepsilon = O(1)$
>   and the KS–KdV Equation 424
>   10.4.4 Viscous and Thermal Effects
>   in a Simple Stratified Fluid Model 427
>   10.4.5 Obukhov Discrete Cascade Systems
>   for Developed Turbulence........................ 435
>   10.4.6 Unpredictability in Viscous Fluid Flow
>   Between a Stationary and a Rotating Disk 439
>   10.5 Some Comments and References 444

References.. 449

Index... 485

Introduction

As a direct consequence of the *Cauchy stress principle*, it follows that [see, for instance, Serrin (1959a; Section B)]

$$\rho \frac{\mathrm{d}\boldsymbol{u}}{\mathrm{d}t} = \rho \boldsymbol{f} + \nabla \cdot \mathbf{T}, \tag{1}$$

which is the equation of motion for the velocity vector $\boldsymbol{u}(t,\mathbf{x})$.

In (1), $\mathrm{d}/\mathrm{d}t = \partial/\partial t + \boldsymbol{u} \cdot \nabla$; the Cartesian components of the nabla (gradient) operator ∇ are $\partial/\partial x_i$, with $i = 1, 2, 3$; $\boldsymbol{f}$ is the body force per unit mass; $\rho(t,\boldsymbol{x})$ is the density; t is the time variable; and $\boldsymbol{x} = (x_i)$ is the position vector. This is the simple and elegant equation of motion discovered by Cauchy (1828), and it is valid for any fluid and for any continuous medium, regardless of the form which the stress tensor $\mathbf{T} = (T^{ij})$ may take. More precisely, for "real" viscous and compressible fluids, the classical theory of (isotropic) viscous compressible fluids is based on the *Cauchy–Poisson* (linear) law, valid for a *Newtonian* fluid, which is a particular class of Stokesian fluids, when linear dependence of the components T^{ij} of the stress tensor on the components d_{ij} of the rate-of-deformation tensor $\mathbf{D}(\boldsymbol{u})$ is assumed. Then,

$$T^{ij} = -p\delta_{ij} + 2\mu d_{ij} + \lambda d_{kk}\delta_{ij}. \tag{2}$$

In (2), p is the (static) pressure, μ and λ are shear/dynamic viscosity and the second coefficient of viscosity; δ_{ij} is the so-called Kronecker symbol with $\delta_{kk} = 1$ and $\delta_{ij} = 0$ if $i \neq j$. In the Chap. 1, the reader can find the main lines of proof of the constitutive equation (2) according to Stokes's *concept of fluidity* by using matrix algebra.

A fully general expression of (2) is given by Poisson (1831), because in Cauchy (1828) the term $-p\delta_{ij}$ in (2) is absent. On the other hand, Saint-Venant (1843) proposed (2) in the special case when dynamic viscosity μ is constant ($=\mu_0$) and for

$$\mu_\mathrm{v} \equiv \lambda + \frac{2}{3}\mu_0 = 0 \Rightarrow \lambda = -\frac{2}{3}\mu_0. \tag{3a}$$

Relation (3a) was also proposed by Stokes (1845) and is called the "Stokes relation." For a discussion the long controversy over the Stokes relation, $3\mu_\mathrm{v} = 0$, in the classical theory of viscous fluids, see the Truesdell's book (1966, p. 113). The viscosity coefficients (shear/dynamic and bulk) and the thermal conductivity coefficient k [see Fourier's law (14) below] are known functions subject to the thermodynamic restrictions [Clausius–Duhem inequalities and concerning the Duhem approach see his paper of 1901/1902 and his book 1911],

$$\mu \geq 0, k \geq 0 \quad \text{and} \quad \mu_v \geq 0. \tag{3b}$$

Thermodynamics based on a work inequality is very well discussed in Man (1989); see also the book edited by Serrin (1986). The dynamic equation that results from (1) with (2) is the classical Navier–Stokes (NS) equation of motion for the Cartesian component u_i of velocity $\boldsymbol{u}$. We derive the following NS equation:

$$\rho \left[\frac{\partial u_i}{\partial t} + u_j \frac{\partial u_i}{\partial x_j} \right] + \frac{\partial p}{\partial x_i}$$
$$= \frac{\partial}{\partial x_j} \left[\mu \left(\frac{\partial u_i}{\partial x_j} + \frac{\partial u_j}{\partial x_i} \right) \right] + \frac{\partial}{\partial x_i} \left[\lambda \left(\frac{\partial u_k}{\partial x_k} \right) \right], \tag{4}$$

$(i = 1, 2 \text{ and } 3)$,

where p is the static pressure; for the Cartesian components d_{ij} of $\mathbf{D}(\boldsymbol{u})$, we have the following relation

$$2d_{ij} = \frac{\partial u_i}{\partial x_j} + \frac{\partial u_j}{\partial x_i}. \tag{5}$$

In Saint-Venant (1843) [and Stokes (1845)!], the resulting dynamic equation for the components of velocity u_i, in place of (4), is

$$\left[\frac{\partial u_i}{\partial t} + u_j \frac{\partial u_i}{\partial x_j} \right] + \frac{\partial p}{\partial x_i} = \mu_0 \left[\Delta u_i - \frac{1}{3} \frac{\partial}{\partial x_i} \left(\frac{d \log \rho}{dt} \right) \right], \tag{6}$$

if we take into account the equation of continuity for an homogeneous but compressible fluid,

$$\frac{\partial u_k}{\partial x_k} = -\frac{d \log \rho}{dt}, \tag{7}$$

and also the Stokes relation (3a). In (6), $\Delta \equiv \nabla^2$ is the three-dimensional Laplace operator.

When we assume that viscosities μ and λ depend on temperature T (and eventually on density ρ), the NS equation (4) with the continuity equation (7) forms a system of only two equations for the functions $\boldsymbol{u}$, p, ρ and T; obviously, it is necessary to associate an equation for T and also an equation of state for p if we want obtain a closed system of equations for our four unknown functions! Usually, in real (viscous) Newtonian fluids, the pressure p is a function of two variables, ρ and T (a so-called trivariate fluid in baroclinic motion), and it is necessary to write an equation of state,

$$p = P(\rho, T), \tag{8}$$

because it is well known from thermodynamics that for each type of matter, a specific relation exists among these three thermodynamic variables.

For real viscous compressible and heat-conducting fluid flows, mechanical energy is converted into heat by viscosity, and heat of compression is diffused by heat conduction. Here, we consider only a homogeneous fluid when the local equation of state is [according to Gibbs's basic postulate (1875/1961); see Truesdell (1952a)]

$$E = E(\rho, S), \tag{9}$$

where E is the specific internal energy and S is the specific entropy. In this case, temperature T and thermodynamic pressure p^* are defined by the two relations,

$$T = \frac{\partial E}{\partial S} \quad \text{and} \quad p^* = -\frac{\partial E}{\partial (1/\rho)}. \tag{10a}$$

Now, the (static) pressure p has appeared already in the NS equation (4) and in the equation of state (8). A characteristic of gas (compressible) dynamics is the further postulate that thermodynamic pressure is equal to static pressure:

$$p^* = p. \tag{10b}$$

For an incompressible fluid, this postulate is automatically satisfied. Also $p^* = p$ if the Stokes relation (3a) is true. Obviously, for a nonviscous (inviscid) fluid, considered recently in Zeytounian (2002a), we have always $p^* \equiv p$.

On the other hand, for any compressible fluid, by differentiating (9) along any curve on the energy surface (characterizing the fluid), we obtain, with $p^* = p$ and (10a),

$$\frac{dE}{dt} = T\frac{dS}{dt} - p\frac{d(1/\rho)}{dt}. \tag{11}$$

But, for any homogeneous medium in motion, conservation of energy is expressed by the total energy equation of Neumann (1894),

$$\rho \frac{dE}{dt} = \mathbf{T} : \mathbf{D} - \nabla \cdot \mathbf{q}, \tag{12a}$$

or, according to (2) and (5),

$$\rho \frac{dE}{dt} + p\left(\frac{\partial u_k}{\partial x_k}\right) = -\frac{\partial q_i}{\partial x_i} + (2\mu d_{ij} + \lambda d_{kk}\delta_{ij})d_{ij}, \tag{12b}$$

where q_i are the Cartesian components of the heat flux vector $\mathbf{q}$.

We observe that, for the special case of a nonviscous incompressible fluid, the energy equation was given by Fourier (1833) and for small motions of a viscous so-called ideal gas, when $p = R\rho T$, with R the constitutive constant of the viscous ideal gas, by Kirchhoff (1868). We observe also that, for a medium undergoing deformation, (11) and (12b) express different and independent assumptions:

the former, the existence of an energy surface characterizing the fluid and the latter, that mechanical and thermal energy are interconvertible.

The terms "first law", "second law," etc., of thermodynamics are rather (!) misleading; for a history of the origin of thermodynamics see, for instance, the book by Poincaré (1892) and Truesdell (1980). Finally, from (11) it follows that, in place of (12b),

$$\rho T \frac{dS}{dt} = -\frac{\partial q_i}{\partial x_i} + (2\mu d_{ij} + \lambda d_{kk}\delta_{ij})d_{ij} , \tag{13}$$

if we use the continuity equation (7). In particular, if heat flux arises solely from thermal conduction, then Fourier's law gives

$$q_i = -k\frac{\partial T}{\partial x_i} , \tag{14}$$

and we observe that k in (14) is the thermal conductivity (or heat conduction) coefficient.

The three equations (4), (7), and (12b), with (14), and the two equations of state,

$$p = R\rho T \quad \text{and} \quad E = C_v T , \tag{15a}$$

valid for a *perfect* (or ideal, but viscous!) *gas*, with constant specific heats C_v and

$$C_p = R + C_v , \tag{15b}$$

considered a compressible, viscous, heat-conducting Newtonian fluid, constitute the so-called Navier–Stokes–Fourier (NSF) system of equations. In this case, it is assumed that the three dissipative coefficients are functions of ρ and T, $\lambda = \lambda(\rho, T)$ and $\mu = \mu(\rho, T)$ in (4) and (12b), and $k = k(\rho, T)$ in (14).

The compact vectorial dimensional form of these NSF equations for an ideal gas is

$$\frac{d\rho}{dt} + \nabla \cdot \boldsymbol{u} = 0 , \tag{16a}$$

$$\rho\frac{d\boldsymbol{u}}{dt} + \nabla p + \rho g\boldsymbol{k} = \nabla \cdot \Pi , \tag{16b}$$

$$\rho C_v \frac{dT}{dt} + p\nabla \cdot \boldsymbol{u} = \nabla \cdot [k\nabla T] + \Phi , \tag{16c}$$

with $p = R\rho T$, where Φ is the dissipation function and gravity $\boldsymbol{g} = -g\boldsymbol{k}$ acts in the negative x_3 direction. In (16b),

$$\Pi = \lambda(\nabla \cdot \boldsymbol{u})\mathbf{I} + 2\mu \mathbf{D}(\boldsymbol{u}) , \tag{17a}$$

where $\mathbf{I} = (\delta_{ij})$, and in (16c),

$$\Phi = 2\,\mathrm{Trace}\,[(\mathbf{D}(\boldsymbol{u}))^2] + \lambda(\nabla \cdot \boldsymbol{u})^2\,, \tag{17b}$$

where

$$\mathrm{Trace}\,[(\mathbf{D}(\boldsymbol{u}))^2] \equiv \mathbf{D}(\boldsymbol{u}) : \mathbf{D}(\boldsymbol{u}) = \frac{1}{4}\left[\frac{\partial u_i}{\partial x_j} + \frac{\partial u_j}{\partial x_i}\right]^2\,. \tag{17c}$$

The NSF equations (16a–c) with (17a–c) are basic (exact!) fluid dynamics equations for the functions $\boldsymbol{u}$, ρ, $p(= R\rho T)$ and T, governing the unsteady motion of a viscous, compressible and heat conducting Newtonian fluid.

In Batchelor (2000) book, which is the last edition of his "Introduction to Fluid Dynamics", the reader can find a traditional exposition of various aspects of fluid flow phenomena. In Chap. 1, we give some complementary information concerning the derivation of the Cauchy equation of motion (1) and also how we obtain the constitutive equation (2) via matrix algebra from the four Stokes postulates (plus linearity condition: components of $\mathbf{T}$ are linear in the components of $\mathbf{D}(\boldsymbol{u})$!). In Chap. 1, the reader can also find a discussion of the initial and boundary conditions for NSF equations (16a–c) to have a *well-posed problem*. With regard to the time-like variable t, it is necessary to consider an initial value or Cauchy problem, where the solution is given in the subspace $t = 0$ as

$$U(0, \boldsymbol{x}) \equiv (\boldsymbol{u}, \rho, T)_{t=0} = U^0(\boldsymbol{x})\,,$$

and is to be determined at subsequent values $t > 0$. But, if the subspace $t = 0$ is bounded by some surface $S(\boldsymbol{x})$, then additional conditions have to be imposed along that surface at all values of $t > 0$, and this defines an initial boundary-value problem. We observe that the pure initial value or Cauchy problem is not well posed for nonhyperbolic (viscous!) problems, and only initial boundary-value problems will be well posed in this case for the NSF system which is an incompletely parabolic system (more precisely, boundary conditions are imposed at all times for $\boldsymbol{u}$ and T on the boundary of the flow domain). The dimensionless form of the NSF equations and the significance of reduced (nondimensional) parameters in the NSF equations and also in conditions characterizing various fluid flow phenomena are also considered in Chap. 1.

In Chap. 2, we discuss some features and derive particular forms of NSF equations (16a–c) (but written in dimensionless form [see (1.30)–(1.33)]). First, the compressible Navier–Stokes (NS) equations for barotropic fluid motion are obtained, when we assume that Stokes relation (3b) is satisfied with a shear/dynamic viscosity coefficient μ as a function of density ρ only and when a specifying relation,

$$p = P(\rho)\,, \tag{18a}$$

between pressure p and density ρ only [in place of (8)], which is not an equation of state, is assumed. In particular, for a polytropic gas (when the specific entropy S is constant),

$$P(\rho) = \text{const}\, \rho^\gamma\,, \quad \gamma = \frac{C_p}{C_v}\,, \tag{18b}$$

and in this case we obtain the *isentropic NS equations*.

Then, the Navier equations for an incompressible and viscous *homogeneous* fluid are obtained, when both the density ρ and the shear/dynamic viscosity coefficient μ are assumed constant. The dimensional form of Navier equations is (with constant density ρ_0 and constant dynamic viscosity μ_0)

$$\rho_0 \frac{d_N \boldsymbol{u}_N}{dt} + \nabla p_N = \mu_0 \Delta \boldsymbol{u}_N\,, \tag{19a}$$

$$\nabla \cdot \boldsymbol{u}_N = 0\,, \tag{19b}$$

with $d_N/dt = \partial/\partial t + \boldsymbol{u}_N \cdot \nabla$, if we take into account the equation of continuity for an homogeneous incompressible (divergenceless) Navier fluid velocity vector $\boldsymbol{u}_N$, where p_N is a pseudo-pressure; in (19a), ∇p_N is the force term that acts on the particles of a homogeneous incompressible Navier fluid and allows them to move as freely as possible, but compatibly with incompressibility constraint (19b). In Chap. 6, we give a consistent asymptotic derivation of the Navier model from the NSF exact model.

For a weakly expansible (pure) liquid, the Oberbeck–Boussinesq (OB) shallow convection equations for the dimensionless velocity components v_{i0B}, the perturbations of temperature θ_{0B}, and pressure π_{0B} are also derived from the full NSF equations; we obtain the following coupled dimensionless system of three equations:

$$\frac{\partial v_{i0B}}{\partial X_i} = 0\,, \tag{20a}$$

$$\frac{d_{0B} v_{i0B}}{d\tau} + \frac{\partial \pi_{0B}}{\partial X_i} - \text{Gr}\, \theta_{0B} \delta_{i3} = \Delta v_{i0B}\,, \tag{20b}$$

$$\frac{d_{0B} \theta_{0B}}{d\tau} = \frac{1}{\text{Pr}} \Delta \theta_{0B}\,, \tag{20c}$$

with $d_{0B}/d\tau = \partial/\partial\tau + v_{i0B} \partial/\partial X_i$, where τ and X_i are time-space reduced dimensionless variables; Gr and Pr are the Grashof and Prandtl numbers, respectively, defined in Sect. 1.6.3 of Chap. 1.

Some simple examples of viscous fluid flows are considered in Chap. 3. For instance, we investigate plane Poiseuille flow and Couette flow between cylinders in relative motion, Bénard thermal convection linear problems with a nondeformable (Rayleigh–Bénard instability related to the buoyancy effect) and deformable (when we take into account the Marangon effect related to the temperature-dependent surface tension) free surface, flow due to a rotating disc, and Rayleigh's problem (in the framework of unsteady-state one-dimensional NSF equations) for a compressible heat-conducting fluid.

In NSF equations written in dimensionless form (see Chap. 1), there are two fundamental, reduced dimensionless parameters, the Reynolds number (Re, which characterizes the viscous effects mainly near the walls) and the Mach number (M, which characterizes compressibility effects), and correspondingly in Chaps. 4 to 6 of this book, the reader can find three approximate theories related, on one hand, to high (Re $\gg$ 1) and low (Re $\ll$ 1) Reynolds numbers, and on the other hand, to the low ($M \ll 1$) Mach number. These three "limit" theories, the large and vanishing viscosity limit and the incompressible limit, are strongly singular. In particular, it is well known that the *high Reynolds number* (Chap. 4) asymptotics of NSF equations is related (via matching) to the *Prandtl boundary-layer model* equations valid near the wall (as a consequence of the no-slip condition on the solid wall). But it appears that in unsteady-state NSF equations, when Re tends to infinity, the associated (with the Euler nonviscous "outer" equations) unsteady-state boundary-layer (BL, inner) equations are *singular close to the initial time* (because the pressure in the BL is independent of the normal coordinate, and consequently the time derivative of the normal component of the velocity is absent in BL unsteady-state equations). Close to the initial time and near the solid wall, it is necessary to consider a thinner viscous layer (than the BL layer), where the *Rayleigh equations* (governing the compressible Rayleigh problem considered in Sect. 3.7) are valid. We observe also that for high Reynolds numbers, when the hierarchical Prandtl BL model breaks down, it is necessary to consider a triple-deck interactive structure, "à la Stewartson" (1969), see, for instance, Stewartson (1974). In this case, the Prandtl classical BL is divided into two parts usually referred to as the lower and main decks, respectively. The third deck of the triplet (the so-called "upper deck") is the region of external nonviscous (Euler) flow which is most significantly affected by rapid changes in the lower deck (where the classical BL equations hold to leading order, but the determination of the pressure is more subtle!) and lies just above the other two decks. Finally, if we want to consider the shock wave problem then in this case (when Re $\gg$ 1, for a vanishing viscosity), it is necessary to introduce a thin Taylor shock layer that describes the shock (viscous) internal structure.

In Chap. 4, the reader can also find a discussion of the inviscid limit with various references relative to mathematical results.

The significant limit equations for low Reynolds number (Chap. 5) asymptotics in steady-state *external* aerodynamics are the *Stokes* (valid close to the wall) and *Oseen* (valid far from the wall) equations. But again, close to the initial time in the unsteady-state, it is necessary to consider an inner time *transient* (unsteady-state adjustment) *region*. In unsteady-state external aerodynamics, it is shown that a *late-wake singular region* also exists which extends all the way downstream to infinity and whose width is comparable to the diameter of the obstacle!

In low Mach number (Chap. 6) asymptotics, we recover (when *time is fixed*) the Navier equations for the velocity vector and perturbation of pressure and also a *Fourier* equation for the perturbation of temperature. Close to initial time and in far field, in place of Navier (outer) equations, are the classical linear *acoustics* equations [see, for instance, Zeytounian (2002a), Chap. 6]. We observe that in low Mach number asymptotics, it is also possible to derive, on the one hand, a hyposonic model for flow in a bounded cavity when the gas is strongly heated, and, on the other hand, a relatively uncommon *compressible* model for a large channel aspect ratio, low-Mach number flow! Finally, in Chap. 6, we derive the viscous nonadiabatic Boussinesq equations.

Chapter 7 is a *miscellany* of various viscous fluid problems and motions. First, some aspects of *oscillatory viscous* incompressible flow are considered and the *steady-state streaming* phenomenon is studied. Then *unsteady-state viscous*, incompressible flow *past a rotating and translating circular cylinder* is considered. A short account of *Ekman* and *Stewartson layers* is given. Some further investigations of *low Reynolds number flow* are presented. An "alternative" is formulated for the Bénard thermal convection problem in connection with the role of the buoyancy and deformable free-surface effects. Some aspects of *nonadiabatic viscous atmospheric flow* are presented. Finally, *miscellaneous topics* are briefly considered: *entrainment of a viscous fluid in a 2-D cavity; unsteady-state boundary-layers*; various topics related to the *boundary-layers equations*; more concerning the *triple-deck theory* and *viscous-inviscid interaction* problem; some problems related to the Navier equations (topology, interfacial flows, Blasius problem with a free-surface, explicit and exact solutions, steady-state flow past an aerofoil with a cavity in its upper surface, rotating Hagen–Poiseuille flow); low and large Prandtl number flows.

In Chap. 8, the reader can find basic tools for *rigorous analysis* of the viscous (mainly Navier) equations. In particular, for definiteness and simplicity, we consider the *Galerkin approximation* and *weak* solutions for the Navier equations, with periodic boundary conditions on $\Omega = [0, L]^d$, in the presence of a body force which is constant in time. Unfortunately, the weak solutions of the Navier equations have not been shown to be unique in any sense! Their construction as the limit of a subsequence of the Galerkin approximation leaves open the possibility that there is more than one *distinct limit*, even for the same sequence of approximations. *Nonunique* evolution would violate the basic tenets of classical Newtonian determinism and would render the Navier equations *worthless* as a predictive model. Naturally, if the weak solutions were smooth enough that all of the terms in the Navier equations made sense as "normal" functions, then we would say that the weak solutions were *"strong"* solutions. As this is noted by Doering and Gibbon (1995):

"To many readers this kind of distinction may seem little more than a mathematical formality of no real consequence or practical importance. The issues involved, however, go straight to the heart of the question of the validity and *self-consistency* of the Navier equations as a hydrodynamical model, and the mathematical difficulties have their source in precisely the same physical phenomena that the equations are meant to describe"!

In Chap. 8, we also give some information concerning recent rigorous mathematical results (existence, regularity, and uniqueness) for incompressible and compressible viscous fluid flow problems.

Chapter 9 is devoted to some aspects of the *stability theory* of viscous fluid flows. The stability theory poses quite a natural question: *Given an evolutionary fluid dynamic equation, we want to know whether a small perturbation of initial conditions produces effects on the solution which are uniformly small in time*! This question is obviously directly related to the third Hadamard condition concerning the *continuous dependence* on initial conditions in the definition of a well-posed fluid flow problem [see, for instance, Zeytounian (2002a), pp. 233–234]. Unfortunately, in general, the perturbation grows exponentially in time, and the main problem is to seek the conditions for which this does not happen! In Chap. 9, we consider linear and weakly nonlinear stability theory and, in particular, using a technique of the *Lyapunov–Schmidt type* derived from bifurcation theory and a perturbation method coupled with a *multiple-scale* process, we present a unified approach to nonlinear hydrodynamic stability according to the paper by de Coninck et al. (1983).

In Chap. 10 the reader can find an approach to a *finite-dimensional dynamical system* and *routes/scenarios to chaos/turbulence*. The turbulent phenomena are not yet completely understood, even from a physical point of view, so that there is no general agreement on what "turbulence" is. In fact, the term "turbulence" describes very different fluid behaviors, and it seems that many notions of turbulence/chaos exist! There is no complete mathematical theory describing turbulent flow, and consequently any exposition must be, at least partially, phenomenological. However, recent progress has been made in understanding why turbulence/chaos develops. We mean the so-called *onset* of turbulence or the *transition from laminar to turbulent flows*, and actually the general theory of dynamical dissipative systems provides a convincing explanation of many features that accompany this transition. This is basically the content of Chap. 10 where, as an particular interesting example of Feigenbaum and Pomeau–Manneville Scenarios, we discuss the routes to chaos in the so-called *deep thermal convection model* problem (formulated in 1989 by Zeytounian) when, in place of (20c) for θ_{0B}, we have rather the following equation for the temperature perturbation Θ in deep thermal convection (see Zeytounian (1989a)):

$$[1 + \delta(1 - X_3)]\frac{d_D \Theta}{d\tau} = \frac{1}{\Pr}\Delta\Theta + \frac{\delta}{2\,\mathrm{Gr}}\left[\frac{\partial V_{iD}}{\partial X_j} + \frac{\partial V_{jD}}{\partial X_i}\right]^2, \qquad (21)$$

where δ is the so-called "deep" dimensionless parameter and V_{iD} the Cartesian components of the velocity vector in the deep thermal convection problem, with $d_D/d\tau = \partial/\partial\tau + V_{iD}\partial/\partial X_i$. The appearance of strange attractors for various viscous fluid flows is also investigated.

1 Navier–Stokes–Fourier Exact Model

1.1 The Transport Theorem

Let $\mathfrak{V} = \mathfrak{V}(t)$ denote an arbitrary volume that is moving with the fluid, and let $F(t, \boldsymbol{x})$ be a function of position vector $\boldsymbol{x}$ and time t.

The volume integral

$$\int_{\mathfrak{V}(t)} F(t, \boldsymbol{x}) \, \mathrm{d}\mathfrak{v}$$

is then a well-defined function of time. Its derivation is given by the following formula

$$\frac{\mathrm{d}}{\mathrm{d}t} \int_{\mathfrak{V}(t)} F(t, \boldsymbol{x}) \, \mathrm{d}\mathfrak{v} = \int_{\mathfrak{V}(t)} \left[\frac{\partial F}{\partial t} + F \operatorname{div} \boldsymbol{u} \right] \mathrm{d}\mathfrak{v} , \tag{1.1}$$

where the velocity vector of a particle of fluid is $\boldsymbol{u} = \mathrm{d}\boldsymbol{x}/\mathrm{d}t = \boldsymbol{u}(t, \boldsymbol{x})$.

Equation (1.1) can be expressed in an alternate way that brings out clearly its kinematical significance (independently of any meaning attached to function F).

By virtue of $\mathrm{d}F/\mathrm{d}t = \partial F/\partial t + \boldsymbol{u} \cdot \operatorname{grad} F$, the integrand on the right side of (1.1) can be written $\partial F/\partial t + \operatorname{div}(F\boldsymbol{u})$, and then by application of the divergent classical theorem, we find that

$$\frac{\mathrm{d}}{\mathrm{d}t} \int_{\mathfrak{V}(t)} F(t, \boldsymbol{x}) \, \mathrm{d}\mathfrak{v} = \frac{\partial}{\partial t} \int_{\mathfrak{V}(t)} F \, \mathrm{d}\mathfrak{v} + \int_{\mathfrak{S}} F\boldsymbol{u} \cdot \boldsymbol{n} \, \mathrm{d}\mathfrak{s} . \tag{1.2}$$

Here, $\partial/\partial t$ denotes differentiation with volume $\mathfrak{V}(t)$ held fixed, and $\boldsymbol{n}$ is the outer normal to the surface $\mathfrak{S}$. Equation (1.2) equates the rate of change of total F across a material volume $\mathfrak{V}(t)$ to the rate of change of total F across the fixed volume that instantaneously coincides with $\mathfrak{V}(t)$ plus the flux of F out of the bounding surface $\mathfrak{S}$.

Of course, the function F is continuously derivable in terms of volume $\mathfrak{V}(t)$ and the surface $\mathfrak{S}$ of this volume is assumed suitably smooth.

1.2 The Equation of Continuity

We suppose that the fluid posesses a density function $\rho = \rho(t, \boldsymbol{x})$, which serves by means of the formula

$$\mathfrak{M} = \int_{\mathfrak{V}(t)} \rho \, d\mathfrak{v}$$

to determine the mass $\mathfrak{M}$ of fluid occupying a region $\mathfrak{V}(t)$. We naturally assume that $\rho > 0$ and assign "*mass per unit volume*" to the physical dimension. According to the principle of conservation of mass, *the mass of fluid in a material volume $\mathfrak{V}(t)$ does not change as $\mathfrak{V}(t)$ moves with the fluid, and consequently,*

$$\frac{d}{dt} \int_{\mathfrak{V}(t)} \rho \, d\mathfrak{v} = 0 \, .$$

From (1.1),

$$\int_{\mathfrak{V}(t)} \left[\frac{\partial \rho}{\partial t} + \rho \operatorname{div} \boldsymbol{u} \right] d\mathfrak{v} = 0 \, , \tag{1.3a}$$

or

$$\frac{d\rho}{dt} + \rho \operatorname{div} \boldsymbol{u} = 0 \, , \tag{1.3b}$$

because $\mathfrak{V}(t)$ is arbitrary [this derivation is substantially attributable to Euler (1755)].

As a consequence of (1.1) and (1.3b), we conclude with the formula that is valid for an arbitrary function $F = F(t, \boldsymbol{x})$,

$$\frac{d}{dt} \int_{\mathfrak{V}(t)} \rho F \, d\mathfrak{v} = \int_{\mathfrak{V}(t)} \rho \frac{dF}{dt} \, d\mathfrak{v} \, . \tag{1.4}$$

1.3 The Cauchy Equation of Motion

According to the principle of conservation of linear momentum, the rate of change of linear momentum of a material volume $\mathfrak{V}(t)$ equals the resultant force on the volume [see Serrin (1959a, p. 134)]:

$$\frac{d}{dt} \iiint_{\mathfrak{V}(t)} \rho \boldsymbol{u} \, d\mathfrak{v} = \iiint_{\mathfrak{V}(t)} \rho \boldsymbol{f} \, d\mathfrak{v} + \iint_{\mathfrak{S}} \boldsymbol{t} \, d\mathfrak{s} \, , \tag{1.5}$$

where $\boldsymbol{f}$ is the extraneous force per unit mass and $\boldsymbol{t}$ is the *stress vector*. The stress vector may therefore be expressed as a linear function of the components of $\boldsymbol{n}$, that is,

$$t^i = n_j T^{ji} \quad \text{where} \quad T^{ji} = T^{ji}(t, \boldsymbol{x}) \ . \tag{1.6}$$

We observe that, according to Cauchy's stress principle (1828), "*upon any imagined closed surface $\mathfrak{S}$ there exists a distribution of stress vector $\boldsymbol{t}$ whose resultant and moment are equivalent to those of the actual forces of material continuity exerted by the material outside $\mathfrak{S}$ upon that inside.*" It is assumed that $\boldsymbol{t} = \boldsymbol{t}(t, \boldsymbol{x}; \boldsymbol{n})$. We observe that this statement of Cauchy's principle is due to Truesdell. In (1.6), the matrix of coefficients T^{ji} obviously forms a tensor called the *stress tensor*, here denoted by $\mathbf{T}$. Each component of $\mathbf{T}$ has a simple physical interpretation, namely, T^{ij} is the j component of the force on the surface element with an outer normal in the i direction.

By using (1.4) and replacing $\boldsymbol{t}$ by $\boldsymbol{n} \cdot \mathbf{T}$, then according to the divergence theorem, we find, in place of (1.5),

$$\int_{\mathfrak{V}(t)} \rho \frac{d\boldsymbol{u}}{dt} d\mathfrak{v} = \int_{\mathfrak{V}(t)} [\rho \boldsymbol{f} + \text{div } \mathbf{T}] d\mathfrak{v} \ ,$$

and because $\mathfrak{V}(t)$ is arbitrary

$$\rho \frac{d\boldsymbol{u}}{dt} = \rho \boldsymbol{f} + \text{div } \mathbf{T} \ . \tag{1.7}$$

1.4 The Constitutive Equations of a Viscous Fluid

Through its appearance in the equation of motion (1.7), the stress tensor $\mathbf{T}(t, \boldsymbol{x})$ governs the dynamic response of a medium. Thus by a relating $\mathbf{T}$ to other kinematic and thermodynamic variables, we effectively define or delimit the type of medium which we study, e.g., fluid, elastic, plastic, and so on.

Such a relation between $\mathbf{T}$ and other flow quantities is called a *constitutive equation*. In Zeytounian (2002a), a particular example of a constitutive equation has already been considered, that of an inviscid (nonviscous) fluid for which $\mathbf{T} = -p\mathbf{I}$, where $\mathbf{I}$ is the unit matrix and p is the *pressure*. When $p > 0$, the stress vector

$$\boldsymbol{t} = -p\boldsymbol{n}$$

acting on a closed surface tends to compress the fluid inside and then p is independent of $\boldsymbol{n}$: $p = p(t, \boldsymbol{x})$.

Our purpose below [we mainly follow the presentation of Serrin (1959a, pp. 230–239)] is to derive a constitutive equation for $\mathbf{T}$ applicable to real (viscous) fluids that exert appreciable tangential (not only normal) stresses.

1.4.1 Stokes's Four Postulates: Stokesian Fluid

Stokes stated the concept of "fluidity" in the following form:
"*That the difference between the pressure on a plane in a given direction passing through any point P of a fluid in motion and the pressure which would exist in all direction about P if the fluid in its neighborhood were in a state of relative equilibrium depends only on the relative motion of the fluid immediately about P; and that the relative motion due to any motion of rotation may be eliminated whthout affecting the difference of the pressures above mentioned.*"

More precisely, Stokes's idea "of fluidity" can be stated as four postulates:

I. $\mathbf{T}$ is a continuous function of the deformation tensor $\mathbf{D}(\boldsymbol{u})$ and is independent of all other kinematic quantities.
II. $\mathbf{T}$ does not depend explicitly on the position vector $\boldsymbol{x}$ (*spatial homogeneity*).
III. There is no preferred direction in space (*isotropy*).
IV. When $\mathbf{D}(\boldsymbol{u}) = 0$, then $\mathbf{T}$ reduces to $-p\mathbf{I}$.

A medium whose constitutive equation satisfies these four postulates is called a *Stokesian fluid* and according to postulates I and II,

$$\mathbf{T} = f(\mathbf{D}(\boldsymbol{u})) \,. \tag{1.8a}$$

The requirement of isotropy (postulate III), according to matrix algebra, is expressed by the condition,

$$\mathbf{S}\,\mathbf{T}\,\mathbf{S}^{-1} = f(\mathbf{S}\,\mathbf{D}(\boldsymbol{u})\,\mathbf{S}^{-1}) \,, \tag{1.8b}$$

for all orthogonal transformation matrices $\mathbf{S}$.

First, relation (1.8b) implies that there is no preferred direction either in fluid or in space and, then, consequently, (1.8a) is invariant under all rectangular coordinate transformations.

We observe that it is tacitly assumed that the stress tensor $\mathbf{T}$ depends on the thermodynamic state through the coefficients α, β, χ in the representation of $\mathbf{T}$ in the following form:

$$\mathbf{T} = \alpha\mathbf{I} + \beta\mathbf{D} + \chi\mathbf{D}^2 \,, \tag{1.9}$$

which is a direct consequence of postulates I to IV.

More precisely, coefficients α, β, χ are functions of the principal invariants of $\mathbf{D}(\boldsymbol{u})$, namely, $I_1 = \text{Trace } \mathbf{D}(\boldsymbol{u}) = \text{div } \boldsymbol{u}$, I_2, I_3, and can be defined as the coefficient of λ in the expression of the determinant

$$\Delta(\lambda) = \det[\lambda\mathbf{I} - \mathbf{D}(\boldsymbol{u})] \,.$$

Thus,

$$\Delta(\lambda) = \lambda^3 - I_1\lambda^2 + I_2\lambda - I_3 \,. \tag{1.10}$$

1.4 The Constitutive Equations of a Viscous Fluid

On the other hand, the principal values d_1, d_2, d_3 of $\mathbf{D}(\boldsymbol{u})$ are the roots of the equation $\Delta(\lambda) = 0$, and they are real because $\mathbf{D}(\boldsymbol{u})$ is a symmetrical tensor. Clearly, d_i ($i = 1, 2, 3$), are functions of I_1, I_2, I_3.

First, the principal directions of $\mathbf{T}$ coincide with the principal directions of $\mathbf{D}(\mathbf{u})$. In other words, any orthogonal transformation $\mathbf{S}$ reducing to $\mathbf{D}(\boldsymbol{u})$ diagonal form likewise reduces $\mathbf{T}$ to diagonal form.

Because $\mathbf{T}$ is diagonal, it follows that the principal values of $\mathbf{T}$, t_i, are functions of d_1, d_2, d_3, that is,

$$t_i = f_i(d_1, d_2, d_3), \quad i = 1, 2, 3, \tag{1.11}$$

and if d_i ($i = 1, 2, 3$) are all different, we can determine multipliers α, β, χ such that

$$t_i = \alpha + \beta d_i + \chi d_i^2, \quad i = 1, 2, 3. \tag{1.12}$$

In fact we observe that the α, β, χ are symmetrical continuous functions of variables d_i and depend solely on the principal invariants div $\boldsymbol{u}$, I_2, I_3. Finally, *α, β, χ are singled-valued, continuous functions of the principal invariants of $\mathbf{D}(\boldsymbol{u})$.* Reverting to matrix notation, (1.12) gives (1.9) again.

1.4.2 Classical Linear Viscosity Theory: Newtonian Fluid

Now, if we add to Stokes's above postulates I to IV the condition that the components of $\mathbf{T}$ be *linear* in the components of $\mathbf{D}(\boldsymbol{u})$, then (1.9) must reduce to the form (*Cauchy–Poisson law*),

$$\mathbf{T} = (-p + \lambda \operatorname{div} \boldsymbol{u})\mathbf{I} + 2\mu \mathbf{D}(\boldsymbol{u}). \tag{1.13}$$

Here, for a compressible (viscous) fluid, p is the thermodynamic pressure, and λ and μ are scalar functions of the thermodynamic state.

A Stokesian fluid whose constitutive equation is given by (1.13) is called a *Newtonian fluid*. The dynamic equation that results from inserting the Cauchy–Poisson law (1.13) into the Cauchy equation of motion (1.7) is known as the *Navier–Stokes* (NS) equation.

The dissipation function that is correspond to the Cauchy–Poisson law (1.13) is

$$\Phi = \lambda (\operatorname{div} \boldsymbol{u})^2 + 2\mu \mathbf{D}(\boldsymbol{u}) : \mathbf{D}(\boldsymbol{u}) \geq 0, \tag{1.14a}$$

and the condition $\Phi \geq 0$ places some restrictions on the possible values of λ and μ. With:

$$3\Phi = (3\lambda + 2\mu)(\operatorname{div} \boldsymbol{u})^2 + 2\mu[(d_1 - d_2)^2 + (d_2 - d_3)^2 + (d_3 - d_1)^2],$$

we obtain as restrictions

$$\mu \geq 0 \quad \text{and} \quad 3\lambda + 2\mu \geq 0. \tag{1.14b}$$

The quantity $(3\lambda + 2\mu)$ also arises in an expression for the difference between pressure p and mean pressure $p^* = -(1/3)\,\mathrm{Trace}\,\mathbf{T}$:

$$3(p - p^*) = (3\lambda + 2\mu)\,\mathrm{div}\,\boldsymbol{u}\,. \tag{1.14c}$$

For various approaches to the rigorous derivation of the constitutive equations for a viscous fluid via matrix algebra, see the articles of Reiner (1945), Rivlin (1947, 1948), and Truesdell (1950, 1951, 1952b, 1953a,b). An interesting point of Noll's (1955) work is his careful treatment of the notion of isotropy. An elegant method [partially used by Serrin (1959a), pp. 230–239] is given by Ericksen and Rivlin (1955), and Ericksen (1954) examined certain restrictions that the physical situation might possibly impose on coefficients α, β, χ. Finally, see also the mathematical theory of Noll (1958) concerning behavior of continuous media.

The dependence of the viscosity coefficients λ and μ in (1.13) on the thermodynamic state is of some importance and is treated at length by Chapman and Cowling (1952). Truesdell (1952c) clarified the nature of the kinetic theory argument used to support the result $\mu \sim T^{1/2}$ and considered the problem from the viewpoint of dimensional analysis.

In conclusion, we observe that the procedure outlined above belongs to the realm of continuum mechanics. Consequently, it is completely independent of the molecular nature of liquids and gases, and accordingly has been criticized as inadequate in some situations. It is claimed, for example, that NS equations cannot be applied to high-altitude flight or to shock layer phenomena because the mean free path then becomes of appreciable size and the Knudsen number [see Zeytounian (2002a), Chap. 1] is not a very small parameter! This argument has considerable force, but ultimately it must stand or fall on the comparison of theory with experiment. Here, the NS equations appear to give satisfactory results whenever it is properly used. In addition, the only alternative seems to be kinetic theory, and this has certainly not yet reached the simplicity or elegance of continuum mechanics. Concerning the validity problem, we note that in the survey paper by Bellomo et al. (1995), the simutaneous solution of the NSF and Boltzmann equations is considered, when these equations are used together but on two different domains. The recent papers of Bardos (1998), Sone et al. (2000) and the book by Bellomo and Pulvirenti (2000), consider the kinetic point of view.

But to complete the system of equations of classical Newtonian fluid mechanics, it is necessary to derive the total energy equation and to express the heat conduction vector $\boldsymbol{q}$ in terms of mechanical and thermodynamical variables.

1.5 The Energy Equation and Fourier's Law

1.5.1 The Total Energy Equation

First, according to fundamental assumptions of the thermodynamics of simple media [see, for instance, Serrin (1959a) pp. 172–177], we postulate *that total energy is conserved*:
the rate of change of the total energy of a material volume is equal to the rate at which work is being done on the volume plus the rate at which heat is conducted into the volume.

Consequently, we obtain the total energy equation,

$$\rho \frac{dE}{dt} = \mathbf{T} : \mathbf{D} - \operatorname{div} \boldsymbol{q} \,. \tag{1.15a}$$

In place of (1.15a), we can write

$$\rho \frac{dE}{dt} + p \operatorname{div} \boldsymbol{u} = \Phi - \operatorname{div} \boldsymbol{q} \,, \tag{1.15b}$$

if we introduce the tangential stresses $\boldsymbol{\Sigma}$ and write $\mathbf{T} = -p\mathbf{I} + \boldsymbol{\Sigma}$. In this case $\Phi = \boldsymbol{\Sigma} : \mathbf{D}(\boldsymbol{u})$. Finally, by using the equation of continuity (1.3b) and the classical thermodynamic relation [see relation (11) in the Introduction],

$$T\, dS = dE + p\, d\left(\frac{1}{\rho}\right) \,,$$

where T is the absolute temperature, S the specific entropy, and p the pressure [defined as a thermodynamic variable and, therefore, in general, $p = P(\rho, T)$], we obtain the following elegant result:

$$\rho T \frac{dS}{dt} = \Phi - \operatorname{div} \boldsymbol{q} \,, \tag{1.16}$$

that expresses the rate of change of entropy of a particle.

We observe that, to be consistent with the laws of thermodynamics, it is necessary that

$$\Phi - \boldsymbol{q} \cdot \left(\frac{\operatorname{grad} T}{T}\right) \geq 0 \Rightarrow \Phi \geq 0 \tag{1.17a}$$

and

$$\boldsymbol{q} \cdot \operatorname{grad} T \leq 0 \,, \tag{1.17b}$$

and these conditions are the mathematical statements of two familiar facts: heat never flows against a temperature gradient, and deformation absorbs energy (converting it to heat), but never releases it. Conversely, if we start with inequalities (1.17b), we may derive the following inequality, which is a direct consequence of simple physical observations (a postulate related to the second law of thermodynamics),

$$\frac{d}{dt} \iiint_{\mathfrak{V}(t)} \rho S \, dv \geq - \iint_{\mathfrak{S}} \left[\frac{(q \cdot n)}{T} \right] d\mathfrak{s} . \tag{1.18}$$

In Truesdell (1977), which is a "rational continuum mechanics" course, the derivation of constitutive equations and also the thermodynamics are presented in an axiomatic approach, inspired mainly by the works of W. Noll [see, for instance, Noll (1958)].

The review paper by Germain (1974) is devoted to the achievements in continuum thermodynamics up to 1974 and contains a list of very pertinents works.

The modern viewpoint seems to be that the thermodynamics of irreversible processes must be handled on an abstract postulated basis, independent of any derivation from presumably more basic assumptions [see, for instance, the classical books by de Groot (1951) and Denbigh (1951)].

In irreversible thermodynamics, one of the basic postulates is known as the "linear law" [see Onsager (1931); an outline of the results will be found in Curtiss (1955)]. Its validity is certainly open to question (this is the opinion of Serrin!), but at least it is worthwhile to examine its consequence in the present case. Stated roughly, it says that a thermodynamic system will tend to equilibrium at a rate linearly dependent on its displacement from equilibrium. In our case, this displacement is measured by $\mathbf{D}(\boldsymbol{u})$ and $\operatorname{grad} T$, hence Σ and $\boldsymbol{q}$ should be linear functions of $\mathbf{D}(\boldsymbol{u})$ and $\operatorname{grad} T$. By virtue of the tensorial difference of these various quantities, it follows that

$$\Sigma = \text{linear function of } \mathbf{D}(\boldsymbol{u}) , \tag{1.19a}$$

and

$$\boldsymbol{q} = \text{linear function of } \operatorname{grad} T . \tag{1.19b}$$

The first, (1.19a), of these leads to the Cauchy–Poisson law of viscosity (1.13), considered previously in Sect. 1.4.2, and the second, (1.19b), is the well-known Fourier's law considered in Sect. 1.5.2.

These consideration must be regarded as merely heuristic, however, and by no means provide an adequate derivation of the constitutive equations of a Newtonian fluid governed by the Navier–Stokes–Fourier equations (for NSF eqs., see Sect. 1.6).

1.5.2 Heat Conduction and Fourier's Law

According to the previous discussion, we postulate that $\boldsymbol{q}$, the heat conduction vector, *is an isotropic function of the temperature gradient and thermodynamic state*. From the condition of isotropy, it is apparent that q must be parallel to $\operatorname{grad} T$, and consequently, we obtain Fourier's law:

$$\boldsymbol{q} = -k \operatorname{grad} T , \qquad (1.20)$$

where k is a scalar function of $|\operatorname{grad} T|$ and of thermodynamic variables.

The thermodynamic condition (1.17b) implies that $k > 0$. The particular functional form of k to be adopted in a given problem is, of course, a matter for experiment and kinetic theory to decide. Here, the evidence favors the relation [see, for instance, Chapman and Cowling (1952), Chap. 13]

$$\Pr = \frac{\mu C_\mathrm{p}}{k} = \mathrm{const.} \qquad (1.21)$$

This is consistent with the frequent assumption that k, μ, and the specific heat $C_\mathrm{p} = T(\partial S/\partial T)_p > 0$ are all constants and in any case, implies that k should not depend on $|\operatorname{grad} T|$. For an ideal gas with constant specific heats, C_p and $C_\mathrm{v} = C_\mathrm{p} - R$, where R is the positive gas constant, the Prandtl number $\Pr = 2/3$ for monoatomic gases, and $\Pr = 0.72$ for air.

For an ideal gas with constant specific heats, C_p and C_v, we obtain two equations of state:

$$\mathrm{d}E = C_\mathrm{v}\,\mathrm{d}T \quad \text{and} \quad \frac{p}{\rho} = RT . \qquad (1.22)$$

For an ideal gas, the specific entropy is given by

$$S = C_\mathrm{v} \log\left(\frac{p}{\rho^\gamma}\right) , \qquad (1.23)$$

and in (1.23) a constant of integration is avoided by choosing the zero point of entropy at $p = \rho = 1$. From (1.23), we obtain

$$p = \rho^\gamma \exp\left(\frac{S}{C_\mathrm{v}}\right) , \qquad (1.24)$$

and the ratio $\gamma = C_\mathrm{v}/C_\mathrm{p} = 5/3$ for a monoatomic gas (according to kinetic theory); for air at moderate temperature, $\gamma = 1.40$. For monoatomic real gases, it is found that C_v is nearly constant across a wide temperature range.

1.6 The Navier–Stokes–Fourier Equations

The *dynamic* equation that results from inserting the Cauchy–Poisson law (1.13) into the Cauchy equation of motion (1.7) is known as the Navier–Stokes equation. If λ and μ are not constant, we obtain the following NS dynamic equation for velocity vector $\boldsymbol{u}$:

$$\rho \frac{\mathrm{d}\boldsymbol{u}}{\mathrm{d}t} = \rho \boldsymbol{f} - \operatorname{grad} p + \operatorname{grad}(\lambda \operatorname{div} \boldsymbol{u}) + \operatorname{div}(2\mu \mathbf{D}(\boldsymbol{u})) . \qquad (1.25\mathrm{a})$$

On the other hand, when the Fourier law (1.20) is substituted in the total energy equation (1.16),

$$\rho T \frac{\mathrm{d}S}{\mathrm{d}t} = \Phi + \mathrm{div}(k \, \mathrm{grad}\, T) \,. \tag{1.25b}$$

This equation (1.25b), the NS equation (1.25a), the equation of continuity [according to (1.3b)],

$$\frac{\mathrm{d}\log}{\mathrm{d}t} + \mathrm{div}\,\boldsymbol{u} = 0 \,, \tag{1.25c}$$

and the thermodynamic equations of state

$$E = \mathfrak{E}(S, 1/\rho) \,, \quad T = \frac{\partial \mathfrak{E}}{\partial S} \,, \quad p = -\frac{\partial \mathfrak{E}}{\partial (1/\rho)} \tag{1.25d}$$

constitute the general set of equations upon which classical Newtonian fluid dynamics is based. Here, this system of equations (1.25a–d) is called "Navier–Stokes–Fourier" (NSF).

1.6.1 The NSF Equation for an Ideal Gas when C_v and C_p are Constants

For an ideal gas with constant specific heats, in place of NSF equations (1.25a–d), we obtain the following system of four equations for $\boldsymbol{u}$, T, ρ, and p:

$$\begin{aligned}
&\frac{\mathrm{d}\log \rho}{\mathrm{d}t} + \mathrm{div}\,\boldsymbol{u} = 0 \,, \\
&p = R\rho T \,, \\
&\rho \frac{\mathrm{d}\boldsymbol{u}}{\mathrm{d}t} = \rho \boldsymbol{f} - \mathrm{grad}\, p + \mathrm{grad}(\lambda \, \mathrm{div}\, \boldsymbol{u}) + \mathrm{div}(2\mu \mathbf{D}(\boldsymbol{u})) \,, \\
&C_\mathrm{v}\rho \frac{\mathrm{d}T}{\mathrm{d}t} + p\,\mathrm{div}\,\boldsymbol{u} = \Phi(\boldsymbol{u}) + \mathrm{div}(k\,\mathrm{grad}\, T) \,,
\end{aligned} \tag{1.26}$$

with

$$\Phi(\boldsymbol{u}) = 2\,\mathrm{Trace}\,[(\mathbf{D}(\boldsymbol{u}))^2] + \lambda(\nabla \cdot \boldsymbol{u})^2 \,. \tag{1.27}$$

If $\lambda = \lambda^0$, $\mu = \mu^0$, and $k = k^0$, are constant and if we assume that

$$\lambda^0 = -\frac{2}{3}\mu^0 \quad \text{(Stokes's relation)}\,,$$

then equations (1.26) simplify to the forms,

$$\frac{\mathrm{d}\log\rho}{\mathrm{d}t} + \nabla \cdot \boldsymbol{u} = 0 \, , \quad p = R\rho T \, ,$$

$$\rho \frac{\mathrm{d}\boldsymbol{u}}{\mathrm{d}t} = \rho \boldsymbol{f} - \nabla p + \frac{1}{3}\mu^0 \nabla(\nabla \cdot \boldsymbol{u}) + \mu^0 \nabla^2 \boldsymbol{u} \, , \tag{1.28}$$

$$C_\mathrm{v}\rho \frac{\mathrm{d}T}{\mathrm{d}t} + p\nabla \cdot \boldsymbol{u} = \Phi(\boldsymbol{u}) + k^0 \nabla^2 T \, ,$$

with

$$\nabla^2 \boldsymbol{u} = \mathrm{grad}(\mathrm{div}\,\boldsymbol{u}) - \mathrm{curl}\,\boldsymbol{\omega} \quad \text{and} \quad \boldsymbol{\omega} = \mathrm{curl}\,\boldsymbol{u} \, , \tag{1.29}$$

for the Laplacian of $\boldsymbol{u}$.

1.6.2 Dimensionless NSF Equations

Below, we work mainly with dimensionless equations. From (1.25a–c), the dimensionless full unsteady-state NSF equations, written with the same notations for the dimensionless velocity $\boldsymbol{u}$ and thermodynamic functions p, ρ, and T, are the following:

$$\mathrm{S}\frac{\mathrm{d}\rho}{\mathrm{d}t} + \rho \nabla \cdot \boldsymbol{u} = 0 \, ; \tag{1.30}$$

$$\rho \mathrm{S}\frac{\mathrm{d}\boldsymbol{u}}{\mathrm{d}t} + \frac{1}{\gamma M^2}\nabla p + \frac{\mathrm{Bo}}{\gamma M^2}\rho \boldsymbol{k}$$
$$= \frac{1}{\mathrm{Re}}\nabla \cdot [2\mu \mathbf{D}(\boldsymbol{u}) + \lambda(\nabla \cdot \boldsymbol{u})\mathbf{I}] \, ; \tag{1.31}$$

$$\rho \mathrm{S}\frac{\mathrm{d}T}{\mathrm{d}t} + (\gamma - 1)p\nabla \cdot \boldsymbol{u} = \frac{\gamma}{\mathrm{Pr}\,\mathrm{Re}}\nabla \cdot [k\nabla T]$$
$$+ \frac{\gamma M^2(\gamma - 1)}{\mathrm{Re}}[2\mu\,\mathrm{Trace}\,(\mathbf{D}(\boldsymbol{u})^2) + \lambda(\nabla \cdot \boldsymbol{u})^2] \, , \tag{1.32}$$

which form a closed set of evolutionary equations, provided that one adds the dimensionless equation of state for an ideal gas

$$p = \rho T \, . \tag{1.33}$$

The reduced dimensionless parameters are defined below in Sect. 1.6.3 via the characteristic time t^0, the characteristic length L^0, and the characteristic velocity U_∞. The thermodynamic characteristic constant values are p^0, ρ^0, and T^0. Finally, μ^0 and k^0 are the characteristic values for the dynamic viscosity and thermal conductivity coefficients (for instance, for $\rho = \rho^0$ and $T = T^0$) and $\nu^0 = \mu^0/\rho^0$ is the characteristic kinematic viscosity coefficient. In (1.30)–(1.32),

$$\mathrm{S}\frac{\mathrm{d}}{\mathrm{d}t} = \mathrm{S}\frac{\partial}{\partial t} + \boldsymbol{u}\cdot\nabla \, . \tag{1.34}$$

Finally, we observe that in place of $\boldsymbol{f}$ in (1.25a), we write $\boldsymbol{g} = -g\boldsymbol{k}$, the gravitational force per unit mass, with $\boldsymbol{k}$ the unit vertical vector directed along the axis Ox_3.

1.6.3 Reduced Dimensionless Parameters

The reduced dimensional parameters in the dimensionless NSF equations (1.30)–(1.32) are

$$S = \frac{L^0}{t^0 U_\infty}, \quad \text{the Strouhal number,} \tag{1.35a}$$

$$\text{Re} = \frac{L^0 U_\infty}{\nu_0}, \quad \text{the Reynolds number,} \tag{1.35b}$$

$$M = \frac{U_\infty}{(\gamma R T^0)^{1/2}}, \quad \text{the Mach number,} \tag{1.35c}$$

$$\text{Bo} = \gamma \left(\frac{M}{\text{Fr}_{L^0}}\right)^2, \quad \text{the Boussinesq number,} \tag{1.35d}$$

$$\text{Pr} = \frac{C_\text{p} \mu_0}{k_0}, \quad \text{the Prandtl number,} \tag{1.35e}$$

where $\gamma = C_\text{p}/C_\text{v}$, and Fr_{L^0} is the Froude number such that

$$\text{Fr}_{L^0} = \frac{U_\infty}{(gL^0)^{1/2}}. \tag{1.36}$$

The Strouhal number takes into account unsteady-state effects, and the Reynolds number the viscous effects. Compressibility is strongly related to the Mach number, and the acoustics effects are closely linked to the following similarity relation:

$$S\,M = M^* = O(1), \tag{1.37}$$

when $S \gg 1$ (a strong unsteady-state effect) and $M \ll 1$ (quasi-incompressibility). A second simlarity relation related to the small Mach number M is

$$\frac{\text{Bo}}{M} = \text{Bo}^* = O(1). \tag{1.38}$$

It gives the possibility of deriving, in the framework of low Mach number asymptotics of NSF equations, the full viscous and heat-conducting Boussinesq equations.

In the Bénard thermal convection problem, when the fluid is an expansible liquid such that the relation

$$\rho = \rho(T) \tag{1.39}$$

1.6 The Navier–Stokes–Fourier Equations

is assumed, we can introduce the constant coefficient of cubical expansion:

$$\alpha^0 = -\frac{1}{\rho^0}\left[\frac{d\rho}{dT}\right]_{T=T^0} \quad (1.40)$$

and in this case, if ΔT^0 is the difference between the two temperatures in a layer of thickness d^0, then the similarity relation

$$\alpha^0 \frac{\Delta T^0}{(\mathrm{Fr}_d)^2} = \mathrm{Gr} = O(1), \quad (1.41)$$

plays a significant role in the derivation of OB equations. In (1.41), the similarity parameter Gr is the so-called Grashof number, and

$$\mathrm{Pr}\,\mathrm{Gr} = \mathrm{Ra} \quad (1.42)$$

is the Rayleigh number. The reduced dimensional parameter

$$\mathrm{Pr}\,\mathrm{Re} = \mathrm{Pe} = \frac{L^0 U_\infty}{(k_0/\rho_0 C_\mathrm{p})} \quad (1.43)$$

is the Péclet number and for $\mathrm{Pr} \ll 1$ with $\mathrm{Re} \gg 1$ but $\mathrm{Pe} = O(1)$, the fluid motion is quasi-nonviscous but thermally conducting, and the so-called "high thermal conductivity" model equations are valid, at least when the Mach number, $M^2 \ll \mathrm{Re}$, is not very high and far from the wall. We observe that when $\mathrm{Pr} \ll 1$, then, $k_0 \gg \mu_0 C_\mathrm{p}$.

Strictly speaking, when $\mathrm{Re} \gg 1$, the viscosity effects remain important mainly in the boundary layer near the wall which has a thickness of the order of

$$H_\mathrm{CL} = \frac{L^0}{(\mathrm{Re})^{1/2}}, \quad (1.44)$$

and in this case we have, again, a similarity relation,

$$\varepsilon_\perp^2 \,\mathrm{Re} \equiv \mathrm{Re}_\perp = 1, \quad \text{where} \quad \frac{H_\mathrm{CL}}{L^0} = \varepsilon_\perp, \quad (1.45)$$

when we assume that

$$\varepsilon_\perp \downarrow 0 \text{ and } \mathrm{Re} \uparrow \infty.$$

$\varepsilon_\perp$ is a "long-wave" parameter and in a viscous fluid, the so-called long-wave approximation is very similar to the classical boundary-layer approximation [according to (1.45)].

For atmospheric viscous and nonadiabatic motion (when, obviously, the Mach number is very low!), we take into account the Coriolis force, which is characterized by the following parameter:

$$f^0 = 2\Omega^0 \sin \phi^0 \,, \tag{1.46}$$

where $\Omega^0 = |\Omega|$; Ω is the angular velocity of rotating Earth's frame and ϕ^0 a reference latitude (for $\phi^0 \approx 45°$, $a^0 = 6300$ km for Earth's radius). In this case, it is helpful to employ spherical coordinates λ, ϕ, r, and for the gradient operator ∇, we can write:

$$\nabla = \frac{1}{r\cos\phi}\frac{\partial}{\partial\lambda}\boldsymbol{i} + \frac{1}{r}\frac{\partial}{\partial\phi}\boldsymbol{j} + \frac{\partial}{\partial r}\boldsymbol{k} \,. \tag{1.47}$$

But it is more convenient to use a right-handed curvilinear coordinates system (x, y, z) that lies on Earth's surface (for flat ground, $r \approx a^0$) at latitude ϕ^0 and longitude $\lambda = 0$:

$$x = a^0 \cos\phi^0 \,,\ y = a^0(\phi - \phi^0) \,,\ z = r - a^0 \,. \tag{1.48}$$

Although x and y are, in principle, new longitude and latitude coordinates in whose terms the basic NSF (for the atmospheric motion) equations may be rewritten without approximation, they are obviously introduced in the expectation that for small

$$\delta^0 = \frac{L^0}{a^0} \tag{1.49}$$

they will be the Cartesian coordinates of the so-called "f^0-plane approximation." A fundamental parameter for atmospheric motion is the so-called Kibel/Rossby number:

$$\mathrm{Ki} = \frac{1}{f^0 t^0} \equiv \mathrm{Ro} = \frac{1/f^0}{L^0/U_\infty} \,, \tag{1.50}$$

when we assume that S = 1 or $t^0 = L^0/U_\infty$.

On the other hand, for atmospheric motions in the troposphere, a very significant characteristic constant vertical altitude is

$$H_\mathrm{s} = \frac{RT^0}{g} \tag{1.51a}$$

and usually (for weather prediction in a hydrostatic approximation),

$$\varepsilon_\mathrm{s} = \frac{H_\mathrm{s}}{L^0} \ll 1 \,. \tag{1.51b}$$

Viscous and nonadiabatic effects in tropospheric motions are limited to a thin layer near the ground – the so-called Ekman layer which is characterized by the following Ekman number:

$$\mathrm{Ek}_\mathrm{s} = \frac{\mathrm{Ki}}{\mathrm{Re}_\mathrm{s}} = \frac{\nu^0}{f^0 H_\mathrm{s}^2} \,, \tag{1.52}$$

where, for Re $\gg 1$,

$$\mathrm{Re}_s = \varepsilon_s^2 \, \mathrm{Re} = O(1) \,. \tag{1.53}$$

The reader can find various asymptotic models for atmospheric motions in Zeytounian (1990), (1991a) books.

Finally, when oscillatory viscous flow (see, for instance, Riley 1967) is induced by a solid body performing harmonic oscillations (assume that the velocity of the body is $U_\infty \cos \omega$) in an unbounded viscous fluid which is otherwise at rest, $t^0 = 1/\omega$ is a typical time. Then, in dimensionless Navier equations for an incompressible and viscous fluid, we have two main parameters:

$$\frac{1}{S} = \alpha = \frac{U_\infty}{\omega L^0} \quad \text{and} \quad \frac{\mathrm{Re}}{\alpha} = \beta^2 \,, \tag{1.54}$$

where

$$\beta^2 = \frac{\omega L^{02}}{\nu^0} \,. \tag{1.55}$$

Riley (1967) considers solely the situation corresponding to $\alpha \ll 1$ (high Strouhal number).

1.7 Conditions for Unsteady-State NSF Equations

Roughly speaking, we can expect that the equations of motion for a viscous fluid are *parabolic*. However, a more detailed analysis of the stucture of (1.25a,c) for u and ρ shows that it seems correct for velocity vector u, but not quite correct for the system (1.25a,c) because the continuity equation with respect to density ρ is *hyperbolic* in the compressible case, even for non-trivial viscosity!

Thus, to be more exact, we can say that a system of Navier–Stokes equations (1.25a,c), with a specifying relation $p = P(\rho)$ and viscosity coefficient $\mu = -(3/2)\lambda(\rho) \equiv \mu^*(\rho)$, is *hyperbolic-parabolic* or *incompletely parabolic*, according to the definition suggested by Belov and Yanenko (1971) and Strikwerda (1977) in rigorous studies of the mathematical properties of these equations.

When the "Fourier" equation (1.25b) is written in dimensionless form [see (1.32)] then, for a small characteristic Mach number (so-called *"hyposonic"* flow), it is also *parabolic* with respect to temperature T.

The reader can find some pertinent information concerning the initial boundary-value problems for Navier–Stokes equations in Kreiss and Lorenz (1989). On the other hand, in Oliger and Sundström (1978), initial boundary-value problems for several systems of partial differential equations from fluid dynamics are discussed. We observe also that both the viscosity and heat conduction coefficients (characterized mainly by the Reynolds number Re) in dimensionless equations (1.31), (1.32) are usually very small (Re $\gg$ 1), but because these terms change the character of the partial differential equations (from parabolic to hyperbolic), we have to reinvestigate the boundary conditions. For the initial condition at initial time $t = 0$, the situation is unchanged relative to hyperbolic Euler unsteady-state compressible nonviscous equations considered in Zeytounian (2002a).

1.7.1 The Problem of Initial Conditions

In NSF equations (1.30)–(1.32), we have five derivatives for components u_i of the velocity $\boldsymbol{u}$, density ρ, and temperature T. Consequently, if we want to resolve a pure initial-value or *Cauchy problem* (in the L^2 norm, for example), it is necessary to have a complete set of initial conditions (data) for $\boldsymbol{u}$, ρ, and T:

$$t = 0 : \boldsymbol{u} = \boldsymbol{u}^0(\boldsymbol{x}), \quad \rho = \rho^0(\boldsymbol{x}), \quad T = T^0(\boldsymbol{x}), \tag{1.56}$$

where $\rho^0(\boldsymbol{x}) > 0$ and $T^0(x) > 0$. Moreover, when we consider a free-boundary problem or unsteady-state flow in a bounded container with a boundary depending on time, an initial condition for the (moving) boundary $\partial \Omega(t)$ has to be specified.

For barotropic/isentropic dynamic flow governed by NS equations, for example, by the first (for ρ) and the third (for $\boldsymbol{u}$) equations of system (1.28) with $p = p(\rho)$, as initial conditions we assume only

$$t = 0 : \boldsymbol{u} = \boldsymbol{u}_b^0(x) \quad \text{and} \quad \rho = \rho_b^0(x) . \tag{1.57}$$

For divergence-free flow when $\nabla \cdot \boldsymbol{u} = 0$, then for the Navier (incompressible viscous) equation for $\boldsymbol{u}$ we have the possibility of writing only one initial condition:

$$t = 0 : \boldsymbol{u} = \boldsymbol{u}_i^0(x) . \tag{1.58}$$

Note that for incompressible, divergence-free flows, it is necessary that

$$\text{the boundary integral } \int \boldsymbol{u} \cdot \boldsymbol{n} \, d\Omega \text{ vanishes,} \tag{1.59a}$$

and

$$\nabla \cdot \boldsymbol{u}_i^0 = 0 . \tag{1.59b}$$

Naturally, this last condition has no analog in compressible (baroclinic or barotropic) flow because of the occurrence of the term $S \, \partial \rho / \partial t$ in the continuity equation (1.30). For incompressible fluids, we have the Navier equation,

1.7 Conditions for Unsteady-State NSF Equations

$$\frac{d\boldsymbol{u}}{dt} = \boldsymbol{f} - \frac{1}{\rho_0}\nabla p + \nu_0 \nabla^2 \boldsymbol{u} , \tag{1.60a}$$

with

$$\rho = \rho_0 = \text{const}, \quad \nabla \cdot \boldsymbol{u} = 0 , \tag{1.60b}$$

and $\nu_0 = \mu_0/\rho_0$ is the constant kinematic viscosity. If one wishes, (1.60a) can be thought of as a special case of an NS equation in system (1.28) which arises when $\nabla \cdot \boldsymbol{u} = 0$.

This view is somewhat superficial, however, because the pressure has a fundamentally different meaning for compressible and incompressible fluids; we prefer, therefore, to write both equations of motion for $\boldsymbol{u}$ [NS equation in (1.28) and the Navier equation (1.60a)] as separate equations.

In Chap. 6, devoted to low Mach number asymptotics of NSF unsteady-state equations, the Navier system (1.60a,b) is consistently derived and in this case a pseudopressure π, in place of p in (1.60a), emerges naturally [see Zeytounian (2000a)]. In this case, we also have the possibility, via an acoustic *adjustment* problem (valid near $t=0$!), of obtaining a Neumann problem for determinating the initial condition $\boldsymbol{u}_i^0(\boldsymbol{x})$.

We observe that the aim of the unsteady-state adjustment problem is clarifying just how a set of initial data for an exact system of fluid flow equations can be related to another set of initial data for simpler, approximate limit fluid flow model equations, which is considered a significant degeneracy of the system of exact equations considered at the start. To solve such a problem, it is necessary to introduce an initial layer in the vicinity of the initial time $t=0$ by distorting the timescale and the unknowns which were initially undefined. For example, when using limit incompressible equations for $\boldsymbol{u}$ and π, one is allowed to specify a set of initial conditions fewer in number (in fact, only one!) than for the full system of compressible baroclinic NSF equations!

This is due to the fact that the limit process (M tends to zero, with t and $\boldsymbol{x}$ both fixed), which leads to the approximate incompressible Navier equation for a divergenceless velocity vector, filters out some time derivatives – those corresponding to acoustic fast waves – because such waves are of no importance for low-speed aerodynamics and various atmospheric and oceanic motions. Due to this, one encounters the problem of deciding what initial condition one may prescribe for $\boldsymbol{u}$, for solving an unsteady-state incompressible viscous Navier equation, and in what way this initial condition is related to the initial conditions for the exact, compressible, baroclinic NSF equations. Note that the exact initial conditions for the full system of compressible equations are not in general consistent with the estimates of basic orders of magnitude implied by the approximate, incompressible limit (without acoustic waves!) Navier equation.

A physical process of temporal evolution is necessary to bring the initial set to a consistent level as far as the order of magnitude is concerned. Such a process is precisely called one of *"unsteady-state adjustment"* of the initial set of data to the approximate structure of incompressible limiting equations under consideration.

The process of adjustment, which occurs in many fields of fluid mechanics, is short on the timescale of approximate simplified equations, and at the end of it, in an asymptotic sense, we obtain values for the set of initial conditions suitable for the simplified equations.

If we consider, for instance, our approximate simplified equations, derived in Chaps. 4 to 6 for Re $\gg 1$, Re $\ll 1$, and $M \ll 1$, then it is necessary, first, to elucidate the problems of the adjustment to the Prandtl boundary layer, to the Stokes and Oseen steady-state, and to the Navier and Boussinesq flows. A number of adjustment problems occur in meteorology for atmospheric motions (adjustment to hydrostatic balance and to geostrophy, for example); the reader can find a deep discussion of these adjustment problems in Zeytounian (1991a; Chap. V). But note that, depending on the physical nature of the problems, we may have two kinds of behavior when the rescaled (short) time goes to infinity. Either one may have a tendency toward a limiting steady state or an undamped set of oscillations [as, for example, the inertial waves in the inviscid problem of spin-up for a rotating fluid; see Greenspan (1990, §2.4)]. For the terminology of the initial layer as adapted to this kind of singular perturbation problem, we refer to Nayfeh (1973, p. 23). Finally, we note that the unsteady-state adjustment to aerodynamic (or meteorological) fields results from the generation, dispersion, and damping of fast internal waves. According to MMAE (Method of Matched Asymptotic Expansions), the initial conditions for model limiting equations are matching conditions between the two asymptotic representations – the main (outer) one (with t fixed) and the local (innertime) one (near $t = 0$).

Finally, for rigorous mathematical results on singular incompressible limits in compressible fluid dynamics, see, for instance, the recent paper by Beirão da Veiga (1994) and also the various references in that paper. More recent papers have been published on the "compressible $\to$ incompressible" limit (by Desjardins, P.-L. Lions, Grenier, Masmoudi, Hagstrom, J. Lorenz, and Iguchi), but here we do not consider these contributions (see the list of References). We believe that this singular limit (in fact, low-Mach asymptotics) deserves a serious fluid dynamics approach via asymptotic modeling, and a review would be appropriate [see, for instance, Zeytounian (2002b), Chaps. 5, 8, and 11].

1.7.2 Boundary Conditions

Several boundary conditions could be considered for different physical situations.

If we consider, as a simple example, the motion of a fluid in a rigid container Ω (with boundary $\partial\Omega$ independent of time t) and a bounded connected open subset of $\boldsymbol{R}^d$ (where $d > 1$ is the physical dimension), the different structure of the *viscous* equations (on the contrary to Euler nonviscous equations) leads to the necessity, when $\mu > 0$ and $\mu_v \equiv \lambda + (2/3)\mu > 0$, to assume the *no-slip* condition:

$$\boldsymbol{u} = 0 \quad \text{on } \partial\Omega . \tag{1.61}$$

As for the (absolute) temperature T, the boundary condition takes different forms in the two alternative cases: $k > 0$ and $k = 0$.

For heat *conductive fluids*, $k > 0$, several boundary conditions have a physical meaning. Just to limit ourselves to the most common cases, we can require that

$$T = T_w \quad \text{on } \partial\Omega \quad \text{(Dirichlet)} \tag{1.62a}$$

$$k\frac{\partial T}{\partial n} = \Xi \quad \text{on } \partial\Omega \quad \text{(Neumann)} \tag{1.62b}$$

$$k\frac{\partial T}{\partial n} + h(T - T_0) = \Xi \quad \text{on } \partial\Omega \text{ (third type)}, \tag{1.62c}$$

where $T_w > 0$ and Ξ are known functions and $h > 0$ is a given constant.

According to a recent paper by Gresho (1992, pp. 47–52], if $\boldsymbol{u}(0, \boldsymbol{x}) \equiv \boldsymbol{u}^0(\boldsymbol{x})$ is the initial (for $t = 0$) velocity field for the Navier unsteady-state equation (1.60a), then it is necessary to impose the incompressibility constraint, $\nabla \cdot \boldsymbol{u}^0(\boldsymbol{x}) = 0$ in domain Ω, and on the boundary, $\partial\Omega = \Gamma$,

$$\boldsymbol{n} \cdot \boldsymbol{u}^0(\boldsymbol{s}) = \boldsymbol{n} \cdot \boldsymbol{w}(0, \boldsymbol{s}) = \boldsymbol{n} \cdot \boldsymbol{w}^0(\boldsymbol{s}) , \tag{1.63}$$

where $\boldsymbol{w}(t, \boldsymbol{s})$ is the specified boundary condition,

$$\boldsymbol{u}(t, \boldsymbol{s}) = \boldsymbol{w}(t, \boldsymbol{s}) \text{ on } \Gamma \text{ for } t > 0 , \tag{1.64}$$

for the unsteady-state Navier velocity vector which satisfies (1.60a).

However, in many situations (inflow-outflow problems), we cannot assume that the velocity vanishes on $\partial\Omega$. This is the case, for instance, for the flow around an airfoil, where an inflow region is naturally present upstream (and an outflow region appears in the wake). In these cases, several different boundary conditions may be prescribed. In the viscous case, a (nonzero) Dirichlet boundary condition on the velocity-field can be imposed everywhere, or, alternatively, only in the inflow region, i.e., the subset of $\partial\Omega$ where: $\boldsymbol{u} \cdot \boldsymbol{n} < 0$, whereas, in the remaining part of the boundary, the conditions,

$$\boldsymbol{u} \cdot \boldsymbol{n} = U^+ > 0 \quad \text{and} \quad (\boldsymbol{n} \cdot \mathbf{D}(\boldsymbol{u})) \cdot \boldsymbol{t} = 0 , \tag{1.65}$$

have to be prescribed [here $\boldsymbol{t}$ is a unit tangent vector on $\partial\Omega$ and $\mathbf{D}(\boldsymbol{u})$ is the rate of strain (deformation) tensor].

Note also that the conditions:

$$u \cdot n = 0 \quad \text{and} \quad (n \cdot D) \cdot t = 0 \,, \tag{1.66}$$

could be considered on the whole $\partial\Omega$; in this case, however, no inflow or outflow regions would be present.

It is more important to analyze the boundary condition for density ρ because now it turns out that it is necessary to prescribe it on the inflow region. In fact, the first-order hyperbolic continuity equation (1.30) can be solved by the theory of characteristics, and the boundary datum for ρ on the inflow region is a (necessary) Cauchy datum for the density on a noncharacteristic surface.

Further information on inflow-outflow boundary-value problems for compressible viscous NS equations can be found in two pertinent papers by Gustafsson and Sundström (1978) and Oliger and Sundström (1978). Note that in numerical, computational, fluid dynamics, these problems of boundary conditions are well discussed.

Another interesting set of boundary conditions appears when we consider the free-boundary problem, i.e. a problem for which the fluid is not contained in a given domain but can move freely. In this case, the vector $n \cdot T$ is prescribed on (interface) $\partial\Omega$, where moreover $u \cdot n$ is required to be zero (stationary case) or equal to the normal velocity of the boundary itself (nonstationary case).

The value of $n \cdot T$ can be zero (free expansion of a fluid in a vacuum), or

$$-p_e n + 2\sigma K n + \nabla_s \sigma \,, \quad \text{at the interface}, \tag{1.67}$$

where p_e is the external pressure, σ is the surface tension (temperature-dependent, when the fluid is an expansible liquid), K is the mean interfacial curvature, and $\nabla_s = \nabla - n(n \cdot \nabla)$ is the surface (projected) gradient at the interface, respectively.

But, in this (viscous) case, it is also necessary to write a heat transfer condition across the interface (for an expansible and thermally conducting fluid); for instance, we write:

$$k\frac{\partial T}{\partial n} + h_s T = \text{prescribed function}, \tag{1.68}$$

and with the coefficient h_s, we can define a Biot number. Because $d\sigma/dT \neq 0$, then for the "film" problem, it is necessary to take into account a Marangoni number proportional to $(d\sigma/dT)_{T=T^0}$, where T^0 is a constant reference temperature. In this case, we consider a thin-film "Bénard–Marangoni" free-surface problem, which is fundamentally different from the classical "Rayleigh–Bénard" instability problem [see, for instance, Velarde and Zeytounian (2002) or the Chapter 10 in Zeytounian (2002b)].

Naturally, we are now imposing one more condition on the interface $\partial\Omega$ because it is an unknown of the problem.

1.7 Conditions for Unsteady-State NSF Equations

For various viscous problems, the viscous equations require additional boundary conditions, and, as an effect, viscous boundary layers may occur at the boundaries. Such boundary layers may sometimes be appropriate, as in the rigid wall situation (for example, the Ekman boundary layer). However, they are inappropriate at open boundaries.

In comparison to flows in interior or exterior domains, there are two new issues when the *boundary extends to infinity*. First, in addition to the usual initial and boundary conditions, there needs to be some prescription of fluxes or pressure drops when the flow domain has several "exits to infinity." Second, the solutions of interest are often estimates developed to deal with this problem. These estimates are called "Saint-Venant type" because the method was first used in the study of Saint-Venant's principle in elasticity. For the behavior of an incompressible fluid velocity field at infinity, we note that a simple method is given in Dobrokhotov and Shafarevich (1996), which makes it possible to determine an upper bound for the decay rate at infinity of an incompressible fluid velocity field of general form, that is, to determine a lower bound for the field itself. This method is based on using simple integral identities that are valid for solving Navier incompressible, viscous equations in an external region that decrease quickly enough. For equations in entire space, some of these identities were obtained by these two authors. The property of "slow decay" or "spreading" of localized fluid flow is a consequence of incompressibility and is not associated with viscosity alone.

To compute the fluid flow in a bounded region modeled by a problem formulated on an infinite domain, one often introduces an artificial boundary Σ and tries to write a "new" problem on the domain $\Omega^* \subset \Omega$ bounded by Γ and Σ whose solution is as close as possible to the original "exact" problem. When the solution of this new problem in Ω^* coincides with the restriction of the original problem, the boundary Σ is said to be transparent. Here, we note only that the reader can find valuable information concerning this approach with applications to viscous fluid flows in recently published papers in the leading journal devoted to numerical fluid dynamics (see, for instance, the recent issues of the *Journal of Computational Physics*, and as example we mention the paper by Tourette (1997)). A less recent, but very pertinent paper is Gustafsson and Kreiss (1979).

Concerning the weak form of the no-slip condition on a moving wall, we believe that the general slip condition,

$$\boldsymbol{n} \cdot (\boldsymbol{u} - \boldsymbol{u}_P) = 0 \,, \tag{1.69}$$

is satisfied in any case for an impermeable solid wall. In (1.69), $\boldsymbol{u}_P$ is the velocity of the moving wall. On the other hand, from the kinetic theory of gases, when the Knudsen number Kn ($= M/\text{Re}$) is small, we obtain

$$\boldsymbol{n} \wedge (\boldsymbol{u} - \boldsymbol{u}_P) = 0 \,, \quad \text{on the moving wall.} \tag{1.70}$$

As consequence of (1.69) and (1.70), we deduce again the classical no-slip condition (but for a moving wall)

$$\boldsymbol{u} = \boldsymbol{u}_{\text{P}}, \quad \text{on the moving wall.} \tag{1.71}$$

The condition (1.70) is the so-called weak form of the no-slip condition on a moving wall.

For the boundary condition for temperature T on a wall, from the kinetic theory of gases, again when the Knudsen number Kn is small, we obtain

$$T = T_{\text{P}} - \beta \boldsymbol{q} \cdot \boldsymbol{n}, \tag{1.72}$$

where β is a scalar function (related to the kinetic Knudsen sub-layer).

An interesting case of a boundary condition is related to the so-called "Prandtl–Batchelor" condition [see, for instance, the papers by Batchelor (1956) and also by Wood (1957)]. For a 2-D incompressible, steady-state Eulerian fluid flow, when $\text{Re} \equiv \infty$, we derive the following Laplace equation for the 2-D stream function $\psi(x,y)$:

$$\nabla^2 \psi = F(\psi),$$

where the function $F(\psi)$ is arbitrary! If the domain Ω where the flow is considered is a bounded connected open subset of $\boldsymbol{R}^2$, then we do not have the possibility of using the behavior condition at infinity to determinate this function $F(\psi)$.

The key of this indeterminacy is strongly related to the vanishing viscosity problem [in fact, to the limiting process $\text{Re} \to \infty$ in the steady-state form of the Navier equation (1.60a) with $\boldsymbol{f} = 0$].

If we assume that the limit streamlines are closed in Ω, then according to Batchelor (1956), we derive the following Prandtl–Batchelor condition for the limit Euler streamline γ^0 (which is the one of the streamlines $\psi^0 = \text{constant}$):

$$\frac{\mathrm{d}F(\psi^0)}{\mathrm{d}\psi^0} = 0 \quad \text{and} \quad F(\psi^0) = F^{00} \equiv \text{const.} \tag{1.73}$$

Consequently, the Eulerian vorticity, $\omega^0 = -(1/2)F(\psi^0)$, for steady-state incompressible 2-D fluid flow is constant in all regions where streamlines are closed. From matching [Wood (1957)] with the corresponding Prandtl boundary layer in the vicinity of $\partial\Omega$, the value of the constant F^{00} is specified.

In Guiraud and Zeytounian (1984a), a process for setting in motion a viscous incompressible liquid inside a 2-D cavity is considered, and it is shown that the basic process occurs for time of the order of $t = O(\text{Re})$, when $\text{Re} \gg 1$, and that the Prandtl–Batchelor flow with constant Euler vorticity is established after time $t \gg \text{Re}$. In this GZ paper, a functional equation is derived that governs the distribution of vorticity in the main stage of interest, and, it is shown that the vorticity in a cylindrical cavity tends toward its own steady-state value exponentially (see Sect. 7.7.1 in Chap. 7).

Finally, note that, when the number of boundary conditions for a given problem is overspecified, the finite-difference approximation (for a numerical calculation) may well be stable. However, the effective boundary conditions that influence the solution are, in general, difficult to determine, especially for problems in several space dimensions. They may well be complicated functions of the conditions given and bear little resemblance to them. An additional complication induced by overspecification is that the underlying solution that is being approximated is not generally continuous. To avoid the problems associated with the proper selection of boundary conditions, the order and type of the differential equations are often raised to obtain a problem that is easier to analyze and approximate. For example, Eulerian equations are usually modified by adding dissipative terms so that the number of boundary conditions is appropriate. Unfortunately, this idea seldom works. If a spurious boundary layer of appreciable size results, the effects are not unlike those for discontinuities (for a system of equations, the errors can propagate away from the discontinuity through other components of the solution) and, unless the dissipative terms are very large, the error introduced at the boundary will again propagate into the interior. Now, if the boundary conditions are underspecified, there are no a priori estimates for the differential equations. For an approximation to be computable, there must be a sufficient number of boundary conditions specified for the approximation. This cannot be fewer than the number required for the differential equation.

Again, the unsteady-state viscous system of equations is essentially parabolic in time and space or hyperbolic-parabolic, whereas the steady-state part is hyperbolic-elliptical due to the hyperbolic character of the compressible continuity equation considered for a known velocity field. For initial boundary-value problems where the flow functions are given at $t = 0$, boundary conditions are imposed at all times for velocity and temperature on the boundary of the flow domain.

The well-posedness of initial boundary-value viscous fluid flow problems follows to some extent from properly formulated initial and boundary conditions. In Chap. 8, some comments are made concerning the various facets (through the existence and uniqueness results) of the solvability of these viscous fluid flow problems.

We note here only that from a mathematical point of view, *no general, global existence theorems for unsteady-state NSF equations with a defined set of initial and boundary conditions can be defined*!

Some partial, local existence theorems have been obtained, and the reader can find information in our recent review papers [Zeytounian (1999, 2001)].

2 Some Features and Various Forms of NSF Equations

2.1 Isentropicity, Polytropic Gas, Barotropic Motion, and Incompressibility

2.1.1 NS Equations

For an adiabatic ($q_i = 0$) and nonviscous (inviscid) fluid ($\mu = 0$ and $\lambda = 0$) from (13) of the Introduction,

$$\frac{\mathrm{d}S}{\mathrm{d}t} = 0 \, . \tag{2.1}$$

In general, the motion is not isentropic [S is different from a constant at each point of the flow and in time], but constant for each particle (along the trajectory). Even if the flow is isentropic, this is valid only up to the first shock front encountered by the particles, after which the isentropic property may well fail.

On the other hand, if in (13) for S, we neglect heating due to viscous dissipation [the term: $(2\mu d_{ij} + \lambda d_{kk}\delta_{ij})d_{ij}$], an ad hoc approximation which seems reasonable (at least from the applied mathematician's point of view), except for hypersonic (high Mach number) flow and assume that thermal conductivity k is vanishing ($k \downarrow 0$), we obtain again the above "Lagrangian" invariant (2.1) for $S(t, \boldsymbol{x})$. An initial condition (at $t = 0$) is necessary for this (2.1). But if S is constant at $t = 0$, (initially $S^0 = $ const), we now expect to deduce $S \equiv S^0 = $ const (we do not expect shocks because μ and λ are both positive). For an ideal gas, the equation of state is also written in the following form:

$$p = \rho^\gamma \exp\left(\frac{S}{C_\mathrm{v}}\right) , \quad \gamma = \frac{C_\mathrm{p}}{C_\mathrm{v}} , \tag{2.2}$$

and consequently for an isentropic motion, we write the following specifying equation relating p and ρ:

$$p = \kappa_0 \rho^\gamma , \tag{2.3}$$

with $\kappa_0 = \exp(S^0/C_\mathrm{v}) = $ const.

The specifying relation (2.3) characterizes a polytropic compressible viscous gas, and consequently, we obtain the following NS equation (with dimensions) of motion for $\boldsymbol{u}$:

$$\frac{\partial \boldsymbol{u}}{\partial t} = -(\boldsymbol{u} \cdot \nabla)\boldsymbol{u} - \gamma \kappa_0 \rho^{\gamma-2} \nabla \rho$$
$$+ \frac{\mu_0}{\rho}\left[\Delta \boldsymbol{u} + \left(\frac{1}{3}\right)\nabla(\nabla \cdot \boldsymbol{u})\right]. \tag{2.4a}$$

When the compressible equation of continuity,

$$\frac{\partial \rho}{\partial t} = -\nabla \cdot (\rho \boldsymbol{u}), \tag{2.4b}$$

is also considered, then we obtain the NS closed system of two equations, (2.4a,b), for the velocity $\boldsymbol{u}$ and density ρ, which eliminates the need for a separate energy or entropy equation. The analoguous system of the above NS equations is considered in §15 of Landau and Lifshitz, *Fluid Mechanics* (1959).

More complete "barotropic NS" equations for the unknowns u_i, p, and ρ are obtained from (see Introduction) (4), with (5), the specifying equation (18a), and continuity equation (17), if we assume that viscosities μ and λ in (4), do not depend on temperature T and are known functions only of density ρ:

$$\rho \frac{du_i}{dt} + \frac{\partial p(\rho)}{\partial x_i} = \frac{\partial}{\partial x_j}\left[\mu(\rho)\left(\frac{\partial u_i}{\partial x_j} + \frac{\partial u_j}{\partial x_i}\right)\right] + \frac{\partial}{\partial x_i}\left[\lambda(\rho)\left(\frac{\partial u_k}{\partial x_k}\right)\right], \tag{2.5a}$$

$$\frac{\partial u_k}{\partial x_k} = -\frac{d \log \rho}{dt}, \tag{2.5b}$$

These NS barotropic equations (2.5a,b) form an ad hoc model for studying viscous gas dynamics motions because, strictly speaking, these NS equations are not a limiting consistent form of full NSF equations for compressible baroclinic (heat-conducting) fluid flow!

2.1.2 Navier System

For an incompressible ($\nabla \cdot \boldsymbol{u} = 0$) and *homogeneous* fluid, when

$$\frac{d\rho}{dt} = 0, \quad \text{with } \rho(t=0, \boldsymbol{x}) = \rho^0 \Rightarrow \rho \equiv \rho_0 = \text{const}, \tag{2.6}$$

we derive the following Navier dynamic equation for $\boldsymbol{u}$ from the NS equation (2.5a):

$$\frac{d\boldsymbol{u}}{dt} + \nabla\left(\frac{p}{\rho_0}\right) = \nu_0 \Delta \boldsymbol{u}, \tag{2.7a}$$

where ν_0 is the constant kinematic viscosity.

2.1 Isentropicity, Polytropic Gas, Barotropic Motion, and Incompressibility   37

This equation (2.7a) was first obtained by Navier (1821/1822), and we observe (again) that the gradient $\nabla(p/\rho_0)$ in (2.7a) is not an unknown quantity of the initial value problem. This force term acts on the particles of an incompressible and homogeneous fluid which moves as freely as possible, but in a way compatible with the incompressibility constraint:

$$\nabla \cdot \boldsymbol{u} = 0 \,. \tag{2.7b}$$

2.1.3 Navier System with Time-Dependent Density

On other hand, let us assume that a (weakly compressible) gas is contained in a container Ω bounded by an impermeable but, eventually, deformable wall, so that the volume occupied by the gas is a given function of time, $V(t)$.

An obvious way of going over is to assume that the density and temperature go to definite limits $\rho_0(\boldsymbol{x},t)$, $T_0(\boldsymbol{x},t)$, when $M \downarrow 0$ with t and $\boldsymbol{x}$ fixed. It is a very easy matter to get an equation satisfied by T_0 from the dimensionless energy equation [written for an ideal gas – see (1.32)]. Such an equation involves the unknown function $p_0(t)$, and it has an obvious simple solution, $T_0 = T_0(t)$, which holds, provided that the two unknown functions, $p_0(t)$ and $T_0(t)$, meet the requirement that $T_0/(p_0)^{\frac{\gamma-1}{\gamma}}$ is independent of time. By simply choosing p^0 and T^0, we may assume that this constant is equal to unity. Consequently, it is then very easy to conclude that

$$\rho_0(t) = [p_0(t)]^{1/\gamma} \,, \tag{2.8}$$

and conservation of mass for all of the gas in the container gives

$$p_0(t) = [V(t)]^{-\gamma} \,. \tag{2.9}$$

So we have found our way out of the indeterminacy of the leading term in the expansion for pressure:

$$p = p_0(t) + M^2 p_N(t,\boldsymbol{x}) + \ldots, \quad M \downarrow 0 \text{ with } t \text{ and } \boldsymbol{x} \text{ fixed.} \tag{2.10}$$

Of course, our argument relies on $T_0(t)$ being independent of space, and we have to discuss the adequacy of that. It is, obviously a matter of conduction of heat within the gas. Such a phenomenon might have two origins. The first is dissipation of energy within the gas, but consideration of the energy equation tells us that this enters at the rate of $O(M^2)$ and is negligible as far as only $T_0(t)$ is concerned. The second origin for the conduction of heat is through variations of temperature on the wall or from heat transfer through it.

For a discussion of the incompressible fluid flow limit, see, for instance, Majda (1985).

Let us come back to NSF dimensionless equations (1.30)–(1.32), with Bo $\equiv 0$, and assume that $\boldsymbol{u}$ goes to $\boldsymbol{u}_N(t,\boldsymbol{x})$ according to the limit:

$$\text{Lim}^N = [M \downarrow 0 \quad \text{with } t \text{ and } \boldsymbol{x} \text{ fixed}] \,. \tag{2.11}$$

We set

$$\frac{p_N}{\gamma \rho_0(t)} = \text{Lim}^N \left\{ \frac{p - p_0(t)}{\gamma M^2 \rho_0(t)} \right\} = \pi, \qquad (2.12)$$

a fictitious pressure, and we get the following dimensionless "modified" Navier equations for the velocity vector $\boldsymbol{u}_N$ and the pressure perturbation π:

$$\nabla \cdot \boldsymbol{u}_N = -\frac{1}{\rho_0(t)} \frac{d\rho_0(t)}{dt}, \qquad (2.13a)$$

$$\left[\frac{\partial}{\partial t} + \boldsymbol{u}_N \cdot \nabla \right] \boldsymbol{u}_N + \nabla \pi = \frac{1}{\text{Re}} \left[\frac{\mu(T_0)}{\rho_0(t)} \right] \Delta \boldsymbol{u}_N, \qquad (2.13b)$$

with

$$\rho_0(t) = \frac{1}{V(t)}; \quad p_0(t) = [V(t)]^{-\gamma}; \quad T_0(t) = [V(t)]^{1-\gamma}. \qquad (2.13c)$$

In (2.13a,b), the Strouhal number S, which characterizes the unsteady-state effect, is assumed equal to one for simplicity and Re is the usual Reynolds number (1.35b). Equations (2.13a,b) with (2.13c) are a slight variant of classical (dimensionless) Navier equations,

$$\left[\frac{\partial}{\partial t} + \boldsymbol{u}_N \cdot \nabla \right] \boldsymbol{u}_N + \nabla \pi = \frac{1}{\text{Re}} \Delta \boldsymbol{u}_N; \qquad (2.14a)$$

$$\nabla \cdot \boldsymbol{u}_N = 0, \qquad (2.14b)$$

where

$$\pi = \text{Lim}^N \frac{p-1}{\gamma M^2}, \qquad (2.15)$$

which consists of a fluid with time-dependent viscosity, but also, rather than a divergenceless motion, one constant in space, variable in time, and divergent. The usual set of Navier dimensionless equations (2.14a,b), with (2.15), is, obviously, obtained for a constant volume container or for external aerodynamics, when the basic thermodynamic state at infinity is constant (in fact, unity in dimensionless form).

2.1.4 Fourier Equation

For a constant volume container or for external aerodynamics, from the dimensionless energy equation (1.32) written for an ideal gas, we can associate to above Navier equations (2.14a,b) a so-called Fourier equation for the temperature perturbation:

$$T_N = \text{Lim}^N \frac{T-1}{M^2}. \qquad (2.16)$$

It is easy to derive the following linear, but nonhomogeneous, Fourier equation (with S = 1):

$$\frac{\partial T_N}{\partial t} + \boldsymbol{u}_N \cdot \nabla T_N - \frac{1}{\Pr \text{Re}} \Delta T_N = (\gamma - 1)\left(\frac{\partial \pi}{\partial t} + \boldsymbol{u}_N \cdot \nabla \pi\right)$$

$$+ \left[\frac{\gamma - 1}{\text{Re}}\right] \text{Trace} \, (\mathbf{D}(\boldsymbol{u}_N)^2) \,. \tag{2.17}$$

In (2.17), $\mathbf{D}(\boldsymbol{u}_N)$ is the rate-of-deformation tensor written for velocity $\boldsymbol{u}_N$. Equation (2.17) for the perturbation of temperature T_N is the only consistent equation associated with the Navier system (2.14a,b) for $\boldsymbol{u}_N$ and π, when $M \downarrow 0$ with t and $\boldsymbol{x}$ fixed.

2.2 Some Interesting Issues in Navier Incompressible Fluid Flow

First, the rotational form of the Navier equation for $\boldsymbol{u}_N$ is relevant and interesting for several reasons. It is

$$\frac{\partial \boldsymbol{u}_N}{\partial t} + \omega_N \wedge \boldsymbol{u}_N + \nabla H_N + \frac{1}{\text{Re}} \nabla \wedge \omega_N = 0 \,, \tag{2.18}$$

where

$$\omega_N = \nabla \wedge \boldsymbol{u}_N \text{ is the (Navier) vorticity} \tag{2.19}$$

and

$$H_N = \pi + \frac{1}{2} q_N^2 \,, \quad \text{with } q_N = |\boldsymbol{u}_N| \,, \tag{2.20}$$

a Bernoulli (or total) pressure that has subsumed a good portion of the nonlinearity of the Navier problem, perhaps most, and possibly all if the flow is irrotational ($\omega_N \equiv 0$), and therefore it is a good portion of the difficulty encountered in numerical simulations. It is interesting to note that a recent article by Frisch and Orszag (1990) mentions that certain coherent structures in turbulent fluid flow seem to have the vorticity vector aligned, or nearly aligned, with the velocity vector (so-called "Beltrami" fluid flow):

$$\omega_N \wedge \boldsymbol{u}_N \approx 0 \text{ or is quite small}, \tag{2.21a}$$

and the helicity density

$$\boldsymbol{u}_N \cdot \omega_N \text{ is then quite large}, \tag{2.21b}$$

even when the vorticity is not, resulting in a momentum equation that almost looks linear. If ω_N is parallel to $\boldsymbol{u}_N$, the advection term

$$(\boldsymbol{u}_N \cdot \nabla)\boldsymbol{u}_N = \frac{1}{2}\nabla q_N^2$$

is irrotational. But we observe that ω_N is always orthogonal to $\boldsymbol{u}_N$ in 2-D fluid flow. Conversely, a large $\omega_N \wedge \boldsymbol{u}_N$ term ($\omega_N \perp \boldsymbol{u}_N$ or nearly so, with concomitant small helicity density) may be correlated with regions of strong turbulence.

Consequently, if the flow is nearly irrotational in large regions of the domain (a common situation in aerodynamics, for example), the Navier equation in the form given by (2.18) becomes approximately,

$$\frac{\partial \boldsymbol{u}_N}{\partial t} + \nabla H_N = 0 \; ; \tag{2.22}$$

i.e., it describes a time-varying potential flow.

An interesting aspect of (2.18) is that no derivatives are higher than first order. If (2.18), with $\nabla \cdot \boldsymbol{u}_N = 0$, and the definition of vorticity, $\omega_N = \nabla \wedge \boldsymbol{u}_N$, are solved simultaneously (for velocity, vorticity, and pressure), the door is open to solution methods that are well suited to dealing only with first-order spatial derivatives.

For the 2-D plane flow, using the identity

$$\nabla \wedge \boldsymbol{\omega}_N = -\boldsymbol{\omega}_N \wedge \nabla \ln |\omega_N| \; , \text{ where } \boldsymbol{\omega}_N = \omega_N \boldsymbol{k} \; ,$$

and $\boldsymbol{k}$ is the unit vector in the third direction (perpendicular to 2-D plane flow), we obtain in place of (2.18),

$$\frac{\partial \boldsymbol{u}_N}{\partial t} + \boldsymbol{\omega}_N \wedge \boldsymbol{V}_N + \nabla H_N = 0 \; , \tag{2.23a}$$

with

$$\boldsymbol{V}_N \equiv \boldsymbol{u}_N - \frac{1}{\text{Re}} \nabla \ln |\omega_N| \; . \tag{2.23b}$$

In the steady-state case, taking the scalar product of the modified velocity vector $\boldsymbol{V}_N$ with the steady-state version of (2.23a), we obtain an extension of Bernoulli's theorem to viscous flow:

$$\boldsymbol{V}_N \cdot \nabla H_N = 0 \; , \tag{2.24}$$

i.e., H_N remains constant along the "streamlines" of $\boldsymbol{V}_N$; however, $\boldsymbol{V}_N$ is not solenoidal! On the other hand, the curl of (2.23a) yields a version of the vorticity transport equation in the plane,

$$\frac{\partial \omega_N}{\partial t} + \nabla \cdot (\omega_N \boldsymbol{V}_N) = 0 \; , \tag{2.25}$$

which, again for the steady-state situation, reveals that vortical transfer occurs along these same streamlines:

$$\omega_N \boldsymbol{V}_N \sigma = \text{const} , \tag{2.26}$$

where σ is the small thickness of a stream tube of $\boldsymbol{V}_N$.

Consequently [see Gresho (1992), p. 55],

"In steady-state plane incompressible flows in the presence of a body force potential, the Bernoulli function remains constant along the lines of vorticity transfer – a valid result for viscous (Navier) or inviscid (Euler) flows."

2.2.1 The Pressure Poisson Equation

Using the incompressibility constraint, $\nabla \cdot \boldsymbol{u}_N = 0$ and operating on Navier equation (2.14a) with the divergence operator, we derive the following Poisson equation for pressure π:

$$\nabla^2 \pi = -\nabla \cdot (\boldsymbol{u}_N \cdot \nabla \boldsymbol{u}_N) \quad \text{in the domain } \Omega , \tag{2.27}$$

where, if desired, the quasi-linear source term can also be expressed, using $\nabla \cdot \boldsymbol{u}_N = 0$, again, as

$$\nabla \cdot (\boldsymbol{u}_N \cdot \nabla \boldsymbol{u}_N) = \nabla \boldsymbol{u}_N : \nabla \boldsymbol{u}_N \equiv \left(\frac{\partial u_{Ni}}{\partial x_j}\right)\left(\frac{\partial u_{Nj}}{\partial x_i}\right) . \tag{2.28}$$

Note that we also have assumed that the acceleration is divergence-free to obtain (2.27).

When the pair (2.14a) for $\boldsymbol{u}$ and (2.27) for π is employed properly [see, Gresho (1992)], it can be used to solve the Navier model equations, i.e., these equations can give the same solution; the resulting velocity field will be divergence-free even though $\nabla \cdot \boldsymbol{u}_N = 0$ is not directly employed in obtaining the solution. The pressure field evolving in time is always in equilibrium [because the "pressure Poisson equation (PPE)" is elliptical] with the time-varying solenoidal velocity field because the speed of sound is infinite ($M \downarrow 0$). We observe also that the influence of local acceleration and the viscous term can only manifest at domain boundaries. Not a bad question is:

"Why form the PPE"?

The answer is obvious: The PPE provides the hope of separating the velocity calculation from the pressure calculation. And that decoupling leads (sometimes) to another:

The separation of one velocity component from another – at least in the viscous terms. Concerning methods for using PPE, see Glowinski (1984).

2.2.2 $\Psi_N - \omega_N$ and $u_N - \omega_N$ Formulations

We consider, first, the following vorticity 2-D transport equation:

$$\frac{\partial \omega_N}{\partial t} + u_N \cdot \nabla \omega_N = \frac{1}{\text{Re}} \nabla^2 \omega_N \,, \tag{2.29}$$

and we introduce the stream function Ψ_N, via

$$u_N = \frac{\partial \Psi_N}{\partial y} \quad \text{and} \quad v_N = -\frac{\partial \Psi_N}{\partial x} \Rightarrow u_N = \nabla \Psi_N \wedge k \,, \tag{2.30}$$

to arrive at the second in the pair called the ($\Psi_N - \omega_N$) formulation, via

$$\omega_N = \frac{\partial v_N}{\partial x} - \frac{\partial u_N}{\partial y} \tag{2.31a}$$

$$\Rightarrow \nabla^2 \Psi_N = -\omega_N \quad \text{in } \Omega \,, \tag{2.31b}$$

an elliptical equation that replaces the elliptical PPE of the other derived variable formulation. Now, the integration of (2.31b) over the 2-D domain Ω yields the following 2-D version of Stokes's circulation theorem:

$$-\int_\Gamma \left(\frac{\partial \Psi_N}{\partial n}\right) d\Gamma = \int_\Omega \omega_N \, d\Omega \,, \tag{2.32}$$

an easy way to obtain the global conservation law for vorticity and where $\partial \Omega = \Gamma$. Relation (2.32) must be satisfied for arbitrary ω_N [for $x = (x,y)$ in Ω a 2-D domain] and arbitrary $\partial \Psi_N / \partial n$ (and for all time $t \geq 0$) in order that (2.31b) possesses a solution. Because (2.31b) always does (it seems) possess a solution of some sort, it is clear that some irregular behaviour of ω_N may well occur, and this does happen near the boundary Γ of the 2-D domain Ω. On the other hand, the basic, $u_N - \omega_N$, equations are (in general) the vorticity transport equation

$$\frac{\partial \omega_N}{\partial t} + u_N \cdot \nabla \omega_N = \omega_N \cdot \nabla u_N + \frac{1}{\text{Re}} \nabla^2 \omega_N \,, \tag{2.33}$$

and the two kinematic equations,

$$\nabla \cdot u_N = 0 \,, \tag{2.34a}$$

$$\omega_N = \nabla \wedge u_N \,. \tag{2.34b}$$

Given appropriate initial and boundary conditions, these can be used to solve for u_N and ω_N. Indeed, the curl of $\nabla \wedge u_N$, using $\nabla \cdot u_N = 0$, leads to another higher order equation – a Poisson equation for velocity:

$$\nabla^2 u_N = -\nabla \wedge \omega_N \,, \tag{2.35}$$

which is still a kinematic equation.

2.2 Some Interesting Issues in Navier Incompressible Fluid Flow

Finally, considering that ω_N is given, an equivalent integral formulation of (2.34a,b) is [see, for instance, Leonard (1985)]

$$\boldsymbol{u}_N = \nabla \phi_N + \boldsymbol{u}_\omega , \qquad (2.36a)$$

where

$$\boldsymbol{u}_\omega = -\beta_d \int \frac{\omega_N(\boldsymbol{x}') \wedge (\boldsymbol{x}' - \boldsymbol{x})}{|\boldsymbol{x}' - \boldsymbol{x}|^d} \, d\Omega(\boldsymbol{x}') , \qquad (2.36b)$$

with

$$\beta_d = \frac{\pi}{4} \text{ for 3-D } (d=3) \quad \text{and} \quad \beta_d = \frac{\pi}{2} \text{ for 2-D } (d=2) .$$

The first component in (2.36a) is determined from

$$\nabla^2 \phi_N = 0 \text{ and the boundary condition } \frac{\partial \phi_N}{\partial n} = \boldsymbol{n} \cdot (\boldsymbol{u} - \boldsymbol{u}_\omega) . \qquad (2.37)$$

The remaining part $\boldsymbol{u}_\omega$ is the vortical (rotational) portion of the velocity field; the velocity $\boldsymbol{u}_N$ from (2.36a,b) will satisfy both (2.34a,b), as well as the appropriate normal boundary conditions on velocity; it may, but is not guaranteed to, and generally will not, also satisfy the tangential velocity conditions [see, for instance, Speziale (1987)].

2.2.3 The Omnipotence of the Incompressibility Constraint

For the Navier (incompressible, everywhere and for all time) model

$$\nabla \cdot \boldsymbol{u}_N = 0 \text{ in } \underline{\Omega} \text{ for } t \geq 0 , \qquad (2.38)$$

where $\underline{\Omega} \equiv \Omega + \Gamma$ and $\partial \Omega = \Gamma$ is the boundary of the domain Ω. Particularly noteworthy are two inclusions:

$$\nabla \cdot \boldsymbol{u}_N = 0 \text{ on } \Gamma \text{ for all } t > 0 \quad \text{and} \quad t = 0 , \qquad (2.39a)$$

and

$$\nabla \cdot \boldsymbol{u}_N = 0 \text{ on } \underline{\Omega} \text{ at } t = 0 , \qquad (2.39b)$$

This Navier model is simple and convenient because it precludes all phenomena associated with compressibility effects and because it is believed that it accurately describes the real behavior of all fluids in many important cases. But we observe that, if we consider the Navier model as a limiting leading-order approximate model derived (for t fixed and small Mach number M) from the full unsteady-state NSF "exact" model, then it is necessary to associate, close to initial time $t = 0$, the classical linear acoustic model that gives the possibility of obtaining asymptotically (via an adjustment process by matching) the initial condition at $t = 0$ for Navier velocity $\boldsymbol{u}_N$ (see, for instance, Chap. 6).

An important consequence of (2.38) is the following relation:

$$\boldsymbol{n} \cdot \boldsymbol{u}_N = \boldsymbol{n} \cdot \boldsymbol{u}_\Gamma \quad \text{on } \Gamma, \tag{2.40}$$

which is just a restatement of $\nabla \cdot \boldsymbol{u}_N = 0$ on Γ and where $\boldsymbol{u}_\Gamma$ is the specified boundary condition for Navier velocity $\boldsymbol{u}_N$.

Consequently, the normal component of the velocity boundary condition is also the realization of (2.38) on Γ. We observe that (2.38) and (2.40) apply for all t – and in particular at $t = 0$, where they become

$$\nabla \cdot \boldsymbol{u}_N^0 = 0 \quad \text{in } \Omega, \tag{2.41a}$$

and

$$\boldsymbol{n} \cdot \boldsymbol{u}_N^0 = \boldsymbol{n} \cdot \boldsymbol{u}_\Gamma(0, \boldsymbol{P}) \equiv \boldsymbol{n} \cdot \boldsymbol{u}_\Gamma^0, \quad \text{on } \Gamma. \tag{2.41b}$$

We note again that $\boldsymbol{u}_N^0(\boldsymbol{x}) \equiv \boldsymbol{u}(0, \boldsymbol{x})$ is the initial (at $t = 0$) Navier velocity field and $\boldsymbol{P} \in \Gamma$. Obviously, (2.41a,b) are significant constraints on the data: *the initial velocity must be divergence-free (well known) and the normal component of the initial velocity must agree pointwise with the normal velocity boundary condition at $t = 0$ (less well known).*

Although the constraint of (2.41b) precludes impulsive starts in the conventional sense (an issue we shall return to in Chap. 6 in the framework of low Mach number asymptotics), it actually permits the following type of start-up:

$$\boldsymbol{u}_N^0 = 0 \text{ in } \Omega, \quad \boldsymbol{n} \cdot \boldsymbol{u}_N^0 = \boldsymbol{n} \cdot \boldsymbol{u}_\Gamma^0 = 0 \text{ on } \Gamma \tag{2.42a}$$

but (in a 2-D case, for simplicity)

$$\boldsymbol{\tau} \cdot \boldsymbol{u}_\Gamma^0 = f(s) \quad \text{on } \Gamma, \tag{2.42b}$$

where $\boldsymbol{\tau}$ is the unit tangent vector.

The function $f(s)$ is such that

$\partial f/\partial s$ *is not required to vanish* $\Rightarrow \nabla \cdot \boldsymbol{u}_N$ *different from zero on* Γ, for the less stringent definition of divergence on Γ.

Finally, note, too, that, in addition to the requirement that the normal component be smooth in $\boldsymbol{x}$ (for $\boldsymbol{x} \to \Gamma$, in particular), we also require smoothness in time (for $t \to 0$)!

Now, if (2.41) is violated, according to Gresho (1992, pp. 51 and 52), a nearby well-posed Navier initial boundary-value problem can be obtained by modifying the (proposed but ill-posed) initial data by solving the following problem (project the initial velocity onto the nearest – in L^2 – divergence-free subspace):

$$\boldsymbol{v} = \boldsymbol{u}_N^0 + \nabla \lambda, \quad \nabla \cdot \boldsymbol{v} = 0 \quad \text{in } \Omega \tag{2.43a}$$

with

$$\boldsymbol{n} \cdot \boldsymbol{v} = \boldsymbol{n} \cdot \boldsymbol{u}_\Gamma^0 = 0 \quad \text{on } \Gamma. \tag{2.43b}$$

Solve

$$\nabla^2 \lambda = -\nabla \cdot \boldsymbol{u}_N^0 \quad \text{in } \Omega, \tag{2.44a}$$

with

$$\frac{\partial \lambda}{\partial n} = \boldsymbol{n} \cdot (\boldsymbol{u}_\Gamma^0 - \boldsymbol{u}_N^0) = 0 \quad \text{on } \Gamma; \tag{2.44b}$$

then compute

$$\boldsymbol{v} = \boldsymbol{u}_N^0 + \nabla \lambda, \text{ and replace } \boldsymbol{u}_N^0 \text{ by } \boldsymbol{v} \text{ in } \boldsymbol{u}(0, \boldsymbol{x}) = \boldsymbol{u}_N^0 \text{ in } \Omega. \tag{2.44c}$$

In general, the data and constant are assumed to be "sufficiently smooth," which we avoid trying to define too carefully, except to insist that $\boldsymbol{n} \cdot \boldsymbol{u}_\Gamma(t, \boldsymbol{P})$ be continuous in time, so that $\partial/\partial t[\boldsymbol{n} \cdot \boldsymbol{u}_\Gamma(t, \boldsymbol{P})]$ exists, and to require that $\boldsymbol{n} \cdot \boldsymbol{u}_N^0$ is a continuous function of the position space vector $\boldsymbol{x}$ as $\boldsymbol{x} \to \Gamma$, where the boundary unit normal $\boldsymbol{n}$ is imagined to be translated (in the $-\boldsymbol{n}$ direction) to form $\boldsymbol{n} \cdot \boldsymbol{u}_N^0$.

We believe also that $\boldsymbol{u}_\Gamma(t, \boldsymbol{P})$ in the boundary condition:

$$\boldsymbol{u}_N(t, \boldsymbol{P}) = \boldsymbol{u}_\Gamma(t, \boldsymbol{P}) \quad \text{on } \Gamma \text{ for } t > 0, \tag{2.45a}$$

is such that

$$\int_\Gamma \boldsymbol{n} \cdot \boldsymbol{u}_\Gamma(t, \boldsymbol{P}) = 0 \quad \text{for } t \geq 0. \tag{2.45b}$$

If any of the three constraints on the data, (2.41b), (2.45b), is violated, the Navier problem is ill-posed, and no solution exists. Obviously, (2.45b) is an independent constraint only for $t > 0$; it is implied by (2.41) at $t = 0$.

The initial data are not required to satisfy the tangential boundary condition (in general: $\boldsymbol{t} \cdot \boldsymbol{u}_N^0 \neq \boldsymbol{t} \cdot \boldsymbol{u}_\Gamma^0$), and in the general case, vortex sheets will therefore exist on Γ at $t = 0$. There are no boundary conditions on pressure, and the Navier solution will determine pseudopressure π only up to an arbitrary additive constant. But it will turn out that the normal component of Navier equation (2.14a) also applies on Γ at $t = 0$ – and helps to set the initial pressure.

2.2.4 A First Statement of a Well-Posed Initial Boundary-Value Problem (IBVP) for Navier Equations

For Navier model equations (2.14a,b), we have a well-posed IBVP if we take into account the no-slip boundary condition (2.45a), with (2.45b), and the initial condition,

$$\boldsymbol{u}_N(0, \boldsymbol{P}) = \boldsymbol{u}_N^0 \quad \text{in } \Omega, \tag{2.46}$$

with (2.41). Gresho (1992, p. 52) asserts:

"It may be argued that the above definition of a well-posed Navier problem is too strict and that we should permit weaker, generalized solutions that do not exact so much smoothness – especially with regard to the constraints (2.41a,b) that result from (2.38)."

For example, Ladyshenskaya (1969, 1975), Temam (1985), and Hopf (1950, 1951) discuss situations that permit much less smoothness by relaxing these compatibility conditions; Hopf discusses cases wherein no spatial or temporal derivatives (in the classical sense) need even exist (!), and Ladyshenskaya (1975) observes:

"... here as in the rest of the paper we indicate the degree of smoothness of neither the data nor the solution under consideration."

But we shall take the stand, following, e.g., Foias (1979), that all of the above boundary and initial conditions and constraints must be satisfied, based on our belief that most (if not all) numerically computable solutions of Navier model equations should be well-posed as defined in the preceding. Heywood (1979), in his discussion of classical solutions, points out some of the difficulties of 'too generalized' solutions.

The article by Chu (1978) and the book by Kreiss and Lorenz (1989) are at least in partial agreement with us; they impose: $\nabla \cdot \boldsymbol{u}_N^0 = 0$ in Ω, but duck the issue of: $\boldsymbol{n} \cdot \boldsymbol{u}_N^0 = \boldsymbol{n} \cdot \boldsymbol{u}_\Gamma(0, \boldsymbol{P}) \equiv \boldsymbol{n} \cdot \boldsymbol{u}_\Gamma^0$ on Γ.

We observe also that for the following Navier "velocity-vorticity" equations in the unbounded domain outside of the domain Σ:

$$\nabla^2 \boldsymbol{u}_N = -\nabla \wedge \omega_N, \tag{2.47a}$$

$$\frac{\partial \omega_N}{\partial t} = \nabla \wedge (\boldsymbol{u}_N \wedge \omega_N) + \frac{1}{\text{Re}} \nabla^2 \omega_N, \tag{2.47b}$$

we have a well-posed IBVP if we assume (for an 'external' flow problem around a finite body Σ with Γ as wall):

$$\omega_N = 0 \text{ at } t = 0, \quad \text{but } \omega_N \neq 0 \text{ on } \Gamma, \tag{2.48a}$$

$$\boldsymbol{u}_N = \boldsymbol{u}_\Gamma(t, \boldsymbol{P}) \text{ on } \Gamma \quad \text{and} \quad \boldsymbol{u}_N \to 0 \text{ far from } \Gamma. \tag{2.48b}$$

In Chap. 6, we give an asymptotic derivation of the initial condition for the Navier (incompressible and viscous) external unsteady-state fluid flow problem.

2.2 Some Interesting Issues in Navier Incompressible Fluid Flow

In Chap. 8, the reader can find some classical and recent rigorous mathematical results related to the existence and uniqueness of solutions of Navier model equations. For these rigorous mathematical results, here, we recommend the recent books by P.-L. Lions (1996) and Doering and Gibbon (1995) and also the recent survey by Temam (2000), where the reader can find some rigorous developments of Navier–Stokes (Navier!) equations in the second half of the twentieth century.

2.2.5 Cauchy Formula for Vorticity

To derive the Cauchy formula for vorticity $\omega = (\omega_i, \ i = 1, 2, 3)$, it is necessary to write the Navier equation with Lagrangian variables, t and a_j, with $j = 1, 2, 3$, such that

$$a_j = a_j(t, x_i) \quad \text{and} \quad a_i \equiv x_i \ , \ \text{when} \ t = 0 \ ; \quad \frac{\mathrm{d}a_j}{\mathrm{d}t} = 0 \ ; \tag{2.49}$$

$$\frac{\partial^2 x_i}{\partial t^2} \frac{\partial x_i}{\partial a_j} = -\frac{\partial \pi}{\partial a_j} + \frac{1}{\mathrm{Re}} \nabla^2 u_{\mathrm{N}i} \frac{\partial x_i}{\partial a_j} \ , \tag{2.50}$$

where $j = 1, 2, 3$ and $\partial x_i / \partial t = u_{\mathrm{N}i}$.

First, we have the following exact differential:

$$\frac{\partial}{\partial t}\left[u_{\mathrm{N}i} \frac{\partial x_i}{\partial a_j} \right] \mathrm{d}a_j - \frac{1}{\mathrm{Re}}\left[\nabla^2 u_{\mathrm{N}i} \frac{\partial x_i}{\partial a_j} \right] \mathrm{d}a_j = \mathrm{d}F(t, a_j)$$

and by integration from $t = 0$ to t, we obtain (because $\partial x_i / \partial t = u_{\mathrm{N}i}$)

$$\left[u_{\mathrm{N}i} \frac{\partial x_i}{\partial a_j} - u_{\mathrm{N}j}^0 - \frac{1}{\mathrm{Re}} \int_0^t \nabla^2 u_{\mathrm{N}i} \frac{\partial x_i}{\partial a_j} \mathrm{d}t \right] \mathrm{d}a_j = \int_0^t F(t, a_j) \mathrm{d}t \ .$$

Consequently, we derive the following relation:

$$\frac{\partial}{\partial a_k}\left[u_{\mathrm{N}i} \frac{\partial x_i}{\partial a_j} - u_{\mathrm{N}j}^0 - \frac{1}{\mathrm{Re}} \int_0^t \nabla^2 u_{\mathrm{N}i} \frac{\partial x_i}{\partial a_j} \mathrm{d}t \right]$$

$$= \frac{\partial}{\partial a_j}\left[u_{\mathrm{N}l} \frac{\partial x_l}{\partial a_k} - u_{\mathrm{N}k}^0 - \frac{1}{\mathrm{Re}} \int_0^t \nabla^2 u_{\mathrm{N}l} \frac{\partial x_l}{\partial a_k} \mathrm{d}t \right] \ . \tag{2.51}$$

But $\partial u_{\mathrm{N}i} / \partial a_k = (\partial u_{\mathrm{N}i}/\partial x_s) \partial x_s / \partial a_k$ and $\partial u_{\mathrm{N}s}/\partial a_j = (\partial u_{\mathrm{N}s}/\partial x_i)\partial x_i/\partial a_j$, so in place of (2.51), we obtain

$$\left(\frac{\partial x_i}{\partial a_j}\right) \frac{\partial x_s}{\partial a_k} \left[\frac{\partial u_{\mathrm{N}i}}{\partial x_s} - \frac{\partial u_{\mathrm{N}s}}{\partial x_i} \right] = \frac{\partial u_{\mathrm{N}j}^0}{\partial a_k} - \frac{\partial u_{\mathrm{N}k}^0}{\partial a_j}$$

$$-\frac{1}{\mathrm{Re}} \int_0^t \left[\frac{D(\nabla^2 u_{\mathrm{N}i}; x_i)}{D(a_j; a_k)} \right] \mathrm{d}t \ , \tag{2.52}$$

where

$$\frac{D(f;g)}{D(a;b)} = \frac{\partial f}{\partial a}\frac{\partial g}{\partial b} - \frac{\partial f}{\partial b}\frac{\partial g}{\partial a}.$$

Finally, because $(\partial x_i/\partial a_j)\partial a_j/\partial x_k \equiv \delta_{ik}$, with $\delta_{kk} = 1$ and $\delta_{ik} = 0$ if $i \neq k$, for t fixed, then from (2.52), we derive the Cauchy formula:

$$2\omega_{is} \equiv \frac{\partial u_{Ni}}{\partial x_s} - \frac{\partial u_{Ns}}{\partial x_i} = 2\omega_{jk}^0 \left(\frac{\partial a_j}{\partial x_i}\right)\frac{\partial a_k}{\partial x_s}$$
$$- \left\{\frac{1}{\mathrm{Re}}\int_0^t \left[\frac{D(\nabla^2 u_{Ni}; x_i)}{D(a_j; a_k)}\right] dt\right\}\left(\frac{\partial a_j}{\partial x_i}\right)\frac{\partial a_k}{\partial x_s}. \qquad (2.53)$$

Now, it is obvious that the classical theorem of Lagrange is not valid because of the viscous term (proportional to $1/\mathrm{Re}$) in (2.53).

Even if $2\omega_{jk}^0 \equiv 0$, the local components of vorticity ω_{is} are different from zero when $t > 0$.

As a consequence, a salient feature of the motion of viscous fluids is the presence of vorticity!

2.2.6 The Navier Equations as an Evolutionary Equation for Perturbations

We consider again the Navier equations for $\boldsymbol{u}_N$ and π, and we write the following decomposition:

$$\boldsymbol{u}_N = \boldsymbol{U} + \boldsymbol{v}, \quad \pi = \Pi + p. \qquad (2.54)$$

For perturbations $(\boldsymbol{v}, p)$, relative to basic flow $(\boldsymbol{U}, \Pi)$, we derive from the Navier system the following equations for perturbations:

$$\frac{\partial \boldsymbol{v}}{\partial t} = \frac{1}{\mathrm{Re}}\nabla^2 \boldsymbol{v} - (\boldsymbol{U}\cdot\nabla)\boldsymbol{v} - (\boldsymbol{v}\cdot\nabla)\boldsymbol{U} - \nabla p - (\boldsymbol{v}\cdot\nabla)\boldsymbol{v}; \qquad (2.55a)$$
$$\nabla\cdot\boldsymbol{v} = 0. \qquad (2.55b)$$

Now, we write for perturbation p, $p = p_1 - p_2$ and, first, we introduce a "projection" operator $\mathcal{P}(\boldsymbol{v})$ such that, when it is applied to $\boldsymbol{v}$, gives (by definition) $\boldsymbol{v} + \nabla p_1$:

$$\mathcal{P}(\boldsymbol{v}) \equiv \boldsymbol{v} + \nabla p_1. \qquad (2.56a)$$

On the other hand, p_1 is such that (again by definition)

$$\nabla\cdot\mathcal{P}(\boldsymbol{v}) \equiv \nabla\cdot\boldsymbol{v} + \nabla^2 p_1 = 0 \Rightarrow \nabla^2 p_1 = 0, \qquad (2.56b)$$

2.2 Some Interesting Issues in Navier Incompressible Fluid Flow

and also

$$\nabla^2 p_1 = 0 \text{ with } \mathcal{P}(\boldsymbol{v}) \cdot \boldsymbol{n} = 0 \text{ on the wall } \Gamma, \qquad (2.56c)$$

where $\boldsymbol{n}$ is the unit vector of the normal at the wall Γ directed to the fluid.

Obviously, the component p_1 in the decomposition $p = p_1 - p_2$ must satisfy the following Neumann (external) problem:

$$\nabla^2 p_1 = 0, \quad \frac{\mathrm{d}p_1}{\mathrm{d}n} = -\boldsymbol{v} \cdot \boldsymbol{n} \quad \text{on the wall } \Gamma, \qquad (2.57a)$$

$$p_1 \to 0 \quad \text{at infinity, far from } \Gamma. \qquad (2.57b)$$

Consequently, when the perturbation of the velocity $\boldsymbol{v}$ is known, then it is possible to determine p_1 from (2.57a,b).

Next, we introduce a *linear* operator in $\boldsymbol{v}$, dependent on the basic velocity $\boldsymbol{U}$ and ∇p_1:

$$\mathcal{L}(\boldsymbol{U})\boldsymbol{v} = \mathcal{P}\left\{-\boldsymbol{U} \cdot \nabla \boldsymbol{v} - \boldsymbol{v} \cdot \nabla \boldsymbol{U} + \frac{1}{\mathrm{Re}} \nabla^2 \boldsymbol{v}\right\}$$

$$\equiv \frac{1}{\mathrm{Re}} \nabla^2 \boldsymbol{v} - (\boldsymbol{U} \cdot \nabla)\boldsymbol{v} - (\boldsymbol{v} \cdot \nabla)\boldsymbol{U} - \nabla p_1, \qquad (2.58)$$

and we have the following divergenceless condition:

$$\nabla \cdot [\mathcal{L}(\boldsymbol{U})\boldsymbol{v}] = 0. \qquad (2.59)$$

Then, a second *quasi-linear, quadratic* operator is introduced that contains ∇p_2:

$$\mathcal{Q}(\boldsymbol{v};\boldsymbol{v}) = -\{\nabla p_2 + (\boldsymbol{v} \cdot \nabla)\boldsymbol{v}\}. \qquad (2.60)$$

Finally, in place of the Navier system of perturbations (2.55a,b), we can write the following single evolutionary equation for the perturbation of velocity $\boldsymbol{v}$:

$$\frac{\partial \boldsymbol{v}}{\partial t} = \mathcal{L}(\boldsymbol{U})\boldsymbol{v} + \mathcal{Q}(\boldsymbol{v};\boldsymbol{v}). \qquad (2.61)$$

We observe that

$$\nabla \cdot \mathcal{Q}(\boldsymbol{v};\boldsymbol{v}) = 0, \qquad (2.62)$$

and, consequently,

$$\nabla^2 p_2 = -\nabla \cdot (\boldsymbol{v} \cdot \nabla)\boldsymbol{v}, \qquad (2.63a)$$

which is a Poisson equation for p_2 when $\boldsymbol{v}$ is known. As boundary conditions for the Poisson equation (2.63a), we write:

$$\frac{\mathrm{d}p_2}{\mathrm{d}n} = -\left(\frac{\partial \boldsymbol{v}}{\partial t} + (\boldsymbol{v} \cdot \nabla)\boldsymbol{v}\right) \cdot \boldsymbol{n} \quad \text{on the wall } \Gamma, \qquad (2.63b)$$

and (for an external problem)

$$p_2 \to 0 \quad \text{at infinity, far from } \Gamma . \tag{2.63c}$$

The evolutionary equation (2.61) for $\boldsymbol{v}$ is a differential equation in a suitable (infinite-dimensional Hilbert) space H, and we recall that H consists of L^2 vector fields (finite, kinetic-energy, vector fields). For (2.61), it is necessary to write an initial condition,

$$\boldsymbol{v} = \boldsymbol{v}^0 \quad \text{at } t = 0 . \tag{2.64}$$

The initial-value (Cauchy) problem, (2.61), (2.64), is the starting point for the stability theory (see Chap. 9 where the reader can find some aspects of the so-called "hydrodynamic stability theory") and also for the conventional theory of turbulence [see, for instance, the recent Note by Foias et al. (2001b) and also the recent book by these authors (2001a)]. The finite-dimensional dynamical system approach and the investigations of the routes/scenarios to chaos/turbulence, considered in Chap. 10, also have the initial-value problem (2.61), (2.64) as a starting point. We observe that the above results are valid only in the framework of the Navier (incompressible viscous) model. Obviously, the situation is, unfortunately, less advanced for compressible NS and NSF models!

2.3 From NSF to Hyposonic and Oberbeck–Boussinesq (OB) Equations

2.3.1 Model Equations for Hyposonic Fluid Flows

In external aerodynamics, when pressure p in dimensionless NSF equations (1.30)–(1.32) with (1.33) tends to unity at infinity, it is judicious again to introduce a reduced pressure:

$$p^* = \frac{p-1}{\gamma M^2} . \tag{2.65a}$$

In this case, from (1.33), we can write the following equation of state for an ideal gas:

$$\rho T = 1 + \gamma M^2 p^* . \tag{2.65b}$$

Next, from (1.32), we obtain the following equation for temperature T in (2.65b):

$$\rho S \frac{DT}{Dt} + (\gamma - 1)\nabla \cdot \boldsymbol{u} = \frac{\gamma}{\Pr \operatorname{Re}} \nabla \cdot [k\nabla T]$$
$$+ (\gamma - 1)\gamma M^2 \left\{ \frac{1}{\operatorname{Re}} [2\mu \operatorname{Tr}(\mathbf{D}(\boldsymbol{u})^2) + \lambda(\nabla \cdot \boldsymbol{u})^2] - (\nabla \cdot \boldsymbol{u})p^* \right\} , \tag{2.65c}$$

without any approximations.

2.3 From NSF to Hyposonic and Oberbeck–Boussinesq (OB) Equations

Then, in place of the continuity equation, we obtain, with (2.65b), the following relation for the divergence of the velocity vector $\boldsymbol{u}$:

$$\nabla \cdot \boldsymbol{u} = \mathrm{S}\frac{\mathrm{d}(\log T)}{\mathrm{d}t} - \gamma M^2 \, \mathrm{S}\frac{\mathrm{d}p^*}{\mathrm{d}t} + O(M^4) \, . \tag{2.65d}$$

Finally, for $\boldsymbol{u}$, we can write the following equation [we do not take into account the influence of the Boussinesq number (Bo $\equiv 0$) because we consider classical aerodynamics]:

$$\rho \mathrm{S}\frac{\mathrm{d}\boldsymbol{u}}{\mathrm{d}t} + \nabla p^* = \frac{1}{\mathrm{Re}}\nabla \cdot [2\mu \mathbf{D}(\boldsymbol{u})]$$
$$+ \frac{1}{\mathrm{Re}}\nabla \cdot \left\{ \lambda \mathrm{S}\left(\frac{\mathrm{d}(\log T)}{\mathrm{d}t}\right)\mathbf{I} \right.$$
$$\left. - \gamma M^2 \lambda \mathrm{S}\left(\frac{\mathrm{d}p^*}{\mathrm{d}t}\right)\mathbf{I} \right\} + O(M^4) \, . \tag{2.65e}$$

This "transformed NSF" system of equations for $\boldsymbol{u}$, ρ, T and p^* is written explicitly in terms up the order of M^2.

Now, if we assume that M^2 is a *small* parameter, then at the leading order, when M^2 *tends to zero with t and $\boldsymbol{x}$ fixed*, we derive, from (2.65c–e), the following so-called "*hyposonic*" NSF system of three equations for $\boldsymbol{u}$, T, and p^*:

$$\nabla \cdot \boldsymbol{u} = \mathrm{S}\frac{\mathrm{d}(\log T)}{\mathrm{d}t} \, ; \tag{2.66a}$$

$$\mathrm{S}\frac{\mathrm{d}(\log T)}{\mathrm{d}t} + (\gamma - 1)\nabla \cdot \boldsymbol{u} = \left(\frac{\gamma}{\mathrm{Pr}\,\mathrm{Re}}\right)\nabla \cdot [k\nabla T] \, ; \tag{2.66b}$$

$$\left(\frac{\mathrm{S}}{T}\right)\frac{\mathrm{d}\boldsymbol{u}}{\mathrm{d}t} + \nabla p^* = \frac{1}{\mathrm{Re}}\nabla \cdot [2\mu \mathbf{D}(\boldsymbol{u})]$$
$$+ \frac{1}{\mathrm{Re}}\nabla \cdot \left\{ \lambda \mathrm{S}\left(\frac{\mathrm{d}(\log T)}{\mathrm{d}t}\right)\mathbf{I} \right\} \, . \tag{2.66c}$$

First, in the hyposonic NSF equations (2.66a–c), the dissipative coefficients μ, λ, and k are functions only of temperature T, and density ρ is related to temperature T by:

$$\rho = \frac{1}{T} \, . \tag{2.67}$$

This hyposonic model system (2.66a–c) is a very adequate approximate system for investigating *low-speed external aerodynamics, when compressibility plays an important role mainly as a consequence of thermal effects*. According to Embid (1987), the hyposonic NSF system with intial and boundary conditions seems well posed.

2.3.2 The Oberbeck–Boussinesq Model Equations

The presence of density gradients in a fluid means that gravitational potential energy can be converted into motion through the action of buoyant forces. Density differences can be induced by heating the fluid and by forcing concentration differences in mixtures like salt water. In the so-called Oberbeck(1879)–Boussinesq(1903) (OB) equations, it is assumed that the fluid has uniform density; density differences are recognized only in those terms that drive motion. For an asymptotic derivation of the OB model equations, it is important to note that one needs to change the pressure by roughly five atmospheres to produce the same change of density as a temperature difference of 1 °C. It follows that even vigorous motions of water will not introduce important buoyant forces other than those from temperature variations [see, for instance, Chap. VIII in Joseph (1976), where the reader can find information about OB equations]. The temperature-induced density changes in such motion can be very small, but they are motive power for the convective part of the motion.

Consequently, for a weakly expansible, "pure" liquid (when we neglect any concentrations), it is sufficient to assume the availability of a specifying equation of the form [see (1.39) – in place of (8) for example]:

$$\rho = \rho(T), \tag{2.68}$$

between the density and the temperature of the liquid. The weakness of the expansion of the liquid is characterized by the small parameter [see (1.40) and (1.41)],

$$\wp = \Delta T^0 \alpha^0 \equiv -\left[\frac{\Delta T}{\rho(T^0)}\right]\left(\frac{\mathrm{d}\rho}{\mathrm{d}T}\right)_{T=T^0} \ll 1, \tag{2.69}$$

with $\Delta T^0 = T_1 - T^0 > 0$, where [as in the classical Bénard (1900/1901) thermal convection problem between two parallel plates] T_1 is the constant temperature at the rigid lower plate ($x_3 = 0$) and T^0 the constant temperature of the rigid upper plate ($x_3 = d^0$).

The coefficient of thermal expansion of the liquid is denoted by α^0. The liquid layer thickness is d^0, and the main objective below is to show that d^0 is not arbitrary but should satisfy some conditions (in fact restrictions!) to validate the asymptotic derivation of the so-called "Oberbeck–Boussinesq" (OB) model equations from the full exact NSF, dimensional equations (16a–c), with (17a–c) and (2.68) that govern the evolution of a weakly expansible liquid. But, with (2.68), the specific internal energy $E = E(T)$, and in place of the term $\rho C_\mathrm{v}\,\mathrm{d}T/\mathrm{d}t$ in the energy equation (16c), we have $\rho C(T)\,\mathrm{d}T/\mathrm{d}t$, where $C(T) = \mathrm{d}E/\mathrm{d}T$ is the specific heat for our liquid.

First, for a weakly expansible liquid (when $\wp \ll 1$), it is possible to derive, in place of full exact NSF equations, a simplified dimensionless system of dominant equations. Let C^0, λ^0, μ^0, k^0, and ρ^0 be reference values of $C(T)$,

2.3 From NSF to Hyposonic and Oberbeck–Boussinesq (OB) Equations

$\lambda(T)$, $\mu(T)$, $k(T)$, and $\rho(T)$ at $T = T^0$ and $\nu_0 = \mu^0/\rho^0$. We scale all distances on the thickness of the liquid layer, d^0. The velocity vector $\boldsymbol{u}$, pressure p, temperature difference $T - T^0$ and time t are referred to scales: ν_0/d^0, $\rho^0 g d^0$, ΔT^0, and $d^{0\,2}/\nu_0$, respectively. As a result, the following dimensionless parameters (Froude, gravity, and Prandtl numbers) arise:

$$\mathrm{Fr}_0 = \frac{(\nu_0/d^0)}{(gd^0)^{1/2}}, \quad \mathcal{G} = \frac{gd^0}{C^0 \Delta T^0}, \quad \mathrm{Pr}_0 = \frac{C^0 \nu_0 \rho^0}{k^0}. \tag{2.70}$$

In this case, from (16a–c), with (17a–c) and (2.68), for dimensionless velocity components, $v_i = u_i/(\nu_0/d^0)$, with $i = 1, 2, 3$, the perturbation pressure

$$\pi = \frac{p - p^0}{\rho^0 g d^0} - \left(1 - \frac{x_3}{d^0}\right), \tag{2.71}$$

where p^0 is a constant pressure (at $T = T^0$, for instance), and the perturbation temperature

$$\theta = \frac{T - T^0}{\Delta T^0}, \tag{2.72}$$

as a function of dimensionless time, $\tau = t/(d^{0\,2}/\nu_0)$, and space, $X_i = x_i/d^0$, variables, we derive a new system of dimensionless *dominant* equations valid with an error of the order of $O(\wp^2)$, as in Zeytounian (1989) and see, for instance, Zeytounian (2002b), Chap. 10.

Now, from these dominant equations, when $\wp \to 0$, it is obvious that if we want to take into account the buoyancy term $-(\wp/\mathrm{Fr}_0^2)\theta\delta_{i3}$, in the equation for v_3, then it is necessary to consider the following OB limit process [according to Zeytounian (1989)]:

$$\wp \to 0, \quad \mathrm{Fr}_0 \to 0, \quad \text{with} \quad \frac{\wp}{\mathrm{Fr}_0^2} = \mathrm{Gr}_0 = O(1), \tag{2.73}$$

where the Grashof number Gr_0 is a similarity parameter for the below OB limit model equations (2.75a–c).

Because $\mathrm{Fr}_0 \to 0$, such that $\mathrm{Gr}_0 = O(1)$, then it is also necessary to associate the following asymptotic expansions with (2.73):

$$v_i = v_{i0\mathrm{B}} + \mathrm{Fr}_0^2 v_{i1} + \dots, \quad \pi = \mathrm{Fr}_0^2 \pi_{0\mathrm{B}} + \mathrm{Fr}^4 \pi_1 + \dots,$$

$$\theta = \theta_{0\mathrm{B}} + \mathrm{Fr}_0^2 \theta_1 + \dots . \tag{2.74}$$

Now, we assume that τ, X_i, Pr_0, $\mathcal{G}$, λ^0, μ^0 *are fixed* [of the order of $O(1)$], when $\mathrm{Fr}_0 \to 0$ with (2.73) and (2.74). Then, from the dominant equations, we derive asymptotically the following model OB (shallow convection) equations for $v_{i0\mathrm{B}}$ ($i = 1, 2, 3$), $\pi_{0\mathrm{B}}$, and $\theta_{0\mathrm{B}}$:

$$\frac{\partial v_{i0B}}{\partial X_i} = 0, \tag{2.75a}$$

$$\frac{d_{0B} v_{i0B}}{d\tau} + \frac{\partial \pi_{0B}}{\partial X_i} - \mathrm{Gr}_0\, \theta_{0B} \delta_{i3} = \Delta v_{i0B}, \tag{2.75b}$$

$$\frac{d_{0B} \theta_{0B}}{d\tau} = \frac{1}{\mathrm{Pr}_0} \Delta \theta_{0B}. \tag{2.75c}$$

In the framework of the classical Bénard shallow convection problem, the boundary conditions relative to X_3 for OB equations (2.75a–c) are

$$v_{i0B} = 0 \quad \text{and} \quad \theta_{0B} = 1 \text{ at } X_3 = 0, \tag{2.76a}$$

$$v_{i0B} = \theta_{0B} = 0 \text{ at } X_3 = 1. \tag{2.76b}$$

In (2.75b,c),

$$\frac{d_{0B}}{d\tau} = \frac{\partial}{\partial \tau} + v_{i0B} \frac{\partial}{\partial X_i} \quad \text{and} \quad \delta_{33} = 1,\ \delta_{13} = \delta_{23} = 0.$$

For the validity of the OB equations (2.75a–c), we note that first, the OB model system is valid only for a weakly expansible liquid when the parameter $\wp$ given by (2.69) is small. For a usual liquid, this is very well satisfied if ΔT^0 is not very large. A second condition is related to the assumption of a low Froude number. From the definition of the Froude number Fr_0, according to (2.70), we obtain

$$d^0 \gg \left[\frac{(\nu_0)^2}{g}\right]^{1/3} \approx 1\,\mathrm{mm}. \tag{2.77}$$

Consequently, for very thin films, when the film thickness d^0 is of the order of millimeters, in place of OB equations [in this case $\mathrm{Fr}_0 = O(1)$, and the buoyancy term in the leading approximate equations (2.75b) is *not* important!], we obtain the classical incompressible model equations for the velocity components and pressure (see, for instance, Chap. 7). The perturbation of temperature is then decoupled from this incompressible (Navier) model, but is present in the free-surface conditions when we assume temperature-dependent surface tension (in this case we take into account the so-called "Marangoni effect"). For an in-depth discussion of this case and the role of the buoyancy in the so-called Bénard–Marangoni problem, when the liquid is bounded by a free-surface with temperature-dependent surface tension (thermocapillary instability problem), see our recent two papers, Zeytounian (1997, 1998), and in Chap. 7, we return to this problem.

Next, the previous OB model equations are valid only if the gravity number $\mathcal{G}$ is bounded [of the order of $O(1)$]. Only for this case, we can neglect the two terms in the dominant equation for θ

$$\wp \mathcal{G}(X_3 - 1)\frac{d\theta}{d\tau} \quad \text{and} \quad \left(\frac{\mathcal{G}\,\mathrm{Fr}_0^2}{2}\right)\left[\frac{\partial v_i}{\partial X_j} + \frac{\partial v_j}{\partial X_i}\right]^2.$$

2.3 From NSF to Hyposonic and Oberbeck–Boussinesq (OB) Equations

Consequently,

$$\mathcal{G} = O(1) \Rightarrow d^0 \approx C^0 \frac{\Delta T^0}{g} . \tag{2.78}$$

But, if

$$\mathcal{G} \gg 1, \quad \text{with} \quad \mathcal{G}\,\mathrm{Fr}_0^2 = \frac{\wp \mathcal{G}}{\mathrm{Gr}_0} = O(1) , \tag{2.79}$$

then for the perturbation of the temperature, we obtain the following, so-called *deep-convection equation*, for dimensionless perturbation of temperature Θ [Zeytounian (1989)]:

$$[1 + \delta(1-X_3)]\frac{\mathrm{d}_\mathrm{D}\Theta}{\mathrm{d}\tau} = \frac{1}{\mathrm{Pr}_0}\Delta\Theta + \frac{\delta}{2\,\mathrm{Gr}_0}\left[\frac{\partial V_{i\mathrm{D}}}{\partial X_j} + \frac{\partial V_{j\mathrm{D}}}{\partial X_i}\right]^2 , \tag{2.80}$$

where

$$\delta = \wp \mathcal{G} \equiv \frac{\alpha^0 g d^0}{C^0} \tag{2.81}$$

and $V_{i\mathrm{D}}$ are the velocity components of the velocity in deep convection with $\mathrm{d}_\mathrm{D}/\mathrm{d}\tau = \partial/\partial\tau + V_{i\mathrm{D}}\partial/\partial X_i$.

For the "deep" thickness of the liquid layer d^0, we obtain the following relation in place of (2.78) because $\delta = O(1)$:

$$d^0 \approx \frac{C^0}{\alpha^0 g} . \tag{2.82}$$

The reader can find rigorous results of the mathematical analysis of the Bénard problem for deep convection in the papers by Charki (1996). For the usual liquids, according to (2.78), the thickness of the liquid layer should be at most of the order of a kilometer! Naturally, this estimate is very good for technological applications of the OB equations, but for geophysical applications, the deep convection equation (2.80) is more convenient. The reader can find some numerical calculations in Errafiy and Zeytounian (1991a,b).

It is interesting to note that the parameter $\mathcal{G}\,\mathrm{Fr}_0^2$ plays the role of the square of a reference Mach number,

$$M^{02} = \frac{(\nu_0/d^0)^2}{C^0 \Delta T^0} , \tag{2.83}$$

based on the temperature difference ΔT^0 between the rigid lower and upper boundaries, and $C^0 \Delta T^0$ plays the role of a characteristic "acoustic speed" for a weakly expansible liquid.

On the other hand, we observe that when $\mathcal{G} = O(1)$, we use, in fact, the small M^{02} expansions (2.74) for the rigorous asymptotic derivation of OB, shallow convection model equations (2.75a–c).

Therefore, the derivation of OB model equations from the full NSF equations for a weakly expansible liquid is really related to the *low Mach number flow* problem considered in Chap. 6.

Now, if Δp^0 is the characteristic pressure fluctuation, then,

$$\frac{\Delta p^0}{\Delta T^0} \ll C^0 \rho^0 , \qquad (2.84)$$

and Δp^0 is always smaller than ΔT^0. This property justifies, to a certain extent, the validity of the specifying equation (2.68), which was adopted at the beginning.

For (2.68), it is necessary to observe also that, from expansion (2.74) for π (proportional to Fr_0^2), it is clear that the presence of pressure p [related to π by (2.71)] in a full baroclinic equation of state, $\rho = \rho(T, p)$, in place of (2.68), does not change [in the limit (2.73) with (2.74)] the form of the derived OB approximate equations (2.75a–c).

Curiously (and obviously erroneously!) a recent paper by Rajagopal et al. (1996), asserts that,

" ... Consequently (the derivation of OB equations), it is free from the additional assumptions usually added on in earlier works in order to obtain the correct (OB) equations"!

Again, the OB model equations (2.75) are singular in the vicinity of the initial time, $\tau = 0$, where the initial conditions are imposed for $\boldsymbol{u}$, ρ, and T for initiating the solution of the NSF exact equations (16a–c). As a consequence, *a detailed analysis in the vicinity of $\tau = 0$ is necessary, which leads to an adjustment problem in the OB state.*

Finally, for the functions U_i, T, and P, such that

$$U_i = \Pr v_{i0\mathrm{B}}, \; i = 1, 2, 3, \quad T = X_3 - 1 + \theta_{0\mathrm{B}}, \qquad (2.85\mathrm{a})$$

and

$$P = \mathrm{Gr}_0 \, X_3 \left[\frac{X_3}{2} - 1\right] + \pi_{0\mathrm{B}} , \qquad (2.85\mathrm{b})$$

in place of (2.75a–c) and (2.76a,b), we obtain the classical Rayleigh–Bénard rigid–rigid shallow convection problem, as in Drazin and Reid (1981). For this problem, see Sect. 3.4 of Chap. 3 where we consider the linear Bénard problem and the corresponding thermal instability via linearized OB equations.

3 Some Simple Examples of Navier, NS and NSF Viscous Fluid Flows

3.1 Plane Poiseuille Flow and the Orr–Sommerfeld Equation

If we consider Navier fluid flow bounded by a channel of width $2L^0$, then for unsteady-state (with Strouhal number $S = 1$) dimensionless Navier model equations [see (2.14a,b) with (2.15)]:

$$\frac{d\boldsymbol{u}_N}{dt} + \nabla\pi = \left(\frac{1}{\text{Re}}\right)\nabla^2 \boldsymbol{u}_N , \tag{3.1a}$$

$$\nabla \cdot \boldsymbol{u}_N = 0 , \tag{3.1b}$$

we can consider plane *Poiseuille* flow as basic motion.

This is a flow under constant pressure (denoted by P) in every cross-section between two fixed planes $z \equiv x_3 = \pm 1$, such that $u_{N2} = u_{N3} = 0$, $u_{N1} = U_P(z)$. In this case, $\partial P/\partial x_3 = \partial P/\partial x_2 = 0$, and from Navier equation (3.1a), only the scalar equation in the x_1 direction remains, and this reduces to

$$\text{Re}\,\frac{dP}{dx_1} = \frac{d^2 U_P(z)}{dz^2} . \tag{3.2}$$

With no-slip boundary conditions on $z = \pm 1$, $U_P(\pm 1) = 0$, the solution of (3.2) is a velocity distribution of parabolic form; using dimensionless quantities, the solution is

$$U_P(z) = (1 - z^2) , \quad -1 \leq z \leq +1 , \tag{3.3}$$

such that

$$-\frac{\partial P}{\partial x_1} = \frac{2}{\text{Re}} = \text{const.}$$

Even though no mathematical restriction has been placed on Re, Poiseuille solution (3.3) is valid only up to a certain critical Reynolds number Re_c (≈ 2300). The Poiseuille flow becomes unstable (in fact turbulent) above this value, when $\text{Re} > \text{Re}_c$.

At higher Reynolds numbers, laminar flow exhibits some kind of instability leading to additional secondary flow, and ultimately a breakdown into turbulent flow (see, for instance, Chaps. 9, 10, where the reader can find a discussion of turbulence and stability).

3.1.1 The Orr–Sommerfeld Equation

Now, we consider the perturbations in basic flow defined by

$$\boldsymbol{u}_{\mathrm{N}} = U_{\mathrm{P}}(z)\boldsymbol{i} + \boldsymbol{u}'(t,\boldsymbol{x}) , \quad \pi = P + p'(t,\boldsymbol{x}) , \tag{3.4}$$

with (3.2). The linearized Navier system of two equations governing the perturbations u'(t, x) and p'(t, x) is

$$\left[\frac{\partial}{\partial t} + U_{\mathrm{P}}(z)\frac{\partial}{\partial x}\right]\boldsymbol{u}' + w'\frac{dU_{\mathrm{P}}}{dz}\boldsymbol{i} = -\nabla p' + \frac{1}{\mathrm{Re}}\nabla^2 \boldsymbol{u} , \tag{3.5a}$$

$$\nabla \cdot \boldsymbol{u}' = 0 , \tag{3.5b}$$

where the coefficients of the former equation depend only on z and $w'(t,\boldsymbol{x})$ is the normal (to $z = \pm 1$) component of the velocity perturbation $\boldsymbol{u}'$. This allows us to seek solutions in the form

$$(\boldsymbol{u}',p') = (\boldsymbol{v}(z),p)\exp[i(kx + my - kct)] , \quad c = c_{\mathrm{r}} + ic_i , \tag{3.6}$$

with the boundary conditions (for the flow between two parallel planes),

$$\boldsymbol{v}(-1) = \boldsymbol{v}(+1) = 0 . \tag{3.7}$$

Substituting (3.6) in (3.5a,b), it follows that the problem leads to an eigenvalue problem with the eigenvalue relation,

$$F\left(k, m, c, \frac{1}{\mathrm{Re}}\right) = 0 . \tag{3.8}$$

However, according to the classical *Squire* theorem [Squire (1933)]:

"in an unstable parallel flow, the disturbance with the greatest rate of amplification $\sigma = ikc$ at a given value of Re is two-dimensional",

to determine the minimum critical Reynolds number, it is sufficient to consider only two-dimensional disturbances. In other words, the three-dimensional perturbation problem has therefore been reduced to an equivalent two-dimensional (x,z) problem (the dependence of $x_2 = y$ is canceled and $x_1 \equiv x$).

For this, we consider a general two-dimensional perturbation by introducing a stream function $\Psi'(t,x,z)$ such that

$$(u',w') = \left(\frac{\partial \Psi'}{\partial z}; -\frac{\partial \Psi'}{\partial x}\right) , \tag{3.9}$$

3.1 Plane Poiseuille Flow and the Orr–Sommerfeld Equation

so that the perturbed vorticity Navier equation can be obtained in the following form:

$$\left[\frac{\partial}{\partial t} + U_P(z)\frac{\partial}{\partial x}\right]\nabla^2\Psi' - \left(\frac{d^2 U_P}{dz^2}\right)\frac{\partial \Psi'}{\partial x} = \frac{1}{\text{Re}}\nabla^4\Psi'. \tag{3.10}$$

Now, we can write

$$\Psi' = \phi(z)\exp[ik(x - ct)], \tag{3.11}$$

so that $\phi(z)$ satisfies the celebrated *Orr–Sommerfeld* equation:

$$\left\{[U_P(z) - c](D^2 - k^2) - \frac{d^2 U_P}{dz^2}\right\}\phi(z) = \left(\frac{1}{ik}\text{Re}\right)(D^2 - k^2)^2\phi(z), \tag{3.12}$$

where $D = d/dz$. The boundary conditions for the flow between two parallel planes are

$$D\phi = \phi = 0 \text{ on } z = -1 \quad \text{and} \quad z = +1. \tag{3.13}$$

The Orr–Sommerfeld equation (3.12), with boundary conditions (3.13), constitutes an eigenvalue problem, and the eigenvalue relation assumes the form,

$$F\left(k, c, \frac{1}{\text{Re}}\right) = 0, \quad \text{with } c = c_r + ic_i, \tag{3.14}$$

where c_i is the exponential decay or amplification factor of the disturbance. Thus, a given flow $U_P(z)$ is stable or unstable according to the linearized theory, provided that $c_i < 0$ or $c_i > 0$. If $c_i = 0$, there is sustained oscillation, and the condition $c_i = 0$ leads to a relation between k and Re or a curve in the (k, Re) plane, as shown in Fig. 3.1.

This curve, usually referred to as the "neutral stability curve". A tangent parallel to the k axis touches it at the minimum critical Reynolds number Re_c so that the neutral curve lies entirely on the right-hand side of $\text{Re} > \text{Re}_c$.

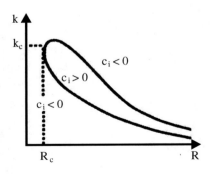

Fig. 3.1. The neutral curve for plane Poiseuille flow

The region of the (k, Re) plane in which basic flow is unstable is separated from the region of stability by the neutral curve, along which $c_i = 0$. For all $\mathrm{Re} < \mathrm{Re_c}$, disturbances of all wave numbers decay with time, whereas for $\mathrm{Re} > \mathrm{Re_c}$, a range of wave numbers k exists for which the corresponding disturbances grow, and then, in turn lead to instability.

3.1.2 A Double-Scale Technique for Resolving the Orr–Sommerfeld Equation

When

$$ik\,\mathrm{Re} = \frac{1}{\varepsilon} \gg 1,$$

a double-scale (in z) technique is very powerful for resolving the Orr–Sommerfeld boundary problem (3.12), (3.13). Because the odd solutions are always stable, it is sufficient to consider only even solutions and then in place of (3.13), we can write as boundary conditions on $z = 0, +1$,

$$D\phi(0) = D^3\phi(0) = \phi(+1) = D\phi(+1) = 0. \tag{3.15}$$

Now, in place of $\phi(z)$, we consider a new function of two vertical variables, z and ζ, and also of ε:

$$\phi^*(z,\zeta,\varepsilon), \quad \text{with} \quad \zeta = \frac{g(z)}{\varepsilon^\beta}. \tag{3.16}$$

For consistency of the double-scale technique, it is necessary to choose

$$\beta = \frac{1}{2} \quad \text{and} \quad g(z) = \int [U_\mathrm{P}(z) - c]^{1/2}\,\mathrm{d}z. \tag{3.17}$$

Then, we can write the following *uniformly* valid asymptotic expansion for $\phi^*(z,\zeta,\varepsilon)$:

$$\phi^* = \phi_0 + \varepsilon^{1/2}\phi_1 + \ldots, \tag{3.18}$$

where

$$\phi_0(z,\zeta) = [U_\mathrm{P}(z) - c]^{-5/4}\left\{a\,\mathrm{e}^{+\zeta} + b\,\mathrm{e}^{-\zeta}\right\}$$
$$\quad + E(z) + F(z) + O(|\varepsilon|^{1/2}). \tag{3.19}$$

Solution (3.19) is valid only if

$$U_\mathrm{P}(z) - c \text{ and } \frac{\mathrm{d}^2 U_\mathrm{P}}{\mathrm{d}z^2} \text{ are both } O(1), \text{ when } z \in [0, +1].$$

Functions $E(z)$ and $F(z)$ in solution (3.19) are solutions of the equation [see Long (1987) for the details]:

$$[U_P(z) - c]\left\{\frac{d^2 f}{dz^2} - k^2 f\right\} = \left(\frac{d^2 U_P}{dz^2}\right) f\,; \quad f = (E, F)\,. \tag{3.20}$$

If, for instance, $U_P(z_c) = c$ for $z = z_c$, then it is necessary to resolve a so-called "turning point" problem and to introduce a new inner variable,

$$z^* = \frac{z - z_c}{\delta}\,, \tag{3.21}$$

and an associated inner asymptotic expansion for $\phi(z)$,

$$\phi(z) = \phi_0^*(z^*) + \delta\phi_1^*(z^*) + \ldots \tag{3.22}$$

Again, consistency imposes

$$\delta = (k\,\mathrm{Re})^{-1/3}\,. \tag{3.23}$$

Finally, for the function $\phi_0^*(z^*)$, we derive an Airy equation,

$$i\frac{d^4\phi_0^*}{dz^{*4}} + \left[\left(\frac{dU_P}{dz}\right)_{z=z_c}\right]z^*\frac{d^2\phi_0^*}{dz^{*2}} = 0\,, \tag{3.24}$$

and for this Airy equation the two nontrivial solutions are Hankel functions.

3.2 Steady Flow Through an Arbitrary Cylinder under Pressure

Now, we consider a steady-state, Navier with no external forces and take axes Ox_1, Ox_2, Ox_3, where Ox_1 is parallel to the generators of the cylinder and Ox_2, Ox_3 are perpendicular thereto. We look for a solution in which the flow is entirely parallel to the generators of the cylinder; thus

$$u_1 = u_1(x_1, x_2, x_3)\,, \quad u_2 = u_3 = 0\,, \tag{3.25}$$

and so the Navier equations become (with dimensions)

$$\frac{\partial u_1}{\partial x_1} = 0\,, \quad \frac{\partial p}{\partial x_2} = 0\,, \quad \frac{\partial p}{\partial x_3} = 0\,, \tag{3.26a}$$

$$\rho_0 u_1 \frac{\partial u_1}{\partial x_1} = -\frac{\partial p}{\partial x_1} + \mu_0 \nabla^2 u_1\,. \tag{3.26b}$$

Equations (3.26a) show that:

$$u_1 = u(x_2, x_3), \quad p = p(x_1),$$

and in place of (3.26b), we obtain:

$$\frac{dp}{dx_1} = \mu_0 \left(\frac{\partial^2 u}{\partial x_2^2} + \frac{\partial^2 u}{\partial x_3^2} \right),$$

where the left-hand side is a function only of x_1, and the right-hand side is a function only of x_2 and x_3. Consequently, we deduce that both are constants (denoted by Π_0), and write

$$\frac{1}{\mu_0} \frac{dp}{dx_1} = \frac{\partial^2 u}{\partial x_2^2} + \frac{\partial^2 u}{\partial x_3^2} = -\frac{\Pi_0}{\mu_0}. \tag{3.27}$$

Equation (3.27) must be solved subject to the boundary conditions that $u = 0$ on the surface of the cylinder. The computation of full pipe flow for arbitrary cross-sectional shapes is based on solving the Poisson differential equation (3.27). Solutions for different cross-sections can be found in Shah and London (1978). We note also that an exact solution of the Navier equations can be given for a pipe of concentric circular cross-section (a so-called *annulus*); see Müller (1936).

3.2.1 The Case of a Circular Cylinder

For a circular cylinder of radius a, we transform into polar coordinates (r, θ, x), and note that the velocity $u(x_2, x_3)$ along the tube will be a function of r alone. Thus,

$$\frac{\partial^2 u}{\partial x_2^2} + \frac{\partial^2 u}{\partial x_3^2} = \frac{1}{r} \frac{\partial}{\partial r} \left(r \frac{\partial u}{\partial r} \right) = -\frac{\Pi_0}{\mu_0},$$

which integrates to give

$$u(r) = A \ln r + B - \frac{1}{4\mu_0} \Pi_0 r^2, \tag{3.28}$$

where A and B are arbitrary constants of integration. The constant A must be zero if the solution is to be physically acceptable along the axis $r = 0$, and B is then determined by the no-slip condition,

$$u = 0 \text{ on } r = a \quad \Rightarrow \quad B = \frac{1}{4\mu_0} \Pi_0 a^2.$$

3.2 Steady Flow Through an Arbitrary Cylinder under Pressure

Thus, the solution is (with dimensions)

$$u(r) = \left(\frac{a^2}{4\mu_0}\right) \Pi_0 \left[1 - \left(\frac{r}{a}\right)^2\right] . \tag{3.29}$$

From this velocity profile, we may deduce the *mass flux* per unit time passing any cross-section of the tube:

$$M = \int_0^a \rho_0 u 2\pi r \, dr = \rho_0 \pi \left(\frac{a^4}{8\mu_0}\right) \Pi_0 . \tag{3.30}$$

This result is known as *Poiseuille's law* – the basis of a method of measuring the viscosity of a fluid. Obviously, solution (3.29) is only valid "far" from the "entry flow" near the entrance to the tube, where the fully developed region with a velocity profile given by (3.29) has not been attained.

3.2.2 The Case of an Annular Region Between Concentric Cylinders

If we consider an annular region between concentric cylinders of radii a and b ($b < a$), then the velocity profile for the flow through this annular region is

$$u(r) = \left(\frac{a^2}{4\mu_0}\right) \Pi_0 \left\{ \left[1 - \left(\frac{r}{a}\right)^2\right] - \left[\frac{\ln(r/a)}{\ln(b/a)}\right] \left[1 - \left(\frac{b}{a}\right)^2\right] \right\} . \tag{3.31}$$

From (3.31), when b tends to zero, we derive (3.29) again. The resulting (3.31) is obtained from the solution (3.28), if we consider no-slip boundary conditions on the cylinders $r = a$ and $r = b$ (in this case the constant A is not zero because the singular axis $r = 0$ is outside the annular region).

3.2.3 The Case of a Cylinder of Arbitrary Section

More generally, for a cylinder of arbitrary section, we may find a solution of (3.27) as follows. We write

$$F(x_2, x_3) = u + \left(\frac{1}{4\mu_0}\right) \Pi_0 (x_2^2 + x_3^2) ,$$

so that

$$\frac{\partial^2 F}{\partial x_2^2} + \frac{\partial^2 F}{\partial x_3^2} = \frac{\partial^2 u}{\partial x_2^2} + \frac{\partial^2 u}{\partial x_3^2} + \frac{\Pi_0}{\mu_0} = 0 . \tag{3.32}$$

The no-slip boundary condition is

$$u = F - \left(\frac{1}{4\mu_0}\right) \Pi_0 (x_2^2 + x_3^2) = 0 , \tag{3.33a}$$

on the boundary of the cylinder; that is, on the boundary of the cylinder,

$$F = \left(\frac{1}{4\mu_0}\right)\Pi_0(x_2^2 + x_3^2) \equiv \left(\frac{1}{4\mu_0}\right)\Pi Z Z^*, \tag{3.33b}$$

where Z is the complex variable $x_2 + ix_3$ and Z^* the complex conjugate.

Finally, the problem has thus been reduced to that of solving Laplace's equation in two dimensions, with F prescribed on the boundary.

As a simple example, we find that the velocity profile for flow through an *elliptical cylinder* with axes $2a$, $2b$, is given by the following solution:

$$u = \left[\frac{a^2 b^2}{a^2 + b^2}\right]\left(\frac{\Pi_0}{2\mu_0}\right)\left[1 - \frac{x_2^2}{a^2} - \frac{x_3^2}{b^2}\right]. \tag{3.34}$$

3.3 Steady-State Couette Flow Between Cylinders in Relative Motion

Now we consider the problem of two concentric circular cylinders of radii r_1, r_2, rotating about their common axis at angular velocities ω_1, ω_2. We take coordinates r, θ, x, with associated velocity components u, v, w and look for a solution in which

$$u = 0, \quad v = V(r), \quad w = 0, \quad p = P(r).$$

From the Navier dimensional equations of motion in cylindrical polar coordinates, we derive simply,

$$-\frac{V^2}{r} = -\frac{1}{\rho_0}\frac{\partial P}{\partial r}, \tag{3.35a}$$

$$0 = \frac{d^2 V}{dr^2} + \frac{1}{r}\frac{dV}{dr} - \frac{1}{r^2}V. \tag{3.35b}$$

Upon integrating explicitly (3.35b), we obtain the following solution for $V(r)$:

$$V(r) = Ar + \frac{B}{r}. \tag{3.36a}$$

Constants A and B are determined by the boundary conditions that

$$V = r_1\omega_1 \text{ when } r = r_1 \quad \text{and} \quad V = r_2\omega_2 \text{ when } r = r_2. \tag{3.36b}$$

The pressure $P(r)$ may then be obtained, if required, from (3.35a). With (3.36b), we obtain for A and B in solution (3.36a)

$$A = \omega_1\left[\frac{\Omega - R^2}{1 - R^2}\right], \tag{3.37a}$$

$$B = \omega_1 r_1^2\left[\frac{1 - \Omega}{1 - R^2}\right], \tag{3.37b}$$

3.3 Steady-State Couette Flow Between Cylinders in Relative Motion

where the nondimensional flow parameters Ω and R are given by

$$\Omega = \frac{\omega_2}{\omega_1} \quad \text{and} \quad R = \frac{r_1}{r_2}. \tag{3.38}$$

3.3.1 The Classic Taylor Problem

We investigate the stability of the flow described by

$$u = w = 0, \quad V(r) = Ar + \frac{B}{r}, \quad \frac{d(P/\rho_0)}{dr} = \frac{V^2}{r}.$$

Following Taylor (1923), we assume perturbed flow in the form

$$\boldsymbol{u} = (u', V(r) + v', w') \quad p = P(r) + p'. \tag{3.39}$$

We substitute these results in the Navier equations and assume that the disturbance is so small that the system can be linearized. We further assume that the perturbations are axisymmetric and independent of θ. By analyzing the disturbance into normal modes, we seek solutions of the resulting equations in the form,

$$(u', v', p') = [u^*(r), v^*(r), p^*(r)] \exp(im\, x + st), \tag{3.40}$$

where m is the axial wave number.

Then, eliminating pressure from the resulting form of the Navier equation yields two disturbance differential equations for $u^*(r)$ and $v^*(r)$; then, if $u^*(r)$ is eliminated, we obtain a single differential equation for $v^*(r)$:

$$\left\{ \frac{d}{dr}\left[\frac{d}{dr} + \left(\frac{1}{r}\right)\right] - m^2 - \left(\frac{\rho_0}{\mu_0}\right)s \right\}^2 \left\{ \frac{d}{dr}\left[\frac{d}{dr} + \left(\frac{1}{r}\right)\right] - m^2 \right\} v^*(r)$$

$$= 4A\left(A + \frac{B}{r^2}\right)\left(\frac{\rho_0 m}{\mu_0}\right)^2 v^*(r). \tag{3.41}$$

The boundary (no-slip) conditions are

$$v^* = 0, \quad \frac{d}{dr}\left[\frac{d}{dr} + \left(\frac{1}{r}\right)\right] v^* = 0, \tag{3.42a}$$

$$\left[\frac{d}{dr} + \left(\frac{1}{r}\right)\right]\left\{\frac{d}{dr}\left[\frac{d}{dr} + \left(\frac{1}{r}\right)\right] - m^2 - \left(\frac{\rho_0}{\mu_0}\right)s\right\} v^* = 0, \text{ on } r = r_1, r_2. \tag{3.42b}$$

It is convenient to simplify problem (3.41), (3.42a,b) further under the "*narrow-gap*" approximation, that is, the gap $d = r_2 - r_1$ between the two cylinders is *small* compared to their mean radius:

$$r_m = \frac{r_1 + r_2}{2}.$$

For this, we introduce the following dimensionless flow variables and parameters:

$$\eta = \frac{r - r_m}{d}, \quad n = md, \quad \sigma = s\rho_0 \frac{d^2}{\mu_0}, \tag{3.43}$$

where $-1/2 \leq \eta \leq +1/2$; n and σ are nondimensional wave number and amplification parameters, respectively. Now, if the terms of the order $\varepsilon = d/r_m$ are neglected, first,

$$\frac{\mathrm{d}}{\mathrm{d}r}\left[\frac{\mathrm{d}}{\mathrm{d}r} + \left(\frac{1}{r}\right)\right] \cong \frac{\mathrm{d}^2}{\mathrm{d}\eta^2},$$

and in view of (3.43), (3.41) and the boundary conditions (3.42a,b) become:

$$\left(\frac{\mathrm{d}^2}{\mathrm{d}\eta^2} - n^2\right)\left(\frac{\mathrm{d}^2}{\mathrm{d}\eta^2} - n^2 - \sigma^2\right)v^*(\eta) = 4A\Phi(\eta)\,\mathrm{d}^4\left(\frac{\rho_0 n}{\mu_0}\right)^2 v^*(\eta), \tag{3.44a}$$

$$v^* = \frac{\mathrm{d}^2 v^*}{\mathrm{d}\eta^2} = \frac{\mathrm{d}}{\mathrm{d}\eta}\left(\frac{\mathrm{d}^2}{\mathrm{d}\eta^2} - n^2 - \sigma^2\right)v^*, \quad \text{on } \eta = \pm\frac{1}{2}, \tag{3.44b}$$

where the function $\Phi(\eta)$ is an angular velocity function of η and may be expanded in powers of ε. Both $\Phi(\eta)$ and A can be expanded in powers of ε and are given, to a first approximation, by

$$\Phi(\eta) = \left(\frac{1}{2}\right)\omega_1(1 + \Omega) + \eta[\omega_1(\Omega - 1)]; \tag{3.45a}$$

$$A = -r_m \omega_1 \frac{1 - \Omega^2}{2d}. \tag{3.45b}$$

3.3.2 The Taylor Number

In the limit $\Omega \to 1$ (the cylinders are rotating in the same direction with their angular velocities nearly equal), in place of the differential equation (3.44a), we obtain the following reduced differential equation:

$$\left(\frac{\mathrm{d}^2}{\mathrm{d}\eta^2} - n^2\right)\left(\frac{\mathrm{d}^2}{\mathrm{d}\eta^2} - n^2 - \sigma^2\right)v^*(\eta) + n^2\,\mathrm{Ta}\,v^*(\eta) = 0, \tag{3.46}$$

3.3 Steady-State Couette Flow Between Cylinders in Relative Motion

where

$$\mathrm{Ta} = -2A(\omega_1 + \omega_2)d^4 \left(\frac{\rho_0}{\mu_0}\right)^2 \tag{3.47}$$

is the *Taylor number*.

If A is replaced by its approximate value (3.45b) for *small* $\varepsilon = d/r_\mathrm{m}$, then the Taylor number becomes

$$\mathrm{Ta} = r_\mathrm{m} d^3 (\omega_1^2 - \omega_2^2) \left(\frac{\rho_0}{\mu_0}\right)^2, \tag{3.48a}$$

and it can be shown that this form of the Taylor number represents the ratio of (destabilizing) centrifugal force to (stabilizing) viscous force.

We observe that in place of (3.48a), we can write

$$\mathrm{Ta} = \left(\frac{\rho_0}{\mu_0}\right)^2 \omega_1^2 d^4 \left[\frac{1-\Omega^2}{\varepsilon}\right], \tag{3.48b}$$

and the ratio $C^0 = (1 - \Omega^2)/\varepsilon$ of two small parameters is assumed to have a *finite* value.

The differential equation (3.46), with boundary conditions (3.44b), constitutes again an eigenvalue problem that yields an eigenvalue relation in the functional form $F(n, \sigma, \mathrm{Ta}) = 0$. This indicates that instability will be possible if Ta is greater than some critical value, $\mathrm{Ta_c}$. This implies that the effect of the centrifugal force would be much larger than that of the viscous force and leads to instability. More precisely, the situation in *neutral* (marginal) stability is given by the value $\sigma = 0$ which is shown on the neutral stability curve in Fig. 3.2. *This result is often known as the principle of exchange of stabilities.* The critical Taylor number is found by solving the equation $F(n, 0, \mathrm{Ta}) = 0$, and the minimum value of Ta can be determined as a function of n. For $\mathrm{Ta} > \mathrm{Ta_c}$, it can be shown that a disturbance within the band of unstable wave numbers leads to *nonoscillatory* amplifying flow.

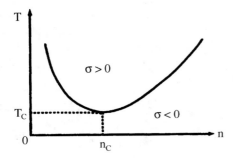

Fig. 3.2. The neutral stability curve for circular Couette flow

3.4 The Bénard Linear Problem and Thermal Instability

As in Sect. 2.3.2, we consider the Bénard problem but in the framework of a linear theory when perturbations relative to a basic equilibrium state are small.

For this, first, we consider the OB equations (2.75a–c) with the new functions U_i, T, and P given by (2.85a,b). The OB system for these functions is written in the following form for $\boldsymbol{v} = (U_1, U_2)$, $w = U_3$, T, and P:

$$\boldsymbol{D} \cdot \boldsymbol{v} + \frac{\partial w}{\partial z} = 0 ; \tag{3.49a}$$

$$\left(\frac{\partial}{\partial \tau} - \nabla^2\right) \boldsymbol{v} + (\boldsymbol{v} \cdot \nabla)\boldsymbol{v} + \Pr \boldsymbol{D} P = 0 ; \tag{3.49b}$$

$$\left(\frac{\partial}{\partial \tau} - \nabla^2\right) w + (\boldsymbol{v} \cdot \nabla)w + \Pr \frac{\partial P}{\partial z} + \operatorname{Ra} T = 0 ; \tag{3.49c}$$

$$\left(\Pr \frac{\partial}{\partial \tau} - \nabla^2\right) T + (\boldsymbol{v} \cdot \nabla)T - w = 0 , \tag{3.49d}$$

where

$$\operatorname{Ra} = \Pr_0 \operatorname{Gr}_0 , \quad \text{and} \quad \nabla = \boldsymbol{D} + \left(\frac{\partial}{\partial z}\right) \boldsymbol{k}$$

with

$$\boldsymbol{D} = \left(\frac{\partial}{\partial X_1}, \frac{\partial}{\partial X_2}\right) \quad \text{and} \quad z = X_3 .$$

The basic equilibrium state, according to (3.49a–d) is given by

$$v = w = 0 , \quad T = P = 0 ,$$

and for the perturbations, $\boldsymbol{v}'$, w', T', and p', after the usual linearization, we derive the following linear OB system:

$$\boldsymbol{D} \cdot \boldsymbol{v}' + \frac{\partial w'}{\partial z} = 0 ; \tag{3.50a}$$

$$\left(\frac{\partial}{\partial \tau} - \nabla^2\right) \boldsymbol{v}' + \Pr \boldsymbol{D} p' = 0 ; \tag{3.50b}$$

$$\left(\frac{\partial}{\partial \tau} - \nabla^2\right) w' + \Pr \frac{\partial p'}{\partial z} + \operatorname{Ra} T' = 0 ; \tag{3.50c}$$

$$\left(\Pr \frac{\partial}{\partial \tau} - \nabla^2\right) T' - w' = 0 , \tag{3.50d}$$

3.4 The Bénard Linear Problem and Thermal Instability

It is possible to eliminate all of the dependent variables except w from (3.50a–d) to obtain a single equation. We obtain, first, an equation between w' and T',

$$\left(\frac{\partial}{\partial \tau} - \nabla^2\right)\nabla^2 w' = \text{Ra}\, \boldsymbol{D}^2 T' , \qquad (3.51)$$

with $\nabla^2 = \boldsymbol{D}^2 + \partial^2/\partial z^2$. Then eliminating T' from (3.50d) and (3.51) yields an equation only for w':

$$\left(\Pr \frac{\partial}{\partial \tau} - \nabla^2\right)\left(\frac{\partial}{\partial t} - \nabla^2\right)\nabla^2 w' = \text{Ra}\, \boldsymbol{D}^2 w' . \qquad (3.52)$$

Next, we analyze, again, the perturbations into normal modes:

$$(w', T') = [W(z), \Theta(z)] f(X_1, X_2) \exp(\sigma \tau) ,$$

where $\sigma = \sigma_{\rm r} + i\sigma_i$.

Then (3.50d) assumes the form

$$\left[\frac{\mathrm{d}^2}{\mathrm{d}z^2} - a^2 - \sigma \Pr\right]\Theta = -W , \qquad (3.53\text{a})$$

provided that

$$(\boldsymbol{D}^2 + a^2)f = 0 , \qquad (3.54)$$

where a^2 is a separation constant and a can be interpreted as a real horizontal wave number of the Bénard problem.

Similarly, using (3.54), (3.51) and (3.52) can be written in the form

$$\left[\frac{\mathrm{d}^2}{\mathrm{d}z^2} - a^2\right]\left[\frac{\mathrm{d}^2}{\mathrm{d}z^2} - a^2 - \sigma\right]W = \text{Ra}\, a^2 \Theta , \qquad (3.53\text{b})$$

$$\left[\frac{\mathrm{d}^2}{\mathrm{d}z^2} - a^2\right]\left[\frac{\mathrm{d}^2}{\mathrm{d}z^2} - a^2 - \sigma\right]\left[\frac{\mathrm{d}^2}{\mathrm{d}z^2} - a^2 - \sigma \Pr\right]W = -\text{Ra}\, a^2 W . \qquad (3.53\text{c})$$

The relevant boundary conditions for the $W(z)$ solution of (3.53c) are [when we take (3.53b) into account]

$$W = \frac{\mathrm{d}W}{\mathrm{d}z} = \left[\frac{\mathrm{d}^2}{\mathrm{d}z^2} - a^2\right]\left[\frac{\mathrm{d}^2}{\mathrm{d}z^2} - a^2 - \sigma\right]W = 0 , \qquad (3.55\text{a})$$

at a *rigid* boundary surface, and

$$W = \frac{\mathrm{d}^2 W}{\mathrm{d}z^2} = \frac{\mathrm{d}^4 W}{\mathrm{d}z^4} = 0 , \qquad (3.55\text{b})$$

at a *free* boundary surface.

Equation (3.53c) for $W(z)$ with the boundary conditions (3.55a) or (3.55b) determines a so-called "self-adjoint" eigenvalue problem for parameters Ra, a, and σ, when Pr is fixed. It can be proven that when Ra is less than a certain critical parameter, all small disturbances of the purely conductive basic equilibrium state *decay* in time (stability). Whereas, when Ra exceeds the critical value Ra_c, instability occurs in the form of convection in cells of a polygonal platform.

These cells are called *Bénard cells*, and the formation of Bénard cells in a weakly expansible liquid layer is one of the most remarkable examples of *bifurcation phenomena* [the bifurcations in a dissipative (dynamical) system are investigated in Chap. 10]. From a physical viewpoint, the fundamental process involved in Bénard instability is the transformation of the potential energy of the gravitational field into the kinetic energy of the convective disturbance. The case $\sigma = 0$ represents, again, neutral (marginal), stability; the neutral stability curve is shown in Fig. 3.3.

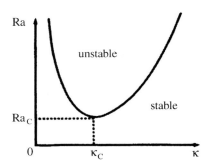

Fig. 3.3. Neutral stability curve for the onset of Bénard thermal convection

Pellew and Southwell (1940) made a comprehensive study of linearized Bénard convection, and they conclusively proved that when the *basic temperature decreases upward*, the only type of disturbance that can appear corresponds to *real σ* so that an *an amplifying wave motion is not possible*.

When $\sigma = 0$, the first variational principle of Pellew and Soutwell leads to an energy-balance relation. This relation establishes a precise balance between the rate of supply of kinetic energy to the velocity field and the rate of dissipation of kinetic energy.

The reader can find examples of bifurcations and instabilities in fluid phenomena in the survey paper by Debnath (1987).

Many excellent experiments have been conducted by a number of authors, including Bénard (1901), Chandra (1938), and Koschmieder (1974). See, also the books by Chandrasekhar (1981) and Drazin and Reid (1981).

3.5 The Bénard Linear Problem with a Free Surface and the Marangoni Effect

In a liquid layer, wavy motion of the free surface open to ambient passive air (nonviscous and at constant temperature T_a and constant pressure p_a) represents one of the most important cases where capillary forces are displayed. The motion induced by tangential gradients of variable surface tension is customarily called the Marangoni effect (after one of the first scientists to explain this effect). Below, we consider mainly thermocapillarity effects and pose an equation of state, $\sigma = \sigma(T)$, for surface tension, as (only) a function of the temperature T:

$$\sigma(T) = \sigma(T^0) - \gamma(T - T^0), \tag{3.56a}$$

with

$$\gamma^0 = -\frac{d\sigma}{dT} = \text{const}, \tag{3.56b}$$

where T^0 is the constant temperature of the free surface in the basic equilibrium state.

We assume (for simplicity) that, in two dimensions (x, z), the equation of a deformable free surface is $z = h(t, x)$.

First, it is necessary to note that it is *not consistent* [from an asymptotic viewpoint, according to Zeytounian (1997, 1998)], *to take into account* the *buoyancy* effect and the *deformation of the free surface simultaneously* in the Bénard thermal convection model problem (for a weakly expansible liquid).

Consequently, here, the model equations are the classical Navier dynamical equations and the Fourier thermal equation, but the boundary conditions at the free surface are very complicated.

In the investigations of Takashima (1981a), the effect of a free-surface deformation on the onset of surface tension driven instability in a horizontal thin liquid layer subjected to a vertical temperature gradient is examined using linear stability theory. Assuming that the neutral state is stationary, the conditions under which instability sets in are determined in detail. In Takashima (1981b), the above linear theory is extended to include the possibility that surface tension driven instability in a horizontal thin liquid layer confined between a solid wall and a deformable free surface can set in a purely oscillatory motion. More recently, linear thermocapillary instabilities with surface deformation in a fluid layer heated from below were studied by Regnier and Lebon (1995).

In linear theory, in place of equations (3.53a,c), we derive the following two ODE for $W(z)$ and $\Theta(z)$:

$$\left\{\sigma - \left[\frac{d^2}{dz^2} - a^2\right]\right\}\left[\frac{d^2 W}{dz^2} - a^2 W\right] = 0, \tag{3.57a}$$

$$\left\{\Pr\sigma - \left[\frac{d^2}{dz^2} - a^2\right]\right\}\Theta = \Pr W, \tag{3.57b}$$

with the following linear boundary conditions:

$$\left.\frac{dW}{dz}\right|_{z=0} = 0, \quad W(0) = \Theta(0) = 0; \tag{3.58a}$$

$$\left.\frac{d^2 W}{dz^2}\right|_{z=1} + a^2 W(1) + a^2 \operatorname{Ma}[\Theta(1) - H^0] = 0; \tag{3.58b}$$

$$[\sigma + 3a^2]\left.\frac{dW}{dz}\right|_{z=1} - \left.\frac{d^3 W}{dz^3}\right|_{z=1} + a^2\left[\left(\frac{1}{\mathrm{Fr}^{02}}\right) + a^2\operatorname{We}\right]H^0 = 0; \tag{3.58c}$$

$$\left.\frac{d\Theta}{dz}\right|_{z=1} + \operatorname{Bi}[\Theta(1) - H^0] = 0; \tag{3.58d}$$

$$W(1) = \sigma H^0, \tag{3.58e}$$

with $H^0 = \mathrm{const}$. These linear boundary (at $z = 0$) and free-surface (at $z = 1$) conditions are the consequence of exact conditions [see, for instance, Zeytounian (1998, §3.3)].

In condition (3.58b), Ma is the Marangoni number that is proportional to surface tension gradient $\gamma^0 = -d\sigma/dT = \mathrm{const}$:

$$\operatorname{Ma} = \gamma^0 d^{03} \frac{\Delta T^0}{\rho^0 \nu_0^2}, \tag{3.59}$$

and in condition (3.58c), $\mathrm{Fr}_0^2 (= (\nu_0/d^0)^2/g d^0)$ is the square of the Froude number defined in (2.70). On the other hand,

$$\operatorname{We} = \sigma(T^0)\frac{d^0}{\rho^0 \nu_0^2} \tag{3.60}$$

is the Weber number in (3.58c). Finally, in condition (3.58d),

$$\operatorname{Bi} = \frac{q^0 d}{k^0} \tag{3.61}$$

is the Biot number, where q^0 is the thermal conductance between a free surface and air, when we write Newton's cooling law of heat transfer at the free surface.

We believe that the basic temperature state describes pure conduction in the liquid at rest, when the free surface (nondeformable) is $z \equiv d^0$.

Equations (3.57a,b) with the boundary conditions (3.58a–e) constitute an eigenvalue (linear) system for a so-called Bénard–*Marangoni* problem.

3.5 Bénard Linear Problem with Free Surface and Marangoni Effect

We note that if, in place of Froude (Fr^0) and Weber (We) numbers, we introduce the crispation (Cr) and Bond (Bn) numbers, according to (where $\mu^0 = \rho^0 \nu_0$)

$$\text{Cr} = \frac{1}{\text{Pr}} \text{We} = \frac{\mu^0 k^0}{\sigma(T^0) d^0}, \tag{3.62a}$$

$$\text{Bn} = \text{Pr} \frac{\text{Cr}}{\text{Fr}^{02}} = \rho^0 g \frac{d^{02}}{\sigma(T^0)}. \tag{3.62b}$$

Then, from this system, we obtain the eigenvalue system considered in Takashima (1981a) and also in Regnier and Lebon (1995), with

$$\omega = \text{Pr}\,\sigma \quad \text{and} \quad w = \text{Pr}\,W,$$

in place of σ and W.

3.5.1 The Case when the Neutral State is Stationary

In this case, $\sigma = 0$ in the above eigenvalue (linear) system. Then, we obtain the following steady-state problem:

$$\left[\frac{d^2 w}{dz^2} - a^2 w\right] = 0, \tag{3.63a}$$

$$\left[\frac{d^2}{dz^2} - a^2\right]\Theta + w = 0, \tag{3.63b}$$

with

$$\left.\frac{dw}{dz}\right|_{z=0} = 0, \quad w(0) = \Theta(0) = 0, \tag{3.64a}$$

$$\left.\frac{d^2 w}{dz^2}\right|_{z=1} + a^2 w(1) + a^2 \text{Ma}[\Theta(1) - H^0] = 0, \tag{3.64b}$$

$$\text{Cr}\left[3a^2 \left.\frac{dw}{dz}\right|_{z=1} - \left.\frac{d^3 w}{dz^3}\right|_{z=1}\right] + a^2[\text{Bn} + a^2] H^0 = 0, \tag{3.64c}$$

$$\left.\frac{d\Theta}{dz}\right|_{z=1} + \text{Bi}[\Theta(1) - H^0] = 0. \tag{3.64d}$$

We observe that the scalar a is a real wave number. The general solutions of (3.63a,b) for $w(z)$ and $\Theta(z)$ can easily be obtained through $\sinh(az)$ and $\cosh(az)$ and, when these solutions are substituted in the boundary conditions (3.64a–d), then, we derive the following eigenvalue relationship:

$$\text{Ma} = \frac{8a[\sinh(a)\cosh(a) - a][a\cosh(a) + \text{Bi}\sinh(a)](\text{Bn} + a^2)}{8\,\text{Cr}\,a^5 \cosh(a) + (\text{Bn} + a^2)[\sinh^3(a) - a^3 \cosh(a)]}. \tag{3.65}$$

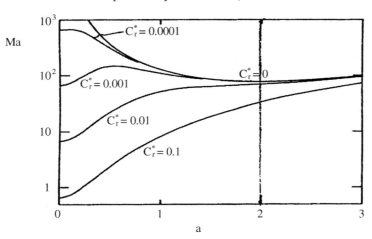

Fig. 3.4. Neutral stability curves for various values of Cr for Bi = 0 and Bn = 0.1. The region below each curve represents the stable state

The result (3.65) was derived 20 years ago by Takashima (1981a). See, also, Regnier and Lebon (1995), where the growth rate of disturbances for the nonzero mode is studied.

For fixed values of Bi, Bn, and Cr, relation (3.65) enables us to plot a neutral stability curve in the (a, Ma) plane (as in Fig. 3.4).

When Bi = 0 is still representative because the value of Bi for a thin layer with which we are concerned is at most 0.1 for most liquids, such a small value of Bi does not affect the results for Bi = 0 appreciably (for a more consistent taking account of the Biot effect see Sect. 7.5). For Bn, if the depth of the layer d^0 is assumed to vary from 0.01 cm to 0.1 cm, for most liquids in contact with air

$$10^{-3} \sim |\mathrm{Bn}| \sim 1 \ . \tag{3.66}$$

The neutral stability curves for Bi = 0 and Bn = 0.1 are shown in Fig. 3.4 as a sample case. The values of Cr are given in the figure.

Because the region below each curve represents a stable state, the lowest point of each curve gives the critical Marangoni number $\mathrm{Ma_c}$ and the corresponding critical wave number a_c.

When Cr → 0, i.e., for a nondeformable free surface, and $\mathrm{Fr}^{02} \to 0$ also but Bn ≠ 0, the result of Pearson (1958),

$$\mathrm{Ma_c} \approx 80 \quad \text{and} \quad a_c \approx 2.0 \ , \tag{3.67}$$

is again obtained. When Cr exceeds 0.00083, a_c changes abruptly from 2.0 to 0, and $\mathrm{Ma_c}$ decreases rapidly as Cr increases. To examine the behavior of Ma analytically in the vicinity of $a = 0$, we expand the right-hand side of relation (3.65) in a power of a.

It has the form:

$$\text{Ma} = \frac{2\,\text{Bn}}{3\,\text{Cr}}(1+\text{Bi})$$
$$+ \frac{2}{3\,\text{Cr}}\left\{(1+\text{Bi})\left[1-\frac{2}{15}\text{Bn}\right] + \frac{1}{3}\text{Bn}\right.$$
$$\left. - \frac{(1+\text{Bi})\,\text{Bn}^2}{120\,\text{Cr}}\right\}a^2 + O(a^4)\,. \qquad (3.68)$$

This result (3.68) of Takashima (1981a) was also derived by Regnier and Lebon (1995). It follows from relation (3.68) that when

$$\text{Cr} > \frac{(1+\text{Bi})\,\text{Bn}^2}{120}\left\{(1+\text{Bi})\left[1-\frac{2}{15}\text{Bn}\right]+\frac{\text{Bn}}{3}\right\},$$

the coefficient of a^2 is positive, and therefore Ma has a minimum value at $a = 0$. It is interesting to note that each curve for Ma_c has a turning point at $\text{Cr} = \text{Bn}/120$ and this turning point is accompanied by a discontinuous change in a_c.

3.5.2 Free-Surface Deformation

When $\text{Cr} < \text{Bn}/120$, the free-surface deformation is not important. On the contrary, when $\text{Cr} > \text{Bn}/120$, the value of Ma_c is strongly dependent on Bn and Cr and the value of a_c is zero. This means that when $\text{Cr} > \text{Bn}/120$, free-surface deformation is important.

According to Takashima (1981), the condition under which free-surface deformation becomes important can also be expressed as

$$d^0 < d^* = \left(\frac{120\nu_0 k^0}{g}\right)^{1/3}, \qquad (3.69)$$

and for water the value of d^* is $0.015\,\text{cm}$.

When $\text{Pr} \approx 1$ ($\nu_0 = k^0$), condition (3.69) is equivalent to the condition derived first in Zeytounian (1997) through an asymptotic analysis of the full exact Bénard problem. In Zeytounian (2002b, Chap. 10), the reader can find asymptotic modeling of thermal convection and interfacial phenomena.

3.6 Flow due to a Rotating Disc

Now, we consider an infinite plane disc, rotating at constant angular velocity Ω_0 in an otherwise unbounded fluid that is at rest apart from the motion induced by the disc. We take cylindrical polar coordinates r, θ, z, with velocity components u, v, w, where $r = 0$ is the axis of rotation of the plane of the disc, $z = 0$.

As boundary conditions, we write

$$u = 0 , \ v = \Omega_0 r , \ w = 0 \text{ on the disc } z = 0 , \tag{3.70a}$$

$$u = v = 0 \text{ when } z \to \infty , \tag{3.70b}$$

and we look for a solution that is independent of θ and time t.

Then, guided partly by dimensional considerations and partly by boundary conditions (3.70a,b), we consider as a solution,

$$u = \Omega_0 r f(z) , \ v = \Omega_0 r g(z) , \ w = (\nu_0 \Omega_0)^{1/2} h(z) ,$$
$$p = \rho_0 (\nu_0 \Omega_0) P(z) . \tag{3.71}$$

Upon substituting these forms in the Navier equations of motion and continuity for an incompressible but viscous fluid flow written in cylindrical polar coordinates, it is found, after some tedious but straightforward algebra, that the functions $f(z)$, $g(z)$, $h(z)$, and $P(z)$ are solutions of a set of ordinary differential equations.

Writing

$$\zeta = z \left(\frac{\nu_0}{\Omega_0} \right)^{1/2} \tag{3.72a}$$

and

$$f\left(\frac{\zeta}{(\nu_0/\Omega_0)^{1/2}} \right) = F(\zeta) , \ g\left(\frac{\zeta}{(\nu_0/\Omega_0)^{1/2}} \right) = G(\zeta) , \tag{3.72b}$$

$$h\left(\frac{\zeta}{(\nu_0/\Omega_0)^{1/2}} \right) = H(\zeta) , \ P\left(\frac{\zeta}{(\nu_0/\Omega_0)^{1/2}} \right) = Q(\zeta) , \tag{3.72c}$$

we derive the following sytem of ODEs:

$$2F + \frac{dH}{d\zeta} = 0 , \tag{3.73a}$$

$$F^2 - G^2 + H \frac{dF}{d\zeta} = \frac{d^2 F}{d\zeta^2} , \tag{3.73b}$$

$$2FG + H \frac{dG}{d\zeta} = \frac{d^2 G}{d\zeta^2} , \tag{3.73c}$$

$$H \frac{dH}{d\zeta} = -\frac{dQ}{d\zeta} + \frac{dH}{d\zeta} , \tag{3.73d}$$

with boundary conditions [for equations (3.73a–c)]

$$F = H = 0, \quad G = 1, \quad \text{when } \zeta = 0, \tag{3.74a}$$

$$F \downarrow 0, \quad G \downarrow 0, \quad \text{as } \zeta \uparrow \infty. \tag{3.74b}$$

We observe that in effect (3.73d) may be regarded as an equation for Q in terms of H.

3.6.1 Small Values of ζ

For small values of ζ, we look for a solution in powers of ζ. By virtue of boundary conditions (3.74a) at $\zeta = 0$ and by using (3.73a–c), we deduce that

$$F = a\zeta - \frac{1}{2}\zeta^2 - \frac{1}{3}b\zeta^3 + \cdots, \tag{3.75a}$$

$$H = -a\zeta^2 + \frac{1}{3}\zeta^3 + \cdots, \tag{3.75b}$$

$$G = 1 + b\zeta + \frac{1}{3}a\zeta^3 + \cdots, \tag{3.75c}$$

where a and b are as yet unknown. To determine these constants, it is necessary to look for a solution of (3.73a–c) with (3.74b) for large values of ζ, far from $\zeta = 0$, and then match both approximate solutions.

3.6.2 Large Values of ζ

For large values of ζ, we note, first of all, that by virtue of boundary conditions (3.74b), (3.73a–c) may be approximated (when ζ is large) by

$$H^\infty \frac{dF}{d\zeta} = \frac{d^2 F}{d\zeta^2}, \quad H^\infty \frac{dG}{d\zeta} = \frac{d^2 G}{d\zeta^2}, \quad H^\infty = \text{const.}$$

Obviously, H^∞ must be negative for consistency, $H^\infty \equiv -c_\infty$, and then for large ζ, we may reasonably look for a solution of the form,

$$F = A_1 e^{-c_\infty \zeta} + A_2 e^{-2c_\infty \zeta} + \cdots, \tag{3.76a}$$

$$G = B_1 e^{-c_\infty \zeta} + B_2 e^{-2c_\infty \zeta} + \cdots, \tag{3.76b}$$

$$H = -c_\infty + C_1 e^{-c_\infty \zeta} + C_2 e^{-2c_\infty \zeta} + \cdots \tag{3.76c}$$

By proceeding as before, it may be shown that (3.73a–c) imply certain relationships between the unknown coefficients in (3.76a–c). In fact, it may be shown that

$$A_2 = -\frac{1}{2c_\infty^2}[A_1^2 + B_1^2], \quad C_1 = \frac{2A_1}{c_\infty},$$

$$C_2 = -\frac{1}{2c_\infty^3}[A_1^2 + B_1^2], \ldots \tag{3.77}$$

where A_1, B_1 and c_∞ are to be determined!

3.6.3 Joining (Matching)

Now, we choose the unknown constants a, b, A_1, B_1, and c_∞ so that F, G, H, and $dF/d\zeta$, $dG/d\zeta$ are continuous where the expansions (3.75a–c) and (3.76a–c) for F, G, H, are "joined." By retaining a sufficient number of terms in each expansion, any desired accuracy can be obtained, and the numerical results are [see, for instance, book by Curle and Davies (1968, p. 163)]:

$$a = 0.510\,,\quad b = -0.616\,,\quad c_\infty = 0.886\,,\quad A_1 = 0.934\,,\quad B_1 = 1.208\,.$$

For all practical purposes, F, G, and H have reached their limiting values when $\zeta \approx 5(\nu_0/\Omega_0)^{1/2}$. For the "disc problem," the Reynolds (local) number is $Re = \Omega_0 r^2/\nu_0$ and we note that the scale normal to the disc is proportional to $r/Re^{1/2}$. For a finite disc of radius a^0, we may calculate the "retarding" torque experienced by the disc, provided we ignore the effects of the edge, in whose vicinity the above theory (and solution) is not valid because the boundary condition, $v = \Omega_0 r$, when $z = 0$, holds only on the surface of the disc, i.e., when $r \leq a^0$. The appropriate retarding shearing stress is

$$-\mu_0\left(\frac{\partial v}{\partial z}\right)_{z=0} = -\mu_0\Omega_0 r \left(\frac{\Omega_0}{\nu_0}\right)^{1/2} \left.\frac{dG}{d\zeta}\right|_{\zeta=0}. \tag{3.78a}$$

Thus, the retarding torque on one side of the finite disc is

$$M = -2\pi\mu_0 \int_0^{a^0} r\left(\frac{\partial v}{\partial z}\right)_{z=0} r\,dr$$

$$= -\left(\frac{\pi}{2}\right)\mu_0\left(\frac{\Omega_0^3}{\nu_0}\right)^{1/2}\left.\frac{dG}{d\zeta}\right|_{\zeta=0} a^{04}. \tag{3.78b}$$

But this result may, alternatively, be expressed in terms of a "moment coefficient" C_m by dividing by $(1/2)\rho^0 a^{03}\Omega_0^2 \Sigma$, where $\Sigma = \pi a^{02}$ is the surface area of the disc, and this yields

$$C_m = -\left(\frac{\nu_0}{\Omega_0 a^{02}}\right)^{1/2}\left.\frac{dG}{d\zeta}\right|_{\zeta=0} = \frac{0.616}{(\mathrm{Re}_{a^0})^{1/2}}, \tag{3.78c}$$

where $\mathrm{Re}_{a^0} = (\Omega_0 a^0)/(\nu_0/a^0) \equiv \mathrm{Re}|_{a^0}$.

3.7 One-Dimensional Unsteady-State NSF Equations and the Rayleigh Problem

Suppose that an infinite flat plate lies *initially at rest* ($t = 0$) on the plane $x_3 \equiv z = 0$ in a viscous compressible heat-conducting fluid (which is an ideal

3.7 1-D Unsteady-State NSF Equations and the Rayleigh Problem

gas with constant γ). At time $t = 0$, it is set *impulsively* into motion in its own plane, moving in the positive x_1 direction at uniform (constant) velocity U_∞, and simultaneously, when $t > 0$, the temperature changes instantaneously on the flat plate $z = 0$ from: T^0 to $T^0 + \Delta T^0$, where $\Delta T^0 = \text{const} > 0$.

Obviously, all x_1 and x_2 derivatives are zero in NSF equations (1.30)–(1.32) and, consequently, the dimensionless equations of our *compressible Rayleigh problem* are one-dimensional unsteady-state NSF equations for the velocity components, u, w, and thermodynamic functions p, ρ, T, dependent on t and z; namely (with $S \equiv 1$ and $Bo = 0$),

$$\frac{\partial \rho}{\partial t} + \frac{\partial (\rho w)}{\partial z} = 0 ; \tag{3.79a}$$

$$\rho \left(\frac{\partial u}{\partial t} + w \frac{\partial u}{\partial z} \right) = \frac{1}{\text{Re}} \frac{\partial}{\partial z} \left(T \frac{\partial u}{\partial z} \right) ; \tag{3.79b}$$

$$\rho \left(\frac{\partial w}{\partial t} + w \frac{\partial w}{\partial z} \right) + \frac{1}{\gamma M^2} \frac{\partial p}{\partial z} = \frac{4}{3 \text{Re}} \frac{\partial}{\partial z} \left(T \frac{\partial w}{\partial z} \right) ; \tag{3.79c}$$

and

$$\rho \left(\frac{\partial T}{\partial t} + w \frac{\partial T}{\partial z} \right) - \left[\frac{\gamma - 1}{\gamma} \right] \left(\frac{\partial p}{\partial t} + w \frac{\partial p}{\partial z} \right)$$
$$= \frac{1}{\text{Pr Re}} \frac{\partial}{\partial z} \left(T \frac{\partial T}{\partial z} \right)$$
$$+ \frac{(\gamma - 1) M^2}{\text{Re}} T \left[\left(\frac{\partial u}{\partial z} \right)^2 + \frac{4}{3} \left(\frac{\partial w}{\partial z} \right)^2 \right], \tag{3.79d}$$

with $p = \rho T$, when we assume that μ and k are proportional to temperature T and use the Stokes relation (3a) in the Introduction.

In dimensionless form, the appropriate initial and boundary conditions are

$$t \leq 0 : u = w = 0 , \quad p = \rho = T = 1 ; \tag{3.80a}$$

$$t > 0 \text{ and } z = 0 : u = 1 , \quad w = 0 , \quad T = 1 + \tau^0 ; \tag{3.80b}$$

$$t > 0 \text{ and } z \uparrow +\infty : u = w = 0 , \quad p = \rho = T = 1 , \tag{3.80c}$$

where

$$\tau^0 = \frac{\Delta T^0}{T^0} . \tag{3.81}$$

From (3.79a), a "particle" function $\phi(t, z)$ exists such that

$$\rho = \frac{\partial \phi}{\partial z} \quad \text{and} \quad w = -\frac{1}{\rho} \frac{\partial \phi}{\partial t} , \tag{3.82a}$$

and it is convenient to rewrite our problem (3.79a–d) with (3.80a–c), via the new coordinate ϕ normal to the flat plate and new time θ:

$$\phi = \int_0^z \rho \, dz' \quad \text{and} \quad \theta \equiv t \,. \tag{3.82b}$$

Obviously,

$$\frac{\partial}{\partial t} = \frac{\partial}{\partial \theta} + \frac{\partial \phi}{\partial t}\frac{\partial}{\partial \phi} \quad \text{and} \quad \frac{\partial}{\partial z} = \rho \frac{\partial}{\partial \phi}\,, \tag{3.82c}$$

and consequently,

$$\frac{\partial}{\partial t} + w\frac{\partial}{\partial z} \equiv \frac{\partial}{\partial \theta}\,. \tag{3.82d}$$

Finally, we derive the following set of equations:

$$w = \frac{\partial z}{\partial \theta}\,, \quad z = \int_0^\phi \left(\frac{1}{\rho}\right) d\phi'\,, \quad \rho = \frac{p}{T}\,; \tag{3.83a}$$

$$\frac{\partial u}{\partial \theta} = \frac{1}{\text{Re}}\frac{\partial}{\partial \phi}\left(p\frac{\partial u}{\partial \phi}\right)\,, \tag{3.83b}$$

$$\frac{\partial p}{\partial \phi} = \gamma M^2 \left\{\frac{4}{3\,\text{Re}}\frac{\partial}{\partial \phi}\left(p\frac{\partial w}{\partial \phi}\right) - \frac{\partial w}{\partial \theta}\right\}\,, \tag{3.83c}$$

$$\frac{\partial T}{\partial \theta} = \frac{1}{\text{Pr}\,\text{Re}}\frac{\partial}{\partial \phi}\left(p\frac{\partial T}{\partial \phi}\right) + \left[\frac{\gamma - 1}{\gamma}\right]\frac{\partial p}{\partial \theta}$$

$$+ M^2\left[\frac{\gamma - 1}{\text{Re}}\right]p\left[\left(\frac{\partial u}{\partial \phi}\right)^2 + \frac{4}{3}\left(\frac{\partial w}{\partial \phi}\right)^2\right]\,, \tag{3.83d}$$

and the initial and boundary conditions are

$$\theta = 0 \text{ and } \phi \uparrow +\infty : u = w = 0\,, \quad p = T = 1\,, \tag{3.84a}$$

$$\theta > 0 \text{ and } \phi = 0 : u = 1\,, \quad w = 0\,, \quad T = 1 + \tau^0\,, \tag{3.84b}$$

because, when $z \uparrow +\infty$, $\rho \to 1$; consequently, $\phi \equiv z$ at the limit, and $z \uparrow +\infty$ is equivalent to $\phi \uparrow +\infty$.

Below, the initial boundary-value problem (3.83a–d), (3.84a,b) is investigated when

$$\text{Pr} \equiv 1\,, \quad \text{Re} = O(1)\,, \quad M^2 \to 0 \text{ and } \tau^0 \to 0\,, \tag{3.85a}$$

such that

$$\tau^0 = \Lambda^0 M^2\,, \tag{3.85b}$$

with

$$\Lambda^0 = \gamma R \frac{\Delta T^0}{U_\infty^2} = O(1)\,. \tag{3.85c}$$

3.7.1 Small M^2 Solution – Close to the Flat Plate but far from the Initial Time

If M^2 is small, the variation in pressure p, density ρ, and temperature T are small, and so is the velocity component normal to the flat plate w. Therefore we write

$$u = u_I + M^2 u' + \ldots ;$$
$$w = +M^2 w' + \ldots ;$$
$$p = 1 + M^4 p' + \ldots ;$$
$$T = 1 + M^2 T' + \ldots ;$$
$$\rho = 1 + M^2 \rho' + \ldots ;$$
$$z = \phi_I + M^2 \phi' + \ldots , \tag{3.86}$$

where the functions with the primes (') are dependent on θ and ϕ and the functions u_I (dependent on θ and $\phi_I \equiv z$), $w_I \equiv 0$, $p_I \equiv 1$ represent the limiting incompressible case. $u_I(\theta, \phi_I)$ is the solution of the following problem:

$$\frac{\partial u_I}{\partial \theta} = \frac{1}{\mathrm{Re}} \frac{\partial^2 u_I}{\partial \phi^2}, \quad u_I(\theta, 0) = 1,$$

$$\lim_{\phi_I \uparrow +\infty} u_I(\theta, \phi_I) = 0 \Leftrightarrow u_I(0, \phi_I) = 0 . \tag{3.87a}$$

With the new coordinate

$$\zeta = \frac{\phi_I}{2\varepsilon \theta^{1/2}}, \quad \varepsilon^2 = \frac{1}{\mathrm{Re}}, \tag{3.87b}$$

the function $u_I \equiv u_I^*(\zeta)$ is the solution of the "incompressible" problem:

$$u_I^{*\prime\prime} + 2\zeta u_I^{*\prime} = 0, \quad u_I^*(0) = 1, \quad u_I^*(+\infty) = 0 . \tag{3.87c}$$

Finally, we obtain the following classical solution of the Rayleigh problem for incompressible viscous fluid flow:

$$u_I^*(\zeta) = 1 - \left(\frac{2}{\pi^{1/2}}\right) \int_0^\zeta \exp(-s^2)\, ds \equiv L_0(\zeta) . \tag{3.87d}$$

We observe that the new (similarity) coordinate ζ is expressed through $z \equiv \phi_I$ according to (3.82b) when $M^2 \downarrow 0$.

Now, a weakly "compressible" effect is taken into account when we consider the problem for $T'(\theta, \phi)$ – see (3.86) – proportional to M^2:

$$\frac{\partial T'}{\partial \theta} = \varepsilon^2 \left\{ \frac{\partial^2 T'}{\partial \phi^2} + (\gamma - 1)\left(\frac{\partial u}{\partial \phi_I}\right)^2 \right\},$$

$$T' = \Lambda^0 \text{ on } \phi = 0, \quad T' = 0 \text{ for } \theta = 0, \quad \lim_{\phi \uparrow +\infty} T' = 0 . \tag{3.88a}$$

The solution of problem (3.88a) for T' is

$$T' \equiv T^{*\prime}(\zeta) = \left(\Lambda^0 + \frac{1}{2}\right) L_0(\zeta) - \frac{1}{2}(\gamma-1)L_0^2(\zeta), \qquad (3.88b)$$

with $\zeta = \phi_I/2\varepsilon\theta^{1/2}$. In the particular case when $\Lambda^0 = (\gamma-1)/2$, we recover a solution obtained by Van Dyke (1952). From expansions (3.86) for p, ρ, and T, when we take into account the dimensionless equation of state for an ideal gas, $p = \rho T$, we deduce a relation between ρ' and T':

$$\rho' = -T',$$

and

$$\frac{1}{\rho} = \left[\frac{1}{1 - M^2 T' + \ldots}\right] \approx 1 + M^2 T' + O(M^4). \qquad (3.88c)$$

Consequently, from (3.83a), we obtain for z an explicit relation (with an error of the order of M^4) between z and ϕ:

$$z = \int_0^\phi \left(\frac{1}{\rho}\right) d\phi' = 2\varepsilon\theta^{1/2} \int_0^\zeta \left(\frac{1}{\rho}\right) d\zeta' = 2\varepsilon\theta^{1/2}\zeta + M^2 \phi' + O(M^4), \qquad (3.89)$$

where

$$\phi' = 2\varepsilon\theta^{1/2} \left\{ \left(\frac{1}{\pi^{1/2}}\right) \left[\Lambda^0 + \frac{1}{2}(\gamma-1)\right] \left[\pi^{1/2}\zeta L_0(\zeta) + 1 - \exp(-\zeta^2)\right] \right.$$
$$+ \left[\frac{(\gamma-1)}{\gamma}\right] \left[\left(\frac{2}{\pi^{1/2}}\right)\left(1 - L_0(2^{1/2}\zeta)\right)\right.$$
$$\left.\left. - \left(\frac{2}{\pi^{1/2}}\right)\left(1 - \exp(-\zeta^2)L_0(\zeta)\right) - \zeta L_0^2(\zeta)\right]\right\}. \qquad (3.90)$$

Then, from relation (3.83a) for $w(= \partial z/\partial\theta)$, we derive the following relation for w':

$$w' = \frac{\partial \phi'}{\partial \theta} = \left(\frac{\varepsilon}{\pi^{1/2}}\right)\left(\frac{1}{\theta^{1/2}}\right) \left\{ \left[\Lambda^0 + \frac{1}{2}(\gamma-1)\right][1 - \exp(-\zeta^2)] \right.$$
$$\left. +(\gamma-1)\left[\left(\frac{1}{2^{1/2}}\right)\left(1 - L_0(2^{1/2}\zeta)\right) - \left(1 - \exp(-\zeta^2)L_0(\zeta)\right)\right]\right\}. \qquad (3.91)$$

From (3.91), we deduce that (when θ and ε are both fixed)

$$w' \to \left(\frac{\varepsilon}{\pi^{1/2}}\right)\left\{\Lambda^0 + \frac{1}{2}(\gamma-1)\left[\frac{2 - 2^{1/2}}{2^{1/2}}\right]\right\}\left(\frac{1}{\theta^{1/2}}\right), \qquad (3.92)$$

when $\zeta \to +\infty$, *and we see that w' does not tend to zero at infinity!*

3.7 1-D Unsteady-State NSF Equations and the Rayleigh Problem

Now, with solution (3.91), we can also obtain a solution for p', according to (3.83c) and expansion (3.86) for p. It is necessary to resolve the following equation:

$$\frac{1}{\gamma}\frac{\partial p'}{\partial \phi} = \frac{4}{3}\varepsilon^2 \frac{\partial^2 w'}{\partial \phi^2} - \frac{\partial w'}{\partial \theta},$$

and we derive as a solution,

$$p' = \gamma\varepsilon^2 \left(\frac{1}{\pi^{1/2}\theta}\right) \left\{ \left[\Lambda^0 + \frac{\gamma-1}{2}\right]\left[\zeta\left(1 + \frac{1}{3}\exp(-\zeta^2)\right)\right] \right.$$

$$+ (\gamma - 1)\left[\left(\frac{1}{3}\right)\zeta \exp(-\zeta^2)(1 - L_0(\zeta))\right]$$

$$\left. + \left(\frac{1}{2^{1/2}}\right)\zeta\left(1 - L_0(2^{1/2}\zeta)\right)\right] \right\} + P(\theta), \qquad (3.93)$$

where $P(\theta)$ is an arbitrary function.

Again, for θ and ε fixed, p' does not tend to zero at infinity (in fact, p' tends to infinity as $z \to +\infty$). We observe that $\zeta \to +\infty$ is equivalent to $z = \phi_I \to +\infty$ when $\theta \equiv t$ is fixed and also to $\theta = 0$ when ϕ_I is fixed.

On the other hand, we note that in (3.87b) and also in relations (3.89)–(3.93), the parameter ε is fixed of the order of $O(1)$.

If we assume that $\varepsilon \ll 1$ (high Reynolds number) such that

$$\varepsilon \equiv M^2 \Rightarrow (\text{Re})^{1/2} M^2 = 1, \qquad (3.94)$$

then, solutions (3.87d) for $u_I^*(\zeta)$ and (3.88b) for $T^{*\prime}(\zeta)$ with the boundary layer coordinate

$$\Phi_I = \frac{\phi_I}{\varepsilon}, \quad \text{and} \quad \zeta = \frac{\Phi_I}{2\theta^{1/2}}, \qquad (3.95)$$

and the solution for w'/ε, according to (3.91) and (3.94), with $p = 1$ and $\rho' = -T'$, give a *boundary layer solution* (a function of θ and Φ_I).

Therefore, expansions (3.86) are *inner expansions* valid only close to the flat plate and far from the initial time!

3.7.2 Small M^2 Solution – Far from a Flat Plate

Far from a flat plate, it is necessary to consider *outer expansions*, and for this we assume that

$$M^2 \to 0 \text{ with } t \text{ and } z^* = Mz \text{ fixed}, \qquad (3.96a)$$

and introduce a new vertical component of the velocity

$$w^* = Mw . \tag{3.96b}$$

Then, for the functions u^*, w^*, p^*, ρ^*, and T^* (dependent on t and z^*), we must consider the following equations in place of (3.79a–d):

$$\frac{\partial \rho^*}{\partial t} + \frac{\partial (\rho^* w^*)}{\partial z^*} = 0 ; \tag{3.97a}$$

$$\rho^* \left(\frac{\partial u^*}{\partial t} + w^* \frac{\partial u^*}{\partial z^*} \right) = \left(\frac{M^2}{\mathrm{Re}} \right) \frac{\partial}{\partial z^*} \left(T^* \frac{\partial u^*}{\partial z^*} \right) ; \tag{3.97b}$$

$$\rho^* \left(\frac{\partial w^*}{\partial t} + w^* \frac{\partial w^*}{\partial z^*} \right) + \frac{1}{\gamma} \frac{\partial p^*}{\partial z^*} = \left(\frac{4M^2}{3\,\mathrm{Re}} \right) \frac{\partial}{\partial z^*} \left(T^* \frac{\partial w^*}{\partial z^*} \right) ; \tag{3.97c}$$

$$\rho^* \left(\frac{\partial T^*}{\partial t} + w^* \frac{\partial T^*}{\partial z^*} \right) - \left[\frac{(\gamma - 1)}{\gamma} \right] \left(\frac{\partial p^*}{\partial t} + w^* \frac{\partial p^*}{\partial z^*} \right)$$

$$= \left(\frac{M^2}{\Pr \mathrm{Re}} \right) \frac{\partial}{\partial z^*} \left(T^* \frac{\partial T^*}{\partial z^*} \right)$$

$$+ M^4 \left[\frac{\gamma - 1}{\mathrm{Re}} \right] T^* \left[\left(\frac{\partial u^*}{\partial z^*} \right)^2 + \left(\frac{4}{3M^2} \right) \left(\frac{\partial w^*}{\partial z^*} \right)^2 \right] , \tag{3.97d}$$

with $p^* = \rho^* T^*$. For an asymptotic solution of (3.97a–d), we write the following outer expansions, in place of (3.86):

$$u^* = M^\alpha u_\mathrm{a} + \ldots ;$$
$$w^* = M^\alpha w_\mathrm{a} + \ldots ;$$
$$p^* = 1 + M^\alpha p_\mathrm{a} + \ldots ;$$
$$T^* = 1 + M^\alpha T_\mathrm{a} + \ldots ;$$
$$\rho^* = 1 + M^\alpha \rho_\mathrm{a} + \ldots , \tag{3.98}$$

and we observe that u_a, w_a, p_a, ρ, and T_a are dependent on t and z^*. The scalar $\alpha > 0$ is determined such that the behavior of w' at $z \to +\infty$, according to (3.92), must be compatible with the behavior of w^* when $z^* \to 0$ according to (3.96b) and (3.98) (matching!):

$$w = \frac{w^*}{M} = M^{\alpha-1} w_\mathrm{a} + \ldots \sim w = M^2 w' + \ldots , \tag{3.99a}$$
$$\text{when } z^* \to 0 \qquad\qquad \text{when } z \to +\infty ,$$

thus

$$\alpha - 1 = 2 \to \alpha = 3 . \tag{3.99b}$$

3.7 1-D Unsteady-State NSF Equations and the Rayleigh Problem

With (3.99b) in (3.98), from (3.97a–d), we derive the system of classical acoustic equations for w_a, p_a, ρ_a, and T_a [we assume that the Reynolds number Re $= O(1)$]:

$$\frac{\partial \rho_a}{\partial t} + \frac{\partial w_a}{\partial z^*} = 0 ; \tag{3.100a}$$

$$\frac{\partial w_a}{\partial t} + \frac{1}{\gamma}\frac{\partial p_a}{\partial z^*} = 0 ; \tag{3.100b}$$

$$\gamma \frac{\partial T_a}{\partial t} - (\gamma - 1)\frac{\partial p_a}{\partial t} = 0 ; \tag{3.100c}$$

$$p_a = \rho_a + T_a , \tag{3.100d}$$

and for w_a, we obtain the acoustic (wave) equation

$$\frac{\partial^2 w_a}{\partial z^{*2}} - \frac{\partial^2 w_a}{\partial t^2} = 0 . \tag{3.101}$$

First, from the initial conditions (3.80a) for w and p, we write as initial conditions for w_a, a solution of (3.101),

$$w_a(0, z^*) = 0 , \quad \left.\frac{\partial w_a}{\partial t}\right|_{t=0} = 0 . \tag{3.102a}$$

Matching with the inner solution for $z \to +\infty$ gives as a boundary condition on $z^* = 0$,

$$w_a(t, 0) = \left(\frac{\varepsilon}{\pi^{1/2}}\right)\left\{\Lambda^0 + \frac{\gamma - 1}{2}\left[\frac{2 - 2^{1/2}}{2^{1/2}}\right]\right\}\left(\frac{1}{t^{1/2}}\right), \text{ for } t > 0 . \tag{3.102b}$$

We see that for the wave equation (3.101), it is also necessary to write a radiation condition "à la Sommerfeld", simply,

$$w_a(t, +\infty) \to 0 . \tag{3.102c}$$

Finally, the solution of (3.101)–(3.102a–c) is

$$w_a(t, z^*) = 0 , \quad \text{when } z^* \geq t > 0 ; \tag{3.103a}$$

$$w_a(t, z^*) = \frac{\varepsilon}{[\pi(t - z^*)]^{1/2}}\left\{\Lambda^0 + \frac{1}{2}(\gamma - 1)(2^{1/2} - 1)\right\} , \tag{3.103b}$$

when $z^* \leq t > 0$.

Now, with the solution (3.103a,b), from (3.100b), we determine $p_a(t, z^*) = 0$, when

$$p_a(t, z^*) = 0 \quad \text{when } z^* \geq t > 0 ; \tag{3.104a}$$

$$p_a(t, z^*) = \frac{\gamma \varepsilon}{[\pi(t - z^*)]^{1/2}} \left\{ \Lambda^0 + \frac{1}{2}(\gamma - 1)(2^{1/2} - 1) \right\} , \tag{3.104b}$$

when $z^* \leq t > 0$,

and when $z^* \to 0$, with $t > 0$, we obtain an expression for $p_a(t, 0)$ which is not zero:

$$p_a(t, 0) = \frac{\gamma \varepsilon}{(\pi t)^{1/2}} \left\{ \Lambda^0 + \frac{1}{2}(\gamma - 1)(2^{1/2} - 1) \right\} , \quad \text{for } t > 0 . \tag{3.104c}$$

Consequently, to match the inner and outer pressure expansions, it is necessary to include a term for p in (3.86) that is proportional to M^3:

$$p = 1 + M^3 p_a(\theta, 0) + M^4 p' + \ldots , \tag{3.105}$$

where $p_a(\theta, 0)$ is given by (3.104c) with θ in place of t. Obviously, it is also necessary to insert the terms proportional to M^3 in the expansions (3.86) for w, ρ, and T. To determine the arbitrary function $P(\theta)$ in (3.93), then it is necessary to consider the terms proportional to M^4 in (3.98), where $\alpha = 3$.

3.7.3 Small M^2 Solution – Close to the Initial Time

Clearly, the procedure used in Sect. 3.7.1 and 3.7.2 yields only an asymptotic solution for time $t = O(1)$. It is not valid near $t = 0$ because solutions (derived above) would imply that an infinite impulse per unit area is required to set the flat plate in motion. The main problem is matching the solution that is valid in the initial transient layer with the above asymptotic solution that is valid for fixed time $t > 0$! In Howarth (1951), the reader can find some information. Howarth investigated the initial stages of motion [see Howarth (1951), §4].

We consider the problem of the initial stages of motion for the large and low Reynolds numbers in Chaps. 4 and 5, respectively, and also in Chap. 6 devoted to low-Mach-number asymptotics of NSF equations.

Here, we note only that to investigate the solution of our problem (3.79a–d)–(3.80a–c) for small M^2 close to initial time $t \leq 0$, it is necessary, first, to introduce a short time adapted to the initial transient layer. For this, we can write:

$$\tau = \frac{t}{M^\beta} , \quad \text{with } \beta > 0 , \tag{3.106a}$$

and with the *short time* τ, we consider the following limiting (initial) process:

$M \to 0$ with τ and z fixed and $\text{Re} = O(1)$. \hfill (3.106b)

Then, with the following initial expansions:

$$w = M^a w_\text{i} + \ldots ;$$

$$p = 1 + M^b p_\text{i} + \ldots ;$$

$$T = 1 + M^c T_\text{i} + \ldots ;$$

$$\rho = 1 + M^d \rho_\text{i} + \ldots , \tag{3.106c}$$

from (3.79a) and (3.79c,d) and the relation $p = \rho T$, we *recover again* the *classical acoustic equations* (3.100a–d) for w_i, p_i, ρ_i and T_i if

$$a = b - \beta , \quad b - 2 + \beta = 0 , \quad c = b , \quad d = c , \tag{3.106d}$$

with $\beta > 0$.

A "plausible" case corresponds to

$$\beta = 1 , \quad a = 0 \quad \text{and} \quad b = c = d = 1 , \tag{3.107}$$

and in this case in the initial expansions for p, ρ, and T, close to the initial time, we have a term proportional to M!

3.8 Complementary Remarks

First, we observe that when we have a solution of Navier equations, then we can associate a solution for the temperature perturbations according to Fourier equation (2.17), and in this case, the boundary condition on the wall for the temperature (assumed for the NSF equations) is taken into account.

Thus, for each viscous incompressible (dynamic) fluid flow, we have the possibility of determining an associated thermal field that takes into account the thermal effects on the wall.

In the recent book by Schlichting and Gersten (first published in German in 1997 – English ed. 2000; see Chap. 5), the reader can also find various exact solutions of viscous fluid dynamics equations for steady-state and unsteady-state flows. Among others, we note plane and axisymmetric stagnation-point flow [extension to the 3-D stagnation point was treated by Gersten (1973)], flow between two disks rotating in opposite directions [a good summary has been given by Zandbergen and Dijkstra (1987)], axisymmetric free jet [see, Sherman (1990) and Schneider (1981)], unsteady-state plane flows (Stokes problems – flow at a wall suddenly set into motion and flow at an oscillating wall, start-up of Couette flow, oscillating channel, and pipe flows).

These exact solutions are obtained mainly in the framework of the boundary layer equations valid near the wall for high Reynolds number flow. In the recent book by Papanastasiou et al. (2000) devoted to *Viscous Fluid Flow*, the reader can also find various exact solutions of Navier equations. The survey of Wang (1991) is especially devoted to exact solutions of steady-state NS equations.

Chapter 4 is devoted just to this asymptotic theory at $Re \gg 1$, and the reader can find, first in this chapter, a simple derivation of the Prandtl boundary-layer equations. Then, we investigate the formal asymptotic structure of unsteady-state NSF equations at $Re \gg 1$ and, in particular, we recover the one-dimensional unsteady-state NSF equations (3.79a–d), considered in Sect. 3.7, which are valid only near the wall for a thin unsteady-state layer close to the initial time. This is the consequence of the singular nature of the unsteady-state Prandtl boundary-layer equations close to the initial time.

In Chap. 7, we consider also some interesting viscous problems and motions and in "Introduction" the reader can find a short comment concerning these viscous flows.

It is also interesting to observe that many examples of paradoxical solutions exist for viscous flows [see, for instance, Birrkhoff (1960, Chap. II) and the review paper by Goldshtik (1990)].

4 The Limit of Very Large Reynolds Numbers

4.1 Introduction

When frictional (viscous) terms are appreciably larger than inertial terms, the limiting case is of considerably greater importance in various practical applications. Air and water have low viscosities, and we can consider the limit of very large (high) Reynolds number, Re $\to \infty$, which is the singular case of vanishing viscosity.

Obviously, if we simply delete the frictional term on the right-hand side of Navier equation (2.14a), then we obtain the Euler nonviscous incompressible equation considered in the companion book devoted to nonviscous fluid flows [Zeytounian (2002a)]. But in this case, unfortunately, because the order of the partial differential equation is reduced, the no-slip boundary condition on the wall must be replaced by the more simple slip boundary condition on the wall.

This question is central in the framework of large Reynolds asymptotics of NSF equations!

The situation is more complicated for NSF these (nonadiabatic) equations, because it is necessary to assume a boundary condition on the wall for the temperature also. On the contrary, for the full compressible Euler equations, which are obtained from NSF equations, when Re $\equiv \infty$, such a temperature equation is not necessary if the slip condition is satisfied for the normal to the wall component of the Eulerian velocity – the temperature is not subject to transport phenomena through the wall.

Another point is important and concerns the position of the points $M(\boldsymbol{x})$ in flow space and the value of time t relative to $t = 0$, where the initial conditions are given for the NSF unsteady-state equations! Eulerian nonviscous fluid flows are derived when the Eulerian limiting process,

$$\text{Re} \to \infty \text{ when } \boldsymbol{x} \text{ and } t \text{ are fixed of the order of } O(1), \qquad (4.1)$$

and the variables $\boldsymbol{x}$ and t are, respectively, dimensionless variables relative to L^0 and L^0/U_∞ [if the Strouhal number S (1.35a) is assumed $= 1$]. When Re $\to \infty$, near the wall, it is necessary to take into account, at least, part (!) of the frictional (dissipative) terms proportional to Re.

For this, we must consider, in place of (4.1), another limiting process (the Prandtl limiting process); in the steady-state,

$$\text{Re} \to \infty \text{ when } \boldsymbol{s}, \eta = n/\delta \text{ and } t \text{ are fixed of the order of } O(1), \qquad (4.2)$$

when we write $\boldsymbol{x} = (\boldsymbol{s}, n)$, with $\boldsymbol{s} = (s_1, s_2)$ and consider: s_1, s_2, and n (with $n = \boldsymbol{x} \cdot \boldsymbol{n}$ where $\boldsymbol{n}$ the unit normal to the wall directed to the fluid) as a *system of curvilinear coordinates* of a point $M^*(\boldsymbol{s}, n)$ near the wall $\eta = 0$. The small parameter in (4.2) is

$$\delta = L^0/l^0, \qquad (4.3)$$

with l^0 the thickness of the thin viscous layer near the wall where a part of the dissipative terms proportional to $1/\text{Re}$ must be retained.

With (4.1) and (4.2), the full NSF (exact) low viscosity flow is divided into two parts:

> the so-called "outer" or "main nonviscous" flow, obtained with the help of the Euler equations of motion, and a so-called "inner" or "local-boundary layer" flow.

By matching, both flows should be matched onto the another so that both flows are valid in an overlap region, and via this matching, we obtain a consistent, rational approximation of the real NSF slightly viscous fluid flows.

In an unsteady-state, the above asymptotic structure (Euler-matching-Prandtl) should be complemented by two initial layers, close to $t = 0$, where the significant times are, respectively,

$$\tau = t/\delta \quad \text{and} \quad \tau^* = t/\delta^2. \qquad (4.4a)$$

It is also necessary to consider a viscous sublayer, in an initial layer characterized by t/δ^2, when in place of $\eta = n/\delta$, we have as a significant, normal to wall, vertical coordinate

$$\eta^* = n/\delta^2 = \eta/\delta. \qquad (4.4b)$$

Naturally, in this unsteady-state, when the Prandtl boundary-layer concept fails close to the initial time. Because, in the Prandtl boundary-layer equations, for compressible, steady-state, incompressible or unsteady-state fluid flows, we always have $\partial p_{\text{BL}}/\partial \eta = 0$, in place of the third equation of motion for the vertical component of the velocity vector (normal to the wall $\eta = 0$). This problem is considered in our last book [Zeytounian (2002b, Chap. 7)] devoted to asymptotic modeling of fluid flow phenomena.

In Sect. 4.3 in the framework of a formal asymptotic structure of unsteady-state NSF equations at $\text{Re} \gg 1$ we again touch on this problem, but the approach is slightly different. Unfortunately, the problem of matching various limiting approximations among themselves is far from resolved!

In particular, matching between the unsteady-state Prandtl BL model and the Rayleigh compressible viscous sublayer (VS-L) model, when both τ^* and η^* tend to infinity (matching at $\eta = 0$ and $t = 0$), is a very difficult problem, and it seems that only via numerical computation of the corresponding "unsteady-state adjustment problem" we have the possibility of understanding this matching process with the unsteady-state Prandtl boundary layer.

In Sect. 4.2 we give an asymptotic derivation of steady-state two-dimensional, incompressible Prandtl boundary equations near a semi-infinite flat plate (which leads to the so-called *"Blasius"* problem), when

$$\text{Re} \to \infty \text{ and } \delta \to 0 \text{ such that } \text{Re}\,\delta^a = R^* = O(1) , \qquad (4.5)$$

where $a > 0$ and R^* is a similarity parameter.

With (4.5) we can understand the Prandtl boundary-layer concept as a double-scale concept and it is possible to derive the classical Prandtl boundary-layer equations via a multiple scale method (MSM).

The Prandtl boundary-layer concept also fails, for example, in the vicinity of an "accident" at the wall that can cause boundary-layer separation. In Sect. 4.4, we consider the simple case of viscous incompressible Navier flow over a flat plate with a small hump situated downstream of the leading edge. The reader can find, in [Zeytounian (2002b, Chap. 12)] a detailed exposition of the triple-deck asymptotic model which leads to *singular coupling* between nonviscous Eulerian flow and viscous boundary-layer flow via an intermediate coupling layer.

On the contrary, the classical theory, Euler-matching-Prandtl, is *hierarchical*, and the outer and inner approximations are derived *step by step*. The coupling is "regular".

In Sect. 4.5, we give complementary information relative to boundary-layer equations in two and three dimensions for a curvilinear body. The problem of the unsteady-state incompressible laminar boundary layer developing from rest is also briefly considered.

Here, we note that our understanding of viscous incompressible (2-D steady-state) flow past a finite plate, at $\text{Re} \gg 1$, has been revolutionized by applying the triple-deck theory to the flow near the trailing edge (in Fig. 4.1, the small parameter $\varepsilon = 1/\text{Re}$). It comprises a region around the trailing edge whose extent is of the order $\text{Re}^{-3/8}$ times the length of the plate and which matches upstream with the classical Blasius solution and its associated outer flow and downstream with the two-layered wake analyzed by Goldstein (1930). There is also a small circular core in which, just as in the corresponding vicinity of the leading edge, the full Navier equations apply. In Fig. 4.1, we have (1) Potential flow; (2) Blasius BL; (3 and 7) Navier region; (4) Main deck, inviscid; (5) Lower deck, viscous; (6) Upper deck, potential flow; (8) Inner Goldstein wake; and (9) Outer Goldstein wake.

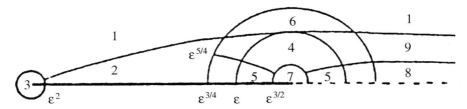

Fig. 4.1. Asymptotic structure near the trailing edge of a finite plate

We have the following formula for the integrated skin friction on one side of the finite plate, as a consequence of the structure discussed above:

$$C_F = 1.328/\text{Re}^{1/2} + 2.654/\text{Re}^{7/8} + \ldots + 2.226/\text{Re} + \ldots \\ - (1.192/\text{Re}^{3/2}) \log \text{Re} - 4.864/\text{Re}^{3/2} + \ldots . \quad (4.6)$$

In (4.6), the dots mean that intermediate terms probably exist that are not yet known; the first term is the classical Blasius drag.

The structure discussed, near the trailing edge, contributes a correction to the Blasius drag that is of the order $\text{Re}^{-7/8}$, and hence is slightly more important than the dispacement effects of the order $1/\text{Re}$ calculated first by Kuo (1953) and Imai (1957a) using a global balance of momentum. The coefficient 2.654 was obtained numerically by Veldman and van de Vooren (1974); in fact, these authors find 2.651 ± 0.003. The coefficient 2.226 was also obtained by a numerical code (for the full Navier equations) by van de Vooren and Dijkstra (1970). For a discussion of the last two terms in (4.6), see Goldstein (1960). In fact, the fourth one comes from the leading order of the second-order approximation of the boundary layer and was given by Imai (1957a) who was unable to find the fifth term because it must include the contribution of the Navier solution right at the leading edge. Note that the last term in (4.6) is given by $-(0.204 + 2C)/\text{Re}^{3/2}$, according to Imai (1957a), and the coefficient $C \approx 2.33$ was obtained numerically by solving the full 2-D steady-state Navier equations near a region around the leading edge whose extent is of the order $1/\text{Re}$ times the length of the plate [see van de Vooren and Dijkstra (1970)]. For the $\text{Re}^{-7/8}$ dependence of the second term in (4.6), we observe that there is no change in the order of the skin friction when going from the upstream boundary layer to the lower deck, but this one term brings in a perturbation to an $O(\text{Re}^{-1/2})$ term integrated over an $O(\text{Re}^{-3/8})$ distance which gives an integrated contribution of $O(\text{Re}^{-7/8})$.

4.2 Classical Hierarchical Boundary-Layer Concept and Regular Coupling

4.2.1 A 2-D Steady-State Navier Equation for the Stream Function

Below we return to the Navier system (2.14a,b) and consider a 2-D steady-state motion in the $(O; x, y)$ plane. From the continuity equation (2.14b), we obtain for the components u_N and v_N of the Navier velocity vector $\boldsymbol{u}_N$

$$\frac{\partial u_N}{\partial x} + \frac{\partial v_N}{\partial y} = 0, \tag{4.7a}$$

and we have the possibility of introducing a stream function $\psi(x, y)$ such that

$$u_N = \frac{\partial \psi}{\partial y} \quad \text{and} \quad v_N = -\frac{\partial \psi}{\partial x}. \tag{4.7b}$$

Then, if we eliminate the pseudopressure π in the two Navier scalar equations of motion written for $u_N(x, y)$ and $v_N(x, y)$,

$$u_N \frac{\partial u_N}{\partial x} + v_N \frac{\partial u_N}{\partial y} + \frac{\partial \pi}{\partial x} = \frac{1}{\mathrm{Re}} \left(\frac{\partial^2 u_N}{\partial x^2} + \frac{\partial^2 u_N}{\partial y^2} \right),$$

$$u_N \frac{\partial v_N}{\partial x} + v_N \frac{\partial v_N}{\partial y} + \frac{\partial \pi}{\partial y} = \frac{1}{\mathrm{Re}} \left(\frac{\partial^2 v_N}{\partial x^2} + \frac{\partial^2 v_N}{\partial y^2} \right),$$

we derive the following single equation for $\psi(x, y)$:

$$\left(\frac{\partial \psi}{\partial y} \frac{\partial}{\partial x} - \frac{\partial \psi}{\partial x} \frac{\partial}{\partial y} \right) \left(\frac{\partial^2 \psi}{\partial x^2} + \frac{\partial^2 \psi}{\partial y^2} \right) = \frac{1}{\mathrm{Re}} \left(\frac{\partial^2}{\partial x^2} + \frac{\partial^2}{\partial y^2} \right) \left(\frac{\partial^2 \psi}{\partial x^2} + \frac{\partial^2 \psi}{\partial y^2} \right). \tag{4.8}$$

As boundary conditions, we assume simply the no-slip condition on the flat plate $y = 0$, when $0 < x < +\infty$ (a semi-infinite flat plate) and the existence of an upstream (at $-\infty$) constant uniform velocity vector $U^0 \boldsymbol{i}$ parallel to the direction $x > 0$:

$$\psi = 0 \quad \text{and} \quad \frac{\partial \psi}{\partial y} = 0 \quad \text{on } y = 0, \tag{4.9a}$$

$$\psi \to y \quad \text{when } x \to -\infty. \tag{4.9b}$$

We note that in dimensionless problem (4.8), (4.9a,b), we have U^0 and L^0 as a reference velocity, and reference length, and in this case, the reference value of ψ is $U^0 L^0$.

Obviously, the dimensionless problem (4.8), (4.9a,b), for $\psi(x, y; \mathrm{Re})$ is significant only when (at least for $\mathrm{Re} \to \infty$) we consider weakly viscous flow for the points which are at distance L^0 from the flat plate $y = 0$.

4.2.2 A Local Form of the 2-D Steady-State Navier Equation for the Stream Function

But, if we want to consider weakly viscous flow for the points which are at distance $l^0 \ll L^0$ from the flat plate $y = 0$, then, first, we must introduce a new "distorted" vertical coordinate,

$$\eta = y/\delta , \qquad (4.10a)$$

and also a new (local) stream function,

$$\psi_{\text{BL}} = \psi/\delta . \qquad (4.10b)$$

Consequently, near the flat plate $\eta = 0$, in place of dimensionless problem (4.8), (4.9a,b), we obtain the following new dimensionless problem for $\psi_{\text{BL}}(x, \eta; \text{Re}, \delta)$:

$$\left(\frac{\partial \psi_{\text{BL}}}{\partial \eta} \frac{\partial}{\partial x} - \frac{\partial \psi_{\text{BL}}}{\partial x} \frac{\partial}{\partial \eta} \right) \left(\delta \frac{\partial^2 \psi_{\text{BL}}}{\partial x^2} + \frac{1}{\delta} \frac{\partial^2 \psi_{\text{BL}}}{\partial \eta^2} \right)$$
$$= \frac{1}{\text{Re}} \left(\frac{\partial^2}{\partial x^2} + \frac{1}{\delta^2} \frac{\partial^2}{\partial \eta^2} \right) \left(\delta \frac{\partial^2 \psi}{\partial x^2} + \frac{1}{\delta} \frac{\partial^2 \psi}{\partial \eta^2} \right) , \qquad (4.11)$$

$$\psi_{\text{BL}} = 0 \text{ and } \frac{\partial \psi_{\text{BL}}}{\partial \eta} = 0 \quad \text{on } \eta = 0 , \qquad (4.12a)$$

$$\psi_{\text{BL}} \to \eta \quad \text{when } x \to -\infty . \qquad (4.12b)$$

We observe that in both problems (4.8), (4.9a,b) and (4.11), (4.12a,b), we have the same horizontal coordinate x (parallel to the semi-infinite flat plate), and $0 < x < +\infty$.

The decoupling of our exact problem into two subproblems is obviously interesting only for large Reynolds numbers when the boundary-layer concept works!

4.2.3 A Large Reynolds Number and "Principal" and "Local" Approximations

For very large Re ($\to \infty$), it is necessary to consider two limiting processes:

(i) $\quad \displaystyle\lim_{\text{Re} \to \infty}^{\text{P}} = [\text{Re} \to \infty \text{ with } x \text{ and } y \text{ fixed}, O(1)] , \qquad (4.13a)$

(ii) $\quad \displaystyle\lim_{\text{Re} \to \infty}^{\text{loc}} = [\text{Re} \to \infty \text{ with } x \text{ and } \eta \text{ fixed}, O(1)] . \qquad (4.13b)$

The first "Principal" limiting process for

$$\lim_{\text{Re} \to \infty}^{\text{P}} \psi = \psi_0(x, y) , \qquad (4.14a)$$

4.2 Classical Hierarchical Boundary-Layer Concept and Regular Coupling

leads to the following equation from (4.8):

$$\left(\frac{\partial \psi_0}{\partial y}\frac{\partial}{\partial x} - \frac{\partial \psi_0}{\partial x}\frac{\partial}{\partial y}\right)\left(\frac{\partial^2 \psi_0}{\partial x^2} + \frac{\partial^2 \psi_0}{\partial y^2}\right) = 0 ,$$

and consequently,

$$\frac{\partial^2 \psi_0}{\partial x^2} + \frac{\partial^2 \psi_0}{\partial y^2} = F(\psi_0) . \tag{4.14b}$$

With the condition $\psi_0 \to y$, when $x \to -\infty$, according to (4.9b), and when we assume that the considered limiting flow is continuous, obviously $F(\psi_0) \equiv 0$, and consequently the classical 2-D Laplace equation for the potential outer flow is derived:

$$\frac{\partial^2 \psi_0}{\partial x^2} + \frac{\partial^2 \psi_0}{\partial y^2} = 0 . \tag{4.14c}$$

But, for Laplace equation (4.14b), we do not have any possibility of accounting for the no-slip boundary condition (4.9a) for ψ_0! Therefore the "P" limit (4.13a) is singular near the flat plate! Consequently, it is necessary to consider (near the flat plate) the "loc" limit (4.13b), with the hope that this second "local" limiting process leads to a different limiting equation for

$$\operatorname*{Lim}_{\mathrm{Re} \to \infty}^{\mathrm{loc}} \psi_{\mathrm{BL}} = \psi_{\mathrm{BL},0}(x, \eta) . \tag{4.15a}$$

For this, we are constrained (!) to assume the existence of a similarity relation between the parameter Re and δ:

$$\mathrm{Re}\,\delta^a = R^* = O(1), \quad \text{with } a > 0 \text{ and } R^* = O(1) , \tag{4.16}$$

if we want to derive a new limit equation [other than (4.14b)] from (4.11) for the local leading stream function $\psi_{\mathrm{BL},0}(x,\eta)$!

From (4.11), with (4.16), we derive first:

$$\left(\frac{\partial \psi_{\mathrm{BL},0}}{\partial \eta}\frac{\partial}{\partial x} - \frac{\partial \psi_{\mathrm{BL},0}}{\partial x}\frac{\partial}{\partial \eta}\right)\frac{\partial^2 \psi_{\mathrm{BL},0}}{\partial \eta^2} = \frac{1}{R^*}\delta^{a-2}\frac{\partial^4 \psi_{\mathrm{BL},0}}{\partial \eta^4} . \tag{4.15b}$$

Only the case

$$\alpha = 2 \Rightarrow \delta = 1/\mathrm{Re}^{1/2} \quad \text{for } R^* = 1 \tag{4.15c}$$

is significant and, when we take into account condition (4.12b) for $\psi_{\mathrm{BL},0}$, we derive, after an integration relative to η, the Prandtl BL equation (with $R^* = 1$):

$$\frac{\partial^3 \psi_{\mathrm{BL},0}}{\partial \eta^3} + \frac{\partial \psi_{\mathrm{BL},0}}{\partial x}\frac{\partial^2 \psi_{\mathrm{BL},0}}{\partial \eta^2} - \frac{\partial \psi_{\mathrm{BL},0}}{\partial \eta}\frac{\partial^2 \psi_{\mathrm{BL},0}}{\partial \eta \partial x} = F_{\mathrm{BL}}(x) \equiv 0 ; \tag{4.17}$$

the arbitrary function $F_{\mathrm{BL}}(x)$ is identically zero as a consequence of the upstream constant uniform velocity parallel to $x > 0$. More precisely, the condition $F_{\mathrm{BL}}(x) \equiv 0$ is strongly related to the matching between the principal and local degeneracies (see Sect. 4.2.4).

Now, for the Prandtl BL equation (4.17), we can write [from (4.12a)] the following two boundary conditions relative to η:

$$\frac{\partial \psi_{\mathrm{BL0}}}{\partial \eta} = 0 \tag{4.18a}$$

and $\psi_{\mathrm{BL0}} = 0$ on $\eta = 0$. (4.18b)

But, it is necessary to add to conditions (4.18a,b), a third condition for $\eta \to \infty$ which is a matching condition with the solution $\psi_0(x, y)$ of the Laplace equation (4.14c) when $y = 0$.

4.2.4 Matching

We observe, first, that we have considered the following two asymptotic expansions via the principal and local limiting processes for stream functions $\psi(x, y; 1/\operatorname{Re})$ with $\delta = 1/\operatorname{Re}^{1/2}$:

$$(\mathrm{P}) : \psi(x, y; 1/\operatorname{Re}) = \psi_0(x, y) + \nu(\delta)\psi_1(x, y) + \ldots , \tag{4.19a}$$

and

$$(\mathrm{loc}) : \psi(x, y; 1/\operatorname{Re}) = \frac{1}{\operatorname{Re}^{1/2}}\psi_{\mathrm{BL0}}(x, \eta) + \mu(\delta)\psi_{\mathrm{BL1}}(x, \eta) + \ldots , \tag{4.19b}$$

according to the MAEM [matched asymptotic expansions method; see Lagerstrom (1988) and Zeytounian (1994)]. We observe that the gauge functions $\nu(\delta)$ and $\mu(\delta)$ are arbitrary but tend to zero as $\delta \downarrow 0$. According to the classical rule of matching [à la Van Dyke (1975)], we write (because $y = \delta\eta$):

$$\psi_0(x, 0) + \delta \left.\frac{\partial \psi_0}{\partial y}\right|_{y=0} + \nu(\delta)\psi_1(x, 0) + \ldots$$
$$= \delta\psi_{\mathrm{BL0}}(x, \infty) + \mu(\delta)\psi_{\mathrm{BL1}}(x, \infty) + \ldots , \tag{4.19c}$$

and at the leading order, we recover the nonviscous slip condition for the Laplace (principal outer) equation (4.14c):

$$\psi_0(x, 0) = 0 . \tag{4.20a}$$

For Laplace equation (4.14c), it is clear (!) that the following behavior upstream condition:

$$\psi_0(-\infty, y) = y , \tag{4.20b}$$

is also satisfied. The solution of the leading-order outer problem (4.14c), (4.20a,b) is

$$\psi_0(x,y) = y \ . \tag{4.20c}$$

Now, as a direct consequence of (4.20c), we find that the arbitrary function $F_{\mathrm{BL}}(x)$ in (4.17) is really zero. When η tends to infinity the local inner equation (4.17) must be identical to the principal outer equation

$$\left(\frac{\partial \psi_0}{\partial y}\frac{\partial}{\partial x} - \frac{\partial \psi_0}{\partial x}\frac{\partial}{\partial y}\right)\left(\frac{\partial^2 \psi_0}{\partial x^2} + \frac{\partial^2 \psi_0}{\partial y^2}\right) = 0 \ ,$$

written at $y = 0$, and this is possible only if $F_{\mathrm{BL}}(x) \equiv 0$.

The third condition for the Prandtl BL equation (4.17) is derived when we write the matching of the vertical (normal to flat plate) component of the velocity:

$$\frac{\partial \psi}{\partial y} \approx \frac{\partial \psi_0(x,\delta\eta)}{\partial y} = \left.\frac{\partial \psi_0}{\partial y}\right|_{y=0} + \delta\eta \left.\frac{\partial^2 \psi_0}{\partial y^2}\right|_{y=0} + \ldots \equiv 1 + 0 \ ,$$

and

$$\frac{\partial \psi}{\partial y} \approx \frac{\partial \psi_{\mathrm{BL0}}(x,y/\delta)}{\partial \eta} = \left.\frac{\partial \psi_{\mathrm{BL0}}}{\partial \eta}\right|_{\eta=\infty} + \ldots \ .$$

Consequently, we derive the following behavior condition:

$$\left.\frac{\partial \psi_{\mathrm{BL0}}}{\partial \eta}\right|_{\eta=\infty} = 1 \ . \tag{4.18c}$$

4.2.5 The Prandtl–Blasius and Blasius BL Problems

The boundary-value problem,

$$\frac{\partial^3 \psi_{\mathrm{BL0}}}{\partial \eta^3} + \frac{\partial \psi_{\mathrm{BL0}}}{\partial x}\frac{\partial^2 \psi_{\mathrm{BL0}}}{\partial \eta^2} - \frac{\partial \psi_{\mathrm{BL0}}}{\partial \eta}\frac{\partial^2 \psi_{\mathrm{BL0}}}{\partial \eta \partial x} = 0 \ , \tag{4.21}$$

$$\psi_{\mathrm{BL0}} = 0 \text{ and } \frac{\partial \psi_{\mathrm{BL0}}}{\partial \eta} = 0 \text{ on } \eta = 0, \ 0 < x < +\infty, \quad \lim_{\eta\uparrow+\infty} \frac{\partial \psi_{\mathrm{BL0}}}{\partial \eta} = 1 \ ,$$

is the *Prandtl–Blasius BL problem* [see, for instance, Prandtl (1904) and Blasius (1908)]. This problem is parabolic and, according to Nickel's (1973) mathematically rigorous theory, it is necessary to give the datum for $\partial \psi_{\mathrm{BL0}}/\partial \eta$ at a particular value of $x > 0$.

Blasius canonical problem. According to (4.3), (4.5)/(4.16), and (4.15c), the thickness of the BL is

$$l^0 = L^0/\mathrm{Re}^{1/2} \ , \tag{4.22}$$

where L^0 is arbitrary for the semi-infinite flat plate Blasius problem! Consequently a "self-similar solution" exists, and this means that the solution (if unique) of the Prandtl–Blasius BL problem (4.21) cannot involve the function $\psi_{\mathrm{BL}0}$ and the variables x, η separately but only in combinations that are invariant under the transformation. From this similarity, it follows that

$$\psi_{\mathrm{BL}0}(x,\eta) = x^{1/2} f_{\mathrm{B}}(\zeta) \quad \text{with } \zeta = \eta/x^{1/2} \,. \tag{4.23a}$$

Reintroducing the original dimensional function and variables at this point shows that the length L^0 disappears from f_{B} and ζ.

Substituting (4.23a) in the BL problem (4.21) gives the following *Blasius problem* for the Blasius function $f_{\mathrm{B}}(\zeta)$:

$$2 f_{\mathrm{B}}''' + f_{\mathrm{B}} f_{\mathrm{B}}'' = 0, \quad f_{\mathrm{B}}(0) = f_{\mathrm{B}}'(0) = 0, f_{\mathrm{B}}'(\infty) = 1 \,. \tag{4.23b}$$

For our purposes the essential results of the numerical solution are contained in the expansions for small and large ζ:

$$f_{\mathrm{B}}(\zeta) = (\alpha/2)(\zeta^2/2!) - (\alpha^2/2)(\zeta^4/4!) + (11\alpha^3/4)(\zeta^8/8!) + \ldots , \tag{4.24a}$$

and

$$f_{\mathrm{B}}(\zeta) \approx \eta - \beta + \gamma \int_\eta^\infty \mathrm{d}\sigma \int_\sigma^\infty \exp[-(\lambda-\beta)^2/4]\,\mathrm{d}\lambda + \ldots , \tag{4.24b}$$

with

$$\alpha = f_{\mathrm{B}}''(0) = 0.332, \ \beta = \lim_{\zeta \uparrow \infty} [\zeta - f_{\mathrm{B}}(\zeta)] = 1.73, \ \gamma = 0.231 \,. \tag{4.25}$$

However, Weyl (1942) points out that this is a "dangerous" business because (originally patched by Blasius) the series (4.24a) has only a modest radius of convergence and the second (4.24b) probably none!

The values of constants (4.25) in the series (4.24a,b) are obtained numerically via the following initial value problem:

$$2 G_{\mathrm{B}}''' + G_{\mathrm{B}} G_{\mathrm{B}}'' = 0, G_{\mathrm{B}}(0) = G_{\mathrm{B}}'(0) = 0, G_{\mathrm{B}}''(0) = 1 \,. \tag{4.26}$$

with

$$\lim_{Z \uparrow \infty} G_{\mathrm{B}}'(Z) = 2 K_0^2 > 0, \text{ exists and } Z = \zeta/2K_0 \,. \tag{4.27}$$

In this case [see, for instance, Meyer (1971)],

$$f_{\mathrm{B}}(\zeta) = (1/K_0) G_{\mathrm{B}}(Z) \tag{4.28}$$

is the solution of the Blasius problem (4.23b). The existence and unicity of the Blasius problem (4.23b) were first proved by Weyl (1942) in the class of functions belonging to $C^\infty[0,\infty)$. For an another point of view on the "uniqueness" of the Blasius solution see also Van Dyke (1975, p. 131).

4.2 Classical Hierarchical Boundary-Layer Concept and Regular Coupling

An important feature of the Blasius BL problem is the

existence of a velocity component normal to the flat plate at the "upper" limit of the Blasius BL:

$$\lim_{\eta\uparrow+\infty} \frac{\partial \psi_{BL0}}{\partial x} = -\frac{\beta}{2}x^{-1/2} . \tag{4.29}$$

Validity. Note that the boundary-layer concept is not valid near the leading edge ($x = y = 0$) of a semi-infinite flat plate in a region of radius $1/\text{Re}$; in this singular region, it is necessary to consider the full Navier equation. More precisely, in this singular region, the significant variables are

$$x^* = \text{Re}\, x, \quad y^* = \text{Re}\, y, \quad \text{and } \psi^* = \text{Re}\,\psi , \tag{4.30}$$

and the function $\psi^*(x^*, y^*)$ is the solution of the equation

$$\left(\frac{\partial \psi^*}{\partial y^*}\frac{\partial}{\partial x^*} - \frac{\partial \psi^*}{\partial x^*}\frac{\partial}{\partial y^*}\right)\mathbf{D}^{*2}\psi^* = \mathbf{D}^{*2}[\mathbf{D}^{*2}\psi^*] , \tag{4.31}$$

where

$$\mathbf{D}^{*2} = \frac{\partial^2}{\partial x^{*2}} + \frac{\partial^2}{\partial y^{*2}} .$$

Obviously, when $x^* \to \infty$, we must recover the Blasius BL flow for the value $x = 0$. Consequently, we have the possibility of obtaining (by matching) the value of

$$\left.\frac{\partial \psi_{BL0}}{\partial \eta}\right|_{x=0} = \lim_{x^* \to \infty} \frac{\partial \psi^*}{\partial y^*} , \tag{4.32}$$

which is necessary for computing the parabolic Prandtl–Blasius BL problem (4.21).

Between the Blasius BL region and the Navier region [where (4.31) is significative], we have an intermediate matching BL region which is characterized by the following scaling:

$$X = \frac{x}{\Delta(1/\text{Re})}, \quad Y = \frac{\text{Re}^{1/2}\,y}{\Delta^{1/2}(1/\text{Re})}, \quad \Psi = \frac{\text{Re}^{1/2}\,\psi}{\Delta^{1/2}(1/\text{Re})} , \tag{4.33}$$

with

$$1/\text{Re} \ll \Delta(1/\text{Re}) \ll 1 .$$

In this intermediate matching region (near the flat plate), the significative problem is the Prandtl–Blasius BL problem (4.21) with the (initial) condition (4.32) for $\psi_{BL0}(x, \eta)$. In the Navier region, for (4.31), it is necessary to assume as boundary conditions,

$$\psi^* = 0, \quad \frac{\partial \psi^*}{\partial y^*} = 0 \text{ on } y^* = 0 \text{ when } x^* > 0 ; \qquad (4.34\text{a})$$

$$\psi^*(-\infty, y^*) = y^* \text{ and } \psi^* = 0 \text{ on } y^* = 0 \text{ when } x^* < 0 . \qquad (4.34\text{b})$$

The existence and uniqueness of the intermediate matching BL problem, (4.21) with (4.32), was proved by Serrin (1967), and consequently when $x^* \to \infty$, this solution must tend to Blasius solution $f_B(\zeta)$.

The Navier problem (4.31), (4.34a,b) was numerically solved by Van Vooren and Dijkstra (1970). These authors also assume that the local vorticity $\Gamma^* = \mathbf{D}^{*2} \psi^*$ satisfies the following behavior:

$$\Gamma^* = 0 \text{ for } x^* < 0 \text{ and } y^* = 0 \text{ (symmetry)} , \qquad (4.35\text{a})$$

$$\Gamma^* \to 0 \text{ when } x^* \to -\infty \text{ and } y^* \to +\infty . \qquad (4.35\text{b})$$

On the other hand, it is assumed that, matching the numerical solution with the BL solution, when $x^* \to +\infty$ and for $y^* \leq O(x^{*1/2})$, and also with the outer potential (uniform), inviscid flow can be realized.

In Fig. 4.2, the reader can find a sketch of four regions necessary for asymptotic study of incompressible fluid flow with vanishing viscosity above a semi-infinite flat plate, when at upstream infinity, the flow is uniform and directed parallel to the flat plate.

The first region (I) is the "Navier" region governed by problem (4.31), (4.34a,b) with (4.35a,b). Region (II) is an "intermediate" matching region governed by the Prandtl–Blasius problem (4.21) with (4.32). Region (III) is

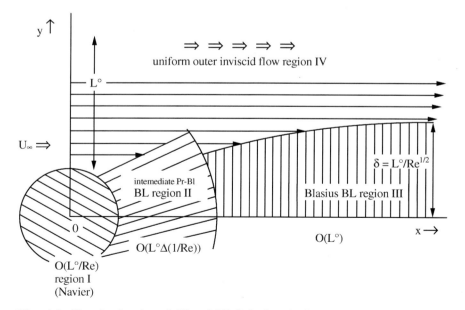

Fig. 4.2. Sketch of regions I, II and IV, linked with the Blasius BL region III

4.2 Classical Hierarchical Boundary-Layer Concept and Regular Coupling

the "Blasius" BL region governed by the classical Blasius problem (4.23b) or (4.26) with (4.27), (4.28). Finally, region (IV) is the "nonviscous," potential region with uniform flow $\psi_0 = y$.

Regular coupling. In the framework of so-called "regular coupling," first, we observe that knowledge of the outer (uniform) inviscid potential flow, $\psi_0 = y$, gives the possibility of formulating the Prandtl–Blasius BL problem (4.21) which, according to (4.23a) and (4.23b), leads to the Blasius BL solution $f_B(\zeta)$ with $\zeta = (\text{Re}/x)^{1/2} y$.

On the other hand, knowledge of the solution of the local Blasius BL problem [at least for large ζ; see (4.24b)] gives the possibility of determining (see below) the inviscid (linearized) solution $\psi_1(x, y)$ in the (outer) principal expansion (4.19a). The following step is determination of the second-order BL solution $\psi_{BL1}(x, \eta)$ in (inner) local expansion (4.19b), and so on!

In Fig. 4.3, we represent schematically the process of regular coupling for the horizontal [longitudinal, see (4.7b)] component of Navier incompressible velocity $\partial \psi / \partial y$.

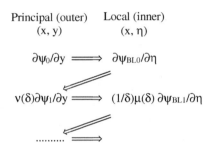

Fig. 4.3. Regular coupling (noninteractive): inviscid (outer) flow – BL (inner) flow

For the moment, we have determined the form of the leading term of the local (inner) expansion (4.19b):

$$\psi \approx (1/\text{Re}^{1/2}) x^{1/2} f_B(\zeta) = (1/\text{Re}^{1/2}) x^{1/2} f_B(\text{Re}^{1/2} x^{-1/2} y)$$
$$= (1/\text{Re}^{1/2}) x^{1/2} [\text{Re}^{1/2} x^{-1/2} y - \beta + \text{Exp}] ,$$
(4.36a)

according to (4.24b), where "Exp" stands for terms that are exponentially small for large y. With the inner variables, in place of (4.36a), we obtain

$$\psi \approx (1/\text{Re}^{1/2}) \eta - (1/\text{Re}^{1/2}) x^{1/2} \beta + \ldots .$$
(4.36b)

But, according to the (outer) principal expansion (4.19a), we can also write, with the inner variables [because $\psi_0(x, y) = y = \eta/\text{Re}^{1/2}$],

$$\psi \approx (1/\text{Re}^{1/2}) \eta + \nu(\delta)[\psi_1(x, 0) + \ldots] ,$$
(4.36c)

102   4 The Limit of Very Large Reynolds Numbers

by matching

$$\nu(\delta) = 1/\operatorname{Re}^{1/2} = \delta, \tag{4.37a}$$

and

$$\psi_1(x, 0) = -\beta x^{1/2}, \tag{4.37b}$$

which is equivalent to (4.29) and has a familiar physical interpretation in terms of the *displacement thickness* of the boundary layer.

Therefore, the effect of the inner boundary layer on inviscid outer flow (at the order $\delta = 1/\operatorname{Re}^{1/2}$) is expressed by linearized inviscid fluid flow around a parabola of nose radius $\beta^2/\operatorname{Re}$. The equation of this parabola is

$$y = (\beta/\operatorname{Re}^{1/2}) x^{1/2} \text{ with } 0 < x < +\infty. \tag{4.38}$$

Condition (4.37b) is then the slip condition on this parabola (tangency condition for the displacement parabola) but linearized according to the thin-airfoil approximation and transferred to the axis of the parabola. Indeed, the slip condition on the parabola (4.38) is

$$\psi = y + \operatorname{Re}^{-1/2} \psi_1(x, y) + \ldots = 0 \text{ on } y = (\beta/\operatorname{Re}^{1/2}) x^{1/2},$$

or

$$\operatorname{Re}^{-1/2} \{(\beta/\operatorname{Re}^{1/2}) x^{1/2} + \psi_1(x, (\beta/\operatorname{Re}^{1/2}) x^{1/2}) + \ldots\} = 0,$$

and when Re tends to infinity,

$$\beta x^{1/2} + \psi_1(x, 0) = 0,$$

which is condition (4.37b).

On the other hand, substituting the outer inviscid expansion $\psi = y + \operatorname{Re}^{-1/2} \psi_1(x, y) + \ldots$, in exact equation (4.8) again leads to a Laplace equation for the function $\psi_1(x, y)$ because

$$\frac{\partial \mathbf{D}^2 \psi_1}{\partial x} = 0 \Rightarrow \mathbf{D}^2 \psi_1 = 0, \tag{4.39a}$$

as a consequence of the upstream infinity (irrotationality) condition.

For (4.39a), we have the following two conditions:

$$\psi_1(x = -\infty, y) = 0, \ \psi_1(x, 0) = 0, \text{ if } x < 0, \tag{4.39b}$$

$$\psi_1(x, 0) = -\beta x^{1/2}, \text{ if } 0 < x < +\infty. \tag{4.39c}$$

The solution of problem (4.39a–c) for $\psi_1(x, y)$ is obvious from the viewpoint of complex variable theory $[i = (-1)^{1/2}]$:

$$\psi_1(x, y) = -\beta \operatorname{Real}[(x + iy)^{1/2}]. \tag{4.40}$$

4.3 Asymptotic Structure of Unsteady-State NSF Equations at Re ≫ 1

Now, substituting the inner expansion (4.19b) in the exact equation (4.8) for ψ, we derive a *homogeneous* equation for $\psi_{BL1}(x,\eta)$. But from matching between (below, we exibit the second term because its coefficient will be different from zero in any other problem, including that of a finite plate)

$$\frac{\partial \psi}{\partial y} \approx 1 + 0/\operatorname{Re}^{1/2} \quad \text{(principal)}$$

and

$$\frac{\partial \psi}{\partial y} \approx 1 + \operatorname{Re}^{1/2} \mu(\delta) \left. \frac{\partial \psi_{BL1}}{\partial \eta} \right|_{\eta=\infty} \quad \text{(local)}$$

gives

$$\operatorname{Re}^{1/2} \mu(\delta) = 1/\operatorname{Re}^{1/2} \Rightarrow \mu(\delta) = 1/\operatorname{Re} \qquad (4.41a)$$

and

$$\left. \frac{\partial \psi_{BL1}}{\partial \eta} \right|_{\eta=\infty} = 0. \qquad (4.41b)$$

With the no-slip condition $\psi_{BL1} = 0$ and $\partial \psi_{BL1}/\partial \eta = 0$ on the flat plate $\eta = 0$, we obtain a *completely homogeneous second-order* linear BL problem for $\psi_{BL1}(x,\eta)$ and consequently,

$$\psi_{BL1} \equiv 0,$$

and the complete *second approximation* is

$$\psi(x,y;\operatorname{Re}) \approx \begin{cases} y - \operatorname{Re}^{-1/2} \beta \operatorname{Real}[(x+iy)^{1/2}], \text{ outer} \\ (1/\operatorname{Re}^{1/2})x^{1/2} f_B(\operatorname{Re}^{1/2} x^{-1/2} y) + 0/\operatorname{Re}, \text{ inner} \end{cases} \qquad (4.42)$$

In Van Dyke (1975, pp. 139–140), the reader can find information concerning the third-order boundary-layer approximation for a semi-infinite flat plate. The correct solution was given by Goldstein (1956) and Imai (1957a) and is described in detail in Goldstein (1960).

In the recent book by Schlichting and Gersten (2000; 8th Revised and Enlarged Edition of the famous *Boundary-Layer Theory*, 7th Edition, by Hermann Schlichting), the reader can find various subjects related to high Reynolds fluid flows. For an "asymptotic" viewpoint on $\operatorname{Re} \gg 1$ flows, see also our recent book, Zeytounian (2002b, Chaps. 7 and 12).

4.3 Asymptotic Structure of Unsteady-State NSF Equations at Re ≫ 1

The simple fact that, in the framework of the boundary-layer approximation valid near the wall, washed by a fluid that has a vanishing viscosity ($\operatorname{Re} \gg 1$), the pressure is constant in the normal to the wall direction [see relation (4.43c)] gives strong motivation in the unsteady-state compressible case for a thorough revision of large Reynolds number asymptotics of NSF equations that govern unsteady-state, compressible, viscous, heat-conducting fluid flow. In particular, the unsteady-state compressible, heat-conducting Prandtl boundary-layer equations, valid near the wall ($n = 0$), for the horizontal and normal (relative to n) components of the velocity vector [$\boldsymbol{u} = (\boldsymbol{v}, w)$], $\boldsymbol{v}_{\mathrm{Pr}}^0$, w_{Pr}^1, and thermodynamic functions, p_{Pr}^0, ρ_{Pr}^0, T_{Pr}^0, which are dependent on the variables t, s_1, s_2, and $N = n/\delta$ can be written in the following form:

$$\mathrm{S}\frac{\partial \rho_{\mathrm{Pr}}^0}{\partial t} + \mathbf{D} \cdot \left(\rho_{\mathrm{Pr}}^0 \boldsymbol{v}_{\mathrm{Pr}}^0\right) + \frac{\partial \left(\rho_{\mathrm{Pr}}^0 w_{\mathrm{Pr}}^1\right)}{\partial N} = 0 \; ; \tag{4.43a}$$

$$\rho_{\mathrm{Pr}}^0 \left[\mathrm{S}\frac{\partial \boldsymbol{v}_{\mathrm{Pr}}^0}{\partial t} + \left(\boldsymbol{v}_{\mathrm{Pr}}^0 \cdot \mathbf{D}\right) \boldsymbol{v}_{\mathrm{Pr}}^0 + w_{\mathrm{Pr}}^1 \frac{\partial \boldsymbol{v}_{\mathrm{Pr}}^0}{\partial N}\right] + \frac{1}{\gamma M^2} \mathbf{D} p_{\mathrm{Pr}}^0$$

$$= \frac{\partial^2 \boldsymbol{v}_{\mathrm{Pr}}^0}{\partial N^2} \; ; \tag{4.43b}$$

$$\frac{\partial p_{\mathrm{Pr}}^0}{\partial N} = 0 \; ; \tag{4.43c}$$

$$\rho_{\mathrm{Pr}}^0 \left[\mathrm{S}\frac{\partial T_{\mathrm{Pr}}^0}{\partial t} + \boldsymbol{v}_{\mathrm{Pr}}^0 \cdot \mathbf{D} T_{\mathrm{Pr}}^0 + w_{\mathrm{Pr}}^1 \frac{\partial T_{\mathrm{Pr}}^0}{\partial N}\right] + (\gamma - 1) p_{\mathrm{Pr}}^0 \left[\mathbf{D} \cdot \boldsymbol{v}_{\mathrm{Pr}}^0 + \frac{\partial w_{\mathrm{Pr}}^1}{\partial N}\right]$$

$$= \frac{\gamma}{\mathrm{Pr}} \frac{\partial^2 T_{\mathrm{Pr}}^0}{\partial N^2} + \gamma(\gamma-1) M^2 \left|\frac{\partial \boldsymbol{v}_{\mathrm{Pr}}^0}{\partial N}\right|^2 \; , \tag{4.43d}$$

with $p_{\mathrm{Pr}}^0 = \rho_{\mathrm{Pr}}^0 T_{\mathrm{Pr}}^0$. The horizontal gradient operator $\mathbf{D}$ is to be formed with $\partial/\partial s_1$ and $\partial/\partial s_2$, and $\boldsymbol{v}_{\mathrm{Pr}}^0$, w_{Pr}^1 are the Cartesian horizontal and normal (relative to N) components of the Prandtl velocity vector. Equations (4.43a–d) are derived from the full unsteady-state NSF equations via the limiting process (4.2) when we take into account the following Prandtl asymptotic representation ($\delta = \mathrm{Re}^{-1/2}$):

$$(\boldsymbol{v}, w) = (\boldsymbol{v}_{\mathrm{Pr}}^0, 0) + \delta(\boldsymbol{v}_{\mathrm{Pr}}^1, w_{\mathrm{Pr}}^1) + \ldots \; , \tag{4.44a}$$

$$(p, \rho, T) = (p_{\mathrm{Pr}}^0, \rho_{\mathrm{Pr}}^0, T_{\mathrm{Pr}}^0) + \delta(p_{\mathrm{Pr}}^1, \rho_{\mathrm{Pr}}^1, T_{\mathrm{Pr}}^1) + \ldots \; . \tag{4.44b}$$

BL equations (4.43a–d) are not valid for unsteady-state ($S \neq 0$) compressible flow, close to the initial time ($t = 0$), where the initial data are given for unsteady-state compressible, "exact" NSF starting equations. As a consequence of this singular initial-time region, it is necessary to formulate an unsteady-state compressible adjustment problem, which gives the possibility, by matching, of deriving the asymptotically correct initial conditions for the unsteady-state Prandtl boundary-layer, compressible, heat-conducting equations (4.43a,b,d). Close to the initial time and near the wall, two singular regions appeared and in fact the consistent asymptotic analysis at large Reynolds number is closely linked with a four-regions structure. In particular, as a result of the below analysis we can give an answer to the following question:

4.3 Asymptotic Structure of Unsteady-State NSF Equations at Re ≫ 1

How are the Euler inviscid, Prandtl boundary-layer, one-dimensional gas dynamics and Rayleigh compressible equations related to the full NSF equations and among themselves?

The initial-time singularity of BL equations was first considered by Zeytounian (1980) in a short Note but did not attract the attention of fluid dynamicists. We observe, again, that this initial-time singularity of BL equations is an unfortunate and direct consequence of the classical feature of the boundary-layer concept, valid near the wall (but far from the initial time), and is related to the constancy of pressure in the normal to the wall direction, within the whole thickness of the boundary layer.

As a consequence of this initial-time singularity, it is necessary to reconsider the classical asymptotic analysis (Euler ↓↑ Prandtl via regular coupling), at high Reynolds number (Re ≫ 1), of NSF equations, that govern unsteady-state, compressible, viscous, heat-conducting fluid flow, which is mainly associated with Euler (inviscid) and Prandtl (boundary-layer) asymptotic models (at least for continuous fluid flow when it is not necessary to take Taylor internal shock structure into account). Or precisely, the loss of the time derivative in the equation for the vertical motion (for w_{Pr}^1) in the boundary layer is the main reason for a fundamental revision of unsteady-state flows at large Reynolds numbers. A first consequence of this revision is closely linked with the resolution of a so-called "adjustment problem" which opens the possibility of determining asymptotically the consistent initial conditions for unsteady-state boundary-layer equations.

4.3.1 Four Significant Degeneracies of NSF Equations

In fluid dynamics, various simplified equations that are approximately valid in certain regions for particular regimes of motion are usually considered in place of the full unsteady-state NSF equations. Below, we are concerned with the following four systems of equations: Euler inviscid, Prandtl boundary-layer, one-dimensional inviscid gas dynamics, and Rayleigh viscous compressible. For simplicity, we consider dimensionless NSF equations (1.30)–(1.32) in a system of rectangular Cartesian coordinates and, consequently, we do not take curvature effects into account.

Euler inviscid equations. When a fluid is assumed inviscid (nonviscous) then, in place of dimensionless NSF equations (1.30)–(1.32), we can consider the following "truncated," inviscid Euler equations for velocity vector $\boldsymbol{u}_E^0$, and thermodynamic functions p_E^0, ρ_E^0, T_E^0 that are dependent on variables $t, x_1, x_2,$ and x_3:

$$S\frac{\partial \rho_E^0}{\partial t} + \boldsymbol{\nabla} \cdot \left(\rho_E^0 \boldsymbol{u}_E^0\right) = 0 \, ; \tag{4.45a}$$

$$\rho_E^0 \left[S\frac{\partial \boldsymbol{u}_E^0}{\partial t} + \left(\boldsymbol{u}_E^0 \cdot \boldsymbol{\nabla}\right) \boldsymbol{u}_E^0\right] + \frac{1}{\gamma M^2}\boldsymbol{\nabla} p_E^0 = 0 \, ; \tag{4.45b}$$

$$\rho_E^0 \left[S\frac{\partial T_E^0}{\partial t} + \boldsymbol{u}_E^0 \cdot \boldsymbol{\nabla} T_E^0\right] + (\gamma-1) p_E^0 \left(\boldsymbol{\nabla} \cdot \boldsymbol{u}_E^0\right) = 0 \, , \tag{4.45c}$$

with $p_E^0 = \rho_E^0 T_E^0$.

Prandtl boundary-layer equations. Near a wall washed by a compressible, viscous, heat-conducting fluid in a thin layer, where the no-slip condition on the wall is assumed for the velocity vector and a thermal condition is written for the temperature, in place of the full NSF equations, we can write the following Prandtl boundary-layer equations for the functions $\boldsymbol{v}_{\text{Pr}}^0$, w_{Pr}^1, p_{Pr}^0, ρ_{Pr}^0, T_{Pr}^0, that are dependent on variables t, x_1, x_2, and $\eta = x_3/\delta$:

$$S\frac{\partial \rho_{\text{Pr}}^0}{\partial t} + \mathbf{D} \cdot \left(\rho_{\text{Pr}}^0 \boldsymbol{v}_{\text{Pr}}^0\right) + \frac{\partial (\rho_{\text{Pr}}^0 w_{\text{Pr}}^1)}{\partial \eta} = 0 \, ; \tag{4.46a}$$

$$\rho_{\text{Pr}}^0 \left[S\frac{\partial \boldsymbol{v}_{\text{Pr}}^0}{\partial t} + \left(\boldsymbol{v}_{\text{Pr}}^0 \cdot \mathbf{D}\right) \boldsymbol{v}_{\text{Pr}}^0 + w_{\text{Pr}}^1 \frac{\partial \boldsymbol{v}_{\text{Pr}}^0}{\partial \eta}\right] + \frac{1}{\gamma M^2}\mathbf{D} p_{\text{Pr}}^0 = \frac{\partial^2 \boldsymbol{v}_{\text{Pr}}^0}{\partial \eta^2} \, ; \tag{4.46b}$$

$$\frac{\partial p_{\text{Pr}}^0}{\partial \eta} = 0 \, ; \tag{4.46c}$$

$$\rho_{\text{Pr}}^0 \left[S\frac{\partial T_{\text{Pr}}^0}{\partial t} + \boldsymbol{v}_{\text{Pr}}^0 \cdot \mathbf{D} T_{\text{Pr}}^0 + w_{\text{Pr}}^1 \frac{\partial T_{\text{Pr}}^0}{\partial \eta}\right] + (\gamma-1) p_{\text{Pr}}^0 \left[\mathbf{D} \cdot \boldsymbol{v}_{\text{Pr}}^0 + \frac{\partial w_{\text{Pr}}^1}{\partial \eta}\right]$$
$$= \frac{\gamma}{\text{Pr}} \frac{\partial^2 T_{\text{Pr}}^0}{\partial \eta^2} + \gamma(\gamma-1) M^2 \left|\frac{\partial \boldsymbol{v}_{\text{Pr}}^0}{\partial \eta}\right|^2 , \tag{4.46d}$$

with $p_{\text{Pr}}^0 = \rho_{\text{Pr}}^0 T_{\text{Pr}}^0$. Here, the operator $\mathbf{D}$ is formed with $\partial/\partial x_1$ and $\partial/\partial x_2$, and $\boldsymbol{v}_{\text{Pr}}^0$, w_{Pr}^1 are the Cartesian horizontal and normal (relative to η) components of the Prandtl velocity vector.

One-dimensional unsteady-state inviscid gas dynamics equations. In (inviscid) gas dynamics, in place of Euler equations, often the following one-dimensional unsteady-state inviscid gas dynamics equations are considered for the functions $\boldsymbol{v}_G^0$, w_G^0, p_G^0, ρ_G^0, T_G^0, that are dependent on variables $\tau = t/\delta$, x_1, x_2, and η:

$$S\frac{\partial \boldsymbol{v}_G^0}{\partial \tau} + w_G^0 \frac{\partial \boldsymbol{v}_G^0}{\partial \eta} = 0 \, ; \tag{4.47a}$$

$$S\frac{\partial \rho_G^0}{\partial \tau} + \frac{\partial (\rho_G^0 w_G^0)}{\partial \eta} = 0 \, ; \tag{4.47b}$$

4.3 Asymptotic Structure of Unsteady-State NSF Equations at Re ≫ 1

$$\rho_G^0 \left[S \frac{\partial w_G^0}{\partial \tau} + w_G^0 \frac{\partial v_G^0}{\partial \eta} \right] + \frac{1}{\gamma M^2} \frac{\partial p_G^0}{\partial \eta} = 0 \; ; \tag{4.47c}$$

$$\rho_G^0 \left[S \frac{\partial T_G^0}{\partial \tau} + w_G^0 \frac{\partial T_G^0}{\partial \eta} \right] + (\gamma - 1) p_G^0 \frac{\partial w_G^0}{\partial \eta} = 0 \; , \tag{4.47d}$$

with $p_G^0 = \rho_G^0 T_G^0$. In (4.47a–d), the horizontal variables x_1, x_2, play the role of parameters, and v_G^0, w_G^0 are the Cartesian horizontal and normal (also relative to η) components of the velocity vector.

Rayleigh viscous compressible equations. Generalization to a viscous and heat-conducting fluid of the one-dimensional, unsteady-state inviscid gas dynamics equations (4.47a–d) are the so-called Rayleigh, viscous, compressible flow equations, used by Howarth (1951), who first investigated some aspects of Rayleigh's classical problem for a compressible, heat-conducting fluid (considered in Chap. 3, Sect. 3.7).

These Rayleigh equations are written for the functions v_R^0, w_R^0, p_R^0, ρ_R^0, T_R^0 which are dependent on variables $\tau^* = t/\delta^2$, x_1, x_2, and $\eta^* = x_3/\delta^2$:

$$\rho_R^0 \left[S \frac{\partial v_R^0}{\partial \tau^*} + w_R^0 \frac{\partial v_R^0}{\partial \eta^*} \right] = \frac{\partial^2 v_R^0}{\partial \eta^{*2}} \; ; \tag{4.48a}$$

$$S \frac{\partial \rho_R^0}{\partial \tau^*} + \frac{\partial (\rho_R^0 w_R^0)}{\partial \eta^*} = 0 \; ; \tag{4.48b}$$

$$\rho_R^0 \left[S \frac{\partial w_R^0}{\partial \tau^*} + w_R^0 \frac{\partial w_R^0}{\partial \eta^*} \right] + \frac{1}{\gamma M^2} \frac{\partial p_R^0}{\partial \eta^*} = \frac{4}{3} \frac{\partial^2 w_R^0}{\partial \eta^{*2}} \; ; \tag{4.48c}$$

$$\rho_R^0 \left[S \frac{\partial T_R^0}{\partial \tau^*} + w_R^0 \frac{\partial T_R^0}{\partial \eta^*} \right] + (\gamma - 1) p_R^0 \frac{\partial w_R^0}{\partial \eta^*}$$
$$= \frac{\gamma}{\Pr} \frac{\partial^2 T_R^0}{\partial \eta^{*2}} + \gamma(\gamma - 1) M^2 \left[\left| \frac{\partial v_R^0}{\partial \eta^*} \right|^2 + \frac{4}{3} \left| \frac{\partial w_R^0}{\partial \eta^*} \right|^2 \right] , \tag{4.48d}$$

with $p_R^0 = \rho_R^0 T_R^0$. We observe that, v_R^0, w_R^0 are the Cartesian horizontal and normal (but relative to η^*) components of the velocity vector.

Now, a basic question is the following:

How are the Euler inviscid, Prandtl boundary-layer, one-dimensional, unsteady-state, inviscid, and Rayleigh equations related to the NSF full equations and among themselves?

It is necessary also to understand

How are the various functions (with indexes "E, Pr, G, and R") and variables (t, τ, τ^* and x_3, η, η^*) related with the original NSF functions and variables (t, x_3) and among themselves?

The present asymptotic analysis works with three various times and three associated vertical (to wall) coordinates.

4.3.2 Formulation of a Simplified Initial Boundary-Value Problem for the NSF Full Unsteady-State Equations

We observe that the experienced reader is aware of the fact that the above set of equations is not sufficient for discussing flow problems, and this seemingly anodyne remark has far-reaching consequences that will be discussed in Sect. 4.3.3.

For a detailed and convincing answer to the basic questions posed at the end of Sect. 4.3.1, it is necessary to formulate an initial boundary-value problem for the starting, full, dimensionless, NSF unsteady-state equations, (1.30)–(1.32), with the ideal (perfect) gas dimensionless equation of state $p = \rho T$.

Boundary conditons. To avoid unnecessary additional technicalities, we consider a very simplified physical test problem. But we observe that the main results of the present asymptotic analysis are also true for a more complicated, realistic, fluid flow problem.

First, we assume (with dimensionless quantities) that the wall washed by a viscous, compressible, heat-conducting fluid is determined at any given instant by the plane $x_3 = 0$, and we presuppose that at $x_3 = 0$ in the bounded area D, the temperature T is a function of time t and Cartesian horizontal coordinates x_1 and x_2. For the temperature T on the wall, $x_3 = 0$, we write $T = T_w(\beta t, x_1, x_2)$ where $\beta = t^0/\Delta t^0$ and Δt^0 is a characteristic time, different from $t^0 = L^0/U_\infty$, when we assume that the Strouhal number $S \equiv 1$. The length scale L^0 (see Sect. 1.6.2) is closely linked with the diameter of the given bounded area D.

As a consequence of the above assumptions we write the following boundary conditions for the starting NSF dimensionless unsteady-state equations (1.30)–(1.32):

$$\text{on } x_3 = 0: \; \boldsymbol{v} = 0, w = 0, \tag{4.49a}$$

$$\text{and } T = T_w(\beta t, x_1, x_2), \tag{4.49b}$$

where $(x_1, x_2) \subset D$. In fact, we assume that (in the framework of a so-called "outbreak of fire" problem!)

$$T_w(0^-, x_1, x_2) \equiv 1 \quad \text{and} \quad T_w(\infty, x_1, x_2) \equiv T_w^\infty(x_1, x_2), \tag{4.50}$$

and when $T_w \equiv 1$, for $t \leq 0^-$, the temperature of the bounded area D is the same as that of the entire plane wall simulated by $x_3 = 0$. On the other hand, when $T = T_w^\infty(x_1, x_2)$, the temperature of the bounded area on $x_3 = 0$ is independent of the time, and this is physically true for $t^0 \gg \Delta t^0$ far from the initial transition time layer.

Initial conditions. Now, for the initial conditions, for $t \leq 0^-$ we assume, first, that

4.3 Asymptotic Structure of Unsteady-State NSF Equations at Re ≫ 1

$$p = \rho = T = 1, \text{ at } t \leq 0^- . \tag{4.51a}$$

Then we write for the NSF velocity vector $\boldsymbol{u} = \boldsymbol{U}^0(\alpha x_3)$, for time $t \leq 0^-$, where $\alpha = L^0/H^0$ and H^0 is the vertical characteristic length scale closely linked with given initial shear velocity $\boldsymbol{U}^0$.

Consequently, as initial conditions,

$$\boldsymbol{v} = \boldsymbol{V}^0(\alpha x_3), \ w = W^0(\alpha x_3) \text{ at } t \leq 0^- . \tag{4.51b}$$

We observe that, when $L^0 \gg H^0$ ($\alpha \gg 1$), at the limit $\alpha \to \infty$,

$$\boldsymbol{V}^0(\infty) \equiv \mathbf{1} \text{ and } W^0(\infty) \equiv 1 . \tag{4.51c}$$

When it is assumed that M^2 and Pr are of order unity, then for large Reynolds numbers, our main small parameter is $\delta = 1/\text{Re}^{1/2}$. For parameters α and β in conditions (4.51b) and (4.49b), we consider in the Sect. 4.3.3, that the following two similarity relations are satified:

$$\alpha \delta^a = 1 \quad \text{and} \quad \beta \delta^b = 1 , \tag{4.52}$$

where a and b are two positive scalars. As explained in Sect. 4.3.3, it is necessary to consider two consistent cases when $a = b = 1$ and $a = b = 2$. Consequently, we must consider two kinds of adjustment problems depending on the initial conditions (4.51b) for the velocity components and boundary condition (4.49b) for the temperature.

Finally, we stress again that the asymptotic analysis in Sects. 4.3.3 and 4.3.4 can be generalized for a more complicated and realistic unsteady-state aerodynamic external problem, when the wall of the body is determined at any given instant t by the function $F(t, \boldsymbol{x}) = 0$. In this case [see, for example, Cheatham and Matalon (2000)], it is necessary to use intrinsic wall coordinates (s_1, s_2) aligned with the principal directions of curvature at each point of the wall to parametrize the wall. On the other hand, it is necessary to take into account the velocity $\boldsymbol{V}_p$ of a point P on the solid wall and the angular velocity W_p of P in the chosen Galilean frame [see, for instance, Germain and Guiraud (1966)]. Our formulation is well adapted to meteorological problems, where the quasi-hydrostatic approximation is singular close to the initial time, as demonstrated by Guiraud and Zeytounian (1982).

4.3.3 Various Facets of Large Reynolds Number Unsteady-State Flow

The systems of approximate equations in Sect. 4.3.1 are derived when $\delta \downarrow 0$. But for each case, it is necessary to define more exactly the relation between the starting NSF variables (t, x_1, x_2, x_3) and the variables (τ, τ^*) and (η, η^*) introduced in the system of approximate equations (4.46), (4.47), and (4.48).

4 The Limit of Very Large Reynolds Numbers

Consequently, it is necessary to define four limiting processes when $\delta \downarrow 0$:

$$\text{LIM}^E = [\delta \downarrow 0, \text{ with } t \text{ and } x_1, x_2, x_3 \text{ fixed}], \tag{4.53a}$$

$$\text{LIM}^{\text{Pr}} = [\delta \downarrow 0, \text{ with } t \text{ and } x_1, x_2 \text{ and } x_3/\delta = \eta \text{ fixed}], \tag{4.53b}$$

$$\text{LIM}^G = [\delta \downarrow 0, \text{ with } t/\delta = \tau \text{ and } x_1, x_2 \text{ and } x_3/\delta = \eta \text{ fixed}], \tag{4.53c}$$

$$\text{LIM}^R = [\delta \downarrow 0, \text{ with } t/\delta^2 = \tau^* \text{ and } x_1, x_2 \text{ and } x_3/\delta^2 = \eta^* \text{ fixed}]. \tag{4.53d}$$

Euler–Prandtl coupling. First, when we pass from NSF starting, full, unsteady-state equations (1.30)–(1.32) (with Bo $\equiv 0$) to Euler inviscid equations (4.45a–c), through the limiting process (4.53a), with the following Euler outer asymptotic representation:

$$\boldsymbol{u} = (\boldsymbol{v}, w) = (\boldsymbol{v}_E^0, w_E^0) + \delta(\boldsymbol{v}_E^1, w_E^1) + \ldots, \tag{4.54a}$$

$$(p, \rho, T) = (p_E^0, \rho_E^0, T_E^0) + \delta(p_E^1, \rho_E^1, T_E^1) + \ldots, \tag{4.54b}$$

we do not have the possibility of applying boundary conditions (4.49a) on the wall $x_3 = 0$; this results from the loss of higher derivatives relative to x_3 in Euler equations (4.45b,c).

Consequently, it is necessary to derive, at least, an approximate consistent (from an asymptotic point of view) system of equations near the wall, which opens the possibility of accounting for the missed boundary conditions on the wall.

The limiting process (4.53b) realizes this task, but it is also necessary to take into account the following Prandtl asymptotic representation:

$$(\boldsymbol{v}, w) = (\boldsymbol{v}_{\text{Pr}}^0, 0) + \delta(\boldsymbol{v}_{\text{Pr}}^1, w_{\text{Pr}}^1) + \ldots, \tag{4.55a}$$

$$(p, \rho, T) = (p_{\text{Pr}}^0, \rho_{\text{Pr}}^0, T_{\text{Pr}}^0) + \delta(p_{\text{Pr}}^1, \rho_{\text{Pr}}^1, T_{\text{Pr}}^1) + \ldots. \tag{4.55b}$$

At this stage, we can deduce that both systems of equations (4.45a–c), for $\boldsymbol{u}_E^0 = (\boldsymbol{v}_E^0, w_E^0)$, p_E^0, ρ_E^0, T_E^0, and (4.46a–d), for $\boldsymbol{v}_{\text{Pr}}^0, w_{\text{Pr}}^1, p_{\text{Pr}}^0, \rho_{\text{Pr}}^0, T_{\text{Pr}}^0$, are derived asymptotically from the full NSF unsteady-state system of equations (1.30)–(1.32), when we consider the limiting processes (4.53a) and (4.53b) with the associated asymptotic representations (4.54a,b) and (4.55a,b). In this case, according to matching condition,

$$\lim_{\eta \uparrow \infty}(\text{LIM}^{\text{Pr}}) = \lim_{x_3 \downarrow 0}(\text{LIM}^E),$$

we derive the following classical relations:

$$w_E^0 = 0 \text{ on } x_3 = 0, p_{\text{Pr}}^0 = p_E^0(t, x_1, x_2, 0) \equiv p_{E0}^0(t, x_1, x_2), \tag{4.56a}$$

$$\lim_{\eta \uparrow \infty} \boldsymbol{v}_{\text{Pr}}^0 = \boldsymbol{v}_E^0(t, x_1, x_2, 0) \equiv \boldsymbol{v}_{E0}^0(t, x_1, x_2), \tag{4.56b}$$

4.3 Asymptotic Structure of Unsteady-State NSF Equations at Re ≫ 1   111

such that in the boundary equation (4.46b) [with $\rho_{E0}^0 \equiv \rho_E^0(t, x_1, x_2, 0)$],

$$\mathbf{D}p_{\mathrm{Pr}}^0 \equiv \gamma M^2 \rho_{E0}^0 \left[S \frac{\partial \boldsymbol{v}_{E0}^0}{\partial t} + (\boldsymbol{v}_{E0}^0 \cdot \mathbf{D}) \boldsymbol{v}_{E0}^0 \right] . \tag{4.56c}$$

For the initial conditions for the Euler inviscid unsteady-state equations (4.45a–c) for functions $\boldsymbol{u}_E^0 = (\boldsymbol{v}_E^0, w_E^0)$, ρ_E^0 and T_E^0 dependent on t, x_1, x_2, and x_3, we can obviously write (because we assume $a = 1$ or $a = 2$ in the first of (4.52)) the following reduced uniform flow conditions as initial conditions:

$$p_E^0 = \rho_E^0 = T_E^0 = 1, \; \boldsymbol{v}_E^0 = 1, \; w_E^0 = 1, \; \text{at } t \leq 0^- . \tag{4.56d}$$

The singular nature of Prandtl unsteady-state boundary-layer equations. Obviously, in the framework of the Prandtl unsteady-state equations (4.46a–d), we have the possibility of taking into account the boundary conditions on $\eta = 0$ for Prandtl velocity components:

$$\text{on } \eta = 0: \; \boldsymbol{v}_{\mathrm{Pr}}^0 = 0 \text{ and } w_{\mathrm{Pr}}^0 = 0 . \tag{4.57a}$$

For the thermal boundary condition (4.49b), we must consider two cases. First, the simple case, $b = 0$, in the second of the similarity relations (4.52) is too degenerate, as a consequence of the matching (relative to time) of the Prandtl limiting process (4.53b) with the limiting process (4.53c):

$$\lim_{t \downarrow 0} (\mathrm{LIM}^{\mathrm{Pr}}) = \lim_{\tau \uparrow \infty} (\mathrm{LIM}^G) .$$

A second case, linked with the second of the similarity relations (4.52), gives the following steady-state thermal condition (according to the second of (4.50)) for both $b = 1$ and $b = 2$ for the unsteady-state Prandtl boundary-layer temperature equation (4.46d):

$$\text{on } \eta = 0: \; T_{\mathrm{Pr}}^0 = T_w^\infty(x_1, x_2) . \tag{4.57b}$$

A second behavior condition for T_{Pr}^0 is derived from the matching condition between Prandtl and Euler:

$$\lim_{\eta \uparrow \infty} T_{\mathrm{Pr}}^0 = T_E^0(t, x_1, x_2, 0) \equiv T_{E0}^0(t, x_1, x_2) , \tag{4.57c}$$

and from (4.45c), we can write the following compatibility equation:

$$\left[\frac{\partial}{\partial t} + \boldsymbol{v}_{E0}^0 \cdot \mathbf{D} \right] \left\{ \log \left[\frac{T_{E0}^0}{(\rho_{E0}^0)^{\gamma-1}} \right] \right\} = 0 . \tag{4.57d}$$

Now, it is also necessary to write initial conditions at $t = 0$ for the unsteady-state Prandtl boundary-layer equations (4.46a,b,d)! For this, it is necessary to resolve, according to the above matching condition in time, $(t \downarrow 0 \approx \tau \uparrow \infty)$, an unsteady-state adjustment problem associated with the one-dimensional unsteady-state gas dynamics equations (4.47a–d).

But unfortunately for this inviscid system (4.47a–d), we can write, the following initial conditions for the velocity components v_G^0 and w_G^0 only when $a = 1$ in the first of similarity relations (4.52):

$$v_G^0 = \mathbf{V}^0(\eta), \ w_G^0 = W^0(\eta), \ \text{at } \tau \leq 0^- . \tag{4.58}$$

When $a = 2$, the initial conditions are the same as those for the unsteady-state Euler inviscid equations:

$$v_G^0 = \mathbf{1}, \ w_G^0 = 1 \text{ at } \tau \leq 0^- . \tag{4.59a}$$

On the other hand, matching in time, $(t \downarrow 0 \approx \tau \uparrow \infty)$ gives

$$w_G^0 = 0, \text{ when } \tau \text{ tends to infinity} \tag{4.59b}$$

because (4.47a–d) are derived from the full NSF equations under the limiting process (4.53c) with the following "gas dynamics" asymptotic representation:

$$(v, w) = (v_G^0, w_G^0) + \delta(v_G^1, w_G^1) + \ldots , \tag{4.60a}$$
$$(p, \rho, T) = (p_G^0, \rho_G^0, T_G^0) + \delta(p_G^1, \rho_G^1, T_G^1) + \ldots . \tag{4.60b}$$

But, unfortunately, we do not have the possibility of taking into account the boundary conditions on $\eta = 0$, in the framework of inviscid one-dimensional unsteady-state gas dynamics equations (4.47a–d). Consequently, it is necessary to consider a new approximate (viscous, heat-conducting) system close to initial time and near the wall!

The role of Rayleigh equations (4.48a–d). The new, viscous, heat-conducting, compressible system of equations close to initial time and near the wall is the Rayleigh system of unsteady-state one-dimensional (in the plane τ^*, η^*), viscous, compressible, heat-conducting equations, which is derived from the full NS-F unsteady-state equations under limiting process (4.53d). More precisely, these Rayleigh equations are derived from the full NSF equations under the limiting process (4.53d) with the following Rayleigh asymptotic representation:

$$(v, w) = (v_R^0, w_R^0) + \delta(v_R^1, w_R^1) + \ldots , \tag{4.61a}$$
$$(p, \rho, T) = (p_R^0, \rho_R^0, T_R^0) + \delta(p_R^1, \rho_R^1, T_R^1) + \ldots . \tag{4.61b}$$

The inviscid compressible one-dimensional unsteady-state gas dynamics equations (4.47a–d) are a degenerate form of the Rayleyh unsteady-state one-dimensional, viscous, compressible, heat-conducting equations (4.48a–d). Close to initial time, the couple of equations (4.48a–d) $\leftrightarrow$ (4.47a–d) plays the same role as the couple NSF $\leftrightarrow$ Euler.

4.3 Asymptotic Structure of Unsteady-State NSF Equations at Re ≫ 1

But it is necessary to observe that Rayleigh equations (4.48a–d) are valid in a layer of the thickness $O(\delta^2)$ relative to time and to the vertical coordinate, whereas the inviscid one-dimensional unsteady-state gas dynamics equations (4.47a–d) are valid in a layer of the thickness $O(\delta)$ relative to time and to the vertical coordinate; the following relation exists between the (τ, η) and (τ^*, η^*) variables: $\tau^* = \tau/\delta$ and $\eta^* = \eta/\delta$, and we can write the following matching conditions:

$$\lim_{\tau^* \uparrow \infty} (\mathrm{LIM}^R) = \lim_{\tau \downarrow 0} (\mathrm{LIM}^G) \tag{4.62a}$$

$$\text{and} \quad \lim_{\eta^* \uparrow \infty} (\mathrm{LIM}^R) = \lim_{\eta \downarrow 0} (\mathrm{LIM}^G) . \tag{4.62b}$$

Consequently, matching between the Rayleigh viscous and heat-conducting solution and the inviscid, one-dimensional, unsteady-state gas dynamics solution opens the possibility (as a rule!) of obtaining a well-posed initial-boundary condition problem for inviscid one-dimensional unsteady-state gas dynamics equations (4.47a–d). But, for this, the following choice must be made in (4.52):

$a = 1$ and $b = 1$.

In such a case, obviously, for the Rayleigh system (4.48a–d) we can write only as initial conditions,

$$v_R^0 = 0, \ w_R^0 = 0, \ \text{and} \ p_R^0 = \rho_R^0 = T_R^0 = 1, \ \text{at} \ \tau^* \leq 0^- . \tag{4.63a}$$

On the other hand, again for this Rayleigh system, we have the following degenerate boundary conditions on the wall:

$$v_R^0 = 0, \ w_R^0 = 0, \ \text{and} \ T_R^0 = 1 \ \text{on} \ \eta^* = 0 . \tag{4.63b}$$

Then, as a consequence of initial (4.63a) and boundary (4.63b) conditions for the Rayleigh viscous and heat-conducting homogeneous equations (4.48a–d), the solution of these homogeneous equations is trivial:

$$v_R^0 \equiv 0, \ w_R^0 \equiv 0 \ \text{and} \ p_R^0 \equiv 1, \ \rho_R^0 \equiv 1, \ T_R^0 \equiv 1 . \tag{4.64}$$

Therefore, when $a = b = 1$, in the framework of the simplified initial boundary-value problem formulated in Sect. 4.3.2, in the leading order the Rayleigh layer is *not* relevant for the adjustment problem.

A relevant Rayleigh problem is derived, if $\underline{a = b = 2}$ and then, the initial and boundary conditions for the Rayleigh equations are

$$v_R^0 = \boldsymbol{V}^0(\eta^*), \ w_R^0 = W^0(\eta^*),$$
$$p_R^0 = 1, \ \rho_R^0 = 1, \ T_R^0 = 1, \ \text{at} \ \tau^* \leq 0^- , \tag{4.65a}$$

and

114   4 The Limit of Very Large Reynolds Numbers

$$v_R^0 = 0, \; w_R^0 = 0, \; \text{on} \, \eta^* = 0 \,, \tag{4.65b}$$

$$T_R^0 = T_w(\tau^*, x_1, x_2), \; \text{when} \; (x_1, x_2) \subset D, \; \text{on} \; \eta^* = 0 \,. \tag{4.65c}$$

But in this Rayleigh relevant case (when $\underline{a = b = 2}$), the initial and boundary conditions for inviscid, one-dimensional, unsteady-state gas dynamics equations are very simplified (naturally, in the framework of the simplified initial boundary-value problem formulated in Sect. 4.3.2). When $a = b = 2$ in (4.52), we obtain the following conditions for (4.47a–d):

$$p_G^0 = 1, \rho_G^0 = 1, T_G^0 = 1, \boldsymbol{v}_G^0 = \mathbf{1}, w_G^0 = 1 \; \text{at} \; \tau \leq 0^- \,. \tag{4.66a}$$

and

$$\boldsymbol{v}_G^0 = 0, \; w_G^0 = 0, \; T_G^0 = T_w^\infty(x_1, x_2) \; \text{on} \; \eta = 0 \,. \tag{4.66b}$$

Then, conditions (4.66b) are behavior conditions at infinity, when $\eta^* \uparrow \infty$, for the Rayleigh viscous solution. Finally, we establish that it is necessary to investigate two different adjustment problems for the initialization of unsteady-state Prandtl BL equations relative to time. The main question is:

What are the initial conditions for unsteady-state boundary-layer equations (4.46a,b,d)?

4.3.4 The Two Adjustment Problems

Adjustment from the Rayleigh problem via intermediate equations. In the framework of the adjustment problem through the Rayleigh problem, we observe that a fifth intermediate region appears between the Rayleigh and Prandtl regions. In this intermediate region, we have a particular sytem of equations which is again derived from the full unsteady-state NS-F equations.

More precisely, the equations in the intermediate region are obtained, on the one hand, when we carry out the classical boundary-layer approximations on the Rayleigh equations (4.48a–d), and, on the other hand, when we carry out the approximation used to derive the Rayleigh equations on the Prandtl equations (4.46a–d).

The intermediate matching unsteady-state equations have the following form:

4.3 Asymptotic Structure of Unsteady-State NSF Equations at Re ≫ 1

$$\rho_I^0 \left[S \frac{\partial v_I^0}{\partial t'} + w_I^k \frac{\partial v_I^0}{\partial z} \right] = \frac{\partial^2 v_I^0}{\partial z^2} \, ;$$

$$S \frac{\partial \rho_I^0}{\partial t'} + \frac{\partial (\rho_I^0 w_I^k)}{\partial z} = 0 \, ;$$

$$\frac{\partial p_I^0}{\partial z} = 0 \, ;$$

$$\rho_I^0 \left[S \frac{\partial T_I^0}{\partial t'} + w_I^k \frac{\partial T_I^0}{\partial z} \right] + (\gamma - 1) p_I^0 \frac{\partial w_I^k}{\partial z}$$

$$= \frac{\gamma}{\Pr} \frac{\partial^2 T_I^0}{\partial z^2} + \gamma(\gamma-1) M^2 \left| \frac{\partial v_I^0}{\partial z} \right|^2 , \tag{4.67}$$

with $p_R^0 = \rho_R^0 T_R^0$. In (4.67) the new intermediate variables t' and z are related to the starting variables t, x_3 by the relations,

$$t' = t/\sigma(\delta) \text{ and } z = x_3/\delta\sigma(\delta)^{1/2} , \tag{4.68a}$$

and

$$w_I^k = (w/\delta)\sigma(\delta)^{1/2}, \ k = 0, 1 . \tag{4.68b}$$

The gauge $\sigma(\delta)$ characterizes the thickness of the initial layer where intermediate equations (4.67) are valid.

Unfortunately, the precise localization of this fifth intermediate matching region is not possible at this stage of the asymptotic analysis.

But we observe that, when

$$\sigma(\delta) = \delta^2, \text{ then } k = 0 \text{ and } t' \equiv \tau^*, \ z \equiv \eta^*, \ w_I^0 \equiv w_R^0 .$$

On the contrary, if $\sigma(\delta) = \delta^0 = 1$, then $k = 1$, and $t' \equiv t$, $z \equiv \eta$, $w_I^1 \equiv w_{\Pr}^1$.

Finally, it seems that the process of unsteady-state adjustment from Rayleigh to Prandtl, via the intermediate matching region, is the more appropriate process provided that for the Rayleigh equations (4.48a–d), we take into account the initial conditions (4.65a), the boundary conditions (4.65b,c) and also the behavior conditions (4.66b) at $\eta^* \uparrow \infty$ (as matching conditions with the solution of inviscid, one-dimensional, unsteady-state gas dynamics equations).

We observe (curiously) that intermediate equations (4.67) are not related to the equations (4.47a–d) for inviscid, one-dimensional, unsteady-state gas dynamics flow!

But a question now remains:

> How can the Rayleigh model equations, close to initial time and near the wall, having only parametrical dependence on horizontal coordinates, possibly settle down to the horizontal coordinate dependent (but time independent for $t \downarrow 0$) Prandtl solution of (4.46a–d)?

Obviously, in the unsteady-state adjustment process, from Rayleigh to Prandtl via an intermediate matching region, the horizontal coordinates play only the role of parameters through the given, on $\eta^* = 0$, thermal field: $T_w(\tau^*, x_1, x_2)$, $(x_1, x_2) \subset D$, in the thermal boundary condition (4.65c), for the Rayleigh temperature T_R^0 that is governed by (4.48d).

As a consequence of matching (4.62a,b), the given thermal field $T_w(\tau^*, x_1, x_2)$, on $\eta^* = 0$, matches the limiting steady-state thermal field $T_w^\infty(x_1, x_2)$ on $\eta = 0$.

It is important to observe, also, that the Rayleigh model settles down with the Prandtl model only after an unsteady-state adjustment process via the intermediate matching unsteady-state model that is governed by (4.67), and this gives the asymptotically consistent initial conditions at $t = 0$ for the Prandtl unsteady-state boundary-layer equations (4.46a,b,d).

In Fig. 4.4 the four-region, unsteady-state singular asymptotic structure at $Re \gg 1$ is sketched. The existence of the intermediate matching region opens the possibility of better understanding the relation between the Rayleigh and Prandtl asymptotic structures.

We stress again that the influence of the Rayleigh structure on the Prandtl structure is realized only through the initial conditions at $t = 0$ for the Prandtl equations.

These initial conditions, at $t = 0$, for the Prandtl equations are the result of the unsteady-state adjustment problem (when $\tau^* \uparrow \infty$) from the Rayleigh problem via the intermediate equations.

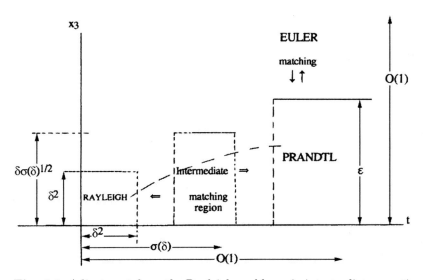

Fig. 4.4. Adjustment from the Rayleigh problem via intermediate equations

4.3 Asymptotic Structure of Unsteady-State NSF Equations at Re ≫ 1

Adjustment from the inviscid, one-dimensional, unsteady-state gas dynamics problem. A second kind of (inviscid!) unsteady-state adjustment problem appears when the Rayleigh solution is trivial. In this particular case, it is necessary to take into account inviscid, one-dimensional, unsteady-state gas dynamics equations (4.47a–d), valid in the thickness of order $O(\delta)$ near the wall and close to the initial time. For these equations, initial conditions are

$$\text{at } \tau \leq 0^- : v_G^0 = \boldsymbol{V}^0(\eta),\ w_G^0 = W^0(\eta)\,, \tag{4.69a}$$

$$p_G^0 = 1,\ \rho_G^0 = 1,\ T_G^0 = 1,\ \text{at } \tau \leq 0^-\,, \tag{4.69b}$$

and as boundary conditions

$$v_G^0 = 0,\ w_G^0 = 0,\ \text{on } \eta = 0\,, \tag{4.69c}$$

$$T_G^0 = T_w(\tau, x_1, x_2),\ \text{when } (x_1, x_2) \subset D,\ \text{on } \eta = 0\,. \tag{4.69d}$$

This adjustment problem (4.47a–d), (4.69a–d), is closely linked with filtering of acoustic waves generated by compressibility at the initial stage of motion.

In this case we observe that for the large Reynolds number at the leading order, *a consistent* (see Fig. 4.5) *"three approximate model problems" approach to the full NSF unsteady-state viscous compressible and heat-conducting problem, is constituted, on the one hand, by the Euler problem which opens the possibility of taking into account the behavior conditions as $\eta \uparrow \infty$ in the framework of the Prandtl unsteady-state boundary-layer problem and, on the other hand, by the one-dimensional, inviscid, unsteady-state gas dynamics problem which gives the possibility of deriving the (correct) initial conditions at $t = 0$ for this Prandtl unsteady-state boundary-layer problem.*

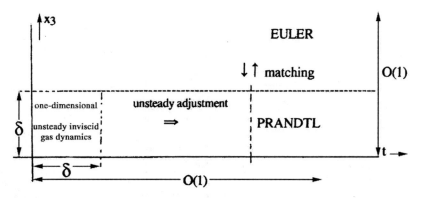

Fig. 4.5. Adjustment from the inviscid, one-dimensional, unsteady-state gas dynamics problem

We observe that, in this case, for the unsteady-state Prandtl boundary-layer equations (4.46a,b,d), it is necessary to write as initial conditions

$$t = 0: \quad \boldsymbol{v}^0_{\text{Pr}} = \boldsymbol{v}^0, \ \rho^0_{\text{Pr}} = \rho^0, \ T^0_{\text{Pr}} = T^0 \,, \tag{4.70}$$

where $\boldsymbol{v}^0$ is different from $\boldsymbol{V}^0$, ρ^0 is different from 1, and T^0 is different from 1, according to (4.51a). In the framework of an unsteady-state adjustment problem, the main question arises:

How are $\boldsymbol{v}^0$, ρ^0, and T^0 related to the given data (4.51a,b)?

We may understand the relation of BL initial conditions (4.70) to NSF initial conditions (4.51a,b). We should consider (4.70) as matching conditions between a BL approximation at $t = 0$ and the initial-time region approximation at $\tau \uparrow \infty$. This matching requires

$$\lim_{\tau \uparrow \infty} (w^0_G, T^0_G, \rho^0_G) = (0, T^0, \rho^0) \,, \tag{4.71a}$$

$$\lim_{\tau \uparrow \infty} \boldsymbol{v}^0_G = \boldsymbol{v}^0 \,. \tag{4.71b}$$

Due to the nonlinearity (quasi-linearity) of the system of equations (4.47b–d), it is difficult to make rigorous statements concerning the behavior of the solution to (4.47b–d) and (4.47a), as $\tau \uparrow \infty$. Nevertheless, it is quite clear on physical grounds that the solution, besides entropy waves ruled by the third of equations (4.47d), consists of acoustic waves radiating toward infinity, leaving behind them a BL structure, which are dependent on η, and horizontal coordinates as parameters, but with pressure independent of the vertical coordinate η.

4.4 The Triple-Deck Concept and Singular Interactive Coupling

The main physical ideas underlying triple-deck theory were laid down by Lighthill (1953) in an article on the interaction between a boundary layer and a supersonic stream arising from expansive steady-state disturbances (either an incident wave or a departure of the wall shape from straight) or from relatively weak compressive disturbances.

The asymptotic foundation of two-dimensional, steady-state triple-deck theory (for incompressible and compressible flows) is reviewed by Nayfeh (1991). Chapter 12 in Zeytounian (2002b) is devoted to "singular coupling and the triple-deck model." See also the pertinent review paper by Meyer (1983).

Below, we give some information concerning the "discovery" of the triple-deck structure, and for this we have taken into account, in particular, the Introduction of the Nayfeh (1991) review paper.

First, Lighthill (1953) treated interaction mathematically by perturbing a parallel flow, linearizing the Navier equations around it, and produced a coherent self-consistent theory of interaction that divides the region of

4.4 The Triple-Deck Concept and Singular Interactive Coupling

interest into three parts: the inviscid flow outside the boundary layer (in which the perturbations are governed by the so-called Prandtl–Glauert equation), the displaced boundary layer in which the perturbations are governed by linearized compressible Euler equations, and an inner part close to the wall in which the perturbations are governed by incompressible boundary-layer equations. Using this theory, Lighthill (1953) indicated how spontaneous changes in the boundary layer could be set up which might lead to separation and that the length of the interaction is of the order of $L^0 \, \text{Re}^{-3/8}$, where the length L^0 characterizes the location of the disturbance. In this theory, Lighthill assumed a streamwise length scale that is $O(L^0 \, \text{Re}^{-1/2})$, which is of the same order as the boundary-layer thickness.

Then, Gadd (1957) developed an approximate theory to extend the Lighthill's analysis to the nonlinear case.

Twelve years later, taking over the basic ideas of Lighthill and Gadd without prior justification, Stewartson and Williams (1969) extended Lighthill's theory to nonlinear interactions and put the approximate method developed by Gadd on a firm basis.

The most important of these ideas is that the intraction length scale is $O(L^0 \, \text{Re}^{-3/8})$ and, in the *triple-deck structure*, the *middle* inviscid rotational deck (without any pressure gradient) has a thickness of the order of $L^0 \, \text{Re}^{-1/2}$, the *upper* irrotational inviscid deck has a thickness of the order $L^0 \, \text{Re}^{-3/8}$, and the *lower rotational viscous* deck has a thickness of the order of $L^0 \, \text{Re}^{-5/8}$.

Neiland (1969) arrived, independently, at the same scalings and the triple-deck structure for the problem of propagating of perturbations upstream of the interaction between a hypersonic flow and a boundary layer.

Messiter (1970) also arrived, independently, at the same scalings and the triple-deck structure for the flow near the trailing edge of a finite flat plate. He showed that Goldstein's (1930) solution of the boundary-layer equations *downstream* from the trailing edge of a finite flat plate (see Fig. 4.1) *breaks down* if the dimensionless distance x/L_0 from the edge, where x/L_0 is the length of the plate, is $O(\text{Re}^{-3/8})$. He also obtained a second-order correction to Goldstein's solution that accounts for the pressure gradient induced locally in the external flow and found that it overtakes the first-order term when x/L_0 is $O(\text{Re}^{-3/8})$. Alternatively, Messiter (1970) assumed the streamwise length scale $X = \text{Re}^a \, x/L_0$ and investigated the limit of the Navier equations as $\text{Re} \uparrow \infty$. Matching the resulting expansion downstream with Goldstein's solution, he found that $a = 3/8$ leads to a distinguished limit and consequently arrived at the same scalings and triple-deck structure. Stewartson (1969) discovered the same triple-deck structure for the trailing edge problem, and Hunt (1971) and Smith (1973), using order of magnitude arguments, investigated the structure of an incompressible flow at high Reynolds number past a hump on an otherwise smooth surface. The triple-deck structure was extended to a plane at an angle of attack of the order of $\text{Re}^{-1/16}$ by Brown and Stewartson (1970), to an airfoil at a finite trailing-edge angle of the order of $\text{Re}^{-1/4}$ by Riley and Stewartson (1969), and to a three-dimensional wing with a pair of eddies at the trailing edge by Guiraud (1974). Various problems involving multistructured boundary layers, were comprehensively surveyed by Stewartson (1974).

An important step in the triple-deck theory is also Sychev's (1972) proposal that gives the possibility of formalizing the triple deck allowing for separation. Smith (1977) completely solved the problem including numerics and asymptotic behavior upstream and downstream. Guiraud and Zeytounian (1979a) formally extended the analysis to three-dimensional separation along a smooth curve, and Riley (1979) considered separation along a ray of a conical body [for Sychev's proposal and the Sychev problem for lower-deck equations, see Meyer (1983; §9) and Zeytounian (2002b; Chap. 12)]. In Sect. 4.4.1, we consider formally the simple case of an accident at distance L_0 from the leading edge of a semi-infinite flat plate (we do not more precise the nature of this accident – for instance, a small hump!).

4.4.1 The Triple-Deck Theory in 2-D Steady-State Navier Flow

The accident is placed near abscissa x_a, and we introduce a local horizontal coordinate

$$x^* = (x - x_a)/\text{Re}^{-a/2} . \tag{4.72}$$

We assume that $a < 1$, which is compatible with a triple-deck structure. The starting Navier 2-D steady-state equations are

$$\frac{\partial u_N}{\partial x} + \frac{\partial v_N}{\partial y} = 0 , \tag{4.73a}$$

$$u_N \frac{\partial u_N}{\partial x} + v_N \frac{\partial u_N}{\partial y} + \frac{\partial \pi}{\partial x} = \frac{1}{\text{Re}} \left[\frac{\partial^2 u_N}{\partial x^2} + \frac{\partial^2 u_N}{\partial y^2} \right] , \tag{4.73b}$$

$$u_N \frac{\partial v_N}{\partial x} + v_N \frac{\partial v_N}{\partial y} + \frac{\partial \pi}{\partial y} = \frac{1}{\text{Re}} \left[\frac{\partial^2 v_N}{\partial x^2} + \frac{\partial^2 v_N}{\partial y^2} \right] . \tag{4.73c}$$

Upstream from point $(x_a, 0)$, we assume that the Prandtl boundary-layer structure is established and consequently, for $0 < x < x_a$ we can write

$$u_N = U_{\text{BL}}(x, \eta) \text{ and } v_N = \delta V_{\text{BL}}(x, \eta) , \tag{4.74a}$$

with $\eta = y/\delta$ and $\delta = \text{Re}^{-1/2}$. \hfill (4.74b)

We observe that, according to Lighthill's (1953) fundamental paper, we assume that the accident leads to a "strong" singular coupling near point $(x_a, 0)$ via a triple-deck structure, and this corresponds just to condition $a < 1$ in (4.72).

4.4 The Triple-Deck Concept and Singular Interactive Coupling

The lower deck in this triple-deck structure is characterized by the vertical coordinate

$$y^* = y/\operatorname{Re}^{-b/2}, \text{ with } b > 1, \tag{4.75}$$

because obviously it is thinner than the Prandtl boundary-layer thickness $\delta = \operatorname{Re}^{-1/2}$.

Main deck. This deck is the natural continuation of the classical Prandtl boundary layer, and it is assumed that our accident generates only a small perturbation of the boundary-layer flow:

$$u_N = U_{\mathrm{BL}}(x_a, \eta) + \operatorname{Re}^{-\alpha/2} u(x^*, \eta) + \ldots + \operatorname{Re}^{-a/2} x^* \left.\frac{\partial U_{\mathrm{BL}}}{\partial x}\right|_{x_a} + \ldots , \tag{4.76a}$$

$$v_N = \operatorname{Re}^{-1/2} \left\{ V_{\mathrm{BL}}(x_a, \eta) + \operatorname{Re}^{-\beta/2} v(x^*, \eta) + \ldots \right.$$

$$\left. + \operatorname{Re}^{-a/2} x^* \left.\frac{\partial V_{\mathrm{BL}}}{\partial x}\right|_{x_a} + \ldots \right\} , \tag{4.76b}$$

$$\pi = 1 + \operatorname{Re}^{-\gamma/2} p(x^*, \eta) + \ldots , \tag{4.76c}$$

where α, β, and γ are three positive real scalars.

Now, first, from the continuity equation (4.73a), we derive the following relation:

$$\beta = \alpha - a \Rightarrow \beta < 0 , \tag{4.77}$$

because $\alpha < a < 1 + \alpha$.

Lower deck. This deck near the flat plate is characterized by the variables x^* and $y^* = y/\operatorname{Re}^{-b/2}$, with $b > 1$ and, obviously, in this deck, the equations are boundary-layer type. We write

$$u_N = \operatorname{Re}^{-\lambda/2} u^*(x^*, y^*) + \ldots , \tag{4.78a}$$

$$v_N = \operatorname{Re}^{-\mu/2} v^*(x^*, y^*) + \ldots , \tag{4.78b}$$

$$\pi = 1 + \operatorname{Re}^{-\gamma/2} p^*(x^*, y^*) + \ldots , \tag{4.78c}$$

because the perturbations of pressure π match continuously.

On the other hand, we observe that

$$\left.U_{\mathrm{BL}}\right|_{x_a} \approx \eta = \operatorname{Re}^{-(b-1)/2} y^* \text{ when } \eta \downarrow 0 \text{ and } y^* \uparrow \infty \Leftrightarrow \eta \downarrow 0 . \tag{4.78d}$$

To derivate boundary-layer type equations in the lower deck from Navier equations (4.73a–c), via expansions (4.78a–c), the following relations must be satisfied:

$$\mu = \lambda + b - a, \ \gamma = 2\lambda, \ \gamma - a + 2b = \lambda + 2 \Rightarrow \mu = 2 - b. \tag{4.79a}$$

From (4.78a) and (4.78d), we also obtain, with (4.79a), the relation

$$\lambda = b - 1 \Rightarrow a = 3(b-1), \tag{4.79b}$$

and consequently, we have the following upstream infinity behavior, for the first term of (4.78a):

$$u^*(x^* \uparrow -\infty, y^*) \to y^*. \tag{4.80}$$

Upper deck. For this inviscid deck, again,

$$\pi = 1 + \mathrm{Re}^{-\gamma/2} P(x^*, Y) + \dots, \tag{4.81a}$$

because the perturbations of pressure π match continuously, where

$$Y = y/\mathrm{Re}^{-c/2}, \ c < 1, \tag{4.82}$$

and the Euler inviscid equations are derived from (4.73a–c) via the expansions with (4.81a),

$$u_N = U_{\mathrm{BL}}(x_a, \infty) + \mathrm{Re}^{-\gamma/2} U(x^*, Y) + \dots, \tag{4.81b}$$

$$v_N = \mathrm{Re}^{-\gamma/2} V(x^*, Y) + \dots, \tag{4.81c}$$

because $\eta = Y/\mathrm{Re}^{-(1-c)/2}$, $1 - c > 0$, and $\eta \uparrow \infty \Leftrightarrow Y \downarrow 0$.

In this case, we derive, first, the following linear inviscid equation between U and P:

$$U_{\mathrm{BL}}(x_a, \infty)\frac{\partial U}{\partial x^*} + \frac{\partial P}{\partial x^*} = 0, \tag{4.83a}$$

and then, with the choice of $c = a < 1$, we also derive two inviscid equations:

$$U_{\mathrm{BL}}(x_a, \infty)\frac{\partial V}{\partial x^*} + \frac{\partial P}{\partial Y} = 0, \tag{4.83b}$$

$$\frac{\partial U}{\partial x^*} + \frac{\partial V}{\partial Y} = 0. \tag{4.83c}$$

The matching of vertical velocity between the upper and main decks [which gives the possibility for the inviscid equations (4.83a–c) of writing a linearized slip condition] is possible only if

$$\gamma = \beta + 1 < 1. \tag{4.84}$$

Then, we can write

$$V(x^*, 0) = \lim_{\eta \uparrow \infty} v(x^*, \eta), \tag{4.85}$$

and this condition is associated with the Laplace equation for vertical velocity $V(x^*, Y)$ which governs upper deck inviscid flow:

$$\frac{\partial^2 V}{\partial x^{*2}} + \frac{\partial^2 V}{\partial Y^2} = 0, \quad (4.86)$$

which is derived from the linear Euler incompressible system (4.83a–c) for U, V, and P, because $U_{\mathrm{BL}}(x_a, \infty) \neq 0$.

Now, with (4.77), (4.79b), and (4.84), we obtain

$$\beta + 1 = 1 + \alpha - a = \gamma = 2\lambda = 2(b - 1) = 2a/3,$$

or

$$\gamma - \alpha = (2/3)a - \alpha = 1 - a > 0. \quad (4.87)$$

A system of equations for main-deck inviscid flow. In the main deck, according to expansions (4.76a–c), we derive the following very simple equations for the perturbed flow (u, v, p):

$$U_{\mathrm{BL}}(x_a, \eta) \frac{\partial u}{\partial x^*} + \left[\frac{\mathrm{d}U_{\mathrm{BL}}(x_a, \eta)}{\mathrm{d}\eta}\right] v = 0,$$

$$\frac{\partial u}{\partial x^*} + \frac{\partial v}{\partial \eta} = 0,$$

$$\frac{\partial p}{\partial \eta} = 0, \quad (4.88)$$

and when $\eta \downarrow 0$, $U_{\mathrm{BL}}(x_a, 0) = 0$. In this case matching between the lower and main decks for horizontal velocity gives

$$\alpha = \lambda = b - 1 = a/3 \Rightarrow \alpha = a/3. \quad (4.89)$$

The solution of (4.88) for u and v is

$$u(x^*, \eta) = A(x^*) \frac{\mathrm{d}U_{\mathrm{BL}}(x_a, \eta)}{\mathrm{d}\eta}, \quad (4.90a)$$

$$v(x^*, \eta) = -U_{\mathrm{BL}}(x_a, \eta) \frac{\mathrm{d}A(x^*)}{\mathrm{d}x^*}, \quad (4.90b)$$

where function $A(x^*)$ is unknown at this stage.

As a consequence of (4.90b) with (4.85), we obtain

$$V(x^*, 0) = -U_{\mathrm{BL}}(x_a, \infty) \frac{\mathrm{d}A(x^*)}{\mathrm{d}x^*}, \quad (4.91)$$

and the velocity $V(x^*, 0)$ can be identified with the perturbation of the blowing velocity used in classical boundary-layer studies [see, for instance, (4.37b)]. $A(x^*)$ is called the displacement function.

Determination of the various scalars and the triple-deck structure.

Finally, from the relations (4.77), (4.79a), (4.79b), (4.84), and (4.87), with (4.89), we obtain the following values for the unknown scalar a, α, b, $\beta + 1$, γ, λ, and, μ:

$$a = 3/4, \ \alpha = 1/4, \ b = 5/4, \ \beta + 1 = 1/2, \ \lambda = 1/4, \ \mu = 3/4. \qquad (4.92)$$

As consequence of (4.92), Navier flow around the accident is divided into three regions as indicated in Fig. 4.6.

The longitudinal and transverse scales (made dimensionless with L_0) of the perturbed region are $\text{Re}^{-3/8}$. Inside the pertubed squared region in the (x, y) plane, three decks are identified. The thickness of the lower deck is $\text{Re}^{-5/8}$, and the viscous effects are important. The thickness of the main deck is $\text{Re}^{-1/2}$, and the viscous effects are negligible; this deck is the prolongation of the classical upstream boundary layer. The thickness of the upper deck is $\text{Re}^{-3/8}$, and the viscous effects are also negligible in it. These three different decks are the different ways the flow can go from the upstream region to the downstream region. These upstream and downstream regions behave like classical Prandtl boundary layers.

Note that the main-deck and the lower deck equations are included in the boundary-layer equations. The lower deck equations (see below) are even the classical boundary layer equations but with *unusual* boundary conditions (see below). By contrast with the classical boundary-layer theory, the sets of viscous flow and inviscid flow equations are strongly coupled in the sense that the solution in the upper deck depends on the solution in the lower and main decks through the displacement function $A(x^*)$, whereas the solution in the lower and main decks depends on the solution in the upper deck through the pressure distribution. The solution in the upper deck cannot be determined independently from the solution in the lower and main decks and vice versa! We observe that the existence of the upper deck is justified by the fact that the vertical component of the velocity in the main deck, according to (4.90b), is different from zero when $\eta \uparrow \infty$ (at the "upper" edge of the main deck or for $Y = 0$ at the "lower" edge of the upper deck).

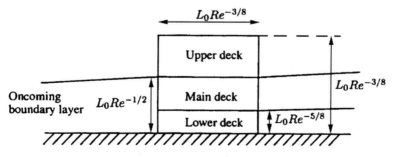

Fig. 4.6. Triple-deck stucture

4.4 The Triple-Deck Concept and Singular Interactive Coupling 125

Boundary-layer problem (4.96) for the lower deck. First, from the main deck solution (4.90a,b), with the main expansions (4.76a–c) and the values (4.92) for α and $\beta + 1$, we obtain the following behavior for u and v, when $\eta \downarrow 0$, or for u^* and v^*, when $y^* \uparrow \infty$:

$$u^*(x^*, y^*) \approx \left.\frac{dU_{\mathrm{BL}}}{d\eta}\right|_{\eta \downarrow 0} [y^* + A(x^*)], \tag{4.93a}$$

$$v^*(x^*, y^*) \approx -\left.\frac{dU_{\mathrm{BL}}}{d\eta}\right|_{\eta \downarrow 0} y^* \frac{dA(x^*)}{dx^*}, \tag{4.93b}$$

because $y^* = \eta/\mathrm{Re}^{1/8}$. In fact, $dU_{\mathrm{BL}}/d\eta|_{\eta \downarrow 0} \equiv \lambda^0 = \mathrm{const}.$

Now, it is obvious that in the upper deck, by analogy with (4.86), we can also obtain a Laplace equation for $P(x^*, Y)$:

$$\frac{\partial^2 P}{\partial x^{*2}} + \frac{\partial^2 P}{\partial Y^2} = 0, \tag{4.94}$$

and, by continuity of p, p^*, and P across the three decks, we can write the following boundary condition for (4.94):

$$P(x^*, 0) = P^*(x^*), \tag{4.95a}$$

with (because $\gamma = 2\lambda = 1/2$)

$$\frac{\partial p^*}{\partial y^*} = 0 \Rightarrow p^* = P^*(x^*) \equiv p(x^*), \tag{4.95b}$$

because, according to (4.88), $\partial p/\partial \eta = 0$.

The function $P^*(x^*)$ plays the role of "pressure" in the *boundary-layer equations* that govern the *lower deck viscous flow*; from (4.78a–c), with $\lambda = \gamma/2 = 1/4$ and $\mu = 3/4$, via Navier equations (4.73a–c), we derive the following equations:

$$\frac{\partial u^*}{\partial x^*} + \frac{\partial v^*}{\partial y^*} = 0, \tag{4.96a}$$

$$u^* \frac{\partial u^*}{\partial x^*} + v^* \frac{\partial u^*}{\partial y^*} + \frac{dP^*(x^*)}{dx^*} = \frac{\partial^2 u^*}{\partial y^{*2}}. \tag{4.96b}$$

The lower deck equations (4.96a,b) are even the classical boundary-layer equations but have the following unusual boundary conditions:

$$y^* = 0 \text{ and } x^* = O(1): u^* = v^* = 0, \tag{4.96c}$$

$$y^* \uparrow \infty: u^* \to \lambda^0[y^* + A(x^*)], \tag{4.96d}$$

$$y^* \uparrow \infty: v^* \to -\lambda^0 y^* \frac{dA(x^*)}{dx^*}, \tag{4.96e}$$

$$x^* \uparrow -\infty: u^* \to \lambda^0 y, \ v^* \to 0, \ P^*(x^*) \to 0, \tag{4.96f}$$

$$\left.\frac{dA(x^*)}{dx^*}\right|_{x^* \uparrow -\infty} \to 0 \text{ and } A(-\infty) = 0. \tag{4.96g}$$

Finally, from the upper equation (4.83b) for $V(x^*, Y)$, we derive

$$\left.\frac{\partial P}{\partial Y}\right|_{Y=0} = -U_{\text{BL}}(x_a, \infty)\frac{\partial V(x^*, 0)}{\partial x^*} = [U_{\text{BL}}(x_a, \infty)]^2 \frac{\mathrm{d}^2 A(x^*)}{\mathrm{d}x^{*2}}, \qquad (4.97)$$

according to relation (4.91). This relation is remarkable because we see that the resolution of the Laplace "upper deck" equation (4.94) with condition (4.95a) gives a functional relation between $A(x^*)$ and $P^*(x^*)$. It is necessary to understand that the function $P^*(x^*)$ in the boundary-layer "lower deck" equation is not a given function via matching with the solution of inviscid fluid flow, as in classical boundary-layer theory!

More precisely, the solution of the upper deck gives (we assume that $U_{\text{BL}}(x_a, \infty) \equiv 1$)

$$P(x^*, 0) = \frac{1}{\pi} \, C\!\!\int\limits_{-\infty}^{+\infty} \left\{ \frac{\mathrm{d}A(x^*)/\mathrm{d}x^*}{x^* - \xi} \right\} \mathrm{d}\xi, \qquad (4.98)$$

where "C" means "Cauchy principal value."

A final remark concerns the wall shear stress that is given by the slope of the velocity profile at the wall in the lower deck. The choice of gauges is such that the perturbation of the wall shear stress introduced by the triple deck has the same order as the wall shear stress of the basic flow. Therefore, the resulting wall shear stress can be negative, i.e., the boundary layer can separate!

4.5 Complementary Remarks

In Chap. 14 of the recent book by Schlichting and Gersten (2000), the reader can find various extensions to the Prandtl boundary-layer theory, in particular, higher order boundary-layer theory [comprehensive descriptions of higher order boundary-layer theory have been given by Van Dyke (1969, 1970)]. We observe here that it is usual to split the solution of second-order boundary-layer (incompressible, but linear) equations into a curvature part and a displacement part. The Re $\gg 1$ asymptotic theory has been extended to three-dimensional flows in numerous studies; these may be the flow at a yawing plane body [Gersten and Gross (1973)] or the flow past a three-dimensional body, such as the 3-D paraboloid, investigated by Papenfuß (1974, 1975). The extension of higher order boundary-layer theory to compressible flow has been carried out by Van Dyke (1962a,b,c), and the displacement and curvature effects appearing in incompressible flow. There are also two further second-order effects: the influence of vorticity in the outer flow which could be, for example, a consequence of curved shock waves in front of the body, and noncontinuum effects. These are effects such as slipping of the flow and a temperature jump at the wall. Such effects are consequences of the corresponding boundary conditions at the outer edge (shock wave) of a supersonic flow or at the wall. For example, Papenfuß (1975) showed that heat transfer at the stagnation point of a paraboloid of revolution at zero incidence, with the free stream Mach number $M = 4$, neglecting the displacement effect, is given by

$$q_w/q_{w\infty} = 1 + [0.236 + 0.514 + 1.002](1/\mathrm{Re}_\infty)^{1/2}, \tag{4.99}$$

where the Reynolds number Re_∞ is formed with the radius of curvature at the stagnation point and the quantities belonging to the free stream. In (4.99), we have, respectively, the curvature 0.236, noncontinuum 0.514 and vorticity 1.002 effects.

Interacting boundary-layer equations. In practice, instead of carrying out the four calculations (first order outer flow, first order inner flow, second order outer flow, and second order inner flow) in succession, only one calculation is performed, and the matching of the inner flow to the outer flow takes place iteratively. Simplified NSF equations, which still contain all terms that contribute to second-order boundary-layer equations are used for the inner boundary-layer solution. For plane (2-D) steady-state compressible flow, these read

$$\frac{\partial \rho u}{\partial s} + \frac{\partial[(1+KN)\rho v]}{\partial N} = 0 \ ;$$

$$\rho\left[\frac{u}{1+KN}\frac{\partial u}{\partial s} + v\frac{\partial u}{\partial N}\right] + \frac{1}{1+KN}\frac{\partial p}{\partial s} = \frac{\partial \tau_{sN}}{\partial N} + 2K\tau_{sN} \ ;$$

$$K\rho u^2 = \frac{\partial p}{\partial N} \ ;$$

$$\rho C_p \left[\frac{u}{1+KN}\frac{\partial T}{\partial s} + v\frac{\partial T}{\partial N}\right] = \frac{\partial}{\partial N}\left(k\frac{\partial T}{\partial N}\right)$$

$$+ K\left(k\frac{\partial T}{\partial N}\right) + \beta T\left[\frac{u}{1+KN}\frac{\partial p}{\partial s} + v\frac{\partial p}{\partial N}\right] + \frac{(\tau_{sN})^2}{\mu}, \tag{4.100}$$

where s and n are the curvilinear coordinates in a system of natural coordinates for a plane (2-D) body and $N = n/\delta$ with $\delta = \mathrm{Re}^{-1/2}$.

$K(s) = L_0/R(s)$ is the dimensionless surface curvature of this body, when the geometry of the body is given by the local radius of curvature $R(s)$ along a meridian. On the other hand,

$$\tau_{sN} = \mu\left(\frac{\partial u}{\partial N} - Ku\right) \quad \text{and} \quad \beta = -\frac{1}{\rho}\frac{\partial \rho}{\partial T}\bigg|_p . \tag{4.101}$$

The interacting boundary-layer equations for axisymmetric compressible flow are given in the book by Schlichting and Gersten (2000; p. 388). The disadvantage of this procedure is that a separate calculation must be carried out for each Reynolds number. For an application to hypersonic and high temperature gas dynamics, see Anderson (1989).

Hypersonic interaction. Frequently, *strong* interactions occur in hypersonic flows past slender bodies, and then the first-order outer flow depends on the behavior of the boundary layer. The manner in which this behaves is itself a consequence of the outer flow (in a *weak* interaction, the first-order boundary layer affects the second-order outer flow, but not the first-order outer flow). The outer flow and the first-order boundary layer, therefore, have a *mutual effect* on each other and must be computed *simultaneously*.

More precisely, high Mach numbers in hypersonic flow ($M > 5$) produce two effects that lead to a strong interaction. On the one hand, there is a considerable increase in the boundary-layer thickness with increasing Mach number, and, on the other hand, as the Mach number increases, the shock angle becomes flatter, i.e., the shock front approaches the body more closely. An analysis of the strong interaction shows that the two solutions (the frictionless outer solution for hypersonic flow past slender bodies and the inner boundary-layer solution) cannot be matched correctly with respect to temperature or density. According to Bush (1966), the flow field behind the shock front consists of three layers. A transition solution in the middle allows matching the two solutions above correctly.

Marginal and "massive" separations. *Marginal separation* always occurs in flow arrangements that are described by one particular (singular) parameter S. This parameter must be able to take on values, where a simple boundary-layer calculation along the entire geometry of the body is possible without the appearance of the Goldstein singularity (attached flows), and values, where the boundary-layer calculation cannot be carried on any further because of the appearance of the Goldstein singularity. The limit between these two values is given by the critical value S_c. For marginal separation, the flow for high Reynolds number is given by this limiting case. The Goldstein singularity has just not appeared! For examples, see Schlichting and Gersten (2000; pp. 404–408).

Massive separation occurs when the boundary layer leaves the wall as one and marks the boundary between the outer flow and a separation region (*backflow* region) as a free shear layer. The separation point is the point where the boundary layer leaves the wall and massive separation takes place when the thickness of the boundary layer in front of the separation point is small compared to the dimension of the separation region perpendicular to the direction of the main flow.

Sychev proposal and F.T. Smith computations. A strong increase in the v component close to separation implies that the reaction on the outer flow is no longer asymptotically small and again an interaction process is present that must be described (in principle?) using the triple-deck theory. However(!) Stewartson (1970) was able to show that the triple-deck theory cannot remove the singularity. It turns out that the positive pressure gradient imposed on the boundary layer, which is modified only in an asymptotically small region in triple-deck theory, is the important obstacle. The positive pressure gradient is a necessary condition for the onset of separation but is simultaneously the origin of the singular behavior of the boundary-layer solution! Sychev (1972) discovered an amazingly simple way out of this "dilemma": close to the separation point, the pressure gradient is assumed to be asymptotically small; it exists only for finite Reynolds numbers. In the limit of infinite Reynolds number (only in this limit can the Goldstein singularity occur), there is no pressure increase directly in front of the separation point and thus no Goldstein singularity. It has not been removed, but has been avoided!

According to Schlichting and Gersten (2000; pp. 408–411), three important aspects determine the asymptotically consistent description of flows with massive separation:

(1) In the limit $1/\mathrm{Re} = 0$, all (thin, at high Reynolds numbers) shear layers "degenerate" to lines. If we start by assuming that the boundary layer leaves the wall at the separation point, then in the limit $1/\mathrm{Re} = 0$, a so-called free-streamline leaves the body (see Fig. 4.7).

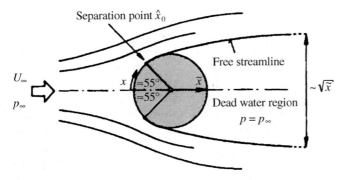

Fig. 4.7. Limiting solution ($1/\mathrm{Re} = 0$) for a circular cylinder, according to the free-streamline theory by Helmholtz (1868)–Kirchhoff (1869)

(2) For $1/\mathrm{Re} \neq 0$, flow must be modified so that a pressure gradient arises that matches the pressure from triple-deck scaling. It follows from triple-deck theory that, at finite Re, the separation point x_0 shifts downstream with $x_0 - \widehat{x}_0 = O(\mathrm{Re}^{-1/16})$ compared to the break-away point $\widehat{x}_0$ of the limiting free streamline. The pressure distribution in front of the separation point x_0 is

$$p - p_0 = -c_0(x_0 - x)^{1/2} + O((x_0 - x)) \text{ for } x < x_0 \text{ and } x \downarrow x_0,$$
(4.102a)

with

$$c_0 = 0.44\Delta^{9/8} - \mathrm{Re}^{-1/16} \tag{4.102b}$$

where Δ describes the wall gradient of the approaching flow boundary layer, see F.T. Smith (1977).

(3) A new inviscid limiting solution has to be determined where the free streamline leaves the body at x_0 (and no longer at $\hat{x}_0$). Cheng and F.T. Smith (1982) also presented an approximate method for computing this limiting solution; see also Cheng and Lee (1986). As yet, there is no complete theory for the asymptotic description of the flow in the so-called far field [but see, for instance, F.T. Smith (1979, 1986)], for example, close to the reattachment point.

In Guiraud and Zeytounian (1979a), steady-state flow with a vortex sheet separating from a smooth body is considered, and a criterion is derived for the location of the separation line. This criterion generalizes that derived (for incompressible irrotational flow) by J.H.B. Smith (1977) and F.T. Smith (1977, 1978) to compressible flow for an ideal gas with constant specific heats. There are two main limitations: the first is that the flow has to be steady-state; the second is that the component of the wall velocity normal to the separation line has to be subsonic. As in the work referred to above, the criterion is derived from the *compatibility of a perfect fluid analysis with a triple-deck simulation of the laminar boundary layer near the separation line (à la Sychev)*; the wall temperature is either adiabatic or at constant temperature.

4.5.1 Three-Dimensional Boundary-Layer Equations

Consider NSF steady-state flow over a solid body with a regular surface (wall) Σ and any point M of the flow. Let us draw through M a normal to the surface Σ meeting it in P_Σ. Then the position of M in 3-D space near Σ, can be expressed in terms of the position of P_Σ and the distance $n = P_\Sigma M$ along the normal to the surface Σ. We assume that $\boldsymbol{n}$ is the unit normal vector to Σ (in the direction of the fluid region) and $\boldsymbol{s} = (s_1, s_2)$ defines a system of Lagrangian curvilinear coordinates on the surface Σ. More precisely, we assume that the intrinsic surface coordinates (s_1, s_2) are aligned with the principal directions of curvature at each point of Σ and parametrize the wall of the solid body considered.

From (x_1, x_2, x_3), we pass to a local dimensionless coordinate system (s_1, s_2, n), and in the steady-state, if $\boldsymbol{M}$ is the position vector of the point M in 3-D space, then, for the *point near the wall* Σ, we write

$$\boldsymbol{M} = \boldsymbol{P}_\Sigma(\boldsymbol{s}) + n\boldsymbol{n}(\boldsymbol{s}) , \tag{4.103}$$

so that (s_1, s_2, n) may be taken as a system of curvilinear coordinates of point M near the wall of the solid body considered. Near the wall Σ, it is convenient (at least in high Reynolds number asymptotics) to introduce, in place of n, the normal to the wall coordinate

$$N = \mathrm{Re}^\alpha\, n, \quad \alpha > 0,\ \mathrm{Re} \equiv 1/\varepsilon^2\ . \tag{4.104}$$

Finally, near the wall Σ of the given solid body, we work with the system of curvilinear coordinates (s_1, s_2, N), and the wall Σ is defined by the position vector $\boldsymbol{P}_\Sigma = \boldsymbol{P}_\Sigma(s_1, s_2)$ for steady-state fluid flow at high Reynolds numbers.

Now, this system of coordinates will be used below to form *steady-state Prandtl boundary-layer* equations, when

$$\varepsilon \to 0,\ \text{with}\, s_1, s_2, N\ \text{fixed}\ , \tag{4.105}$$

from the steady-state NSF (dominant) equations written with $(\boldsymbol{s}, N)$ in place of Cartesian coordinates (x_1, x_2, x_3).

To derive Prandtl limiting steady-state approximate equations, it is more convenient to introduce the following representation for velocity vector $\boldsymbol{u}$:

$$\boldsymbol{u} = \boldsymbol{v} + w\boldsymbol{n},\ \boldsymbol{v} \cdot \boldsymbol{n} = 0\ , \tag{4.106a}$$

and then, we have the following no-slip condition:

$$\boldsymbol{v} = 0\ \text{and}\ w = 0\ \text{on}\ \Sigma\ . \tag{4.106b}$$

Again, near the wall Σ, it is convenient (in fact, "necessary") to introduce a new vertical component of the velocity in place of w:

$$w^* = \mathrm{Re}^\beta\, w,\quad \beta > 0\ . \tag{4.107}$$

Now, we consider the field $U^* = [\boldsymbol{v}^*, w^*, p^*, \rho^*, T^*] = U^*(\boldsymbol{s}, N; \varepsilon)$, and a simple calculation in the steady-state gives,

$$S\, D/Dt \Rightarrow \boldsymbol{v}^* \cdot \mathbf{D} + \varepsilon^{2(\beta-\alpha)} w^* \frac{\partial}{\partial N} + O(\varepsilon^{2\alpha})\ , \tag{4.108}$$

where $\mathbf{D}$ is the *gradient operator on the wall* Σ and it is necessary to define $\mathbf{D}$.

Dominant NSF equations near the wall Σ. Now, if we work with the above gradient operator $\mathbf{D}$ defined on wall Σ, then it can be shown that

$$\boldsymbol{\nabla} \cdot \boldsymbol{u} = \mathbf{D} \cdot \boldsymbol{v}^* + \varepsilon^{2(\beta-\alpha)} \frac{\partial w^*}{\partial N} + O(\varepsilon^{2\alpha})\ ; \tag{4.109a}$$

$$(\boldsymbol{\nabla}\boldsymbol{u}) + (\boldsymbol{\nabla}\boldsymbol{u})^{\mathrm{T}} - \frac{2}{3}(\boldsymbol{\nabla} \cdot \boldsymbol{u})\mathbf{1}$$
$$= \frac{1}{\varepsilon^{2\alpha}}\left[\frac{\partial \boldsymbol{v}^*}{\partial N} \otimes \boldsymbol{n} + \boldsymbol{n} \otimes \frac{\partial \boldsymbol{v}^*}{\partial N}\right] + \mathbf{O}(1) + \mathbf{O}[\varepsilon^{2(\beta-\alpha)}]\ , \tag{4.109b}$$

where the dyadic product $(\boldsymbol{f} \otimes \boldsymbol{g})$ of two vectors in (4.109b) generates a second-order tensor and $\mathbf{O}[\varepsilon^{2(\beta-\alpha)}]$ is a "small" second-order tensor that tends to zero, as $\varepsilon \to 0$.

132   4 The Limit of Very Large Reynolds Numbers

In conclusion, with the above relations, we derive the following "dominant" NSF steady-state equations (with Bo = 0), valid for high Reynolds number *steady-state flow* near a 3-D solid body wall Σ:

$$\boldsymbol{v}^* \cdot \mathbf{D}\rho^* + \rho^* \mathbf{D} \cdot \boldsymbol{v}^* + \varepsilon^{2(\beta-\alpha)} \left[\rho^* \frac{\partial w^*}{\partial N} + w^* \frac{\partial \rho^*}{\partial N} \right] = O(\varepsilon^{2\alpha}) ; \quad (4.110\mathrm{a})$$

$$\rho^*(\boldsymbol{v}^* \cdot \mathbf{D})\boldsymbol{v}^* + \varepsilon^{2(\beta-\alpha)} \rho^* w^* \frac{\partial \boldsymbol{v}^*}{\partial N} + \frac{1}{\gamma M^2} \mathbf{D} p^* - \varepsilon^{2(1-2\alpha)} \frac{\partial}{\partial N} \left(\mu^* \frac{\partial \boldsymbol{v}^*}{\partial N} \right)$$

$$= \mathbf{O}(\varepsilon^{2\beta}) + \mathbf{O}(\varepsilon^{2\alpha}) + \mathbf{O}(\varepsilon^{2(1-\alpha)}) ; \quad (4.110\mathrm{b})$$

$$\frac{\partial p^*}{\partial N} = \mathbf{O}(\varepsilon^{2\alpha}) + \mathbf{O}(\varepsilon^2) ; \quad (4.110\mathrm{c})$$

$$\rho^* \boldsymbol{v}^* \cdot \mathbf{D} T^* + (\gamma - 1) p^* \left[\mathbf{D} \cdot \boldsymbol{v}^* + \varepsilon^{2(\beta-\alpha)} \frac{\partial w^*}{\partial N} \right] + \varepsilon^{2(\beta-\alpha)} \rho^* w^* \frac{\partial T^*}{\partial N}$$

$$- \varepsilon^{2(1-2\alpha)} \left[\gamma(\gamma-1) M^2 \mu^* \left| \frac{\partial \boldsymbol{v}^*}{\partial N} \right|^2 \right] + \frac{\gamma}{\mathrm{Pr}} \frac{\partial}{\partial N} \left(k^* \frac{\partial T^*}{\partial N} \right)$$

$$= \mathbf{O}(\varepsilon^{2\alpha}) + \mathbf{O}(\varepsilon^{2(1-\alpha)}) ; \quad (4.110\mathrm{d})$$

$$p^* - T^* \rho^* = 0 . \quad (4.110\mathrm{e})$$

From the above "dominant" NSF steady-state equations (4.110a–e), we can now elucidate the Prandtl degeneracy of these steady-state "dominant" equations in relation to the two positive scalars, α and β.

The Prandtl (inner) limit and Prandtl (BL) equations. First, from the equation of continuity (4.110a), it is clear that significant degeneracy corresponds to

$$\alpha = \beta . \quad (4.111\mathrm{a})$$

Next, from the equations (4.110b,d), we derive viscous and thermally conducting significant degeneracy (near the wall) if and only if

$$\alpha = 1/2 . \quad (4.111\mathrm{b})$$

Local degeneracy corresponding to $\alpha = \beta = 1/2$ is the so-called "*Prandtl degeneracy*," and consequently, we derive the following *Prandtl inner boundary-layer equations*:

$$\mathbf{D} \cdot (\rho_p \boldsymbol{v}_p) + \frac{\partial(\rho_p w_p)}{\partial N} = 0 ;$$

$$\rho_p \left[(\boldsymbol{v}_p \cdot \mathbf{D}) \boldsymbol{v}_p + w_p \frac{\partial \boldsymbol{v}_p}{\partial N} \right] + \frac{1}{\gamma M^2} \mathbf{D} p_e = \frac{\partial}{\partial N} \left(\mu_p \frac{\partial \boldsymbol{v}_p}{\partial N} \right) ;$$

$$\frac{\partial p_p}{\partial N} = 0, \; p_p \equiv p_e(\boldsymbol{s}) \; ;$$

$$\rho_p \left[\boldsymbol{v}_p \cdot \mathbf{D} T_p + w_p \frac{\partial T_p}{\partial N} \right] + (\gamma - 1) p_p \left[\mathbf{D} \cdot \boldsymbol{v}_p + \frac{\partial w_p}{\partial N} \right]$$

$$= \gamma(\gamma - 1) M^2 \mu_p \left| \frac{\partial \boldsymbol{v}_p}{\partial N} \right|^2 + \frac{\gamma}{\Pr} \frac{\partial}{\partial N} \left(k_p \frac{\partial T_p}{\partial N} \right) ,$$

$$p_e = T_p \rho_p , \tag{4.112}$$

where $p_e(\tau, \boldsymbol{s})$ is given by matching with the Euler outer flow.

The above Prandtl inner boundary-layer, limiting steady-state equations (4.112) can be derived from the system of "dominant" NSF equations (4.110a–e) by the following *Prandtl inner (s, N fixed) steady-state limit*:

$$\mathrm{Lim}^P = [\varepsilon \to 0; \; \boldsymbol{s}, N, M, \Pr, \gamma, \text{ fixed } = O(1)] , \tag{4.113a}$$

and in this case,

$$U_p = [\boldsymbol{v}_p, w_p, p_p, \rho_p, T_p] = \mathrm{Lim}^P U^*(\boldsymbol{s}, N; \varepsilon) , \tag{4.113b}$$

and $\mu_p = \mu(T_p)$, $k_p = k(T_p)$. More precisely, in the boundary layer, we consider the following inner (with $\boldsymbol{s}$ and $N = n/\varepsilon$ fixed) asymptotic expansions:

$$\boldsymbol{v} = \boldsymbol{v}_p + O(\varepsilon), \; w = \varepsilon w_p + o(\varepsilon) , \tag{4.113c}$$

$$(p, \rho, T) = (p_p, \rho_p, T_p) + O(\varepsilon) , \tag{4.113d}$$

where $[\boldsymbol{v}_p, w_p, p_p, \rho_p, T_p] = U_p$ are functions of $\boldsymbol{s}$ and N and they satisfy the Prandtl, leading-order boundary-layer equations (4.112). We note that $O(\varepsilon) \to 0$ as $\varepsilon \to 0$ and $o(\varepsilon)/\varepsilon \to 0$ as $\varepsilon \to 0$.

The Prandtl BL equations in a system of curvilinear coordinates locally orthogonal on wall Σ. For many practical purposes, it is convenient to choose an orthogonal curvilinear coordinates s_1, s_2 system on surface Σ. Note that the three surfaces $s_1 = $ const, $s_2 = $ const, and $N = $ const are *not* necessarily orthogonal everywhere in space. Of course, they are *locally orthogonal* at Σ but only when the curves $s_1 = $ const and $s_2 = $ const on Σ coincide with the *lines of curvatures* of Σ do the *three families of surfaces become orthogonal everywhere*. However, the boundary layer is a local phenomenon on the surface of Σ, and it is quite sufficient for our purpose to have the system of curvilinear coordinates locally orthogonal at Σ [see Rosenhead (1963; pp. 412–414)]. Therefore, when the curves $s_1 \equiv \xi = $ const and $s_2 \equiv \eta = $ const are orthogonal on Σ, we write

$$\boldsymbol{v}_p = u_P \boldsymbol{e}_1 + v_p \boldsymbol{e}_2 \,, \tag{4.114}$$

where

$$\boldsymbol{e}_1 = \frac{1}{h_1}\frac{\partial \boldsymbol{P}_\Sigma}{\partial \xi}, \quad \boldsymbol{e}_2 = \frac{1}{h_2}\frac{\partial \boldsymbol{P}_\Sigma}{\partial \eta} \,, \tag{4.115a}$$

$$h_1 = \left|\frac{\partial \boldsymbol{P}_\Sigma}{\partial \xi}\right|, \quad h_2 = \left|\frac{\partial \boldsymbol{P}_\Sigma}{\partial \eta}\right| \,. \tag{4.115b}$$

In (4.114) $\boldsymbol{e}_1$ and $\boldsymbol{e}_2$ are unit tangential vectors, respectively, to the two parametric curves on Σ, and $(\boldsymbol{e}_1, \boldsymbol{e}_2, \boldsymbol{n})$ form an orthogonal triad of unit vectors. Because Σ is supposed to be *smoothly curved*, it follows that h_1, h_2, their derivatives with respect to ξ and η, and the derivatives of $\boldsymbol{e}_1$, $\boldsymbol{e}_2$, and $\boldsymbol{n}$ are all $O(1)$. If we assume that $\boldsymbol{P}_\Sigma$ is independent of time, $\boldsymbol{P}_\Sigma = \boldsymbol{P}_\Sigma(\xi, \eta)$, then all of these quantities are, of course, functions only of ξ and η only. In the boundary-layer equations (4.112), with (4.114), (4.115a,b), the gradient operator on Σ takes the form,

$$\mathbf{D} = \frac{1}{h_1}\frac{\partial}{\partial \xi}\boldsymbol{e}_1 + \frac{1}{h_2}\frac{\partial}{\partial \eta}\boldsymbol{e}_2 \tag{4.116a}$$

and with ζ in place of N,

$$\boldsymbol{v}_p \cdot \mathbf{D} + w_p\frac{\partial}{\partial N} \equiv \frac{u_p}{h_1}\frac{\partial}{\partial \xi} + \frac{v_p}{h_2}\frac{\partial}{\partial \eta} + w_p\frac{\partial}{\partial \zeta} \,. \tag{4.116b}$$

Finally, in terms of the local coordinates ξ, η, ζ, we derive the following boundary-layer equations for u_p, v_p, w_p, ρ_p, T_p, in place of system (4.112):

$$\frac{1}{h_1 h_2}\frac{\partial(\rho_p u_p h_2)}{\partial \xi} + \frac{1}{h_1 h_2}\frac{\partial(\rho_p v_p h_1)}{\partial \eta} + \frac{\partial(\rho_p w_p)}{\partial \zeta} = 0 \,,$$

$$\frac{u_p}{h_1}\frac{\partial u_p}{\partial \xi} + \frac{v_p}{h_2}\frac{\partial u_p}{\partial \eta} + w_p\frac{\partial u_p}{\partial \zeta} + \left[\frac{1}{h_1 h_2}\frac{\partial h_1}{\partial \eta}\right] u_p v_p - \left[\frac{1}{h_1 h_2}\frac{\partial h_2}{\partial \xi}\right](v_p)^2$$

$$= -\frac{1}{\gamma M^2}\frac{1}{\rho_p h_1}\frac{\partial p_e}{\partial \xi} + \frac{1}{\rho_p}\frac{\partial}{\partial \zeta}\left(\mu_p \frac{\partial u_p}{\partial \zeta}\right) \,,$$

$$\frac{u_p}{h_1}\frac{\partial v_p}{\partial \xi} + \frac{v_p}{h_2}\frac{\partial v_p}{\partial \eta} + w_p\frac{\partial v_p}{\partial \zeta} - \left[\frac{1}{h_1 h_2}\frac{\partial h_1}{\partial \eta}\right](u_p)^2 + \left[\frac{1}{h_1 h_2}\frac{\partial h_2}{\partial \xi}\right] u_p v_p$$

$$= -\frac{1}{\gamma M^2}\frac{1}{\rho_p h_2}\frac{\partial p_e}{\partial \eta} + \frac{1}{\rho_p}\frac{\partial}{\partial \zeta}\left(\mu_p \frac{\partial v_p}{\partial \zeta}\right) \,,$$

$$\rho_p\left[\frac{u_p}{h_1}\frac{\partial T_p}{\partial \xi} + \frac{v_p}{h_2}\frac{\partial T_p}{\partial \eta} + w_p\frac{\partial T_p}{\partial \zeta}\right]$$

$$+ (\gamma-1)p_e\left[\frac{1}{h_1 h_2}\left\{\frac{\partial(u_p h_2)}{\partial \xi} + \frac{\partial(v_p h_1)}{\partial \eta}\right\} + \frac{\partial w_p}{\partial \zeta}\right]$$

$$= \gamma(\gamma-1)M^2 \mu_p\left[\left(\frac{\partial u_p}{\partial \zeta}\right)^2 + \left(\frac{\partial v_p}{\partial \zeta}\right)^2\right] + \frac{\gamma}{\Pr}\frac{\partial}{\partial \zeta}\left(k_p \frac{\partial T_p}{\partial \zeta}\right) \,, \tag{4.117a}$$

$$p_e = T_p \rho_p \,, \tag{4.117b}$$

where $p_p \equiv p_e(\xi, \eta)$.

Unlike the two-dimensional case ($v_p = 0$ and $\partial/\partial\eta = 0$), the boundary-layer equations (4.117) contain terms $\partial h_1/\partial \eta$ and $\partial h_2/\partial \xi$ that depend explicitly on the *curvature* of the coordinate system. We may notice that if it is possible to choose a coordinate system such that both $\partial h_1/\partial \eta$ and $\partial h_2/\partial \xi$ *vanish everywhere on the surface*, then the *Gaus*sian *curvature* K of the surface Σ which is given by

$$K = -h_1 h_2 \left\{ \frac{\partial}{\partial \xi} \left[\frac{1}{h_1} \frac{\partial h_2}{\partial \xi} \right] + \frac{\partial}{\partial \eta} \left[\frac{1}{h_2} \frac{\partial h_1}{\partial \eta} \right] \right\} \,,$$

vanishes, and so surface Σ must be *developable*. Conversely [see Howarth (1959, pp. 309 and 310)], if surface Σ is developable, the parametric curves can be chosen so that $h_1 = h_2 = 1$, and the curvature terms *disappear* from the equations of motion. This is a point of some practical importance.

Euler–Prandtl matching. To complete the formulation of the system of Prandtl boundary-layer equations (4.117), it is necessary to specify appropriate boundary conditions.

Obviously, at the wall $\zeta = 0$ (which is the inside of the domain of validity of the boundary-layer equations), assumed impermeable (and neglecting slip effects),

$$u_p = v_p = w_p = 0 \quad \text{and} \quad T_p = T_w \text{ on } \zeta = 0 \,, \tag{4.118a}$$

where $T_w > 0$ is a known function of (ξ, η) (a prescribed wall temperature). A permeable wall means that w_p takes a value (usually prescribed) there. Slip effects are strictly outside the scope of continuum mechanics and are related to the so-called Knudsen layer.

The appropriate boundary conditions, as $\zeta \to \infty$, require a little care! According to (4.113a,b) and (4.114), we consider the following inner (with ξ, η and $\zeta = n/\varepsilon$ fixed) asymptotic expansions in the boundary layer:

$$(u, v) = (u_p, v_p) + O(\varepsilon), \quad w = \varepsilon w_p + o(\varepsilon) \,,$$
$$(p, \rho, T) = (p_p, \rho_p, T_p) + O(\varepsilon) \,,$$

where u_P, v_P, w_P, p_p, ρ_p, T_p are functions of ξ, η, and $\zeta = n/\varepsilon$.

In inviscid Euler flow, we consider the following outer (with ξ, η, and n fixed) asymptotic expansions:

$$(u, v, w, p, \rho, T) = (u_E, v_E, w_E, p_E, \rho_E, T_E) + O(\varepsilon) \,,$$

where u_E, v_E, w_E, p_E, ρ_E, T_E are functions of ξ, η, and n. These functions (with subscript "$_E$") satisfy the Euler equations written in curvilinear coordinates ξ, η, and n. Now, according to the Prandtl asymptotic matching principle, it is necessary, first, to calculate the limiting values of u_E, v_E, w_E, p_E, ρ_E, and T_E, when $n \to 0$. Consequently, we obtain the values of inviscid flow "above" (just outside) the boundary layer:

$$\lim_{n \to 0}(u_E, v_E, w_E, p_E, \rho_E, T_E) = (u_e, v_e, w_e, p_e, \rho_e, T_e),$$

and these limiting values $(u_e, v_e, w_e, p_e, \rho_e, T_e)$ are functions only of ξ and η. But, as a consequence of the above inner asymptotic expansion for w $(= \varepsilon w_p + o(\varepsilon))$, it is obvious that

$$w_e = 0 \Rightarrow w_e = 0 \text{ on } n = 0. \tag{4.119}$$

Boundary condition (4.119) is the steady-state slip condition on the wall, $n = 0$, for the inviscid Euler equations (written in curvilinear coordinates, ξ, η, and n). Then, the following outer expansion for w is obtained: $w = w_e + \varepsilon w_1 + \ldots$, and

$$w_p \to w_1|_{n=0}, \text{ as } \zeta \to \infty;, \tag{4.118b}$$

where w_1 is the second-order, outer, vertical velocity (in the direction of the normal $\bm{n}$ to the body wall Σ) in the outer expansion (a function of ξ, η, and n). The matching condition (4.118b) gives an "interactive" relation between the Prandtl boundary-layer solution and the second-order linear Eulerian outer flow. Because the pressure $p_p = p_e$ in the boundary layer is independent of ζ, then it is determined by its value at $n = 0$ according to the inviscid solution,

$$\frac{u_e}{h_1}\frac{\partial u_e}{\partial \xi} + \frac{v_e}{h_2}\frac{\partial u_e}{\partial \eta} + \left[\frac{1}{h_1 h_2}\frac{\partial h_1}{\partial \eta}\right] u_e v_e - \left[\frac{1}{h_1 h_2}\frac{\partial h_2}{\partial \xi}\right](v_e)^2$$
$$= -\frac{1}{h_1 \gamma M^2}\frac{1}{\rho_e}\frac{\partial p_e}{\partial \xi},$$
$$\frac{u_e}{h_1}\frac{\partial v_e}{\partial \xi} + \frac{v_e}{h_2}\frac{\partial v_e}{\partial \eta} - \left[\frac{1}{h_1 h_2}\frac{\partial h_1}{\partial \eta}\right](u_e)^2 + \left[\frac{1}{h_1 h_2}\frac{\partial h_2}{\partial \xi}\right] u_e v_e$$
$$= -\frac{1}{h_2 \gamma M^2}\frac{1}{\rho_e}\frac{\partial p_e}{\partial \eta},$$

with

$$\rho_p T_p = T_e \rho_e.$$

Thus, at "infinity", we write the following conditions for the Prandtl boundary-layer equations (4.117) for u_p, v_p, T_P,, ρ_p (as a consequence of matching with the Euler solution):

$$(u_p, v_p, T_p, \rho_p) \to (u_e, v_e, T_e, \rho_e) \text{ as } \zeta \to +\infty. \tag{4.118c}$$

We note that if

$$(u_p, v_p, T_p, \rho_p) = (u_p, v_p, T_p, \rho_p)^\infty + \mathrm{Exp}(\zeta), \ \zeta \to +\infty\ , \tag{4.120a}$$

where

$$\mathrm{Exp}(\zeta)/O(1/\zeta) \to 0 \text{ when } \zeta \to +\infty\ ,$$

then from the Van Dyke (intermediate) matching principle, we derive again matching conditions (4.118c). The slip condition (4.119) for the Euler equations is also a consequence of the intermediate matching condition. Finally, we note that we derive the following "compatibility" equation from the Euler equation for T_e:

$$\left\{ \frac{u_e}{h_1} \frac{\partial}{\partial \xi} + \frac{v_e}{h_2} \frac{\partial}{\partial \eta} \right\} \left[\frac{T_e}{(\gamma - 1)\rho_e^{(\gamma-1)}} \right] = 0\ . \tag{4.121}$$

Now, if we take into account the following behavior (at infinity) of the vertical velocity component in the boundary layer [as a consequence of the continuity equation in (4.117)]:

$$w_p = w_{p,1}\zeta + (w_p)^\infty + \mathrm{Exp}(\zeta), \text{ when } \zeta \to \infty\ , \tag{4.120b}$$

then the limit boundary-layer functions $(u_p, v_p, T_p, \rho_p)^\infty$, in (4.120a), also satisfy an equation analogous to (4.121).

Again, we derive the following relation from the continuity equation [first equation in (4.117)] for the Prandtl boundary layer and (4.120b):

$$(\rho_p w_p)^\infty = \mathbf{D} \cdot \int_0^\infty [(\rho_p \boldsymbol{v}_p)^\infty - \rho_p \boldsymbol{v}_p]\, \mathrm{d}\zeta\ , \tag{4.122}$$

where $\boldsymbol{v}_P = u_p \boldsymbol{e}_1 + v_p \boldsymbol{e}_2$ and $\mathbf{D}$ is defined by (4.116a). Relation (4.122) is closely linked with the displacement thickness of the leading-order (Prandtl) boundary layer.

Concerning the second-order BL effects for compressible 3-D stagnation-point flow see Papenfuß (1977). Finally, in Eichelbrenner (1973) the reader can find a survey concerning the 3-D boundary layers.

4.5.2 Unsteady-State Incompressible Boundary-Layer Formulation

In Zeytounian "ONERA" Technical Note (1968a), the reader can find a "contribution to the study of the tridimensional incompressible laminar boundary layer in the unsteady-state regime". More precisely, we consider the temporal evolution of a laminar incompressible boundary layer from rest. The external inviscid flow is assumed known as a power series expansion of $t^{1/2}$ and in this case, correspondingly, the solution of the unsteady-state boundary layer is also determined in a series of $t^{1/2}$. According to the so-called "prevalence" principle, developed by Eichelbrenner (1959) and adapted for the study of unsteady-state boundary layer equations by Eichelbrenner and Askovic (1967), we can write the following initial boundary-value problem for velocity components u_{BL}, v_{BL}, w_{BL} in the unsteady-state boundary layer:

$$\frac{\partial u_{\text{BL}}}{\partial t} + u_{\text{BL}}\frac{\partial u_{\text{BL}}}{\partial s} + v_{\text{BL}}\frac{\partial u_{\text{BL}}}{\partial N} = \frac{\mathrm{d}f(t)}{\mathrm{d}t}u_e + f(t)^2 u_e \frac{\partial u_e}{\partial s} + \frac{\partial^2 u_{\text{BL}}}{\partial N^2};$$

$$\frac{\partial u_{\text{BL}}}{\partial s} + \frac{\partial v_{\text{BL}}}{\partial N} - K_2 u_{\text{BL}} = 0;$$

$$\frac{\partial w_{\text{BL}}}{\partial t} + u_{\text{BL}}\frac{\partial w_{\text{BL}}}{\partial s} + v_{\text{BL}}\frac{\partial w_{\text{BL}}}{\partial N} + K_1 u_{\text{BL}}^2 - K_2 w_{\text{BL}} u_{\text{BL}}$$

$$= K_1 f(t)^2 u_e^2 + \frac{\partial^2 w_{\text{BL}}}{\partial N^2}; \qquad (4.123\text{a})$$

$$N = 0 \text{ and } t > 0 \;:\; u_{\text{BL}} = w_{\text{BL}} = v_{\text{BL}} = 0,$$
$$N \uparrow \infty \;:\; u_{\text{BL}} \to f(t)u_e \text{ and } w \to 0,$$
$$t = 0 \text{ and } N \geq 0 \;:\; u_{\text{BL}} = w_{\text{BL}} = v_{\text{BL}} = 0, \qquad (4.123\text{b})$$

where $f(t)u_e$ is the given velocity of the outer flow (at the edge of the boundary layer) and $f(0) = 0$. Usually, it is judicious to introduce the new variable,

$$y = N/2\tau \text{ with } \tau = t^{1/2}, \qquad (4.124\text{a})$$
$$\text{and } f(t) = f(\tau^2) \equiv f^*(\tau). \qquad (4.124\text{b})$$

Then, for the functions u^*, w^*, and v^* such that

$$u_{\text{BL}} = u^*(s,y,\tau), \; w_{\text{BL}} = w^*(s,y,\tau) \qquad (4.125\text{a})$$
$$\text{and } v_{\text{BL}} = (1/2)\tau v^*(s,y,\tau), \qquad (4.125\text{b})$$

we derive, the following system of equations for u^*, w^* and v^* in place of the boundary-layer equations (4.123a):

$$\Lambda_0(u^*) = 2\tau \frac{\partial u^*}{\partial \tau} - 2u_e \tau \frac{\mathrm{d}f^*}{\mathrm{d}\tau} + 4\tau^2 \left[u^* \frac{\partial u^*}{\partial s} + v^* \frac{\partial u^*}{\partial y} - f^*(\tau)^2 u_e \frac{\partial u_e}{\partial s} \right],$$
$$\qquad (4.126\text{a})$$

$$\frac{\partial u^*}{\partial s} + \frac{\partial v^*}{\partial y} - K_2 u^* = 0, \qquad (4.126\text{b})$$

$$\Lambda_0(w^*) = 2\tau \frac{\partial w^*}{\partial \tau} + 4\tau^2 \left[u^* \frac{\partial w^*}{\partial s} + v^* \frac{\partial w^*}{\partial y} - K_2 w^* u^* \right.$$
$$\left. + K_1 \left(u^{*2} - f^*(\tau)^2 u_e^2 \right) \right], \qquad (4.126\text{c})$$

where the differential operator $\Lambda_0(g)$ is given by

$$\Lambda_m(g) = g'' + 2yg' - 2mg ,\quad (4.127)$$

when $m = 0$, and g is considered as a function of y.

With (4.124a,b), in place of initial and boundary conditions (4.123b), we obtain for the equations (4.126a–c),

$$y = 0 : u^* = w^* = v^* = 0 ; \quad (4.128a)$$
$$y \uparrow \infty : u^* \to f^*(\tau)u_e \text{ and } w^* \to 0 , \quad (4.128b)$$

and we observe that $y \uparrow \infty$ is equivalent to $\tau = 0$ (for N fixed). The initial condition for v_{BL} is satisfied according to (4.125b) when $|v^*(s,y,0)|$ is bounded. In Zeytounian (1968), the reader can find a careful investigation of solutions of the following problem:

$$\Lambda_m(g) = \phi(y),\ y = 0:\ g(0) = a^0 \text{ and } y \uparrow \infty :\ g(\infty) = 0 , \quad (4.129)$$

where the function $\phi(y)$ is expressed in terms of functions $\mathcal{L}_s(y)$, ($s = 1, 2, \ldots, p$). These $\mathcal{L}_s(y)$ functions are solution of the equation,

$$\Lambda_s(g) = 0 \text{ with } g(0) = 1 \text{ and } g(\infty) = 0 . \quad (4.130)$$

Various solutions corresponding to the given function $f^*(\tau)$ have been obtained and also a simple application for a circular cylinder. In Chap. 7, Sect. 7.7.2, we give some complementary information concerning these unsteady-state boundary layers developing from rest. Chapter 13 in the recent book by Schlichting and Gersten (2000) is devoted to unsteady-state boundary layers, and the reader can find various applications [similar and semi-similar solutions, solutions for small times (high frequencies), separation, start-up processes, oscillation of bodies in a fluid at rest, and compressible unsteady-state boundary layers. Finally, we note that Duck and Dry (2001) identified and investigated a number of aspects of the effects of a class of three-dimensional, unsteady-state disturbances, defined by

$$u_p = U(\xi,\varsigma,t),\ v_p = V(\xi,\varsigma,t),\ w_p = \eta V(\xi,\varsigma,t),\ p_p = p_e(\xi,t) , \quad (4.131)$$

that is, with a linearly growing cross flow in the cross flow direction, on a base state possessing the same spanwise variation. This form of disturbance was chosen deliberately by the authors partly to ensure that the analysis was completely rational and also to reduce the dimension of the problem by one. In that paper, the reader can find recent references related to studies of diverse facets of unsteady-state boundary layers (see also Chap. 7, Sect. 7.7.2, devoted to some complementary remarks relative to features of unsteady-state boundary-layer problems). To unsteady-state BL is devoted also the proceedings of the IUTAM Symposium 1971 (Quèbec, Canada), edited by Eichelbrenner (1972).

4.5.3 The Inviscid Limit: Some Mathematical Results

An interesting but difficult mathematical problem arising quite naturally is strongly related to the behavior of Navier, incompressible, viscous solutions in the vanishing viscosity limit

$$\nu_0 \to 0 \,!$$

It is a well-known question whether, as $\nu_0 \to 0$, the solution of the Navier problem, $\boldsymbol{u}_N$, for viscous incompressible fluid flow tends to the solution $\boldsymbol{u}_E$ of the corresponding problem for Euler incompressible nonviscous equations, in which the viscous term is omitted, and the Navier no-slip boundary condition $[\boldsymbol{u}_N(t, \boldsymbol{P}) = 0$, with $\boldsymbol{P} \in \partial\Omega]$ is weakened to Euler slip condition:

$$\boldsymbol{n} \cdot \boldsymbol{u}_E(t, \boldsymbol{P}) = 0, \quad \boldsymbol{P} \in \partial\Omega, \tag{4.132}$$

where $\boldsymbol{n}$ is the outer normal to $\partial\Omega$. In the absence of a boundary, it is easy to prove rigorously that $\boldsymbol{u}_N \to \boldsymbol{u}_E$ in L^2 (L^2 convergence) when $\nu_0 \to 0$. The solutions of the Navier and Euler problems $\boldsymbol{u}_N$ and $\boldsymbol{u}_E$, respectively, are both associated with the fixed initial value (for $t = 0$) $\boldsymbol{u}^0$. We consider the inviscid limit, that is, we fix time t and let the control parameter (Reynolds number, Re) tend to infinity. Let us focus on the Navier equation (written with dimensionless quantities),

$$\frac{\partial \boldsymbol{u}}{\partial t} + (\boldsymbol{u} \cdot \boldsymbol{\nabla})\boldsymbol{u} + \boldsymbol{\nabla}p = \frac{1}{\mathrm{Re}} \Delta \boldsymbol{u} \,. \tag{4.133}$$

When $\Omega = R^3$ and for an interval of time depending on the initial values but not the viscosity, the problem was solved by Swann (1971), and Kato (1972). See also Kato (1975), Marchioro and Pulvirenti (1984), and Esposito et al. (1988). In the last two papers, the proof is based on the convergence of stochastic processes describing the Navier flow.

But the problem with boundaries remains a famous challenge. Obviously, the behavior of the solution in the vanishing viscosity limit when boundaries are present is much more complicated (this problem is singular!) and not yet completely mathematically understood. A rigorous mathematical theory of the boundary-layer problem is far from being achieved, and an analysis of the existing results is beyond the purposes of this chapter. It seems that the key to the analysis is to construct solutions of the Navier equations incorporating an appropriate boundary approximation of the Prandtl type [see, for instance, Asano (1989)]. Thus, the result is important as a step in the rigorous justification of classical boundary-layer theory.

However, the problem still remains open for more general boundaries because Asano's construction assumes that the boundary is a plane. On the other hand, according to a short but pertinent paper by Constantin (1995; pp. 661–662), the first question is: what are the limiting equations? In realistic closed systems where boundary effects are important, unstable boundary layers drive the system, and the inviscid limit is not well understood. When there are no boundaries (periodic solutions or solutions decaying at infinity), the issue becomes one of smoothness and rates of convergence. In 2-D, if the initial data are very smooth, then the limit is the Euler incompressible 2-D equation, and the difference between Navier solutions and corresponding Euler solutions is optimally small $[O(1/\operatorname{Re})]$. However, if the initial data are not that smooth, for instance, in the case of vortex patches, then the situation changes. Vortex patches are solutions whose vorticity (the antisymmetric part of the gradient) is a step function. They are the building blocks for the phase space of an important statistical theory [see, for example, the papers by Miller (1990) and Robert (1991)]. When one leaves the realm of smooth initial data, the inviscid limit becomes more complicated. Internal transition layers form because the smoothing effect present in the Navier solution is absent in the Eulerian solution. In the case of vortex patches with smooth boundaries, the inviscid limit is still the Euler equation, but there is a definite price to pay for rougher data:

the difference between solutions (in L^2) is only $O[(1/\operatorname{Re})^{1/2}]$,

as shown in Constantin and Wu (1994). This drop in the rate of convergence actually occurs. Exact solutions exist providing lower bounds. The question of the inviscid limit for the whole phase space of the statistical theory of Miller (1990) and Robert (1991) seems open! If the initial data are more singular, then even the classic notion of weak solutions for the Euler equation might need revision [see, Diperna and Majda (1987)], except when the vorticity is of one sign [see, Delort (1991)].

In many interesting flows, the generation of vorticity occurs on very small sets, typically at the surface of an obstacle, if the viscosity is low, as time goes on, the support of the vorticity remains small, as for wakes or detached boundary layers. It is therefore natural [according to Cottet (1992)] to study the Navier equation with an initial vorticity concentrated on a set of Lebesgue measure zero. However, when the viscosity tends to zero, we have to deal with a singular perturbation problem. The a priori estimates available for the Navier equation obviously fail to be uniform with respect to viscosity. If one has available a model for generating vorticity in the limit case of vanishing viscosity and expects the Euler equations to provide a good model for the evolution of this vorticity, the question of the well-posedness of these equations for singular initial data arises. In particular, if one starts with a vorticity field concentrated on a curve, it is important to know whether more severe singularities can develop. This problem has recently received a lot of interest from applied mathematicians because it meets a fundamental question in fluid dynamics and requires sophisticated tools from applied analysis. In Cottet (1992), the reader can find some essential features of the concept of concentration and details of the key steps in the proof of the global existence (in 2-D incompressible fluid flow) of positive vortex sheet. See also Constantin and Wu (1996).

For the 3-D fluid flow and smooth initial data, the inviscid limit is the Euler equation, as long as the corresponding solution of the Euler equation is smooth [see, for example, Constantin (1986)]. This might be a true limitation because of the possibility of finite time blowup! The blowup problem for the Euler ($1/\operatorname{Re} = 0$) equation is,

> do smooth data (with smooth, rapidly decaying initial velocity) guarantee smooth solutions for all time?

The answer is known to be yes only for 2-D and not known for 3-D! The main difference between 2-D and the 3-D is now clear. On the right-hand side of the classical vorticity equation for an incompressible fluid [see, for instance, (9.39a) in Chap. 9, p. 245, of Zeytounian (2002a)], the second and the third terms are zero, but the first term (for 3-D flow) is different from zero! It is important to note that a valid blowup scenario requires the formation of a singular object in physical space. Such an object might be as simple as conical "elbows" in a pair of vortex tubes, and such structures might have been observed in numerical simulations [see, for instance, Kerr (1993)].

In the recent Note by Caflisch and Sammartino (1997), the authors are concerned with constructing the solution of incompressible 2-D unsteady-state Navier equations outside a circular domain and the zero viscosity limit [these authors generalize the result reported by Asano (1989) and cited above]. Under suitable hypotheses and by explicit construction of the solution, it is proved that, when the viscosity goes to zero, the solution converges to the Euler solution ($\boldsymbol{u}_E$) outside the boundary layer and to the Prandtl solution ($\boldsymbol{u}_p$) inside the boundary layer, which exponentially decays outside the boundary layer. It is stated that the existence and uniqueness of Euler and Prandtl solutions can be shown by using some abstract Cauchy–Kovalevskaya theorems. The Euler and Prandtl solutions together with a small corrective term [proportional to $(1/\operatorname{Re}^{1/2})$] are used to construct the Navier solution (as a composite asymptotic expansion):

$$\boldsymbol{u} = \boldsymbol{u}^E + \boldsymbol{u}^P + \left(\frac{1}{\operatorname{Re}}\right)^{1/2} \boldsymbol{w}, \tag{4.134}$$

and it turns out that the norm of $\boldsymbol{w}$ in the appropriate functional space, remains bounded for a time T which is independent of the viscosity. These authors, imposing analyticity on the initial data, solve two problems on a half space:

(1) to prove the existence and uniqueness of the solution $\boldsymbol{u}^P$;
(2) to prove that the Euler and Prandtl equations correctly describe the behavior of the solution of the Navier equations in the zero viscosity limit.

Vanishing viscosity in the initial boundary-value problem (for 2-D unsteady-state flow in a bounded domain) for Navier equations is also considered by

Khatskevich (1996). Under some assumptions (that include, in particular, potential external forces), a positive answer is obtained by the above author to the problem of convergence of solutions to the initial boundary-value problem for Navier equations to the solution of the corresponding initial boundary-value problem for the Euler equations, as the viscosity coefficient tends to zero. The inviscid limit of 2-D incompressible fluids with bounded vorticity in the whole space is analyzed by Chemin (1996).

Finally, we note that in the theory of viscous flows, the double passage to the limit,

$$\nu_0 \to 0 \text{ and } t \to \infty \tag{4.135}$$

is quite intriguing and its elucidation would shed light on many important questions. Naturally, the order in which the limits are taken is crucial. In general, the inviscid limit ($\nu_0 \to 0$) and the temporal limit ($t \to \infty$) do not commute. The appropriate order for closed systems is taking the temporal limit first, but taking the inviscid limit before the temporal limit might be appropriate for systems in which forcing fluctuates rapidly in time.

If we consider, for simplicity, a linear model, then due to the linearity of the model, there is no boundary layer on the steady-state limiting solution and, accordingly, we cannot understand from this simplified model, how the growth of the thickness of the boundary layer settles down to a finite $O(\sqrt{\nu_0})$ value, as it should for a true flow problem. On the other hand, in a sufficiently long time, $t = O(1/\nu_0)$, diffusion has invaded the whole domain and it is only through its pervading action that, ultimately, a steady-state solution is established. For a true flow, nonlinearity drastically changes the situation, but we may safely conjecture that the pervading action of viscosity contributes significantly to the ultimate setting of the steady-state solution, on a time-scale, $t = O(\nu_0)$. The main difference with the linear model is that, probably, due to nonlinearity, the steady-state solution is built up much more rapidly and is almost completed by the time the pervading effect of viscosity is felt. Accordingly, this ultimate behavior should be controlled by linear equations.

In a short Note, Guiraud and Zeytounian (1984b) consider, as a test (linear) problem, the long-term behavior of the solution of the initial boundary-value problem for the heat equation by using a double scale technique and, in this case, it is shown how a Rayleigh type layer can penetrate the whole domain in the long term.

In a short paper by Constantin (1995a), cited above, the reader can find a pertinent discussion of the temporal limit [but Constantin calls the case in which the infinite time limit is taken first the *"temporal limit"* and the case in which one takes the control parameter (the Reynolds number Re, for instance) to infinity, first, the *"inviscid limit"*].

Finally, in a recent paper by Temam and Wang (1998), the zero viscosity limit for Navier viscous, incompressible equations is also rigorously investigated. See also Temam and Wang (1996a,b).

4.5.4 Rigorous Results for the Boundary-Layer Theory

On the mathematical side in the context of modern functional analysis for incompressible viscous fluids, there is a more limited literature. Oleinik (1963) addresses the mathematical theory of the Prandtl equations themselves; see also the more recent book by Oleinik and Samokhin (1999). Vishik and Lyusternik (1957) and J.L. Lions (1973) address many boundary-layer issues in the context of other areas of sciences and engineering. In his article, J.L. Lions introduces the concept of correctors, which was used recently by Temam and Wang (2000). The point of view is slightly different from the traditional point of view in boundary-layer theory. See also Gues (1995), Grenier and Gues (1998), and Xin (1998) for various results related to viscous boundary layers. In Temam and Wang (2001) a rigorous result regarding the boundary layer associated with incompressible Newtonian channel flow with injection and suction is presented.

5 The Limit of Very Low Reynolds Numbers

5.1 Large Viscosity Limits and Stokes and Oseen Equations

5.1.1 Steady-State Stokes Equation

First, we consider the Navier equation in the form $\left(\frac{D}{Dt} \equiv \frac{\partial}{\partial t} + \boldsymbol{u}_N \cdot \boldsymbol{\nabla}\right)$:

$$\text{Re}\,\frac{D\boldsymbol{u}_N}{Dt} + \boldsymbol{\nabla} p_s = \boldsymbol{\nabla}^2 \boldsymbol{u}_N \tag{5.1}$$

when we assume that the gravitational force $\boldsymbol{g}$ is zero and define the (Stokes) pressure p_s such that

$$p_s = \text{Re}\,\frac{p - p_\infty}{\rho_0 (U_\infty)^2}, \tag{5.2}$$

with p_∞ a constant reference pressure.

Now, if one lets Reynolds number Re tend to zero (with time-space dimensionless variables $(t, \boldsymbol{x})$ fixed), one obtains the following steady-state Stokes classical equation:

$$\boldsymbol{\nabla}^2 \boldsymbol{u}_s = \boldsymbol{\nabla} p_s, \tag{5.3}$$

with

$$\boldsymbol{\nabla} \cdot \boldsymbol{u}_s = 0 \text{ and } \lim{}^S \boldsymbol{u}_N = \boldsymbol{u}_s(t, \boldsymbol{x}), \tag{5.4a}$$

where

$$\lim{}^S = [\text{Re} \downarrow 0, \text{ with } p_s, t \text{ and } \boldsymbol{x} \text{ fixed}]. \tag{5.4b}$$

One would expect the same boundary conditions for the Stokes equation (5.3) as those for the full Navier equation (5.1). It was noticed by Stokes (1851) himself that solutions do not exist for stationary 2-D flow past a solid that satisfies the boundary conditions at the solid (no-slip) as well as at infinity where the behavior condition is

$$\text{uniform flow}: \boldsymbol{u}_N \to U_\infty \boldsymbol{i} \text{ at infinity}. \tag{5.5}$$

The Stokes steady-state flow in an *unbounded* domain, *exterior* to a solid body, is an *inner flow* valid only *near the wall* of this body.

5.1.2 Unsteady-State Oseen Equation

At a large distance from a wall, near infinity, when

$$\mathrm{Re}\,|\boldsymbol{x}| \equiv |\boldsymbol{x}_0| = O(1)\,,$$

it is necessary to derive another consistent [outer, so-called Oseen (1910)] equation, in place of the Stokes equation (5.3)! Using Oseen time-space variables,

$$t_0 = \mathrm{Re}\,t \text{ and } \boldsymbol{x}_0\,, \tag{5.6a}$$

and also Oseen pressure,

$$p_0 = \frac{p_s}{\mathrm{Re}}\,, \tag{5.6b}$$

we derive the following dimensionless Navier equation (à la Oseen) in place of Navier (à la Stokes) equation (5.1):

$$\frac{D_0 \boldsymbol{u}_N}{Dt_0} + \boldsymbol{\nabla}_0 p_0 = \boldsymbol{\nabla}_0^2 \boldsymbol{u}_N\,, \tag{5.6c}$$

where

$$\frac{D_0}{Dt_0} = \frac{\partial}{\partial t_0} + \boldsymbol{u}_N \cdot \boldsymbol{\nabla}_0 \text{ and } \boldsymbol{\nabla}_0 = \left(\frac{\partial}{\partial \boldsymbol{x}_0}\right)\,.$$

But, near infinity at the "Oseen limit",

$$\lim{}^0 = [\mathrm{Re} \downarrow 0, \text{ with } p_0, t_0 \text{ and } \boldsymbol{x}_0 \text{ fixed}]\,, \tag{5.6d}$$

a finite body shrinks to a point that cannot cause a finite disturbance in a viscous fluid!

As a consequence, we can write the following asymptotic expansion for the outer Oseen region:

$$\boldsymbol{u}_N = \boldsymbol{i} + \delta(\mathrm{Re})\boldsymbol{u}_0(t_0, \boldsymbol{x}_0) + \dots\,, \tag{5.7}$$

when we assume that (with dimensionless quantities) $\boldsymbol{u} \to \boldsymbol{i}$ at infinity. For the "Oseen" velocity vector, $\boldsymbol{u}_0(t_0, \boldsymbol{x}_0) = \lim{}^0 \boldsymbol{u}$, we derive the following unsteady-state Oseen equation from Navier equation (5.6c):

$$\left[\frac{\partial}{\partial t_0} + \boldsymbol{\nabla}_0 \cdot \boldsymbol{i}\right]\boldsymbol{u}_0 + \boldsymbol{\nabla}_0 p_0 = \boldsymbol{\nabla}_0^2 \boldsymbol{u}_0\,. \tag{5.8}$$

In the outer Oseen asymptotic expansion (5.7) for $\boldsymbol{u}$, the gauge function $\delta(\mathrm{Re}) \downarrow 0$, with $\mathrm{Re} \downarrow 0$, and only by matching (outer to inner) do we have the possibility of determining this gauge. The matching conditions are somewhat more complicated than in the high Reynolds number case [see, for instance, Lagerstrom (1964, pp. 163–167)].

5.1.3 Unsteady-State Stokes and Steady-State Oseen Equations

Near $t = 0$, it is necessary to take into account an *unsteady-state Stokes equation* in place of the steady-state Stokes equation (5.3):

$$\frac{\partial \boldsymbol{u}_s^*}{\partial t_s} + \boldsymbol{\nabla}^2 \boldsymbol{u}_s^* = \boldsymbol{\nabla} p_s \,, \tag{5.9a}$$

where

$$t_s = t/\operatorname{Re} \text{ and } \lim{}^{S*} \boldsymbol{u} = \boldsymbol{u}_s^*(t_s, \boldsymbol{x}) \,, \tag{5.9b}$$

when the following limit is considered:

$$\lim{}^{S*} = [\operatorname{Re} \downarrow 0, \text{ with } p_s, t_s \text{ and } \boldsymbol{x} \text{ fixed}] \,. \tag{5.9c}$$

On the other hand, far from $t = 0$, we have a *steady-state Oseen equation* in place of unsteady-state Oseen equation (5.8):

$$(\boldsymbol{\nabla}_0 \cdot \boldsymbol{i}) \boldsymbol{u}_0^* + \boldsymbol{\nabla}_0 p_0 = \boldsymbol{\nabla}_0^2 \boldsymbol{u}_0^* \,, \tag{5.10}$$

with

$$\boldsymbol{u} = \boldsymbol{i} + \delta^*(\operatorname{Re}) \boldsymbol{u}_0^*(t, \boldsymbol{x}_0) + \ldots \,, \text{ and } \delta^*(\operatorname{Re}) \downarrow 0, \text{ with } \operatorname{Re} \downarrow 0 \,. \tag{5.11}$$

5.1.4 Unsteady-State Matched Stokes–Oseen Solution at $\operatorname{Re} \ll 1$ for the Flow Past a Sphere

The reader can find an asymptotic analysis in Sano (1981), by the MMAE, for a low Reynolds number unbounded uniform flow past a sphere, which gives the matching between the Stokes and Oseen steady-state and unsteady-state regions. Only in 1978, Bentwich and Miloh (1978) considered the unsteady-state matched Stokes–Oseen solution for flow past a sphere.

The solution obtained by these two authors represents the entire process of transition from stagnancy to the steady-state state envisioned by Proudman and Pearson (1957).

But Sano (1981) showed that the matching procedure proposed by Bentwich and Miloh (1978) is incomplete. They divided the (r,t) plane (see Fig. 5.1) into two regions: one is the L-shaped region adjacent to the r and t axes (inner domain = small-time domain + large-time inner domain), and the other is the rectangular region far from the axes (outer domain = large-time outer domain).

The L-shaped region (see Fig. 5.1), suggested by Bentwich and Miloh (1978), must be subdivided into two domains: one is a small-time domain where the time $t = O(1)$, and the other a large-time inner domain where the time $t = O(1/\operatorname{Re}^2)$ and $r = O(1)$.

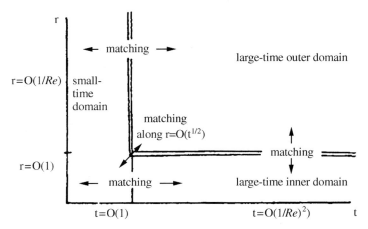

Fig. 5.1. Sketch of the three-region structure for the unsteady-state problem at $\mathrm{Re} \ll 1$ of flow past a sphere

Consequently, including a large-time outer domain where the time $t = O(1/\mathrm{Re}^2)$ and $r = O(1/\mathrm{Re})$, the unsteady-state problem, for Navier fluid flow past a sphere at low Re, has a three-region structure. A spherical coordinate system (r, θ, ϕ), centered at the sphere (where $r = 0$) and with $\theta = 0$ in the direction of the undisturbed stream, is chosen. The unsteady-state motion (time-dependent) is assumed to be axially symmetrical, and no swirling motion occurs.

In the small-time domain, the vorticity layer is confined to the inner region near the surface $[r = O(1)]$, and consequently, the assumption that the nonlinear inertia terms in the unsteady-state vorticity Navier equation [written with the spherical coordinates (r, θ) for the stream function $\psi(t, r, \theta)$], are negligible, when $\mathrm{Re} \ll 1$, is valid throughout the flow field.

In the large-time inner domain, the $\partial/\partial t$ term in the unsteady-state vorticity Navier equation is also small for $\mathrm{Re} \ll 1$, along with the nonlinear terms, and the motion in this $(T = \mathrm{Re}^2 t, r)$ domain is quasi-steady-state.

The requirement of the outer domain in r in the large-time region is due to the fact that, as t increases, the vorticity layer diffuses into the outer region where all terms in the momentum equation are of the same order of magnitude and the (large-time) inner solution fails. In the large-time inner domain, the representative variables are the time $T = \mathrm{Re}^2 t$ and $r = \mathrm{Re}\, r$.

The two asymptotic expansions in the large-time region are constructed so that

(i) the inner expansion satisfies the boundary conditions on the surface (on $r = 1$, for $0 \leq \theta \leq \pi$);
(ii) the outer expansion satisfies the boundary condition at infinity (when $r \to \infty$);

(iii) the two expansions match identically in the overlapping domain in space (in r) where both expansions are valid and also match with the small-time expansion at small values of T (when t tends to infinity).

The matching along $r = O(t^{1/2})$, proposed by Bentwich and Miloh (1978), is incomplete to obtain the higher order terms in the corresponding asymptotic expansions, and this way of dividing the (r, t) plane is correct only if the expansion in the L-shaped region contains one term.

In Zeytounian (2002b, §9.3, Chap. 9), the reader can find a presentation of unsteady-state Navier flow past a sphere with the analysis of three regions and matching.

5.2 Low Reynolds Number Flow due to an Impulsively Started Circular Cylinder

The steady-state low Reynolds number flow of an incompressible, viscous (Navier) fluid past a circular cylinder was studied by Stokes (1851). He linearized the governing equations about a state of rest and found that the resulting boundary-value problem had no solution. Oseen (1910) observed that this *Stokes paradox* was due to incorrect treatment of the flow far from the cylinder and could be avoided by linearizing about the flow at infinity (see Sects. 5.1.1–5.1.3). Lamb (1945, Art. **343**) solved the resulting Oseen equations to obtain a first approximation for the flow and the drag. Much later, Proudman and Pearson (1957) and Kaplun (1957) showed that this (singular perturbation) problem could be solved to any order by systematic use of the MMAE. The inner expansion involved Stokes linearization, and the outer expansion involved Oseen linearization. The method and results are very well described in Van Dyke book (1975, Chap. VIII).

The worst of the matter is that both the inner and the outer expansions involve a series of positive powers of $1/\log \text{Re}$, and in addition, there are terms in powers of Re and in powers of Re multiplied by powers of $1/\log \text{Re}$. These terms are smaller than all positive powers of $1/\log \text{Re}$, so they are said to be "beyond all orders" of $1/\log \text{Re}$, or to be "transcendentally small terms." Skinner (1975) showed how to calculate a few of these terms. He pointed out, however, that they are negligible compared to the terms of orders $(1/\log \text{Re})^4$ and higher, which were unknown. Only the first three terms in powers of $1/\log \text{Re}$ had been found, the last one by Kaplun (1957). A hybrid asymptotic numerical method for calculating low Reynolds numbers flow past symmetrical cylindrical bodies was developed recently by Kropinski et al. (1995). In the recent paper by Keller and Ward (1996), the authors (who previously showed how to determine the sum of all terms in powers of $1/\text{Re}$), show (by using a hybrid method that combines numerical computation and asymptotic analysis) now how to go beyond all of those terms to find the sum of all terms containing Re times a power of $1/\log \text{Re}$. The first sum gives the drag coefficient and represents symmetrical flow in the Stokes region near the cylinder. The second tem reveals the asymmetry of the flow near the body.

5.2.1 Formulation of the Steady-State Problem

Let $\psi(r, \theta; \text{Re})$ be the dimensionless stream function for the steady-state flow of an incompressible, viscous fluid around a circular cylinder. The Reynolds number Re is based upon the cylinder radius, and r is measured in units of this radius. Then, ψ satisfies the following (exact) problem:

$$\Delta^2 \psi + \text{Re}\, J(\psi; \Delta\psi) = 0, \ r > 1 , \tag{5.12}$$

$$\psi(1, \theta; \text{Re}) = 0, \ \left.\frac{\partial \psi}{\partial r}\right|_{r=1} = 0, \ \psi(r, \theta; \text{Re}) \sim r \sin\theta, \text{ as } r \uparrow \infty . \tag{5.13}$$

Here, $J(f; g) = (1/r)[\partial f/\partial r\, \partial g/\partial\theta - \partial f/\partial\theta\, \partial g/\partial r]$ is the Jacobian.

The *inner* (Stokes) expansion of ψ is

$$\begin{aligned}\psi(r, \theta; \text{Re}) &= a(\delta)[r \log r - (r/2) + (1/2r)] \sin\theta \\ &+ \text{Re}\{(1/32)a^2(\delta)[2r^2 \log^2 r - r^2 \log r + (1/4)r^2 - (1/4r^2)] \\ &+ [b(\delta)/8][r^2 - 2 - (1/r^2)]\} \sin 2\theta + O(\text{Re}^2) .\end{aligned} \tag{5.14a}$$

To determine the two constants $a(\delta)$ and $b(\delta)$ with

$$\delta \equiv [\log 4 - \gamma - \log \text{Re} + (1/2)]^{-1} = [\log(3.7026/\text{Re})]^{-1} , \tag{5.15}$$

where γ is Euler's constant, we first introduce the outer (Oseen) variable $R = \text{Re}\, r$ and rewrite (5.14a) in terms of it:

$$\begin{aligned}\text{Re}\,\psi(R/\text{Re}, \theta; \text{Re}) &= \{a(\delta)R \log R - a(\delta)[\log \text{Re} + (1/2)]R + O(\text{Re}^2)\} \sin\theta \\ &+ \{(1/16)a^2(\delta)(R \log R)^2 - (1/8)a^2(\delta)[\log \text{Re} + (1/4)]R^2 \log R \\ &+ (1/32)a^2(\delta)[2 \log^2 \text{Re} + \log \text{Re} + (1/4)]R^2 + O(\text{Re}^2)\} \sin 2\theta .\end{aligned} \tag{5.14b}$$

Expansion (5.14b) is used to match the inner expansion to an outer, and by doing so we can find $a(\delta)$ and $b(\delta)$.

To obtain the outer (Oseen) expansion, we set

$$\Psi(R, \theta; Re) = \text{Re}\,\psi(R/\text{Re}, \theta; \text{Re}) , \tag{5.16a}$$

and we expand

$$\Psi(R, \theta; \text{Re}) = \Psi_0(R, \theta) + \text{Re}^2\, \Psi_1(R, \theta) + \ldots . \tag{5.16b}$$

From the starting, exact problem (5.12), (5.13), we derive the following (outer) Oseen equation for the leading term $\Psi_0(R, \theta)$ of the outer expansion (5.16b), namely:

$$\Delta^{*2}\Psi_0 + \text{Re}\, J(\Psi_0; \Delta^*\Psi_0) = 0 , \tag{5.17a}$$

where Δ^* is written relative to variable $R = \text{Re}\, r$.

For (5.17a), we have the following behavior condition:

$$\Psi_0(R,\theta) \sim R\sin\theta, \quad \text{as } R \uparrow \infty, \tag{5.17b}$$

and also the matching condition

$$\Psi_0(R,\theta) \sim a\, R\log R \sin\theta. \quad \text{as } R \downarrow 0, \tag{5.17c}$$

because matching with (5.14b) shows that $\Psi_0(R,\theta)$ must have singularity! Conditions (5.17b,c) determine a unique solution Ψ_0 of the Oseen equation (5.17a) for each choice of a; this was computed numerically in Kropinski et al. (1995). To match that solution with (5.14b), Keller and Ward (1996) computed its two Fourier sine coefficients

$$\Psi_{0j}(R) = \frac{2}{\pi} \int_0^\pi \Psi_0(R,\theta)\sin j\theta\, \mathrm{d}\theta, \quad \text{for } j = 1 \text{ and } 2. \tag{5.18}$$

From (5.14b), we see that these coefficients must have the following forms, as $R \downarrow 0$:

$$\Psi_{01}(R) \sim aR\log R + C_{00}R, \quad \text{as } R \downarrow 0, \tag{5.19a}$$
$$\Psi_{02}(R) \sim C_{01}(R\log R)^2 + C_{02}R^2\log R + C_{03}R^2, \quad \text{as } R \downarrow 0. \tag{5.19b}$$

For each choice of a, Ψ_0 can be computed, $\Psi_{0j}(R)$ evaluated, and then C_{0j} can be found. The $C_{0j}(a)$ are functions of a because the solution $\Psi_0(R,\theta)$ is determined by a.

Now, matching requires that $\Psi_{01}(R)$ match with the coefficient of $\sin\theta$ in (5.14b) and $\Psi_{02}(R)$ match with the coefficient of $\sin 2\theta$. The first match yields the equation

$$C_{00}(a) = -a[\log\text{Re} + (1/2)], \tag{5.20a}$$

and determines a as function of $\log\text{Re}$ and therefore as a function of $\delta = [\log(3.7026/\text{Re})]^{-1}$. Then the conditions

$$C_{01}(a) = a^2/16 \quad \text{and} \quad C_{02}(a) = -(a^2/8)[\log\text{Re} + (1/4)], \tag{5.20b}$$

must be identities. The fourth matching condition gives an equation for b whose solution is

$$b(\delta) = 8C_{03}(a) - (1/4)a^2[2\log^2\text{Re} + \log\text{Re} + (1/4)], \tag{5.20c}$$

where $\log\text{Re} = \log 3.7026 - (1/\delta)$.

Finally, in this way, Keller and Ward (1996) completed the determination of the term of order Re in the inner expansion (5.14a). This term is "beyond all orders" of δ. All orders of δ are contained in the first term $a(\delta)$, and (5.20c) determines $b(\delta)$ to all order. Kaplun (1957) obtained the first two nonzero terms in the expansion for $a(\delta)$, and Skinner (1975) obtained the first two nonzero terms in $b(\delta)$.

For $\delta \downarrow 0$, they found that

$$a(\delta) \sim \delta - 0.9669\delta^3 + \ldots \quad \text{and} \quad b(\delta) \sim -(1/2) + (1/4)\delta + \ldots . \quad (5.21)$$

Equation (5.21) becomes inaccurate for $a(\delta)$ when Re ≈ 0.5 and for $b(\delta)$ when Re ≈ 0.2!

We observe that the parameter

$$\mu(\text{Re}) = -\text{Re}\, b(\delta(\text{Re}))/a(\delta(\text{Re})), \quad (5.22)$$

gives a measure of the upstream–downstream asymmetry of the flow field near the cylinder. Finally, the drag coefficient C_D is given by

$$C_D = (4\pi/\text{Re})\{a(\delta) + (1/32)\text{Re}^2[1 - \delta/2] + O(\delta^2 \text{Re}^2)\}. \quad (5.23)$$

The numerical result obtained by Keller and Ward (1996) for the drag coefficient is better than that in Kropinski et al. (1995) because a precise procedure was used, via the Fourier coefficients, to obtain the sum from the numerical solution, rather than the cruder method of Kropinski et al. (1995). In the Appendix of Keller and Ward (1996, pp. 264–265), the authors summarize the numerical method of Kropinski et al. (1995) used to solve the hybrid problem (5.17a,b) with the singularity condition (5.17c) and also describe the improved method for computing $C_{00}(a)$ in (5.19a) and the extension of the method of Kropinski et al. (1995) to calculate C_{03} in (5.19b); Re and b are calculated from (5.20a) and (5.20c), respectively.

5.2.2 The Unsteady-State Problem

Bentwich and Miloh (1982) considered the solution for unsteady-state flow due to an impulsively started cylinder. This solution is expressed in terms of three expansions that represent the flow in three different space-time subdomains: one that holds "early" in the process "throughout" the exterior of the cylinder, and two that hold later in the process, a "late inner" and a "late outer" expansion.

However, although it was tacitly assumed by the authors that the later expansion represents the flow everywhere beyond the immediate vicinity of the obstacle, it was subsequently found that this assumption is incorrect.

In Bentwich and Miloh (1984), it is shown that a fourth "late-wake" subdomain exists and an appropriate additional expansion is developed and matched with the other three. This later expansion represents the flow late in the process in a wake region. This wake extends all the way downstream to infinity, and its width is comparable to the diameter of the obstacle.

We observe that the question concerning the validity of the three-expansion solution proposed in Bentwich and Miloh (1982) is, whether potential flow is a valid representation of a high Reynolds number flow past an obstacle.

Thus, just as the irrotational solution holds throughout the domain exterior to the obstacle but not on its surface, the three-expansion solution proposed in Bentwich and Miloh (1982) holds almost but not quite throughout to the "later outer" space-time subdomain under discussion.

Furthermore, just as the inadequacy of the classical irrotational flow can be mended by appending a boundary layer, the discontinuity of the "late outer" expansion along the half-plane behind the cylinder can be mended by constructing an additional fourth expansion.

It expresses the flow over this plane and in its immediate vicinity late in the process. In the "late-wake" subdomain, we derive a fourth-order signficant equation in the coordinate normal to the flow direction.

Two-dimensional flow past a cylindrical body of arbitrary profile at small Reynolds numbers was studied theoretically in Tamada et al. (1983); they assumed that $|\log \text{Re}|^{-1}$ is *not small*, whereas Re is still small. In this case, the outer flow field is governed by the full Navier equations and with the *aid of numerical techniques*, the authors fix the asymptotic flow field which matches with the Stokeslet flow at the origin as well as the uniform flow at infinity (an immersed body is equivalent to a Stokeslet at a large distance).

The result is used to determine the relation between the force experienced by the cylinder and the Reynolds number for a cylinder of arbitrary cross section.

5.3 Compressible Flow

For compressible flow, when $\text{Re} \to 0$, it is also necessary to specify the role of the Mach number M. It is necessary to pose the problem concerning the behavior of solutions of the full dimensionless NSF equations (1.30)–(1.32) with (1.33), when simultaneously,

$$\text{Re} \to 0 \text{ and } M \to 0 . \tag{5.24a}$$

Naturally, for the validity of the NSF equations, it is obvious that it is assumed that the limiting compressible flow at low Reynolds and Mach numbers remains a continuous medium, and this implies that the Knudsen (dimensionless) number is a small parameter:

$$\text{Kn} = M/\text{Re} \ll 1 . \tag{5.24b}$$

For a discussion of the Knudsen number Kn, see Zeytounian (2002a, Chap. 1). Consequently, the double limit process (5.24a) must be done with the following similarity relation:

$$M = R^*(\text{Re})^{1+a}, \text{ with } R^* = O(1) \text{ and } a > 0 \text{ when } \text{Re} \to 0 . \tag{5.25}$$

First, the Stokes (compressible - proximal) limit is denoted by Lim^S and defined by (with Bo = 0)

$$\text{Lim}^S = [\text{Re} \to 0, M \to 0; \|\boldsymbol{x}\|, t, R^*, S, \text{Pr}, \gamma \text{ fixed } = O(1)] . \quad (5.26\text{a})$$

Similarly, the Oseen (compressible - distal) limit is denoted by Lim^0 and defined by (again for Bo = 0)

$$\text{Lim}^0 = [\text{Re} \to 0, M \to 0; \|\boldsymbol{x}_0\|, t, R^*, S, \text{Pr}, \gamma \text{ fixed } = O(1)] , \quad (5.26\text{b})$$

where $\boldsymbol{x}_0$ is the Oseen space variable $\boldsymbol{x}_0 = \text{Re}\,\boldsymbol{x}$.

5.3.1 The Stokes Limiting Case and Steady-State Compressible Stokes Equations

First, we consider the Stokes limit of steady-state NSF equations. If we take into account the similarity relation (5.25), then, when $\text{Re} \to 0$, it is necessary to study the approximate solutions of the NSF equations in the following asymptotic form (when Bo = 0):

$$\boldsymbol{u} = \boldsymbol{u}_s + \ldots; \; p = 1 + (\text{Re})^{1+2a}[p_s + \ldots]; \quad (5.27\text{a})$$
$$\rho = \rho_s + \ldots; T = T_s + \ldots, \quad (5.27\text{b})$$

where the "Stokes" limiting functions (with "s" as subscript) depend only on the position (Stokes inner) variable $\boldsymbol{x} = (x_i)$, $i = 1, 2, 3$.

Then, for $\boldsymbol{u}_s$, p_s, ρ_s, and T_s, we derive an "à la Stokes" steady-state compressible system which is written in three parts:

$$\boldsymbol{\nabla} \cdot [k(T_s)\boldsymbol{\nabla} T_s] = 0 , \quad (5.28\text{a}_1)$$
$$T_s = T_w(\boldsymbol{P}, \tau) \equiv 1 + \tau \Xi(\boldsymbol{P}), \boldsymbol{P} \in \partial\Omega \text{ on } \partial\Omega; \quad (5.28\text{a}_2)$$

$$\rho_s = 1/T_s ; \quad (5.28\text{b})$$

$$\boldsymbol{\nabla} \cdot \boldsymbol{u}_s = \boldsymbol{u}_s \cdot \boldsymbol{\nabla} \log T_s , \quad (5.28\text{c}_1)$$
$$\boldsymbol{\nabla} p_s = \gamma(R^*)^2 \boldsymbol{\nabla} \cdot [2\mu(T_s)\boldsymbol{D}_s\boldsymbol{u}_s + \lambda(T_s)(\boldsymbol{\nabla} \cdot \boldsymbol{u}_s)\boldsymbol{I}] , \quad (5.28\text{c}_2)$$
$$\boldsymbol{D}_s\boldsymbol{u}_s = (1/2)\left[(\boldsymbol{\nabla}\boldsymbol{u}_s) + (\boldsymbol{\nabla}\boldsymbol{u}_s)^{\text{T}}\right] , \quad (5.28\text{c}_3)$$
$$\boldsymbol{u}_s = 0 \text{ on } \partial\Omega . \quad (5.28\text{c}_4)$$

Equation (5.28a$_1$) with the associated boundary condition on $\partial\Omega$ (5.28a$_2$) determines T_s, as soon as the temperature behavior at infinity is specified. Equation (5.28b) is a relation (a limiting form of the equation of state) between T_s and ρ_s and determines ρ_s when T_s is known. Finally, (5.28c$_1$,c$_2$) give a closed system for determining (with (5.28c$_4$)) the velocity vector $\boldsymbol{u}_s$ and the perturbation of the pressure p_s. For an external aerodynamics problem, when the compressible flow at low Reynolds and Mach numbers is investigated in an infinite domain Ω, it is also necessary to write a condition at infinity for $\boldsymbol{u}_s$.

When the rate of temperature fluctuation τ on $\partial\Omega$ tends to zero, as Re $\to 0$, then, we can obtain a particularly simple solution for (5.28a,b):

$$T_s = 1, \ \rho_s = 1 \ ,$$

and (5.28c_1,c_2) with (5.28c_3) reduce to the classical Stokes equations for an incompressible fluid:

$$\boldsymbol{\nabla} \cdot \boldsymbol{u}_s = 0, \ \boldsymbol{\nabla} p_s = \gamma(R^*)^2 \mu(1) \boldsymbol{\nabla}^2 \boldsymbol{u}_s \ . \tag{5.29}$$

We note that the Stokes equations (5.29) for incompressible flow may be obtained either by linearization or by letting Re tend to zero.

That these two procedures give the same result in the incompressible case is fortuitous; for compressible fluids, the low Reynolds number Stokes equations are nonlinear, as shown by the system of equations (5.28a–c). In solving (5.28a–c), one first finds T_s from the energy equation (5.28a_1):

$$\boldsymbol{\nabla}^2 T_s + T_s \left(\frac{\mathrm{d}\log k(T_s)}{\mathrm{d}T_s} \right) \boldsymbol{\nabla} T_s = 0 \ , \tag{5.30}$$

and then ρ_s from the equation of state.

The continuity and momentum equations then become linear equations whose variable coefficients involve the known function T_s.

5.3.2 The Oseen Limiting Case and Steady-State Compressible Oseen Equations

Now, an alternative form of the steady-state dimensionless NSF equations, using the Oseen (outer) space variable $\boldsymbol{x}_0 = \mathrm{Re}\,\boldsymbol{x}$ is (when Bo $= 0$):

$$\begin{aligned}
& \boldsymbol{u} \cdot \boldsymbol{\nabla}^0 \rho + \rho \boldsymbol{\nabla}^0 \cdot \boldsymbol{u} = 0 \ ; \\
& \rho(\boldsymbol{u} \cdot \boldsymbol{\nabla}^0) \boldsymbol{u} + (1/\gamma M^2)\boldsymbol{\nabla}^0 p = \boldsymbol{\nabla}^0 \cdot [2\mu \mathbf{D}^0 \boldsymbol{u} + \lambda(\boldsymbol{\nabla}^0 \cdot \boldsymbol{u})\mathbf{I}] \ ; \\
& \rho \boldsymbol{u} \cdot \boldsymbol{\nabla}^0 T + (\gamma - 1)p\boldsymbol{\nabla}^0 \cdot \boldsymbol{u} = (\gamma/\mathrm{Pr})\boldsymbol{\nabla}^0 \cdot [k\boldsymbol{\nabla}^0 T] \\
& \quad + M^2\gamma(\gamma-1)\left[2\mu\,\mathrm{Tr}(\mathbf{D}^{02}) + \lambda(\boldsymbol{\nabla}^0 \cdot \boldsymbol{u})^2\right] \ ; \\
& p = \rho T \ ,
\end{aligned} \tag{5.31}$$

where the gradient vector $\boldsymbol{\nabla}^0$ is formed with respect to the Oseen space variable $\boldsymbol{x}_0 = (x_{0i})$, $i = 1, 2, 3$, and we have the following relations for the gradient operator and the rate of strain tensor:

$$\boldsymbol{\nabla} = \mathrm{Re}\,\boldsymbol{\nabla}^0 \quad \text{and} \quad \mathbf{D} = \mathrm{Re}\,\mathbf{D}^0 \ .$$

The unknown functions, $\boldsymbol{u}$, p, ρ, T, and the rate of strain tensor $\mathbf{D}^0$ depend on (we consider steady-state flow) all of the Oseen space variables x_{0i}. In the steady-state NSF equations (5.31), the Reynolds number Re has been eliminated! However, Re will reappear in the boundary conditions. For example, if the finite solid body is a sphere of diameter L^0, then the boundary of the body is given in Oseen coordinates x_{0i} by

$$(x_{01})^2 + (x_{02})^2 + (x_{03})^2 = \text{Re}/2 \, ,$$

and when $\text{Re} \to 0$, *with x_{0i} ($i = 1, 2$ and 3) fixed*, the body *shrinks to a point*! For steady-state flow past a solid finite body at rest and when the domain is infinite, the following uniform conditions have to be added:

$$\boldsymbol{u} = U_\infty \boldsymbol{i}, \; p = \rho = T = 1, \text{ at infinity} \, . \tag{5.32}$$

Now, we assume for $\boldsymbol{u}$, p, ρ, and T the following asymptotic (outer) expansions for small Re:

$$\boldsymbol{u} = U_\infty \boldsymbol{i} + \mu_1(\text{Re})\boldsymbol{u}^0(\boldsymbol{x}_0) + \ldots \, , \tag{5.33a}$$

$$p = 1 + (\text{Re})^{2+2a}[\mu_1(\text{Re})p^0(\boldsymbol{x}_0) + \ldots] \, , \tag{5.33b}$$

$$\rho = 1 + \mu_1(\text{Re})\rho^0(\boldsymbol{x}_0) + \ldots \, , \tag{5.33c}$$

$$T = 1 + \mu_1(\text{Re})T^0(\boldsymbol{x}_0) + \ldots \, , \tag{5.33d}$$

where $\mu_1(\text{Re})$ is some suitable gauge function of Re which tends to zero as $\text{Re} \downarrow 0$. In the Oseen, distal outer region, a finite 3-D body *shrinks to a point that cannot cause a finite disturbance in Oseen fluid flow* and hence the values of $\boldsymbol{u}$, p, ρ, and T at any distant fixed point in the Oseen region will tend to the free-stream values (5.32). The expansions (5.33a) take into account this property of Oseen outer flow.

If one then inserts these expansions (5.33) into the NSF equations (5.31) as written in Oseen variables and retains only terms of the order of $\mu_1(\text{Re})$, it is found that $\boldsymbol{u}^0(\boldsymbol{x}_0)$, $p^0(\boldsymbol{x}_0)$, $\rho^0(\boldsymbol{x}_0)$, and $T^0(\boldsymbol{x}_0)$ satisfy the following Oseen (linear outer) limiting equations:

$$\boldsymbol{\nabla}^0 \cdot \boldsymbol{u}^0 = U_\infty \frac{\partial T^0}{\partial x_{01}} \, ;$$

$$U_\infty \frac{\partial \boldsymbol{u}^0}{\partial x_{01}} + \frac{1}{\gamma R^{*2}} \boldsymbol{\nabla}^0 p^0 = \mu(1) \left[\boldsymbol{\nabla}^{02} \boldsymbol{u}^0 + \frac{U_\infty}{3} \boldsymbol{\nabla}^0 \left(\frac{\partial T^0}{\partial x_{01}} \right) \right] \, ;$$

$$U_\infty \frac{\partial T^0}{\partial x_{01}} = \frac{k(1)}{\text{Pr}} \boldsymbol{\nabla}^{02} T^0 \, ;$$

$$\rho^0 = -T^0 \, , \tag{5.34}$$

if we assume that $\lambda(1) = -(2/3)\mu(1)$ and we take into account similarity relation (5.25).

We can write the following boundary conditions for Oseen equations (5.34):

$$\boldsymbol{u}^0, \; p^0, \; T^0, \text{ and } \rho^0 \to 0 \text{ at infinity} \, . \tag{5.35}$$

When T^0 is identically zero, we recover the equations of classical steady-state incompressible Oseen flow for $\boldsymbol{u}^0$ and p^0:

$$\boldsymbol{\nabla}^0 \cdot \boldsymbol{u}^0 = 0; \; U_\infty \frac{\partial \boldsymbol{u}^0}{\partial x_{01}} + \frac{1}{\gamma R^{*2}} \boldsymbol{\nabla}^0 p^0 = \mu(1) \boldsymbol{\nabla}^{02} \boldsymbol{u}^0 \; . \tag{5.36}$$

Therefore, the fundamental problem for the compressible Oseen equations (5.34), with (5.35) and matching with Stokes inner flow when $\boldsymbol{x}_0 \to 0$, is the study of the behavior of the temperature fluctuation $T^0(\boldsymbol{x}_0)$, which is a solution of the linear equation

$$[\frac{\partial^2}{\partial x_{01}^2} + \frac{\partial^2}{\partial x_{02}^2} + \frac{\partial^2}{\partial x_{03}^2}] T^0 - \beta \frac{\partial T^0}{\partial x_{01}} = 0 \; , \tag{5.37}$$

when

$$(x_{01})^2 + (x_{02})^2 + (x_{03})^2 \to 0 \; ,$$

according to matching with the Stokes equations (5.28a$_1$). In (5.37),

$$\beta = U_\infty / (k(1)/\Pr) \; .$$

A consequence of incompressible Oseen equations (5.36) is that p^0 is a harmonic function of $\boldsymbol{x}_0$:

$$\boldsymbol{\nabla}^{02} p^0 = 0 \; . \tag{5.38}$$

Surprisingly, this property is also true for the compressible Oseen equations (5.34), if we assume that $\Pr = (3/4)[k(1)/\mu(1)]$.

In incompressible flow, the Oseen equations (5.36) are actually uniformly valid because the inner (Stokes) limit of these equations gives the classical incompressible Stokes equations (5.29).

This is a fortuitous coincidence closely linked with the fact that the Stokes (proximal) equations for incompressible flow are linear.

The Oseen equations for compressible flow are not uniformly valid near a body.

In solving a problem of low Reynolds number compressible flow past a solid finite body, it is not sufficient to use only the Oseen solution; the outer (distal-Oseen) solution must be matched with an inner solution, i.e., a solution of the compressible Stokes equations.

The method of matching is, in principle, the same as that in incompressible flow, although the computational difficulties are, of course, considerably greater for compressible flow.

To obtain a "crude estimate" of the solution, one may solve the compressible Oseen equations as if they were uniformly valid!

Actually, matching in compressible low Reynolds flow is an open problem!

5.4 Film Flow on a Rotating Disc: Asymptotic Analysis for Small Re

For a disc of arbitrary large diameter which rotates in its own plane at a steady angular velocity Ω_0, we obtain the following three dimensionless equations in place of the three steady-state dimensionless equations (3.73a–c) of Sect. 3.6, in Chap. 3:

$$2F + \frac{\partial W}{\partial \zeta} = 0 , \tag{5.39a}$$

$$\text{Re}^2 \left[\frac{\partial F}{\partial \tau} + F^2 + W \frac{\partial F}{\partial \zeta} \right] - G^2 = \frac{\partial^2 F}{\partial \zeta^2} , \tag{5.39b}$$

$$\text{Re}^2 \left[\frac{\partial G}{\partial \tau} + 2FG + W \frac{\partial G}{\partial \zeta} \right] = \frac{\partial^2 G}{\partial \zeta^2} , \tag{5.39c}$$

for the three functions $F(\tau,\zeta)$, $G(\tau,\zeta)$, and $W(\tau,\zeta)$, related to the components of the velocity vector. The Reynolds number Re is

$$\text{Re} = {h^0}^2 \Omega_0 / \nu_0$$

where h^0 is the initial film thickness.

Below, we restrict our attention to flows that maintain a film thickness that is independent of the radial coordinate r and azimuthal coordinate θ.

Using dimensionless variables, we have the following kinematic condition at the free surface $\zeta = H(\tau)$:

$$\frac{dH}{d\tau} = W(\tau, H) , \tag{5.40}$$

and we assume that at initial time

$$\tau = 0 \ : \ H(0) = 1 \ \text{ and } \ \frac{dH}{d\tau} = 0 . \tag{5.41a}$$

For (5.39a–c), the initial conditions are

$$\tau = 0 \ : \ F = 0, \ G = 0, \text{ and } W = 0 . \tag{5.41b}$$

The boundary conditions for (5.39a–c) at the free surface are

$$\zeta = 0 \ : \ F = 0, G = 1, W = 0 ; \tag{5.42a}$$

$$\zeta = H(\tau) \ : \ \frac{\partial F}{\partial \zeta} = 0, \ \frac{\partial G}{\partial \zeta} = 0 , \tag{5.42b}$$

and also the free surface kinematic condition (5.40). Problem [(5.39a–c), (5.40), (5.41a,b) and (5.42a,b)] was considered by Higgins (1986) and below we follow the main line of that paper.

5.4.1 Solution for Small Re ≪ 1: Long-Time Scale Analysis

To construct an asymptotic solution for Re ≪ 1, we assume the following forms for the expansions of functions F, G, W, and H:

$$F(\tau, \zeta; \text{Re}) = F_0(\zeta, H) + \text{Re}^2 F_1\left(\zeta, \frac{dH}{d\tau}\right) + O\left(\text{Re}^4\right),$$

$$G(\tau, \zeta; \text{Re}) = G_0(\zeta, H) + \text{Re}^2 G_1\left(\zeta, \frac{dH}{d\tau}\right) + O\left(\text{Re}^4\right),$$

$$W(\tau, \zeta; \text{Re}) = W_0(\zeta, H) + \text{Re}^2 W_1\left(\zeta, \frac{dH}{d\tau}\right) + O\left(\text{Re}^4\right),$$

$$H(\tau; \text{Re}) = H_0(\tau) + \text{Re}^2 H_1(\tau) + O\left(\text{Re}^4\right). \tag{5.43}$$

Obviously, expansions (5.43) are not valid near the initial time $\tau = 0$! These are only "long-time" expansions, and "short-time" expansions will be required (see Sect. 5.4.2) to satisfy initial conditions (5.41a,b).

Now, in the long-time limit

$$\text{Re} \downarrow 0 \text{ with } \tau \text{ and } \zeta \text{ fixed}, \tag{5.44}$$

the zero-order problem becomes:

$$2F_0 + \frac{\partial W_0}{\partial \zeta} = 0, \quad \frac{\partial^2 F_0}{\partial \zeta^2} + G_0^2 = 0, \quad \frac{\partial^2 G}{\partial \zeta^2} = 0, \tag{5.45a}$$

$$\zeta = 0 \;:\; F_0 = 0, \; G_0 = 1, \; W_0 = 0, \tag{5.45b}$$

$$\zeta = H(\tau) \;:\; \frac{\partial F_0}{\partial \zeta} = 0, \; \frac{\partial G_0}{\partial \zeta} = 0, \tag{5.45c}$$

and initial conditions (5.41a,b) are discarded.

The solution of problem (5.45a–c) is very simple:

$$F_0(\zeta, H) = -\frac{1}{2}\zeta^2 + H\zeta; \quad G_0(\zeta, H) = 1, \tag{5.46a}$$

$$W_0(\zeta, H) = -\frac{1}{3}\zeta^3 + H\zeta^2. \tag{5.46b}$$

The first-order problem at $O(\text{Re}^2)$ is

$$2F_1 + \frac{\partial W_1}{\partial \zeta} = 0, \tag{5.47a}$$

$$\frac{\partial^2 F_1}{\partial \zeta^2} = \frac{\partial F_0}{\partial \tau} + F_0^2 + W_0 \frac{\partial F_0}{\partial \zeta} - 2G_0 G_1, \tag{5.47b}$$

$$\frac{\partial^2 G_1}{\partial \zeta^2} = \frac{\partial G_0}{\partial \tau} + 2F_0 G_0 + W_0 \frac{\partial G_0}{\partial \zeta}, \tag{5.47c}$$

$$\zeta = 0 \;:\; F_1 = 0, \; G_1 = 0, \; W_1 = 0; \tag{5.47d}$$

$$\zeta = H(\tau) \;:\; \frac{\partial F_1}{\partial \zeta} = 0, \; \frac{\partial G_1}{\partial \zeta} = 0. \tag{5.47e}$$

160   5 The Limit of Very Low Reynolds Numbers

First, from (5.47c) for G_1, because the right-hand side is known from (5.46a,b), we obtain the function G_1 by integration. Then, the solution for G_1 together with (5.46a,b) can be substituted in (5.47b) to give the function F_1 again by integration. Equation (5.47a) gives W_1 after integration. Finally, when we use the boundary conditions (5.47d,e), we obtain the following solution:

$$G_1 = -\frac{1}{12}\zeta^4 + \frac{1}{3}H\zeta^3 - \frac{2}{3}H^3\zeta,$$

$$F_1 = \frac{\partial H}{\partial \tau}\left[\frac{1}{6}\zeta^3 - \frac{1}{2}H^2\zeta\right] + \frac{1}{360}\zeta^6 - \frac{1}{60}H\zeta^5$$
$$+ \frac{2}{9}H^3\zeta^3 - \frac{3}{5}H^5\zeta,$$

$$W_1 = \frac{\partial H}{\partial \tau}\left[\frac{1}{2}H^2\zeta^2 - \frac{1}{12}\zeta^4\right] - \frac{1}{1260}\zeta^7 + \frac{1}{180}H\zeta^6$$
$$- \frac{1}{9}H^3\zeta^4 + \frac{3}{5}H^5\zeta^2. \tag{5.48}$$

Now, it is necessary to take into account the kinematic condition (5.40):

$$\frac{dH}{d\tau} \approx W_0(\tau, H) + \text{Re}^2 W_1(\tau, H)$$
$$= -\frac{2}{3}H^3 + \text{Re}^2\left[\frac{5}{12}H^4\frac{\partial H}{\partial \tau} + \frac{311}{630}H^7\right], \tag{5.49}$$

and with (5.43) for H, we derive two equations:

$$\frac{dH_0}{d\tau} = -\frac{2}{3}H_0^3, \tag{5.50a}$$

$$\frac{dH_1}{d\tau} = -2H_0 H_1^2 + \frac{5}{12}H_0^4\frac{dH_0}{d\tau} + \frac{311}{630}H_0^7. \tag{5.50b}$$

Equations (5.50a,b) are readily integrated to give:

$$H_0 = [3/(4\tau + A^0)]^{1/2}, \tag{5.51a}$$

$$H_1 = B^0 [3/(4\tau + A^0)]^{3/2} - \frac{17}{105}[3/(4\tau + A^0)]^{5/2}, \tag{5.51b}$$

and an inner, "short-time" expansion is required (near $t = 0$) which must then be matched to "long-time" solution (5.51a,b) to determine A^0 and B^0!

5.4.2 Solution for Small Re ≪ 1: Short-Time Scale Analysis

In short-time-scale analysis, only the time variables are stretched:

$$t^* = \tau/\text{Re}^2, \tag{5.52}$$

5.4 Film Flow on a Rotating Disc: Asymptotic Analysis for Small Re

and the unknown functions F^*, G^*, W^*, H^* depend on variables (t^*, ζ^*) and $\text{Re} \ll 1$, where ζ^* is the variable ζ (see (3.72a)) but is significant near the initial time. Then, we derive the following system of equations for the corresponding functions F^*, G^*, and W^* in place of (5.39a–c):

$$2F^* + \frac{\partial W^*}{\partial \zeta^*} = 0 ; \tag{5.53a}$$

$$\frac{\partial F^*}{\partial t^*} + \text{Re}^2 \left[F^{*2} + W^* \frac{\partial F^*}{\partial \zeta^*} \right] - G^{*2} = \frac{\partial^2 F^*}{\partial \zeta^{*2}} ; \tag{5.53b}$$

$$\frac{\partial G^*}{\partial t^*} + \text{Re}^2 \left[2F^* G^* + W^* \frac{\partial G^*}{\partial \zeta^*} \right] = \frac{\partial^2 G^*}{\partial \zeta^{*2}} , \tag{5.53c}$$

with

$$\zeta^* = 0 \;:\; F^* = 0,\; G^* = 1,\; W^* = 0 , \tag{5.53d}$$

$$\zeta^* = H^*(t^*) \;:\; \frac{\partial F^*}{\partial \zeta^*} = 0,\; \frac{\partial G^*}{\partial \zeta^*} = 0,\; \frac{dH^*}{dt^*} = \text{Re}^2 W^* , \tag{5.53e}$$

$$t^* = 0 \;:\; F^* = 0,\; G^* = 0,\; W^* = 0,\; H^* = 1,\; \frac{dH^*}{dt^*} = 0 . \tag{5.53f}$$

Now, we consider in place of (5.44), a short-time limit:

$$\text{Re} \downarrow 0 \text{ with } t^* \text{ and } \zeta^* \text{ fixed} , \tag{5.54}$$

and we expand the functions F^*, G^*, W^*, and H^* relative to a power of Re^2. At zero order, the film thickness can be determined directly from the kinematic condition (5.53e), and the result is

$$\frac{dH_0^*}{dt^*} = 0 \Rightarrow H_0^* = 1 , \tag{5.55}$$

and as a consequence at zero-order approximation, the axial coordinate is specified in the range

$$0 \leq \zeta^* \leq 1 .$$

Now, for G_0^*, we derive the following leading equation from (5.53c):

$$\frac{\partial G_0^*}{\partial t^*} - \frac{\partial^2 G_0^*}{\partial \zeta^{*2}} = 0,\; 0 \leq \zeta^* \leq 1,\; t^* > 0 , \tag{5.56a}$$

with the conditions

$$\zeta^* = 0 \;:\; G_0^* = 1,\; \zeta^* = 1 \;:\; \frac{\partial G_0^*}{\partial \zeta^*} = 0,\; t^* = 0;\; G_0^* = 0 . \tag{5.56b}$$

The solution for G_0^* is given by

$$G_0^* = 1 - 2 \sum_{\substack{n>0 \\ n=1,3,5,\ldots}} (2/n\pi) \sin(n\pi\zeta^*/2) \exp\left[-(n\pi\zeta^*/2)^2 t^*\right] . \tag{5.57}$$

For F_0^*, we derive from (5.53b) the problem

$$\frac{\partial F_0^*}{\partial t^*} - \frac{\partial^2 F_0^*}{\partial \zeta^{*2}} = G_0^{*2}, \ 0 \leq \zeta^* \leq 1, \ t^* > 0, \tag{5.58a}$$

$$\zeta^* = 0 \ : \ F_0^* = 0, \quad \zeta^* = 1 \ : \ \frac{\partial F_0^*}{\partial \zeta^*} = 0, \ t^* = 0 \ : \ F_0^* = 0, \tag{5.58b}$$

and the problem (5.58a,b) is solved by Higgins (1986, p. 3526), via a finite Fourier transform; the solution for $F_0^*(t^*, \zeta^*)$ is rather awkward-looking. Then, once F_0^* is found, (5.53a) can be integrated to find W_0^*. To determine A^0 and B^0, it is also necessary to compute the film thickness $H_1^*(t^*)$ of order Re^2 from (5.53e),

$$H_1^*(t^*) = \int_0^{t^*} W_0^*(t^{*\prime}, 1) \, dt^{*\prime} . \tag{5.59}$$

Matching is realized via an intermediate time-scale $T = \mathrm{Re}^{2\alpha} t^*$, with a scalar $\alpha > 0$ such that $\alpha < 1$, that matches the expansions for $H(\tau)$ and $H^*(t^*)$ in their region of overlap. It is required that the coefficient for each power of Re equal yields $A^0 = 3$ and $B^0 = 0.8227$ and this last value for B^0 is computed thanks to an explicit solution for film thickness $H_1^*(t^*)$, derived from (5.59); see, for instance, relation (56) in the paper by Higgins (1986).

A single composite uniform expansion for the transient film thickness is obtained by adding the long-time-scale solution (5.51a,b) to the short-time-scale solution and subtracting the result of their common part:

$$H^c(\tau; \mathrm{Re}) = H_0(\tau) + \frac{2}{3}\tau + \mathrm{Re}^2 \left[H_1^*(\tau/\mathrm{Re}^2) + H_1(\tau) - 0.6609 \right]. \tag{5.60}$$

The zero-order term $H_0(\tau)$ is shown in Fig. 5.2 as the dashed curve A. This plot is identical to what Emisle et al. (1958) found for the film thickness when they took the initial height of the fluid layer to be independent of the coordinate r.

The two-term long-time-scale solution is shown as the upper dashed curve B. As expected for short times ($\tau < 0.2$ for $\mathrm{Re} = 0.5$), this expansion becomes singular and predicts a film thickness greater than unity. As noted previously, the singular nature of the long-time-scale expansion occurs because the velocity field for the long-time-scale solution does not satisfy the initial conditions.

Interestingly, the second-order correction $H_1(\tau)$ is always positive; hence $H_0(\tau)$ represents a lower bound for the transient film thickness. Note that the choice of t as the independent time variable in Fig. 5.2 suppresses the Reynolds number dependence of $H_0(\tau)$.

When this dependence is made explicit, one obtains the expected result that with decreasing Reynolds number, the relative resistance of the film to

5.4 Film Flow on a Rotating Disc: Asymptotic Analysis for Small Re 163

thinning increases, and a longer spinning time is required to achieve a given reduction in the initial film thickness.

As expected, the difference between curve A, the lower bound for $H(\tau)$, and curve B diminishes with increasing time and with decreasing Re. For example, at Re = 0.5 and $\tau = 0.5$, curve B predicts a film thickness about 11% higher than that predicted by $H_0(\tau)$. At $\tau = 3$, the difference in predicted film thickness is only about 3%.

The composite uniform expansion given by (5.60) – curve D in Fig. 5.2 – merges with the two-term long-time-scale expansion around $\tau = 0.5$ and becomes essentially indistinguishable from it for $\tau > 0.6$.

For $\tau < 0.5$, transient effects due to the initial acceleration of the disc become important, and these effects are accounted for in the composite expansion through the short-time-scale solution, curve C in Fig. 5.2.

Finally, it should be pointed out that the above asymptotic theory is based on the assumption that the interface remains flat during spinning.

The composite expansion predicts a film of zero thickness when the spinning time becomes indefinitely large, a result that is obviously physically unreasonable! Not accounted for in this analysis are interfacial effects that eventually dominate the hydrodynamics at the rim of the disc when the film becomes sufficiently thin.

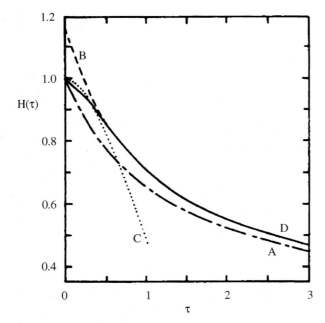

Fig. 5.2. Transient film thickness $H(t)$ for low Reynolds number flow (Re = 0.5)

5.5 Some Rigorous Mathematical Results

For rigorous mathematical results related to low Reynolds number flow, we mention the papers by Finn (1965), Babenko (1976), and Vasiliev (1977).

It is also interesting to note that within the context of Proudman and Pearson's (1957) scheme, the solution for the flow past a cylinder is not unique (!), unless steadiness is postulated.

According to Bentwich (1985), it seems impossible to obtain a proof of uniqueness for unsteady-state two-dimensional flow past a cylinder that satisfies the requirement that the disturbance flow at infinity be finite.

The reader can find a comprehensive study of this problem of uniqueness and various references in the review paper by Ladyzhenskaya (1975).

On the other hand, the reader can find in Pogu and Tournemine (1982) a study by a variational approach of 2-D biharmonic equations acting in unbounded domains. In particular, this study is applied to the problem of low Reynolds number viscous fluid flows. For an *arbitrary profile*, the authors obtain an *existence* and *uniqueness* theorem for the leading term of the *Stokes inner* expansion.

Curiously, in the paper by Temam and Wang (1996a), the reader can find an asymptotic (rigorous) analysis of Oseen type equations in a channel at low viscosity!

6 Incompressible Limit: Low Mach Number Asymptotics

6.1 Introduction

Low Mach number flow plays a dominant role in a number of flow situations, even if this is not established as firmly as it is for high or low Reynolds number flows. We shall emphasize that at least three important parts of fluid dynamics: incompressible (internal and external) viscous unsteady-state aerodynamics, nonlinear acoustics and nonadiabatic atmospheric flow, are concerned with the theory of low Mach number flow.

The smallness of the Mach number is particularly significant when it occurs simultaneously with the smallness of one or several other basic dimensionless parameters (through various similarity relations) such as Reynolds, Boussinesq (Froude), Rossby, or Strouhal numbers.

The study of low Mach number flow occurs in the general context of asymptotic modeling of fluid flow. In our recent book [Zeytounian (2002b)] we have argued the significance of asymptotic modeling of flow at low Mach number as an effective way of building up consistent models amenable to numerical simulation (it is well-known that time-dependent compressible flow schemes often become ineffective at low Mach numbers).

It has long been known [Janzen (1913), Rayleigh (1916)] that this is a sound way of motivating incompressible aerodynamics in low Mach number flow. A comprehensive discussion of it was given by Imai (1957b) in his Technical Note for steady-state flow of a perfect (inviscid) fluid. A rigorous mathematical treatment has even been provided by Klainerman and Majda (1982) and also by Ebin (1982), although the physical situation dealt with is rather simple because the physical space is devoid of material boundaries!

The example of external steady-state (inviscid) aerodynamics, which was mainly considered by Imai (1952, 1957b), is rather straightforward conceptually and allows a high degree of achievement. For instance, Hoffman (1974) used a computer program for generating power series expansions with respect to M^2 up to M^{16} in inviscid flow over a cylinder.

This is somewhat misleading and might give the erroneous impression that low Mach number expansions are straightforward!

For example, in the Proceedings of a Symposium in Honour of Professor J.P. Guiraud [held at the Université Pierre et Marie Curie, Paris, France, 20–22 April 1994; see Bois et al. (1995, p. 5)], in the paper by M. Van

Dyke, "Growing up with Asymptotics", the reader can find the following statement: "On the other hand, the other similarity parameter most familiar to fluid mechanicians, the Mach number, is the ratio of two speeds – and its square (in which form it often appears) is the ratio of two energies – but in no reasonable way can be regarded as the ratio of two lengths. Thus an asymptotic expansion for large or small Mach number is assuredly a regular perturbation (the latter is the 'Janzen-Rayleigh' or 'M^2' expansion)."

Some subtleties are involved when we consider unsteady-state weakly compressible flow at $M \ll 1$ [see, for instance, Zeytounian (1983)]. They concern [according to Zeytounian and Guiraud (1984)] both unsteady-state external aerodynamics, which leads to sound generation by low-speed flow [Blokhintsev (1981), Lighthill (1952, 1954), Lauvstad (1968), Crow (1970), Viviand(1970), Obermeier (1977)], and internal aerodynamics, which raises a number of interesting questions relating to long-time persistence of acoustic oscillations (for instance, in a cavity with a deformable wall) that have scarcely been investigated [Zeytounian and Guiraud (1980a, 1980b, 1984)]. The resonance phenomena and the formation of shock waves in closed tubes which have been investigated so much are also mainly related to low Mach number asymptotics.

In Chap. 6 of Zeytounian (2002a), the reader can find, for example, a sketch of various regions related to the limiting process $M \downarrow 0$.

But the most intriguing features are found when one deals with low Mach number flow in the atmosphere. It is somewhat discouraging to realize [Zeytounian (1990, Chap. 12)] that the low Mach number approximation looks so good from the outset but that very few useful results have been obtained by using it. The pioneering work of Monin (1961) and Drazin (1961) on so-called quasi-nondivergent flow involves some intriguing features tied to the phenomenon of blocking well known in stratified flow at low Froude number, and this might explain why they have not been used in weather prediction [see, for instance, Phillips (1970)]. Blocking is avoided whenever the low Mach number approximation is coupled with another one [Zeytounian and Guiraud (1984)]. When both Rossby and Mach numbers are low, one gets the useful quasi-geostrophic approximation [Kibel (1940)] which has been systematically investigated (from the viewpoint of asymptotic modeling) by Guiraud and Zeytounian (1980) and Zeytounian (1990; Chap. 11).

When both Boussinesq and Mach numbers are simultaneously low, Zeytounian (1974) has shown that an asymptotic expansion produces a useful model, the Boussinesq equations. An obvious interpretation (for Eulerian flow) is that in this way we recover the so-called Long's (1953) model for lee waves generated by a relief in the troposphere with a length scale $O(M)$ in comparison with the height of the troposphere [Zeytounian (1979), Guiraud and Zeytounian (1979b)].

Note also the singular feature of low Mach number expansions in the vicinity of the initial time. Consequently, it is necessary to derive the appropriate initial layer expansions which may deal with adjustment processes. Furthermore, it is also necessary to investigate the singular nature of the far field in external inviscid aerodynamics – the process of going to infinity cannot be exchanged, without caution, with that of letting the Mach number go to zero. The acoustic model may be extracted from the equations for the initial and far fields.

An important problem is matching between the various expansions and issues related to it, the powers of the Mach number that occur in these expansions.

A very difficult problem is the multiple scaling approach to the persistence of initial acoustic oscillations in a deformable cavity. This problem of the long life time of acoustic oscillations generated within a cavity by the starting process and how one can predict their evolution when the shape of the cavity has been deformed a great deal from the original shape, is briefly discussed in Sect. 6.2.2, Chap. 6, of Zeytounian (2002a).

We observe, also, that a number of papers are devoted to so-called "nearly incompressible hydrodynamics." NIHD represents the interface between compressible and incompressible "hydrofluid" descriptions in the subsonic regime. On the basis of a singular expansion technique, modified systems of fluid equations are derived for which the effects of compressibility are admitted only weakly in terms of the different possible incompressible solutions. But here it is not our intention to provide an analysis of these papers [see, for instance, Zank and Matthaeus (1990, 1991, 1992), Bayly, Levermore and Passot (1992), and Ghosh and Matthaeus (1992)].

For numerical simulations of low Mach number flow, we mention the following papers: Majda and Sethian (1985), Fröhlich and Peyret (1992), Choi and Merkle (1986), Dwyer and Yam (1989), Horibata (1992), Volpe (1993), Pletcher and Chen (1993), Klein (1995), and Geratz et al. (1996). In Müller (1996) and Voizat (1998), the reader may find a large number of references to numerical computations and simulations of a low Mach number. Naturally, the ineffectiveness of time-dependent compressible flow schemes occurs because a wide disparity exists (stiff problem!) between the time-scales for slow convection and fast propagating acoustic waves.

In this respect, one should notice that filtering of acoustic waves in compressible fluid flow equations drastically changes the type of the equations and the (initially) well-posed Navier–Stokes–Fourier compressible viscous, heat-conducting equations often become "ill-posed" [see, for instance, Oliger and Sunström (1978) paper and Kreiss and Lorenz (1989) book].

Finally, for recent rigorous mathematical results on the singular incompressible limits in compressible fluid dynamics, see, first, Beirão da Veiga (1994a,b), where the reader can find various pertinent references. For more recent rigorous mathematical results on singular incompressible and acoustic limits in slightly compressible fluid dynamics, see the papers by Iguchi (1997), Lions and Masmoudi (1998), Hagstrom and Lorenz (1998), and the more recent by Desjardins et al. (1999). Unfortunately, in these rigourous papers, the mathematical analysis is carried out almost exclusively from consideration of equations without any boundary conditions corresponding to a physical fluid dynamics problem!

But the experienced reader is aware of the fact that fluid dynamics partial differential equations are not sufficient for discussing flow problems, and this seemingly anodine remark has far-reaching consequences. In particular, applied mathematicians do not clearly distinguish between exterior and interior aerodynamics problems and do not seem to understand well the different role played by acoustic waves generated in the initial stage of motion, even if they consider the initial acoustic layer [see, the recent paper by Iguchi (1997), for example]. Low Mach number asymptotics is used, also, by Kreiss et al. (1991) for the Navier–Stokes (Navier!) to prove the convergence of the compressible flow solutions to incompressible flow solutions for $M \downarrow 0$ under certain conditions. In Sect. 8.4.1, we give information relative to the mathematical aspects of incompressible limits.

6.2 Navier–Fourier Asymptotic Model

In Chap. 2, Sect. 2.1.3, from "time-dependent density" Navier model equations (2.13a,b), with (2.13c), we derived the classical Navier incompressible and viscous system, (2.14a,b), for the divergence-free velocity vector $\boldsymbol{u}_N$ and the pseudopressure π_N given by

$$\mathrm{Lim}^N[(p-1)/M^2] \ .$$

For this, it is necessary to consider an *external aerohydrodynamics* problem. When far from a bounded body – in an unbounded fluid – we write the following (dimensionless) conditions at "infinity" for the NSF starting equations:

$$|\boldsymbol{x}| \to \infty : \rho \to 1, T \to 1, p \to 1 \ . \tag{6.1}$$

More precisely, this classical Navier incompressible viscous model is

$$\boldsymbol{\nabla} \cdot \boldsymbol{u}_N = 0 \ , \tag{6.2a}$$

$$\frac{\mathrm{d}_N \boldsymbol{u}_N}{\mathrm{d}t} + \boldsymbol{\nabla}\left(\frac{\pi_N}{\gamma}\right) = \frac{1}{\mathrm{Re}}\boldsymbol{\nabla}^2 \boldsymbol{u}_N \ , \tag{6.2b}$$

where $\mathrm{d}_N/\mathrm{d}t = \partial/\partial t + \boldsymbol{u}_N \cdot \boldsymbol{\nabla}$ and the Strouhal number S = 1.

Indeed, Navier equations (6.2a,b) are derived from the full unsteady-state NSF system (1.30)–(1.32), with (1.33) and S = 1, when we assume the existence of the following "low Mach number" *outer* asymptotic expansions:

6.2 Navier–Fourier Asymptotic Model

$$u = u_N(t, x) + M^2 u'(t, x) + \ldots ,$$
$$p = 1 + M^2[\pi_N(t, x) + M^2 \pi'(t, x) + \ldots] + \ldots ,$$
$$T = 1 + M^2 T_N(t, x) + \ldots ,$$
$$\rho = 1 + M^2 \rho_N(t, x) + \ldots , \tag{6.3}$$

which are associated with an "outer" limiting process:

$$\mathrm{Lim}^N = [M \downarrow 0, \text{ with } t \text{ and } x \text{ fixed}] . \tag{6.4}$$

Conditions at infinity. Being concerned with the motion of a fluid in a domain extending to a whole neighborhood of infinity, we need some conditions relative to it. First, for the NSF full equations (1.30)–(1.32), it seems obvious that we require, for any finite time, the behavioral conditions (6.1) for p, ρ, and T and also

$$|u| \to 0, \text{ as } |x| \to \infty , \tag{6.5}$$

for the velocity vector. When $M|x| = x^* = O(1)$, we know that low Mach number flow approximates an acoustic field. On the other hand, (6.1) and (6.5) raise some difficulties which come from the wake trailing behind the body, but we leave aside this peculiarity. When we go far from the body, excluding the region within the wake, we know that the effects of viscosity and heat conductivity die out more rapidly than inviscid effects; the perturbations are expected to decay like $O(|x|^2)$. Finally, according to (6.1) and (6.5), we may expect that:

$$\pi_N \to 0 \text{ and } |u_N| \to 0, \text{ as } |x| \to \infty , \tag{6.6}$$

provided this conclusion is not invalidated by a different (far downstream) asymptotic expansion near infinity!

Initial conditions. Here we make the choice that the fluid starts impulsively from a state of rest, at constant density and temperature, so that for the NSF motion,

$$T = 1, \quad \rho = 1, \quad \text{and} \quad u = 0, \text{ at } t = 0 , \tag{6.7}$$

in the whole domain occupied by the fluid.

We assume that the fluid is set into motion by the displacement (and eventually the deformation) of a body and the fluid pervades the entire domain Ω, complementary to this body.

Surprisingly enough, *we cannot use* the initial condition for the limiting Navier velocity vector, (6.7), to set $u_N(0, x) = 0$! Rather, we have to set

$$u_N(0, x) = u_N^0(x) , \tag{6.8a}$$

and, obviously, $u_N^0(x)$ should be divergence-free:

$$\nabla \cdot u_N^0 = 0 \,. \tag{6.8b}$$

It is somewhat puzzling that the initial value of u_N is not zero as one might expect from (6.7), and on the other hand, from what has been said, we don't have any indication of how $u_N^0(x)$ might be obtained. Nevertheless, we have the indication that the body is set impulsively into motion. This problem of impulsive motion has long been known, and the reader will find in Lamb (1932/1945, Sect. 11), a treatment for an inviscid incompressible liquid. Of course, here we have to deal with an asymptotic model, and we should go beyond Lamb's treatment to understand how one goes from the initial value $u(0, x) = 0$ for the NS-F problem to the initial value $u_N^0(x)$ for the asymptotic Navier model equations (6.2a,b), corresponding to the (main outer, Navier) limit (6.4), when we assume implicitly that the other parameters are fixed and $O(1)$ in this outer limiting process.

The conclusion that $\nabla \cdot u_N = O(M^2)$ is valid, only if we assume that $\partial \rho/\partial t$ is of the same order as ρ, that is, it is small. When a discontinuity in velocity occurs at $t = 0$, we may suspect that, close to $t = 0$, for $t > 0^+$, the order of magnitude of $\partial \rho/\partial t$ is not the same as the order of ρ_N itself.

Through the outer limit process (6.4), the derivative $\partial \rho/\partial t$ is lost, and, from what we know about asymptotic expansions, this a clue that we need a local inner expansion in the vicinity of $t = 0$.

Thus, *the divergence-free character of u_N is directly tied to this loss of $\partial \rho/\partial t$*. In Sect. 6.3, we derive the initialization adjustment problem for the Navier equations (6.2a,b), via equations of acoustics and for this, in place of (6.4), it is necessary to consider the following local in time limiting process:

$$\mathrm{Lim}^{\mathrm{Acoust}} = [M \downarrow 0, \text{ with } \tau = t/M, \text{ and } x \text{ fixed}], \tag{6.9}$$

Boundary conditions. We set $\Gamma(=\partial\Omega)$ as the boundary of Ω, n the unit vector normal to Γ, pointing toward the fluid, and Γ is the wall of the body in motion. The spatial derivative of ρ in NSF continuity equation (1.30) enters only in combination with the derivative $\partial \rho/\partial t$, via $\mathrm{d}\rho/\mathrm{d}t = \partial\rho/\partial t + u \cdot \nabla \rho$, when S = 1, and the knowledge of partial differential equations suggests that no condition has to be enforced for ρ on Γ, except if some mass of fluid were transpiring from the inside of the body through Γ. We rule out this possibility so that we need no condition relative to ρ on Γ. On the other hand, we need one condition for u and one for T on Γ. For u, we make the usual assumption that the fluid adheres to the wall (Γ) and this amounts to stating that at each point on the wall, we attach a dimensionless velocity U_w, depending on the time and on the position P on Γ such that

$$u = U_w(t, P) \equiv H(t) u_w(P), \text{ all along } \Gamma \,, \tag{6.10}$$

where $H(t)$ is the so-called Heaviside (or unit) function. This means that we focus our interest on situations when the body starts its motion impulsively.

Of course, this is not very realistic because it means that one should impart an infinite impulse to the body at time $t = 0$ to realize such an impulsive motion! A poor justification for working this way is that this is a classical problem in inviscid incompressible fluid dynamics and that it is worthwhile trying to elucidate the behavior of the motion for the NSF equations. Perhaps a more convincing argument is that, through (6.10), we are somewhat mimicking a catapulting process.

For T, we use the following dimensionless boundary condition [see, for instance, (1.62c)]:

$$\kappa^0 \frac{\partial T}{\partial n} + \text{Bi}^0 (T - 1) = T_w, \text{ all along } \Gamma, \qquad (6.11)$$

where Bi^0 and κ^0 are dimensionless numbers (Bi^0 is the Biot number), and $T_w = T_w(t, \boldsymbol{P})$ is a known function of time t and position $\boldsymbol{P}$ on Γ. We remind the reader that (6.11) simply means that the heat flux from the fluid through Γ goes inside the body at a rate that is proportional to the difference between the actual temperature of the wall and a given temperature. This temperature condition (6.11) is a "third type" condition, which is a mixture of Dirichlet (corresponding to $\kappa^0 \to 0$) and Neumann (when $\text{Bi}^0 \to 0$) conditions. The justification for such a condition (6.11) relies on the assumption that heat conduction within the body is so much faster than within the fluid, that the heat flux on Γ, considered from inside the body, may be approximated by such a difference in temperatures. For a more detailed derivation of condition (6.11), see Joseph (1976).

Finally, because the time $t > 0$ is fixed when M tends to zero, according to (6.4), for the Navier limit system (6.2a,b), we can write the following no-slip boundary condition for the Navier velocity:

$$\boldsymbol{u}_N = \boldsymbol{u}_w(\boldsymbol{P}), \text{ all along } \Gamma \text{ and when } t > 0. \qquad (6.12)$$

The Navier equations (6.2a,b) for $\boldsymbol{u}_N$ and π_N with the initial condition (6.8a) and boundary condition (6.12) form the so-called Navier incompressible and viscous model. But, it is necessary to determine the initial condition $\boldsymbol{u}_N^0(\boldsymbol{x})$ for $\boldsymbol{u}_N$.

6.2.1 The Initialization Problem and Equations of Acoustics

Now, our main goal is to derive a limiting initial boundary-value problem, issuing from $M \downarrow 0$ (but for time near $t = 0$ and $\boldsymbol{x}$ fixed), such that the time derivative in the NSF continuity equation (1.30) remains after obtaining the limiting, low Mach number form of this equation. Due to the impulsive character of the motion of the body, we expect that changes occur within a small interval of time after $t = 0$. Although it is not necessarily obvious, nevertheless, we may expect that during this short time interval, the perturbations of thermodynamic functions, p, ρ, and T, all remain small. On the other hand, we should not expect such behavior for velocity $\boldsymbol{u}$, because the smallness of such a velocity with respect to the speed of sound has already been taken care of within the non-dimensionalization of the original NSF equations.

From inspection, we guess that, with the limiting local in time process (6.9), the following changes:

$$\boldsymbol{u} = \boldsymbol{u}_a , \tag{6.13a}$$
$$(p, \rho, T) = 1 + M(p_a, \rho_a, T_a) , \tag{6.13b}$$

where the "acoustics" functions $\boldsymbol{u}_a$ and p_a, ρ_a, T_a depend on $\tau, \boldsymbol{x}$, and M, will work close to initial time $t = 0$.

It is very easy to check that substituting (6.13a,b) in NSF equations (1.30)–(1.33) gives, neglecting terms which are $O(M^2)$,

$$\frac{\partial \rho_a}{\partial \tau} + \boldsymbol{\nabla} \cdot \boldsymbol{u}_a + M\left[\boldsymbol{u}_a \cdot \boldsymbol{\nabla}\rho_a + \rho_a \boldsymbol{\nabla} \cdot \boldsymbol{u}_a\right] = 0 ; \tag{6.14a}$$

$$\frac{\partial \boldsymbol{u}_a}{\partial \tau} + \boldsymbol{\nabla}(\frac{p_a}{\gamma}) + M\left\{\rho_a\frac{\partial \boldsymbol{u}_a}{\partial \tau} + \boldsymbol{u}_a \cdot \boldsymbol{\nabla}\boldsymbol{u}_a\right.$$
$$\left. - \frac{1}{\text{Re}}\left[\boldsymbol{\nabla}^2\boldsymbol{u}_a + \left[\frac{1}{3} + \sigma^0\right]\boldsymbol{\nabla}(\boldsymbol{\nabla}\boldsymbol{u}_a)\right]\right\} = O(M^2) ; \tag{6.14b}$$

$$\frac{\partial T_a}{\partial \tau} + (\gamma - 1)\boldsymbol{\nabla} \cdot \boldsymbol{u}_a + M\left\{\boldsymbol{u}_a \cdot \boldsymbol{\nabla}T_a + \rho_a\frac{\partial T_a}{\partial \tau}\right.$$
$$\left. + (\gamma - 1)p_a\boldsymbol{\nabla} \cdot \boldsymbol{u}_a - \frac{\gamma}{\Pr\text{Re}}\boldsymbol{\nabla}^2 T_a\right\} = O(M^2) ; \tag{6.14c}$$

$$p_a - \rho_a - T_a - MT_a\rho_a = 0 . \tag{6.14d}$$

Consequently, through the local inner, initial, limiting process (6.9), we derive the following set of leading-order local equations:

$$\frac{\partial \rho_{a,0}}{\partial \tau} + \boldsymbol{\nabla} \cdot \boldsymbol{u}_{a,0} = 0 ,$$
$$\frac{\partial \boldsymbol{u}_{a,0}}{\partial \tau} + \boldsymbol{\nabla}\left(\frac{p_{a,0}}{\gamma}\right) = 0 ,$$
$$\frac{\partial T_{a,0}}{\partial \tau} + (\gamma - 1)\boldsymbol{\nabla} \cdot \boldsymbol{u}_{a,0} = 0 ,$$
$$p_{a,0} = \rho_{a,0} + T_{a,0} . \tag{6.15}$$

The first consequence of this is that, because none of the time derivatives has been lost in the local inner, initial, limit process (6.9), we may apply the initial conditions (6.7) to the system (6.15) and get, according to (6.13a,b),

$$\tau = 0 : \boldsymbol{u}_{a,0} = 0, \ p_{a,0} = \rho_{a,0} = T_{a,0} = 0 , \tag{6.16}$$

everywhere outside the body. Now we run into a problem, this time with the boundary conditions. Equations (6.15) are the dimensionless form of the equations of (linear) acoustics in a homogeneous gas at rest. We know that, for those equations, the only condition that might be applied on the boundary is one of slip of the gas with respect to the wall. We have to come back to (6.10) and observe that $H(t) = H(\tau)$, provided $\tau > 0$. Such a statement necessitates proof, but we may argue physically, and this will be sufficient for our purpose. Then we get the desired boundary condition:

$$\boldsymbol{u}_{a,0} \cdot \boldsymbol{n} = \boldsymbol{u}_w(\boldsymbol{P}) \cdot \boldsymbol{n} \equiv w_w(\boldsymbol{P}), \text{ all along } \Gamma, \tag{6.17}$$

and we observe that $w_w(\boldsymbol{P})$ does not depend on τ.

Let us now leave aside the matter of the boundary conditions which have been lost (on Γ) in the process (6.9) and concentrate on the solution of the acoustics problem (6.15)–(6.17). We observe, first, that due to (6.8b), subtracting $\boldsymbol{u}_N^0$ from $\boldsymbol{u}_N$ does not change anything in the set of (6.15). It is then very easy to check that the following formulae:

$$\boldsymbol{u}_{a,0} = \boldsymbol{u}_N^0(\boldsymbol{x}) + \boldsymbol{\nabla} \phi_{a,0}(\tau, \boldsymbol{x}), \tag{6.18a}$$

$$\rho_{a,0} = -\frac{\partial \phi_{a,0}}{\partial \tau}, \ p_{a,0} = -\gamma \frac{\partial \phi_{a,0}}{\partial \tau}, \tag{6.18b}$$

$$T_{a,0} = (1-\gamma)\frac{\partial \phi_{a,0}}{\partial \tau}, \tag{6.18c}$$

solve (6.15), provided that $\phi_{a,0}$ is a solution of d'Alembert's dimensionless equation of acoustics,

$$\frac{\partial^2 \phi_{a,0}}{\partial \tau^2} - \boldsymbol{\nabla}^2 \phi_{a,0} = 0, \tag{6.19}$$

where the speed of sound is replaced by unity due to the choice in the process of putting the NSF equations in dimensionless form (1.30)–(1.33).

From the initial conditions (6.16), we derive

$$\tau = 0: \ \boldsymbol{\nabla} \phi_{a,0} = -\boldsymbol{u}_N^0(\boldsymbol{x}), \ \frac{\partial \phi_{a,0}}{\partial \tau} = 0, \tag{6.20}$$

everywhere outside the body, and, from the boundary condition (6.17), we get a boundary condition which (because the only restriction put on $\boldsymbol{u}_N^0(\boldsymbol{x})$ until now is (6.8b)) is written in the following form:

$$\boldsymbol{u}_N^0 \cdot \boldsymbol{n} = w_w(\boldsymbol{P}), \text{ all along } \Gamma, \tag{6.21a}$$

so that the boundary condition for ϕ_a on the body wall is

$$\frac{\partial \phi_{a,0}}{\partial n} = 0, \text{ all along } \Gamma. \tag{6.21b}$$

Because (6.8b) and (6.21a) are the only restrictions put on $\boldsymbol{u}_N^0$, we are somewhat short of a condition at infinity for the complete determination of $\phi_{a,0}$. But we may get rid of this slight difficulty by setting

$$\boldsymbol{u}_N^0 = \boldsymbol{\nabla}\psi_N^0 ,\tag{6.22}$$

which is allowed by the first equation of (6.15), and observing that we may then determine ψ_N^0 through the following problem:

$$\boldsymbol{\nabla}^2 \psi_N^0 = 0, \text{ everywhere outside the body },$$
$$\frac{\partial \psi_N^0}{\partial n} = w_w(\boldsymbol{P}), \text{ all along } \Gamma ,$$
$$\psi_N^0 \to 0, \text{ when } |\boldsymbol{x}| \to \infty ,\tag{6.23}$$

which is a straightforward *Neumann problem for the Laplace equation*.

Then, instead of (6.20), we obtain:

$$\tau = 0 \;:\; \phi_{a,0} = -\psi_N^0, \; \frac{\partial \phi_{a,0}}{\partial \tau} = 0 ,\tag{6.24}$$

everywhere outside the body.

Now, the acoustic wave equation (6.19), with (6.21b) and (6.24), leads to a well-posed problem for $\phi_{a,0}$, provided we add

$$\phi_{a,0} \to 0 \text{ as } |\boldsymbol{x}| \to \infty ,\tag{6.25}$$

which amounts to added information that no perturbations come from infinity toward the body, and one must consider that such information is of a physical rather than a mathematical character. We are not actually interested in getting $\phi_{a,0}$; all that we want to know is that

$$\tau \to \infty \;:\; \phi_{a,0} \to 0 ,\tag{6.26}$$

which is guaranteed by the mathematical theory of d'Alembert's equation [see, for instance, Wilcox (1975)]. This provides, according to (6.18a):

$$\tau \to \infty \;:\; \boldsymbol{u}_{a,0} \to \boldsymbol{u}_N^0 ,\tag{6.27}$$

so that, *matching* the Navier solution with the present (acoustics) solution, we get

$$\lim_{t \to +0} \boldsymbol{u}_N(t, \boldsymbol{x}) \sim \lim_{\tau \to \infty} \boldsymbol{u}_{a,0}(\tau, \boldsymbol{x}) ,\tag{6.28}$$

and, finally, we have found what was missing to achieve a complete Navier initial boundary-value problem (Navier IBVP): (6.18a), where $\phi_{a,0}$ is the solution of the acoustics problem, (6.19), (6.21b), (6.24), and (6.25) and $\boldsymbol{u}_N^0$ is completely determined by (6.23) with (6.22).

This omnipotence of the divergenceless velocity constraint and its relation to the initial condition for Navier equation (6.2b), which guarantees the well-posedness of the Navier initial boundary-value problem, is thoroughly discussed in the review paper by Gresho (1992, pp. 47–52). Scrutinizing what is stated there, we see that there is a close relation to Lamb (1932/1945, Sect. 11).

6.2.2 The Fourier Model

A second problem in the framework of the Navier-Fourier asymptotic model is related to the limiting form of the energy equation (1.32) for the temperature T under the outer limiting process (6.4). This problem was solved in Sect. 2.1.4, Chap. 2, and we derived the following consistent Fourier equation:

$$\frac{\partial T_N}{\partial t} + \boldsymbol{u}_N \cdot \boldsymbol{\nabla} T_N - \frac{\gamma}{\Pr \operatorname{Re}} \boldsymbol{\nabla}^2 T_N$$
$$= (\gamma - 1) \left\{ \frac{2\gamma}{\operatorname{Re}} \operatorname{Trace} \left[\mathbf{D}(\boldsymbol{u}_N)^2 \right] - \boldsymbol{\nabla} \cdot \boldsymbol{u}' \right\} . \tag{6.29}$$

An obvious consequence of (1.33) according to (6.3) is

$$\rho_N = \pi_N - T_N . \tag{6.30}$$

Of course, the full determination of T_N necessitates that we provide initial and boundary conditions. Warned by the discussion concerning the initial value of $\boldsymbol{u}_N$, we must be cautious concerning the initial value for θ_N; so we simply write

$$T_N(0, \boldsymbol{x}) = T_N^0(\boldsymbol{x}) , \tag{6.31}$$

leaving aside, for the moment, the problem of determining $T_N^0(\boldsymbol{x})$. For the boundary condition, we look at (6.11), and we immediately see an inconsistency, unless we assume that κ^0 and Bi^0 are large numbers such that

$$\kappa^0 = \frac{\kappa^*}{M^2} \quad \text{and} \quad \mathrm{Bi}^0 = \frac{\mathrm{Bi}^*}{M^2} , \tag{6.32a}$$

with

$$\kappa^* = O(1), \ \mathrm{Bi}^* = O(1) .$$

When (6.32a) holds, we get the full boundary condition for T_N:

$$\kappa^{0*} \frac{\partial T_N}{\partial n} + \mathrm{Bi}^* T_N = T_w, \text{ all along } \Gamma . \tag{6.32b}$$

But, if to the contrary,

$$\kappa^{0*} = 0 \quad \text{or} \quad \mathrm{Bi}^* = 0 , \tag{6.33a}$$

we have, rather, a degenerate (Dirichlet/Neumann) form,

$$T_N = \frac{1}{\mathrm{Bi}^*} T_w, \text{ or } \frac{\partial T_N}{\partial n} = \frac{1}{\kappa^{0*}} T_w, \text{ all along } \Gamma . \tag{6.33b}$$

When κ^0 and Bi^0 are both $O(1)$, we do not escape the conclusion that T_N is no longer of the order of M^2, but we should not be disturbed by that conclusion. In this case, it is necessary to assume that in the "exact" condition (6.11), the known function T_w is proportional (as is the perturbation of temperature T) to M^2. We observe that usually the Biot number is always a small parameter.

As we have already mentioned, there are two causes of heating (or cooling) a gas. One comes from viscous dissipation, and it causes T to grow like M^2. The second cause is through heat transfer at the wall Γ, and such a heat transfer process gives rise to (6.11).

If the heat transfer process at the wall is not consistent with either (6.32a) or (6.33a) and if $T_w \neq M^2 T_w^*$, with $T_w^* = O(1)$, then we need reconsider the expansion (6.3) for T. See, for instance, Zeytounian (1977) and Zank and Matthaeus (1990), for information concerning the case when T_N is no longer of the order of M^2. If κ^0, Bi, and θ_w are $O(1)$, we might expect that T is also $O(1)$. But, when π is $O(M^2)$, $\rho + T + T\rho = 0$, and we run into difficulties because such a relation is inconsistent with two equations derived from (1.30) and (1.32) for ρ and T.

Considering the Fourier equation (6.29), with (6.31) [assuming that we have been able to compute $T_N^0(\boldsymbol{x})$; see below] and either (6.32b) or (6.33b), we get a well-posed "Fourier" initial boundary-value problem (Fourier IBVP). We may state that, under the main outer limit (6.4), with (6.3) and either (6.32a) or (6.33a), both limit problems,

the Navier IBVP and the Fourier IBVP are decoupled from each other.

This conclusion is well known for liquids. Except when strong heating or cooling occurs at the wall, the motion may be computed first using the Navier incompressible equations, and then the temperature is obtained from the Fourier equation and appropriate conditions. Here, we find a corresponding situation for gases. Of course, we know that, when the viscosity of a fluid is very sensitive to temperature, it may happen that heating by viscous dissipation influences the motion through variation in viscosity. This happens, for example, in lubrication theory, but then, the Reynolds number is also a small parameter, as assumed in Chap. 5.

Some remarks concerning the determination of $T_N^0(\boldsymbol{x})$. Going back to equations (6.14a–d), we remind the reader of our warning about $T_N^0(\boldsymbol{x})$. Let us try to examine this issue, assuming that $\boldsymbol{u}_a$, ρ_a, p_a, T_a are expanded in the following form:

$$\boldsymbol{u}_a = \boldsymbol{u}_{a,0} + M\boldsymbol{u}_{a,1} + \ldots , \tag{6.34a}$$

$$(\rho_a, p_a, T_a) = (\rho_{a,0}, p_{a,0}, T_{a,0}) + M(\rho_{a,1}, p_{a,1}, T_{a,1}) + \ldots . \tag{6.34b}$$

Then, for the functions $\boldsymbol{u}_{a,1}$, $\rho_{a,1}$, $p_{a,1}$, $T_{a,1}$, we derive the following second-order inhomogeneous acoustic equations from (6.14a)–(6.14d):

$$\frac{\partial \rho_{a,1}}{\partial \tau} + \nabla \cdot \boldsymbol{u}_{a,1} = \frac{\partial Q_{a,0}}{\partial \tau},$$

$$\frac{\partial \boldsymbol{u}_{a,1}}{\partial \tau} + \nabla \left(\frac{p_{a,1}}{\gamma} \right) = \nabla P_{a,0},$$

$$\frac{\partial T_{a,1}}{\partial \tau} + (\gamma - 1) \nabla \cdot \boldsymbol{u}_{a,1} = \frac{\partial R_{a,0}}{\partial \tau},$$

$$p_{a,1} - (\rho_{a,1} + T_{a,1}) = (\gamma - 1) \left(\frac{\partial \phi_{a,0}}{\partial \tau} \right)^2, \tag{6.35}$$

where

$$Q_{a,0} = \nabla \psi_N^0 \cdot \nabla \phi_{a,0} + \frac{1}{2} (\nabla \phi_{a,0})^2 + \frac{1}{2} \left(\frac{\partial \phi_{a,0}}{\partial \tau} \right)^2;$$

$$P_{a,0} = \frac{1}{2} \left(\frac{\partial \phi_{a,0}}{\partial \tau} \right)^2 - \frac{1}{2} |\nabla \psi_N^0 + \nabla \phi_{a,0}|^2 + \frac{1}{2\,\mathrm{Re}} \left(\frac{4}{3} + \sigma^0 \right) \nabla^2 \phi_{a,0};$$

$$R_{a,0} = (\gamma - 1) \left[\nabla \psi_N^0 \cdot \nabla \phi_{a,0} + \frac{1}{2} (\nabla \phi_{a,0})^2 \right.$$

$$\left. + \frac{\gamma}{2} \left(\frac{\partial \phi_{a,0}}{\partial \tau} \right)^2 - \left(\frac{\gamma}{\mathrm{Pr}\,\mathrm{Re}} \right) \nabla^2 \phi_{a,0} \right]. \tag{6.36}$$

It is easily checked that the following formulae solve (6.35) with (6.36):

$$\boldsymbol{u}_{a,1} = \nabla \frac{\partial \phi_{a,1}}{\partial \tau}, \quad \rho_{a,1} = -\nabla^2 \phi_{a,1} + Q_{a,0},$$

$$T_{a,1} = -(\gamma - 1) \nabla^2 \phi_{a,1} + R_{a,0},$$

$$p_{a,1} = -\gamma \frac{\partial^2 \phi_{a,1}}{\partial \tau^2} + \gamma P_{a,0}, \tag{6.37}$$

provided that $\phi_{a,1}$ is a solution of the d'Alembert's inhomogeneous equation:

$$\frac{\partial^2 \phi_{a,1}}{\partial \tau^2} - \nabla^2 \phi_{a,1} = P_{a,0} - \frac{1}{\gamma}[Q_{a,0} + R_{a,0}] - \frac{\gamma - 1}{\gamma} \left(\frac{\partial \phi_{a,0}}{\partial \tau} \right)^2, \tag{6.38}$$

to which we must add

$$\tau = 0 : \phi_{a,1} = 0, \quad \frac{\partial \phi_{a,1}}{\partial \tau} = 0, \text{ everywhere outside the body,} \tag{6.39a}$$

$$\frac{\partial \phi_{a,1}}{\partial n} = 0, \text{ all along } \Gamma. \tag{6.39b}$$

From the value of $P_{a,0}$, $Q_{a,0}$, and $R_{a,0}$, according to (6.36), we conclude that the right-hand side of d'Alembert's inhomogeneous equation (6.38) tends to zero when $|\boldsymbol{x}| \to \infty$ at τ fixed. Relying on the same physical argument as for (6.25), we may add

$$\phi_{a,1} \to 0, \text{ when } |\boldsymbol{x}| \to \infty, \tag{6.39c}$$

and get, with (6.39a–c), a well-posed initial boundary-value problem for (6.38) relative to the function $\phi_{a,1}$.

Now, matching the Fourier IBVP solution with the solution of the inhomogeneous acoustics problem (6.38), (6.39a–c), gives the following relation for the initial condition for the Fourier IBVP:

$$T_N^0(\boldsymbol{x}) = -(\gamma - 1)\boldsymbol{\nabla}^2 \left[\lim_{\tau \to \infty} \phi_{a,1}\right]. \tag{6.40}$$

A rough argument suggests (!), that the solution of equation (6.38) for $\phi_{a,1}(\tau, \boldsymbol{x})$ tends to zero as $\tau \to \infty$, and, consequently,

$$T_N^0(\boldsymbol{x}) = 0. \tag{6.41}$$

Such a statement necessitates a rigorous proof, and as a consequence the result (6.41) is only a conjecture for the moment!

But, from known results concerning d'Alembert's equation (6.19), for $\phi_{a,0}$, we may be more precise about the behavior (6.26). Thanks again to the Wilcox (1975) paper, in Zeytounian (2002b, Chap. 5, Sect. 5.3), the reader can find a tentative justification of conjecture (6.41).

6.2.3 Influence of Weak Compressibility: Second-Order Equations for $\boldsymbol{u}'$ and $\boldsymbol{\pi}'$

From continuity equation (1.30) for the compressible NSF model, retaining terms factorized by M^2, we get the following equation according to expansion (6.3):

$$\boldsymbol{\nabla} \cdot \boldsymbol{u}' = -\left[\frac{\partial \rho_N}{\partial t} + \boldsymbol{u}_N \cdot \boldsymbol{\nabla}\rho_N\right]. \tag{6.42a}$$

We emphasize that the right-hand side of the Fourier equation (6.29) is known beforehand, once the solution of the Navier model system (6.2a,b) is known and $\boldsymbol{\nabla} \cdot \boldsymbol{u}'$ is expressed from (6.42a) through the density perturbation ρ_N, which satisfies (6.30).

Now, as a useful by-product of the solution of the Fourier equation (with initial and boundary conditions), we may get an equation of motion for $\boldsymbol{u}'$. From (1.31) with the outer asymptotic expansions (6.3), we derive the following equation for $\boldsymbol{u}'$ at order M^2:

$$\frac{\partial \boldsymbol{u}'}{\partial t} + \boldsymbol{u}' \cdot \nabla \boldsymbol{u}_N + \boldsymbol{u}_N \cdot \nabla \boldsymbol{u}' + \nabla\left(\frac{\pi'}{\gamma}\right) - \frac{1}{\text{Re}} \nabla^2 \boldsymbol{u}'$$
$$= -\rho_N \left[\frac{\partial \boldsymbol{u}_N}{\partial t} + \boldsymbol{u}_N \cdot \nabla \boldsymbol{u}_N\right] + \frac{1}{\text{Re}}\left\{2 \left.\frac{d\mu}{dT}\right|_{T=1} \nabla \cdot [\mathbf{D}(\boldsymbol{u}_N) T_N]\right.$$
$$\left.-\left[\frac{1}{3} + \sigma^0\right] \nabla \left[\frac{\partial \rho_N}{\partial t} + \boldsymbol{u}_N \cdot \nabla \rho_N\right]\right\}, \tag{6.42b}$$

where σ^0 is the ratio of viscosities ($= 0$ if we take into acount the Stokes hypothesis).

Equation (6.42b) for $\boldsymbol{u}'$ is linear with a known right-hand side. Thus, we obtain a second-order closed system (6.42a,b) for $\boldsymbol{u}'$ and π'.

Of course, we need initial and boundary conditions. It seems obvious that on the body wall,

$$\boldsymbol{u}' = 0, \text{ all along the wall } \Gamma, \tag{6.43a}$$

but what should be enforced at *infinity* and also at *initial time* is less obvious, but probably,

$$\boldsymbol{u}' = 0, \text{ at infinity and for } t = 0. \tag{6.43b}$$

For concerning the initial condition, it is necessary to investigate again the acoustic region near $t = 0$ and the adjustment problem up to the term $O(M^2)$ for the "acoustic" velocity!

An interesting application of the above results is to compute the companion solution T_N of the Fourier problem (6.29), with (6.32b)/(6.33b) and (6.41) from a known solution $(\boldsymbol{u}_N, \pi_N)$, of the Navier model equations (6.2a,b). Then, the second-order equations (6.42a) and (6.42b) for $\boldsymbol{u}'$ and π', with (6.30), and conditions (6.43a,b) make it possible to take into account the effects of weak compressibility.

From the Navier-Fourier asymptotic model, with all known incompressible and viscous (Navier) solutions, we can associate a solution of the Fourier equation for the temperature perturbation and also take into account the effects of weak compressibility to obtain a weakly compressible solution of the physical problem under consideration.

6.2.4 Concluding Remarks

In conclusion, we note, first, that it is clear that the properties of the initial data influence the type of "nearly incompressible" flow properties obtained asymptotically at a low Mach number. The flow with small initial density fluctuations (thus one excludes acoustic waves in the initial data) remains nearly incompressible for the duration of the simulation, and in agreement with nearly incompressible theory, the density fluctuations remain scaled to the square of a (small) Mach number.

There is no suggestion that the nearly incompressible simulations ever evolve toward strongly compressive scalings.

A second remark concerns the case when the body is set in movement *rapidly* during an interval of time proportional to M or else *progressively* during a period of time $O(1)$.

In the *first case*, the previous result [problem (6.23) with (6.22)] for the initial (Navier) value $\boldsymbol{u}_N^0$ remains true because

$$\boldsymbol{U}_w(t, \boldsymbol{P}) = \boldsymbol{U}_w(\tau, \boldsymbol{P}),$$

and according to matching condition (6.28), with (6.26) and (6.27), in place of $w_w(\boldsymbol{P})$ in condition (6.21b), we can write

$$\boldsymbol{U}_w(\infty, \boldsymbol{P}) \cdot \boldsymbol{n} = W_w(\boldsymbol{P}).$$

On the contrary, in the *second case*, the corresponding Neumann problem (6.23) has only the *trivial zero solution* because in this case the displacement velocity of a material point $\boldsymbol{P}$ of the wall Γ of the body,

$$\boldsymbol{U}_w(t, \boldsymbol{P}) = \boldsymbol{U}_w(M\tau, \boldsymbol{P}) \to \boldsymbol{U}_w(0, \boldsymbol{P}) \equiv 0,$$

when $M \downarrow 0^+$ with τ fixed [local inner limit (6.9)], and consequently, $w_w \equiv 0$, in the boundary condition of the problem (6.23). In this second case, the acoustic region near $t = 0$ plays a "passive role" and does not have any influence on the leading-order Navier limit problem.

Now, a short comment concerning the formulation of a well-posed initial boundary-value problem for the Navier equations, according to Gresho (1992, pp. 47–52) and discussed in Chap. 2, Sect. 2.2.

The analogy with the above result is disconcerting.

For $\boldsymbol{u}_0 \equiv 0$, our ψ_N^0 is the Gresho λ, but our result is only true for the unsteady-state external aerodynamics problem, when, for the acoustics problem (6.19), (6.21b), and (6.24), we have the behavior (6.25)! For the unsteady-state internal aerodynamics problem, this behavior (6.25) seems (!) true only when the wall of the body is set in motion progressively during an interval of time $O(1)$.

A final remark concerning the singular nature of the limit process (6.9), which gives the linear equations of acoustics (6.15) with slip condition (6.17). In the *vicinity of the wall and near the initial time*, it is necessary to consider a new limit process in place of (6.15) and to derive a new set of consistent (so-called *Rayleigh*) equations, in place of (6.15).

In this case, the slip condition (6.17) appears as a matching condition (relative to the coordinate normal to the wall Γ). This problem has been investigated in Sect. 4.3 of Chap. 4, in the framework of high Reynolds number asymptotics for the initialization of the classical Prandtl boundary-layer equations (the initial condition for the tangential component of the boundary-layer velocity vector must be asymptotically derived through an adjustment problem).

6.3 Compressible Low Mach Number Models

6.3.1 Hyposonic Model for Flow in a Bounded Cavity

The so-called "hyposonic" model equations (2.66a–c), derived in Sect. 2.3.1, Chap. 2, for the velocity u, temperature T and modified pressure p^*, according to (2.65a), are a "compressible, low Mach number" model. Below, we generalize this model for flow in a bounded cavity.

When we consider the flow in a bounded cavity Ω, then it is possible to derive a slightly more general model, if we take into account that, in place of (2.65a), we must write

$$p = p_0(t) + M^2 p^*(t, x) + o(M^2) , \qquad (6.44a)$$

because we do not have the possibility in a bounded domain of assuming that $p_0(t) = 1$! With (6.44a), we also consider the expansions,

$$u = u_0(t, x) + o(1), \quad \rho = \rho_0(t, x) + o(1) , \qquad (6.44b)$$
$$T = T_0(t, x) + o(1) . \qquad (6.44c)$$

This might occur if a gas is strongly heated (as a consequence of the boundary condition for the temperature on the wall $\partial\Omega$) and then, for velocity $u(t, x)$ and thermodynamic functions $p^*(t, x)$, $T_0(t, x)$, we derive, when Bo = 0, the following equations from the full NSF equations (1.30)–(1.32):

$$\boldsymbol{\nabla} \cdot \boldsymbol{u}_0 = -S \frac{d(\log \rho_0)}{dt} ; \qquad (6.45a)$$

$$\rho_0 S \frac{d_0 T_0}{dt} + (\gamma - 1) p_0(t) \boldsymbol{\nabla} \cdot \boldsymbol{u}_0 = \frac{\gamma}{\Pr \operatorname{Re}} \boldsymbol{\nabla} \cdot [k(T_0) \boldsymbol{\nabla} T] ; \qquad (6.45b)$$

$$\rho_0 S \frac{d_0 \boldsymbol{u}_0}{dt} + \boldsymbol{\nabla} \left(\frac{p^*}{\gamma} \right)$$
$$= \frac{1}{\operatorname{Re}} \boldsymbol{\nabla} \cdot [2\mu(T_0) \mathbf{D}(\boldsymbol{u}_0)] - \frac{1}{\operatorname{Re}} \boldsymbol{\nabla} \cdot \left\{ \lambda S \left(\frac{d(\log \rho_0)}{dt} \right) \boldsymbol{I} \right\} , \qquad (6.45c)$$

involving $\rho_0(t, x)$ and $p_0(t)$. Here, $p_0(t)$ is ruled by the constancy in time of the quantity

$$\mathcal{M}^0 = p_0(t) \int_\Omega \frac{1}{T_0(t, x)} \, dx , \qquad (6.46)$$

as a consequence of the conservation of overall mass of the bounded cavity Ω, where the integral is over the whole domain occupied by the gas. Of course $\rho_0(t, x)$ should not be considered one of the basic unknown functions because,

182   6 Incompressible Limit: Low Mach Number Asymptotics

according to equation of state (1.33), it is related to $T_0(t, \boldsymbol{x})$ and $p_0(t)$ by the equation:

$$\rho_0(t, \boldsymbol{x}) = \frac{p_0(t)}{T_0(t, \boldsymbol{x})} \ . \tag{6.47}$$

On the other hand, (6.45a) cannot be considered an equation that rules the evolution of $\rho_0(t, \boldsymbol{x})$; it must be considered as a constraint imposed on the velocity field $\boldsymbol{u}_0$ necessary for determining p^*/γ. A significant feature of the above system is that $\mu(T_0)$ and $k(T_0)$ may no longer be considered spatially constant. To some extent, this is reminiscent of the situation that occurs when a constant density liquid is heated and has viscosity and conductivity coefficients which depend, sufficiently strongly, on temperature. In Fig. 6.1, the reader can find the results of a computation performed by Viozat (1998; p. 39) with the above hyposonic (low Mach number) equations, (6.45a–c), (6.46), and (6.47). Viozat considered an adiabatic two-dimensional square cavity (with $\partial T_0/\partial s = 0$ on the wall, with $s = x$ or $s = y$) on the (x, y) plane, and in this case $p_0 = 1$. The upper part of the wall simulated by $y = 1$ and $0 \le x < 1$ moves at a nearly uniform velocity, and the no-

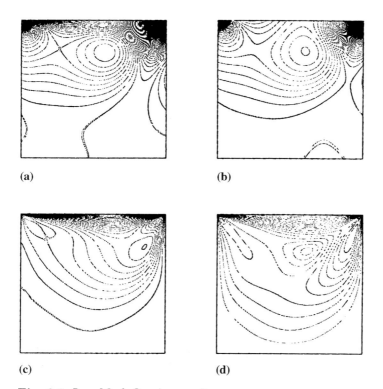

Fig. 6.1. Low Mach flow in a cavity

slip condition is assumed on the wall. Concerning the initial conditions at $t=0$, the fluid is assumed at rest, $T_0 = \rho_0 = 1$ and $\boldsymbol{u}_0 = \boldsymbol{u}_i$, with $\boldsymbol{u}_i$ the steady-state solution of the incompressible problem. The fact that the initial conditions verify the conditions on the wall is important, because the uniqueness and existence theorems of Embid (1987) are proven with this hypothesis. In Fig. 6.1, isovalues are shown for the fluctuations of density (a), pressure (b), and temperature (c) and for the module of the velocity (d) for $Re = 100$, $Pr = 0.7$, and $\gamma = 1.4$.

6.3.2 Large Channel Aspect Ratio, Low Mach Number, Compressible Flow

Another low Mach number compressible flow (LMNF) model, which is a relatively uncommon regime of fluid dynamics, has been recently derived by Shajii and Freidberg (1996). A number of new and somewhat surprising results, which follow from the analysis of this LMCF model, are summarized in Shajii and Freidberg (1996). The specific applications discussed in the paper cited involve steady-state two-dimensional fully developed laminar flow in a straight circular channel. Under these conditions, the LCMF model reduces to [according to Shajii and Freidberg (1996; p. 135)]:

$$\frac{1}{r}\frac{\partial(\rho r u)}{\partial r} + \frac{\partial(\rho w)}{\partial z} = 0 \; ;$$

$$\frac{\partial p}{\partial z} = \frac{1}{r}\frac{\partial}{\partial r}\left[r\mu\left(\frac{\partial w}{\partial r}\right)\right] \; ;$$

$$\frac{1}{r}\frac{\partial}{\partial r}\left[rk\left(\frac{\partial T}{\partial r}\right)\right] = 0; \; p = R\rho T \; ;$$

$$\frac{3}{2}\rho R\left[u\frac{\partial T}{\partial r} + w\frac{\partial T}{\partial z}\right] + p\left[\frac{1}{r}\frac{\partial(\rho r u)}{\partial r} + \frac{\partial(\rho w)}{\partial z}\right]$$

$$= \frac{1}{r}\frac{\partial}{\partial r}\left[rk\left(\frac{\partial T_1}{\partial r}\right)\right] + \mu\left(\frac{\partial w}{\partial r}\right)^2 \; . \tag{6.48}$$

In (6.48), the velocity vector is $\boldsymbol{u} = u\boldsymbol{e}_r + w\boldsymbol{e}_z$, and the funtions u, w, ρ, T, and T_1 are functions of (r, z). The pressure $p = p(z)$ is determined by means of an integrability condition on the last equation of (6.48). System (6.48) describes low Mach number compressible flow according to a derivation by Shajii and Freidberg (1996). Unfortunately, it seems that this derivation via a "basic ordering scheme" that defines low Mach number compressible flow, and the relations between the Mach, Reynolds, and the aspect ratio of the flow,

$$\Lambda = \frac{\text{length}}{\text{diameter}} \gg 1 \; ,$$

which has an enormous value ($\Lambda = 10^6$), is not fully consistent with the spirit of rigorous and logical asymptotic modeling. This remark concerns especially obtaining the "first higher order nonvanishing contribution," the last equations of (6.48) and also the equation for p_2 (written for $\mu = \text{const}$):

$$\frac{\partial p_2}{\partial r} = \mu \frac{1}{r}\frac{\partial(r\partial u/\partial r)}{\partial r} + \frac{1}{3}\mu \frac{\partial}{\partial r}\left[\frac{1}{r}\frac{\partial(\rho r u)}{\partial r} + \frac{\partial(\rho w)}{\partial z}\right]. \quad (6.49)$$

We observe that in (6.48) and (6.49), friction (due to viscosity and thermal conductivity) dominates inertia and even so, because of the large aspect ratio Λ, finite pressure, temperature, and density gradients are required, implying that compressibility effects are also important.

The analysis of the solution of (6.48) shows that for forced channel flow, steady-state-state solutions exist only below a critical value of heat input.

6.4 Viscous Nonadiabatic Boussinesq Equations

In the companion book, Zeytounian (2002a), we derived the Boussinesq inviscid equations in Sect. 5.3, Chap. 5. Below, the viscous, nonadiabatic case is considered.

6.4.1 The Basic State

For the analysis of motions under the influence of the force of gravity, when the Boussinesq number Bo is different from zero in the full NSF equations, (1.30)–(1.32) with (1.33), it is useful to postulate the existence of a basic state which is assumed to exist in the form of a thermodynamic reference situation. Below, we work with dimensionless quantities, and p_S, ρ_S, T_S denote the thermodynamic basic state functions dependent only on the basic vertical coordinate $z_S = \text{Bo}\, z$. In fact, if the relative velocities are small, then the "true" pressure will be disturbed only slightly from the static value $p_S(z_S)$, defined by the relations:

$$\frac{\mathrm{d}p_S}{\mathrm{d}z_S} + \rho_S = 0 \quad \text{and} \quad \rho_S = \frac{p_S(z_S)}{T_S(z_S)}, \quad (6.50a)$$

where $T_S(z_S)$ is assumed to be a solution (in nonadiabatic motion) of the equation

$$k(T_S)\frac{\mathrm{d}T_S}{\mathrm{d}z_S} + \sigma^0 R(T_S) = 0, \quad (6.50b)$$

where

$$\frac{\mathrm{d}R(T_S)}{\mathrm{d}z_S} = \rho_S Q(T_S), \quad (6.50c)$$

where σ^0 is a constant characteristic parameter for the rate of radiative heat transfer. Usually, it is assumed that the influence of the rate of radiative heat

6.4 Viscous Nonadiabatic Boussinesq Equations   185

transfer on the NSF equation (1.32) for temperature is essential, even with the basic state and doing this, we consider only a mean heat source and ignore variations therefrom. For our purpose here, this modeling will be sufficient.

Consequently, the equation for temperature (1.32) is written in the following dimensionless form:

$$\rho S \frac{dT}{dt} + (\gamma - 1)p\boldsymbol{\nabla} \cdot \boldsymbol{u}$$
$$= \frac{\gamma}{\Pr \text{Re}} \boldsymbol{\nabla} \cdot [k\boldsymbol{\nabla}T] + \frac{\gamma M^2(\gamma - 1)}{\text{Re}} \phi + \frac{\gamma}{\Pr \text{Re}} \text{Bo}\,\sigma^0 \frac{dR(T_S)}{dz_S} \,. \quad (6.51)$$

Below, we consider the case with μ and k constants to describe the motions which represent departures from the static basic state, and we introduce the perturbation of pressure π, the perturbation of density ω, and the perturbation of temperature θ, defined by the relations,

$$p = p_S(\text{Bo}z)(1 + \pi), \quad \rho = \rho_S(\text{Bo}z)(1 + \omega), \quad T = T_S(\text{Bo}z)(1 + \theta) \,. \tag{6.52}$$

Then, in place of NSF equation (1.31), we can write the following two dimensionless (exact) equations for the dimensionless velocity components $(\boldsymbol{v}, w)$, with $\boldsymbol{v} = (u, v)$:

$$(1 + \omega)\,\text{S}\,\frac{D\boldsymbol{v}}{Dt} + \frac{T_S(\text{Bo}z)}{\gamma M^2}\boldsymbol{D}\pi$$
$$= \frac{1}{\rho_S(\text{Bo}z)}\frac{1}{\text{Re}}\left[\frac{\partial^2 \boldsymbol{v}}{\partial z^2} + \boldsymbol{D}^2\boldsymbol{v} + \frac{1}{3}\boldsymbol{D}(\boldsymbol{\nabla} \cdot \boldsymbol{u})\right], \quad (6.53\text{a})$$

$$(1 + \omega)\,\text{S}\,\frac{Dw}{Dt} + \frac{T_S(\text{Bo}z)}{\gamma M^2}\frac{\partial \pi}{\partial z} - (1 + \omega)\frac{\text{Bo}}{\gamma M^2}\theta$$
$$= \frac{1}{\rho_S(\text{Bo}z)}\frac{1}{\text{Re}}\left[\frac{\partial^2 w}{\partial z^2} + \boldsymbol{D}^2 w + \frac{1}{3}\frac{\partial}{\partial z}(\boldsymbol{\nabla} \cdot \boldsymbol{u})\right], \quad (6.53\text{b})$$

where

$$\text{S}\frac{D}{Dt} = \text{S}\frac{\partial}{\partial t} + \boldsymbol{v} \cdot \boldsymbol{D} + w\frac{\partial}{\partial z}, \quad \boldsymbol{\nabla} = \boldsymbol{D} + \left(\frac{\partial}{\partial z}\right)\boldsymbol{k}, \quad \boldsymbol{D} = \left(\frac{\partial}{\partial x}, \frac{\partial}{\partial y}\right) \,.$$

As reference characteristic constant values for the thermodynamic functions, we choose $p^0 = p_S^*(0)$, $\rho^0 = \rho_S^*(0)$, $T^0 = T_S^*(0)$, where the * denotes dimensional functions, and in place of equation of state (1.33), we can write the following equation of state for the thermodynamic perturbations:

$$\pi = \omega + (1 + \omega)\theta \,. \tag{6.53c}$$

In place of continuity equation (1.30), we obtain

$$S\frac{D\omega}{Dt} + (1+\omega)\left(\mathbf{D}\cdot\mathbf{v} + \frac{\partial w}{\partial z}\right)$$

$$= (1+\omega)\frac{\mathrm{Bo}}{T_S(\mathrm{Bo}z)}\left[1 + \frac{\mathrm{d}T_S(\mathrm{Bo}z)}{\mathrm{d}z_S}\right]w; . \tag{6.53d}$$

Finally, in place of the energy equation (6.51), when the conductivity coefficient is constant, and according to the thermal balance basic state equation (6.50b), with (6.50c), we derive,

$$(1+\omega)S\frac{D\theta}{Dt} - \frac{\gamma-1}{\gamma}S\frac{D\pi}{Dt}$$

$$+ (1+\pi)\frac{\mathrm{Bo}}{T_S(\mathrm{Bo}z)}\left(\frac{\gamma-1}{\gamma} + \frac{\mathrm{d}T_S(\mathrm{Bo}z)}{\mathrm{d}z_S}\right)w$$

$$= \frac{1}{\rho_S(\mathrm{Bo}z)}\frac{1}{\Pr\mathrm{Re}}\left(\frac{\partial^2\theta}{\partial z^2} + \mathbf{D}^2\theta + 2\frac{\mathrm{Bo}}{T_S(\mathrm{Bo}z)}\frac{\mathrm{d}T_S(\mathrm{Bo}z)}{\mathrm{d}z_S}\frac{\partial\theta}{\partial z}\right.$$

$$\left. + \frac{\mathrm{Bo}^2}{T_S(\mathrm{Bo}z)}\frac{\mathrm{d}^2 T_S(\mathrm{Bo}z)}{\mathrm{d}z_S^2}\theta\right) + \frac{M^2(\gamma-1)}{\mathrm{Re}}\left[\frac{1}{\rho_S(\mathrm{Bo}z)}T_S(\mathrm{Bo}z)\right]\phi . \tag{6.53e}$$

Note that $H_S = RT_S^*(0)/g$ is a characteristic length scale for the vertical coordinate z_S in the static basic state ($z_S = z_S^*/H_S$), and, consequently, $z_S = \mathrm{Bo}z$, with $\mathrm{Bo} = H^0/H_S$, where H^0 is the vertical scale for z in the thermodynamically perturbed motion. The Mach number M is formed with the characteristic value of the speed of sound:

$$a^{*0} = (\gamma R T_S^*(0))^{1/2} .$$

We stress again that dimensionless atmospheric equations (6.53a–e), for $\mathbf{v}$, w, π, ω and θ, are a set of *exact NSF equations*, and this remark is important for a consistent derivation of the Boussinesq model equations.

6.4.2 Asymptotic Derivation of Viscous, Nonadiabatic Boussinesq Equations

Now, we consider exact NSF atmospheric equations, (6.53a–e), when

$$M \ll 1 \text{ and } \mathrm{Bo} \ll 1, \text{ such that } \frac{\mathrm{Bo}}{\mathrm{Ma}} = B^* = O(1). \tag{6.54}$$

Then we assume that, under constraint (6.54), the asymptotic solution of (6.53a–d) has the following form:

$$(u, v, w) = (u_B, v_B, w_B) + \dots , \tag{6.55a}$$

$$(\omega, \theta) = M^\alpha(\omega_B, \theta_B) + \dots , \quad \pi = M^\beta \pi_B \dots . \tag{6.55b}$$

From a straightforward analysis, we easily see that

$$\alpha = 1 \quad \text{and} \quad \beta = 2 \tag{6.55c}$$

is the only significant case.

When M tends to zero, with t, x, y, z, and S, γ, B^* fixed , (6.56)

from the exact NSF atmospheric equations, (6.53a–e), for limiting functions u_B, v_B, w_B, ω_B, θ_B, and π_B, we obtain the following set of Boussinesq viscous non-adiabatic equations:

$$\mathbf{D} \cdot \boldsymbol{v}_B + \frac{\partial w_B}{\partial z} = 0 ; \tag{6.57a}$$

$$S \frac{D_B \boldsymbol{v}_B}{Dt} + \mathbf{D}\frac{\pi_B}{\gamma} = \frac{1}{\mathrm{Re}} \left[\frac{\partial^2 \boldsymbol{v}_B}{\partial z^2} + \mathbf{D}^2 \boldsymbol{v}_B \right] ; \tag{6.57b}$$

$$S \frac{D_B w_B}{Dt} + \frac{\partial(\pi_B/\gamma)}{\partial z} - \frac{B^*}{\gamma}\theta_B = \frac{1}{\mathrm{Re}}\left[\frac{\partial^2 w_B}{\partial z^2} + \mathbf{D}^2 w_B\right] ; \tag{6.57c}$$

$$S \frac{D_B \theta_B}{Dt} + B^*\left[\frac{\gamma-1}{\gamma} - \Gamma_S(0)\right] w_B = \frac{1}{\mathrm{Pr}\,\mathrm{Re}}\left[\frac{\partial^2 \theta_B}{\partial z^2} + \mathbf{D}^2\theta_B\right] ; \tag{6.57d}$$

and

$$\omega_B = -\theta_B . \tag{6.57e}$$

In (6.57b–d)

$$S \frac{D_B}{Dt} = S\frac{\partial}{\partial t} + \boldsymbol{v}_B \cdot \mathbf{D} + w_B \frac{\partial}{\partial z} .$$

In dimensionless form, obviously, $T_S(0) \equiv 1$, but in general $\Gamma_S(0) = -(\mathrm{d}T_S(\mathrm{Boz})/\mathrm{d}z_S)_0$ is different from zero.

6.5 Some Comments

First, some new features occur in *internal aerodynamics*. For an *inviscid fluid*, the reader can find a multiple scale technique in Zeytounian (2002a), based on the following change, for the time derivative (because, when $M \ll 1$, we must build a multiplicity of time-scales into the structure of the solution):

$$\frac{\partial f}{\partial t} = \frac{\partial f^*}{\partial t} + \frac{1}{M}\mathrm{d}f^*, \text{ with } f^* = (\boldsymbol{u}^*, p^*, \rho^*) , \tag{6.58a}$$

where $\partial f^*/\partial t$ stands for the time derivative computed when all fast times are maintained constant and $\frac{1}{M}\mathrm{d}f^*$ is the dimensionless time derivative occurring through the fast times. The representation (6.58a) gives the possibility of investigating the long-time evolution of rapid oscillations of a bounded container with a deformable (in time) wall Σ. An average process also gives the possibility of deriving a system of model equations for the slow variation.

Now, if we deal with a *slightly viscous flow*, then we must start from the full NSF equations in place of the Euler equations and bring into the asymptotic analysis a *second small parameter* $1/\text{Re} = \varepsilon^2$, the inverse of a characteristic Reynolds number. Then we must expect (?) that the oscillations are damped out with time. A precise analysis of this damping phenomenon appears to be a difficult problem and raises many questions. Actually, only the case when $M\,\text{Re} \gg 1$ has been considered, and here we shall be content with an order of magnitude estimate of the damping time. For this, it is necessary to change the relation (6.58a); we write

$$\frac{\partial f}{\partial t} = \frac{\partial f^*}{\partial t} + \frac{1}{M}\,\mathrm{d} f^* + \delta T f^* \,, \tag{6.58b}$$

with $\delta \ll 1$, where the derivative operator T is related to damping short times. Then we may carry out an analysis analogous to that in Zeytounian (2002a, Sect. 6.2.2 for the slow variation) but of order $O(\delta M)$. It appears that the secular terms related to the derivative operator Tf_0^* cannot be eliminated unless the condition on Σ for the velocity is inhomogeneous! Such an inhomogeneous condition may be provided by a boundary-layer analysis of the NSF equations for $\text{Re} \gg 1$ ($\varepsilon^2 \ll 1$) near Σ. For our problem, with $M \ll 1$ and $\varepsilon \ll 1$, the BL is a Stokes layer of thickness

$$O(\varepsilon M^{1/2}) = O((M/\text{Re})^{1/2}) \,. \tag{6.59a}$$

On the other hand, it is found that the flux away from the Stokes layer is also of order $O[(M/\text{Re})^{1/2}]$ and must be equilibrated with the Tf_0^* terms of order δM. Finally, we find the required estimate,

$$\delta t \sim (M\,\text{Re})^{1/2} \,, \tag{6.59b}$$

because

$$\delta M \sim (M/\text{Re})^{1/2} \Rightarrow \delta t = 1/(M\,\text{Re})^{1/2} \text{ and } \delta t = O(1) \,.$$

Another interesting problem in the framework of *small M and large* Re, is related to the *combined effect* of vanishing viscosity and very low compressibility! In this case it is possible to consider three limiting processes:

(i) Re fixed, $M \downarrow 0$, and then $\text{Re} \uparrow \infty$, (6.60a)

(ii) M fixed, $\text{Re} \uparrow \infty$, and then $M \downarrow 0$, (6.60b)

(iii) $M \downarrow 0$ *and* $\text{Re} \uparrow \infty$, such that $1/\text{Re} = \lambda^0 M^b$, (6.60c)

with $b > 0$ and λ^0 a smilarity parameter. It is obvious that limiting process (iii) is the more significant and gives the possibility of taking into account simultaneously two small effects which, via coupling, lead to an effect of the order $O(1)$.

On the other hand, limiting process (iii) is the more significant because limiting process (i) is rediscovered when $\lambda^0 \uparrow \infty$, and limiting process (ii) is rediscovered when $\lambda^0 \downarrow 0$. For instance, if we consider the classical Blasius problem for slightly compressible flow with a vanishing viscosity, then a consistent asymptotic theory leads to

$b = 4$ and with $\lambda^0 = 1 \Rightarrow \operatorname{Re} M^4 = 1$,

according to Godts and Zeytounian (1990). Finally, we note that in Boris and Fridlender (1981) the slow motion of a gas near a strongly heated or cooled sphere is considered.

7 Some Viscous Fluid Motions and Problems

In this chapter, the reader can find some viscous fluid motions that have not been discussed in Chaps. 4–6 in the framework of fluid flows at $\text{Re} \gg 1$, $\text{Re} \ll 1$, and $M \ll 1$.

7.1 Oscillatory Viscous Incompressible Flow

7.1.1 Acoustic Streaming Effect

The so-called *"acoustic streaming"* effect is considered here as a particular flow at a *large Strouhal number*. But only the Navier equations are considered, because I do not have knowledge of any work on compressible flow! Therefore, the problem of investigating the streaming effect, resulting from oscillatory viscous *compressible* (and eventually *heat-conducting*!) flow, is open.

This streaming effect is simply derived in the framework of periodic boundary-layer flow when we consider the boundary layer at a body that moves back and forth in the form of a harmonic oscillation with small amplitude in a fluid at rest.

High frequency oscillations in a fluid initially at rest cause a steady-state secondary flow through the action of viscosity in the boundary layer.

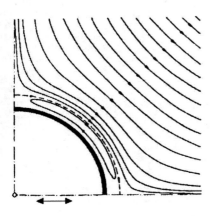

Fig. 7.1. Streaming close to an oscillating circular cylinder

192   7 Some Viscous Fluid Motions and Problems

This secondary flow is such that the entire fluid close to the oscillating body is set into steady-state motion, although the motion of the body is purely periodic. This phenomenon is called "streaming" or "acoustic streaming", and effects of this kind come into play in the formation of dust patterns in a Kundt tube. Figure 7.1 [the calculation is due to Schlichting (1932)] shows the streamline portrait of this steady-state additional flow for an oscillating circular cylinder.

For a circular cylinder, it is convenient to work with the stream function $\psi(t, r, \theta)$, such that, for the components of the velocity vector $u = (u, v)$, we write

$$u = \frac{1}{r}\frac{\partial \psi}{\partial \theta} \text{ and } v = -\frac{\partial \psi}{\partial r}; . \tag{7.1}$$

In this case, it is necessary to consider the following Navier problem:

$$\frac{\partial \nabla^2 \psi}{\partial t} + \alpha \frac{1}{r}\left[\frac{\partial \psi}{\partial \theta}\frac{\partial}{\partial r} - \frac{\partial \psi}{\partial r}\frac{\partial}{\partial \theta}\right]\nabla^2 \psi$$
$$= \alpha \varepsilon^2 \nabla^2(\nabla^2 \psi) ,$$
$$r = 1 : \psi = 0 \text{ and } \frac{\partial \psi}{\partial r} = 0 ,$$
$$r \uparrow \infty : \psi \to -r\sin\theta\cos t . \tag{7.2}$$

For the Laplace operator $\Delta = \nabla^2$, we have the following form:

$$\nabla^2 = \frac{\partial^2}{\partial r^2} + \frac{1}{r}\frac{\partial}{\partial r} + \frac{1}{r^2}\frac{\partial^2}{\partial \theta^2}$$
$$\text{and } \alpha = \frac{U_\infty}{\omega^0 L^0}, \quad \varepsilon^2 = \frac{1}{\text{Re}} = \frac{\nu^0}{U_\infty L^0} , \tag{7.3}$$

when we assume that the reference characteristic time $t^0 = 1/\omega^0$ and here, α plays the role of a *frequency parameter* which is the inverse of a Strouhal number.

Outer flow. When we assume that α and ε^2 are both small parameters

$$\alpha \to 0 \text{ and } \varepsilon^2 \to 0, \text{ with } t, r \text{ and } \theta \text{ fixed} , \tag{7.4}$$

in an "outer" limiting process, then we derive a very degenerate leading-order equation,

$$\frac{\partial \nabla^2 \psi_0}{\partial t} = 0 , \tag{7.5a}$$

and we do not have the possibility of satisfying the no-slip boundary conditions at $r = 1$. Consequently, we obtain only an *outer solution*:

7.1 Oscillatory Viscous Incompressible Flow

$$\psi_0 = -\left[r - \frac{1}{r}\right]\sin\theta \cos t, \tag{7.5b}$$

and from (7.5b) we obtain $\psi_0(t, r = 1, \theta) = 0$, but

$$\left(\frac{\partial \psi_0}{\partial \theta}\right)_{r=1} = 2\sin\theta \cos t \equiv u_e(t, \theta). \tag{7.5c}$$

Local flow: Stokes layer. Obviously, it is necessary to consider a viscous layer close to $r = 1$. For this, in place of r, we introduce the new radial coordinate

$$\eta = \frac{r-1}{(2\alpha\varepsilon^2)^{1/2}} \quad \text{with } \psi^*(t, \eta, \theta) = \frac{\psi}{(2\alpha\varepsilon^2)^{1/2}}. \tag{7.6a}$$

Now, if we consider the following "local" limiting process:

$$\alpha \to 0 \text{ and } \varepsilon^2 \to 0, \text{ with } t, \eta \text{ and } \theta \text{ fixed}, \tag{7.6b}$$

then, in place of the equation for ψ in starting problem (7.2), we derive the following equation for $\psi^*(t, \eta, \theta)$:

$$\frac{\partial^3 \psi^*}{\partial t \partial \eta^2} - \frac{1}{2}\frac{\partial^4 \psi^*}{\partial \eta^4} + 2(2\alpha\varepsilon^2)^{1/2}\left[\frac{\partial^2 \psi^*}{\partial t \partial \eta} - \frac{1}{2}\frac{\partial^3 \psi^*}{\partial \eta^3}\right]$$
$$+ \alpha\left[\frac{\partial \psi^*}{\partial \theta}\frac{\partial}{\partial \eta} - \frac{\partial \psi^*}{\partial \eta}\frac{\partial}{\partial \theta}\right]\frac{\partial^2 \psi^*}{\partial \eta^2} + O\left(\alpha\varepsilon^2\right) + O\left(\alpha(\alpha\varepsilon^2)^{1/2}\right). \tag{7.6c}$$

For (7.6c) as a boundary conditions relative to η, we write

$$\eta = 0: \psi^* = 0, \frac{\partial \psi^*}{\partial \eta} = 0 \quad \text{and for } \eta \uparrow \infty: \text{ matching}. \tag{7.6d}$$

As a consequence of local limiting process (7.6b), from (7.6d), we derive the Stokes equation for ψ_0^*:

$$\frac{\partial^3 \psi_0^*}{\partial t \partial \eta^2} - \frac{1}{2}\frac{\partial^4 \psi_0^*}{\partial \eta^4} = 0, \tag{7.6e}$$

which must be solved with the conditions

$$\eta = 0: \psi_0^* = 0, \frac{\partial \psi_0^*}{\partial \eta} = 0, \tag{7.6f$_1$}$$

$$\eta \uparrow \infty: \frac{\partial \psi_0^*}{\partial \eta} \to -2\sin\theta \cos t. \tag{7.6f$_2$}$$

The solution of (7.6e,f) for $\partial \psi_0^*/\partial \eta$ is

$$\frac{\partial \psi_0^*}{\partial \eta} = 2\sin\theta[\exp(-\eta)\cos(t-\eta) - \cos t]. \tag{7.6g}$$

194   7 Some Viscous Fluid Motions and Problems

The thickness of the Stokes layer, where solution (7.6g) is valid, is the same (in the order of magnitude) as that of the Rayleigh layer (considered in Sect. 3.7, Chap. 3, and in Sect. 4.3.3, Chap. 4) after a time interval equal to the period. But an essential difference between these two viscous layers exists: the Stokes layer has an unlimited life, whereas the Rayleigh layer has a life time only during a finite lapse and its behavior for large times raises a fundamental problem (relative to matching with the unsteady-state Prandtl boundary layer)!

The leading composite approximation. With the outer (7.5b) and local (7.6g) leading-order solutions, we can write a leading composite approximation in the form,

$$\psi_0^c = -\sin\theta \left\{ \left(r - \frac{1}{r}\right) \cos t + 2(2\alpha\varepsilon^2)^{1/2} \mathfrak{G}\left(t, \frac{r-1}{(2\alpha\varepsilon^2)^{1/2}}\right) \right\}$$

with $\dfrac{\partial \mathfrak{G}}{\partial \eta} = -\exp(-\eta)\cos(t-\eta)$, $\mathfrak{G}(t,0) = 0$. (7.7)

Second-order local approximation: steady streaming effect. We consider now the equation of starting problem (7.2) written in the following form

$$\frac{\partial \nabla^2 \psi}{\partial t} + \alpha \frac{1}{r}\left[\left(\frac{\partial \psi}{\partial \theta}\right)\frac{\partial}{\partial r} - \left(\frac{\partial \psi}{\partial r}\right)\frac{\partial}{\partial \theta}\right] \nabla^2 \psi$$

$$= \frac{\alpha^2}{\mathrm{Re}_S} \nabla^2(\nabla^2 \psi) ,$$ (7.2a)

where

$$\frac{\alpha}{\varepsilon^2} = \alpha\,\mathrm{Re} = \mathrm{Re}_S = \frac{U_\infty^2}{\nu^0 \omega^0} .$$ (7.8a)

In the limiting process,

$$\mathrm{Re}\uparrow\infty \text{ and } \alpha\downarrow 0 \text{ with } \mathrm{Re}_S \text{ fixed, and then } \mathrm{Re}_S\uparrow\infty ,$$ (7.8b)

the term proportional to $2(2\alpha\varepsilon^2)^{1/2}$ is neglected in comparison to the term proportional to α. Then, with

$$\psi = \psi_0^* + \alpha\psi_1^* + o(\alpha) ,$$ (7.8c)

we derive the following non-homogeneous equation for the function ψ_1^* [since $\alpha \gg (2\alpha\varepsilon^2)^{1/2}$]:

$$\frac{\partial^3 \psi_1^*}{\partial t \partial \eta^2} - \frac{1}{2}\frac{\partial^4 \psi_1^*}{\partial \eta^4} = \mathrm{Real}\left[\exp(2it)H_1(\eta) + H_0(\eta)\right] \sin\theta \cos\theta .$$ (7.8d$_1$)

7.1 Oscillatory Viscous Incompressible Flow

For (7.8d$_1$) as a boundary condition relative to η, we write

$$\eta = 0 : \psi_1^* = 0, \quad \frac{\partial \psi_1^*}{\partial \eta} = 0 \quad \text{and for } \eta \uparrow \infty : \text{ matching} . \tag{7.8d$_2$}$$

For the leading-order local term ψ_0^*, we can write

$$\psi_0^* = 2\,\text{Real}\,[\exp(2\mathrm{i}t)F_0(\eta)]\sin\theta , \tag{7.8e}$$

and according to (7.6g), we obtain

$$\frac{\mathrm{d}F_0(\eta)}{\mathrm{d}\eta} = \exp[(1+\mathrm{i})\eta] - 1, \quad F(0) = 0 .$$

Consequently, for the functions $H_1(\eta)$ and $H_0(\eta)$ in (7.8d$_1$), we obtain the following expressions:

$$H_0(\eta) = -2\,\text{Real}\left[F_0\frac{\mathrm{d}^3 F_0^*}{\mathrm{d}\eta^3} - \frac{\mathrm{d}F_0}{\mathrm{d}\eta}\frac{\mathrm{d}^2 F_0^*}{\mathrm{d}\eta^2}\right] , \tag{7.8f$_1$}$$

$$H_1(\eta) = -2\left[F_0\frac{\mathrm{d}^2 F_0}{\mathrm{d}\eta^2} - \frac{\mathrm{d}F_0}{\mathrm{d}\eta}\frac{\mathrm{d}^2 F_0}{\mathrm{d}\eta^2}\right] , \tag{7.8f$_2$}$$

where F_0^* is the complex conjugate of F_0.

Finally, the solution of (7.8d$_1$), with the conditions (7.8d$_2$), is

$$\psi_1^* = \{\text{Real}[\exp(2\mathrm{i}t)K_1(\eta)] + K_0(\eta)\}\sin\theta\cos\theta , \tag{7.8g}$$

where

$$\frac{\mathrm{d}^4 K_0(\eta)}{\mathrm{d}\eta^4} = -2H_0(\eta), \quad K_0(0) = 0, \quad \left(\frac{\mathrm{d}K_0(\eta)}{\mathrm{d}\eta}\right)_{\eta=0} = 0 , \tag{7.8h$_1$}$$

$$\frac{\mathrm{d}^4 K_1(\eta)}{\mathrm{d}\eta^4} - 4\mathrm{i}\frac{\mathrm{d}^2 K_1(\eta)}{\mathrm{d}\eta^2} = -2H_1(\eta), \quad K_1(0) = 0 ,$$

$$\left(\frac{\mathrm{d}K_1(\eta)}{\mathrm{d}\eta}\right)_{\eta=0} = 0 , \tag{7.8h$_2$}$$

and the unique solution of problem (7.8h$_2$) for $K_1(\eta)$, which does not increase at infinity as an exponential, must also satisfy the condition

$$\lim_{\eta\uparrow\infty}\frac{\mathrm{d}K_1(\eta)}{\mathrm{d}\eta} = 0 .$$

But, unfortunately, we have as a solution for $K_0(\eta)$,

$$-K_0(\eta) = \frac{13}{8} - \frac{3}{4}\eta - \frac{1}{8}\exp(-2\eta) - \frac{3}{2}\exp(-\eta)\cos\eta$$

$$- \exp(-\eta)\sin\eta - \frac{1}{2}\eta\exp(-\eta)\sin\eta + A^0\eta^2 + B^0\eta^3 , \tag{7.8i}$$

where the constants A^0 and B^0 must be determined by matching.

As a consequence of the solution (7.8i) for $K_0(\eta)$, a *steady-state streaming effect* appears in the solution (7.8g) for ψ_1^*. On the one hand, $K_0(\eta)$ is not oscillating and, on the other hand, $K_0(\eta)$, according to (7.8i), increases at infinity at least as η. This clearly shows that

a steady-state streaming flow emerges outside of the Stokes layer!

7.1.2 Study of the Steady-State Streaming Phenomenon

A general solution of the equation that governs starting problem (7.2) must be written in the following form:

$$\psi = \psi^{\text{st}}(r,\theta) + \sum \{F_n(r,\theta)\cos(nt) + G_n(r,\theta)\sin(nt)\}, \; n \geq 1, \qquad (7.9)$$

where the functions ψ^{st} and F_n, G_n are, it is also assumed, dependent on both small parameters α and ε^2.

By substituting (7.9) in the equation governing the starting problem (7.2) and taking into account only the terms that are independent of time, we obtain:

$$\frac{1}{r}\left[\frac{\partial \psi^{\text{st}}}{\partial \theta}\frac{\partial \nabla^2 \psi^{\text{st}}}{\partial r} - \frac{\partial \psi^{\text{st}}}{\partial r}\frac{\partial \nabla^2 \psi^{\text{st}}}{\partial \theta}\right] - \varepsilon^2 \nabla^2(\nabla^2 \psi^{\text{st}})$$
$$= \frac{1}{2r}\sum \left\{\frac{\partial F_n}{\partial \theta}\frac{\partial \nabla^2 F_n}{\partial r} - \frac{\partial F_n}{\partial r}\frac{\partial \nabla^2 F_n}{\partial \theta}\right.$$
$$\left.+ \frac{\partial G_n}{\partial \theta}\frac{\partial \nabla^2 G_n}{\partial r} - \frac{\partial G_n}{\partial r}\frac{\partial \nabla^2 G_n}{\partial \theta}\right\}, \; n \geq 1. \qquad (7.10a)$$

We have in (7.10) the emergence of a steady-state flow, which is typical in nonlinear oscillating problems.

In the Stokes layer, we have, according to (7.6a), (7.8c), (7.8e), and (7.8g),

$$\psi^{\text{st}}(r,\theta) = \alpha(2\alpha\varepsilon^2)^{1/2} K_0(\eta) \sin\theta \cos\theta. \qquad (7.10b)$$

Now, if we take into account the behavior of the function $K_0(\eta)$ for large values of η $[= (r-1)/(2\alpha\varepsilon^2)^{1/2} \gg 1 \Rightarrow r-1 \gg (2\alpha\varepsilon^2)^{1/2} \ll 1]$, then we can write

$$\psi^{\text{st}}(r,\theta) = \alpha\left\{\frac{3}{4}(r-1) - \frac{B^0}{8}(2\alpha\varepsilon^2)^{1/2} + A^0(2\alpha\varepsilon^2)^{1/2}(r-1)^2\right.$$
$$\left.+ B^0(2\alpha\varepsilon^2)^{-1}(r-1)^3\right\}\sin\theta\cos\theta. \qquad (7.10c)$$

In this case, we can see that the functions F_n in (7.9) verify the Laplace equation $\nabla^2 F_n = 0$. Consequently, we derive the following equation for $\psi^{\text{st}}(r,\theta)$:

$$\frac{\partial \psi^{\text{st}}}{\partial \theta}\frac{\partial \nabla^2 \psi^{\text{st}}}{\partial r} - \frac{\partial \psi^{\text{st}}}{\partial r}\frac{\partial \nabla^2 \psi^{\text{st}}}{\partial \theta} = \frac{\varepsilon^2}{r}\nabla^2\left(\nabla^2 \psi^{\text{st}}\right), \qquad (7.11a)$$

with

$$\frac{1}{r}\psi^{\mathrm{st}} \downarrow 0 \text{ at infinity, when } r \uparrow \infty, \text{ and matching for } r = 1 \,. \quad (7.11\mathrm{b})$$

Because A^0 and B^0 are unknowns, the order of ψ^{st} is a priori also unknown for $r - 1 = O(1)$! Let $\psi^{\mathrm{st}} = \nu\psi_*^{\mathrm{st}}$, with ν a gauge, and in this case, it is necessary to consider three cases:

(i) $\dfrac{\varepsilon^2}{\nu} \gg 1$, (ii) $\dfrac{\varepsilon^2}{\nu} = 1$, (iii) $\dfrac{\varepsilon^2}{\nu} \ll 1$,

and the choice is related to matching for $r = 1$.

On the other hand, according to (7.10c), four options are possible for ν:

(a) $\nu = \alpha(\alpha\varepsilon^2)^{-1}$, (b) $\nu = \alpha(\alpha\varepsilon^2)^{-1/2}$, (c) $\nu = \alpha$, (d) $\nu = \alpha(\alpha\varepsilon^2)^{1/2}$.

Different cases and options have been analyzed in detail in Zeytounian (1987, pp. 273–275), and the conclusion is that the good choice is (and in this case $A^0 = B^0 = 0$):

$$\nu = \alpha, \ \gamma = \varepsilon\alpha^{-1/2} = \mathrm{Re}_S^{-1/2}, \quad (7.11\mathrm{c})$$

where γ is a gauge such that, near $r = 1$ in a boundary layer, we have the expansion

$$\psi_*^{\mathrm{st}} = \gamma\phi_0 + \ldots, \text{ with } r = 1 + \gamma\zeta \,. \quad (7.11\mathrm{d})$$

Finally, the function $\phi_0(\zeta, \theta)$ is a solution of the following problem:

$$\frac{\partial \phi_0}{\partial \theta}\frac{\partial^3 \phi_0}{\partial \zeta^3} - \frac{\partial \phi_0}{\partial \zeta}\frac{\partial^3 \phi_0}{\partial \theta \partial \zeta^2} = \frac{\partial^4 \phi_0}{\partial \zeta^4},$$

$$\zeta = 0 \,:\, \phi_0 = 0, \ \frac{\partial \phi_0}{\partial \zeta} = \frac{3}{4}\sin\theta\cos\theta,$$

$$\zeta \uparrow \infty \,:\, \frac{\partial \phi_0}{\partial \zeta} = 0 \,. \quad (7.12)$$

We observe that problem (7.12) is valid in a viscous boundary layer of thickness $\gamma = \varepsilon\alpha^{-1/2}$, greater than the thickness of the Stokes layer, and it is necessary to determine ψ_*^{st} outside of this viscous boundary layer and justify the behavior condition, $\zeta \uparrow \infty \,:\, \partial \phi_0/\partial \zeta = 0$! Problem (7.12) is parabolic in θ, and consequently, an initial condition is necessary for $\theta = \pi/2$. This last initial condition is derived automatically via a homogeneous solution of the form,

$$\phi_0(\zeta, \theta) = [(\pi/2) - \theta]f_0(\zeta) + o[(\pi/2) - \theta] \,, \quad (7.13\mathrm{a})$$

and $f_0(\zeta)$ is a solution of the following problem:

$$\frac{d^3 f_0}{d\zeta^3} + f_0 \frac{d^2 f_0}{d\zeta^2} - \left(\frac{df_0}{d\zeta}\right)^2 = 0 , \tag{7.13b}$$

$$f_0(0) = 0, \quad \left(\frac{df_0}{d\zeta}\right)_{\zeta=0} = \frac{3}{4}, \quad \left(\frac{df_0}{d\zeta}\right)_{\zeta=\infty} = 0 . \tag{7.13c}$$

The solution of problem (7.13a–c) is

$$f_0(\zeta) = (3^{1/2}/2)\left[1 - \exp\left(-(3^{1/2}/2)4\zeta\right)\right] . \tag{7.14}$$

Outside of the viscous boundary layer of thickness γ, we can write

$$\psi^{st} = \nu\gamma\chi_0(r,\theta) ,$$

and $\chi_0(r,\theta)$ is a solution of the problem,

$$\nabla^2 \chi_0 = 0; \quad r = 1 : \chi_0 = \phi_0(\infty,\theta), \quad r \uparrow \infty : \chi_0 = o(r) . \tag{7.15}$$

Matching the above condition for $r = 1$, also gives the last behavior condition, $\zeta \uparrow \infty : \partial\phi_0/\partial\zeta = 0$, in (7.12), because the tangential velocity (at $r = 1$) is $O(\nu\gamma)$ in the flow outside of the viscous boundary layer of thickness γ, whereas in this viscous boundary layer, this tangential velocity is $O(\nu/\gamma) \gg O(\nu\gamma)$.

7.1.3 The Role of Parameters $\alpha\,\mathrm{Re} = \mathrm{Re}_S$ and $\mathrm{Re}/\alpha = \beta^2$

First, when $\alpha \ll 1$, if we assume in addition that $(\mathrm{Re}/\alpha)^{1/2} = \beta \gg 1$, then the first-order harmonically fluctuating vorticity, created at the surface of the body, is confined to a thin boundary layer or Stokes shear-wave of thickness $O(\nu^0/\omega^0) \ll L^0$. It is well established that second-order steady-state streaming with velocities $O(\alpha U_\infty)$ persists outside this shear-wave (Stokes) layer. The second parameter Re_S is associated with this steady-state streaming, and it is necessary to determine the structure of this steady-state streaming that persists outside the Stokes layer, when $\beta \gg 1$. The parameter Re_S does play a role analogous to the role of the conventional Reynolds number Re for steady-state flow past a solid body.

Thus, if $\mathrm{Re}_S \gg 1$, the outer region, in which the steady-state velocity is adjusted to zero, is of boundary-layer character with thickness $O(L^0/\mathrm{Re}_s^{1/2}) = O[(\nu^0/\omega^0)^{1/2}/\alpha]$ whereas if $\mathrm{Re}_S \ll 1$, the flow is Stokes-like and this adjustment takes place over a much wider region.

On the other hand, because we suppose that the results are valid in the limit, as $\alpha \downarrow 0$, it is convenient to distinguish between the two categories: (1) $\lim \beta = $ finite and (2) $\lim \beta = \infty$, when $\alpha \downarrow 0$.

In Riley (1967), three cases are considered when $\alpha \downarrow 0$:
(I) $M = O(1)$ or $\lim \beta$ finite, when $\alpha \downarrow 0$; (II) $\mathrm{Re} = O(1)$ or $\lim \mathrm{Re}$ finite, when

$\alpha \downarrow 0$; (III) $\text{Re}_S = O(1)$ or $\lim \text{Re}_S$ finite, when $\alpha \downarrow 0$. In each case with $\beta \gg 1$, a thin boundary layer in the neighborhood of the body is involved, and we can expand the solution with α as a perturbation parameter in two complementary series, one associated with the shear-wave region, the other with the flow outside. These series must match at each stage of the expansion. Riley (1967) proceeds in a formal manner, including terms in the solution from "higher order" boundary-layer theory which takes account of body curvature, where appropriate. It is shown that in cases (I) and (II), the steady-state streaming outside the shear-wave is governed by the biharmonic equation corresponding to Stokes flow. In case (III), the dependence of the steady-state streaming on parameter Re_S is revealed. Riley (1967) also discusses, but only very briefly, the nature of the flow when $\beta \ll 1$ and gives a survey of earlier works on oscillatory viscous flows.

Here, we note that the full importance of parameter Re_S for the problem under consideration was first realized by Stuart (1963, 1966). His ideas, set out in a review article (1963), include the notion of an outer boundary layer of thickness $O(L^0/\text{Re}_s^{1/2})$ to which steady-state streaming is confined when the parameter $\text{Re}_s \gg 1$. Detailed results of this aspect of the flow are presented in a subsequent paper (1966), although the work had been completed some years earlier. In Fig. 7.2, according to Stuart (1966), outer layer and steady-state flow due to an oscillating cylinder are presented.

In Fig. 7.3, are presented steady-state streamlines of the mean flow according to the experiment of Masakazu Tatsuno [see Van Dyke (1982, p. 23, Fig. 31)] and reproduced in the recent paper by Badr, Dennis and Kocabiyik (1995).

The Stuart double-boundary layer solution shows that a fluid is dragged steadily in the direction A_1 to B_1 and B_2 and A_2 to B_1 to B_2. Fluid moves in radially to balance this. At the B_1 points and B_2, the outer boundary layers from the two sides impact, and the details of the solution in this region are unknown (in 1966!). After impact, the outer layer fluid motion moves in the direction B_1D_1 and, B_2D_2; at large distances from the cylinder and

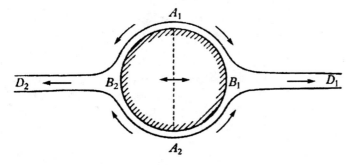

Fig. 7.2. Outer layer and steady-state flow due to an oscillating cylinder

Fig. 7.3. Steady-state streamlines of the mean flow caused by an infinitely long cylinder placed in an unbounded viscous fluid oscillating in a direction normal to the cylinder axis, which is at rest

in the absence of other boundaries nearby, it is plausible(!) to conjecture that the flow tends to the two-dimensional, Bicley (1937), jet solution of the Navier equations [see the book by Rosenhead (1963, pp. 254–256) and also the paper by Davidson and Riley (1972)]; at such distances, the main effect of the cylinder's movement is to provide a source of momentum for the flow. The structure of the outer boundary layer when $Re_S \gg 1$ was also considered by Riley (1965). Other work when $Re_S = O(1)$ includes the papers by Wang (1965), although he does not introduce that parameter explicitly, and Riley (1966). Each considers the problem of an oscillating sphere in an unbounded viscous fluid. We observe, again, that when $Re_S = O(1)$, the Navier equations for an incompressible and viscous fluid in steady-state flow are required, and, in particular, that the Reynolds stresses make no direct contribution to the outer streaming, which is induced indirectly from the action of such stresses in the Stokes shear layer. The purpose of the investigation by Kelly (1966) is to examine the steady-state flow induced by a cylinder which performs *two different types of oscillation simultaneously*; it has been shown that the effect of the second oscillation is greatest when the frequencies of the two oscillations coincide. The extent to which change occurs is beyond the inner, Stokes-type boundary layer. In subsequent work, Davidson and Riley (1972) carried out a theoretical and experimental study of the boundary layers and jets that form on, and in the neighborhood of, a vibrating cylinder when $Re_S \gg 1$.

In Bertelsen et al. (1973), the nonlinear streaming effects associated with oscillating cylinders are considered (and the regime of validity of the theory is discussed) and, in particular, for the one-cylinder model, new numerical results based on the theory by Holtsmark et al. (1954) for $\mathrm{Re}_S \ll 1$ are presented; the authors compare these with some earlier experimental observations by Holtsmark et al. (1954). An experimental study and further observations by Bertelsen (1974), also for large $\mathrm{Re}_S \gg 1$, concentrates on flow in the boundary layer.

Riley (1975), using higher order boundary-layer theory, attempts to reconcile the measured boundary-layer profiles of Bertelsen with those predicted theoretically.

Further such attempts were made by Duck and Smith (1979) and Haddon and Riley (1979) who solved the Navier equations in a bounded annular region for finite values of Re_S. Finally, we note that all of the above studies possess a high degree of symmetry, about each of two perpendicular axes. Wang (1972), in a more asymmetrical situation, considers the streaming that is induced when an acoustic line source is placed close, and parallel, to a circular cylinder.

He analyzes the flow in both the Stokes layer and outside it, although he restricts his attention to $\mathrm{Re}_S \ll 1$.

But as Lighthill (1978) points out, all really noticeable acoustic streaming motions are associated with $\mathrm{Re}_S \gg 1$.

Wang (1982, 1984) also considered the streaming induced when an acoustic source is placed close to a sphere or a plane boundary.

In Riley (1987), the author again considers the configuration of Wang (1972), but when Re_S is *large* and in this case, the asymmetrical feature of the outer flow is further emphasized by the streamline pattern of the mean flow (see Fig. 7.4) in the flow beyond the boundary layer.

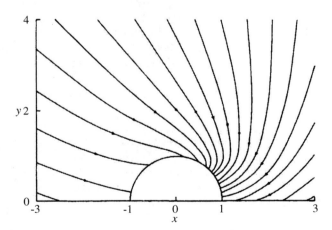

Fig. 7.4. The outer flow mean streamlines for an acoustic line placed at a distance $l^0/a^0 = 1.5$, where a^0 is the radius of the cylinder

7.1.4 Other Examples of Viscous Oscillatory Flow

The laminar boundary layer over a semi-infinite rigid plane $y = 0$, with $x > 0$, was examined, when the free-stream velocity takes the form,

$$u_e(t, x) = U^0(x)(1 + a^* \sin \omega t), \text{ where } 0 \leq a^* < 1 \text{ and } U^0(x) \propto x^n,$$

with $0 \leq n \leq 1$, by Pedley (1972). The principal difference between the Pedley (1972) work and that of earlier authors is that scalar a^* is not required to be small.

In 1961 (during the years 1957–1966, I worked in the Hydrometeorological Center in Moscow, in the department of Prof. I.A. Kibel), I generalized the Schlichting (1932) work to a periodic thermal free-convection problem over a curvilinear wall [see Zeytounian (1968b)], and in Noe (1981) this problem was reconsidered and investigated according to the "double boundary layer" Stuart-Riley theory [see, for instance, Zeytounian (1987, pp. 279–286)].

The object of the paper by Bestman (1983) was to investigate low Reynolds number flow of a purely oscillatory stream past a slowly rotating sphere. This analysis can be combined with that of Singh (1975), if the need be, for pulsatile flow past a spinning sphere. The problem is worthy of note in that purely oscillatory flow past curved bodies exhibit steady-state streaming velocity components, and these steady-state streaming solutions have generated a lot of mathematical and physiological interest in recent times.

Viscous oscillatory flow about a circular cylinder at a small to moderate Strouhal number was considered more recently by Badr, Dennis, Kocabiyik and Nguyen (1995). In that paper, the transient flow field caused by an infinitely long cylinder placed in an unbounded viscous fluid oscillating in a direction normal to the cylinder axis, which is at rest, is considered. The flow, it was assumed, started suddenly from rest and remained symmetrical about the direction of motion. The comparison between viscous and inviscid flow results shows better agreement for higher values of Reynolds and Strouhal numbers. The mean flow for large times is calculated and found in good agreement with previous predictions based on boundary-layer theory (see, for instance, Fig. 7.3).

Two transient problems in slow viscous flow are considered in the paper by S.H. Smith (1997), where the corresponding steady-state behavior leads to the paradoxical results of Stokes and Jeffery (1922). We believe that Jeffery (1922) discovered that when two equal circular cylinders rotate with equal but opposite angular velocities, the solution of the biharmonic equation indicates a uniform stream at large distances, rather than zero velocity!

The resolution of this paradox also required understanding that Jeffery's expression is an inner solution (as in the well-known paradox first detected by Stokes in 1851), requiring an outer domain to complete the description. The paper by Watson (1995) clarified this point.

The reader must not confound this Jeffery's paradox with the Jeffreys (1930) paradox, proposed by H. Jeffreys to emphasize the nontrivial nature of the ideal limit of Newtonian fluid motion. More precisely, the ideal fluid (inviscid fluid in incompressible motion) and bodies are capable only of rigid translation; none of the bodies can even be made to rotate (!). For a proof, see, for instance, Meyer (1971, pp. 87–89). Naturally, the situation is quite different for a viscous flu id (in this case the well-known Poiseuille's flow emerges).

In the recent paper by Misra et al. (2001), an analysis is made of the unsteady-state flow of an incompressible viscous fluid in the entrance region of a parallel-sided channel whose walls pulsate in a prescribed manner. It is interesting to note that the motivation behind studying this problem is related to physiological flow problems (e.g., blood flow through arteries, but in this case, it is necessary to consider a *non*-Newtonian fluid).

Finally, it is also possible to consider the acoustic oscillations in a bounded medium with a small viscosity for a compressible, heat-conducting fluid, when governing NSF equations are linearized (see, for instance, the book by Vasilieva and Boutousov (1990, in Russian, §17). Concerning compressible oscillating boundary layers, see the paper by Telionis and Gupta (1977).

7.2 Unsteady-State Viscous, Incompressible Flow past a Rotating and Translating Cylinder

7.2.1 Formulation of the Governing Problem

Badr and Dennis (1985) made a numerical study of the development with time of the 2-D flow of a viscous, incompressible fluid around a cylinder that suddenly starts rotating about its axis at constant angular velocity ω^0 and translating at right angles to this axis at constant speed U^0. For the solution, a suitable frame of reference is used in which axes translating with the cylinder but fixed in direction are taken in the plane of a circular cross section with origin 0 at the center (we assume that the circular cylinder is "infinitely long"). Modified polar coordinates (ξ, θ) are used, where

$\xi = \ln(r/a^0)$ and a^0 is the radius of the cylinder .

Thus, the cylinder is situated at $\xi = 0$, and the domain of the solution is

$\xi \geq 0$, and $0 \leq \theta \leq 2\pi$,

with $\theta = 0$ in the downstream direction. Dimensionless stream function ψ and scalar vorticity ζ are used. For the dimensionless radial u and transverse v components of velocity, obtained by dividing the physical components by U^0, we write

$$u = e^{-\xi}\frac{\partial \psi}{\partial \theta}, \quad v = -e^{-\xi}\frac{\partial \psi}{\partial \xi}, \quad \zeta = e^{-\xi}\left(\frac{\partial u}{\partial \theta} - \frac{\partial v}{\partial \xi} - v\right) . \quad (7.16)$$

204   7 Some Viscous Fluid Motions and Problems

The Navier equation of motion can be expressed as two equations:

$$e^{2\xi}\frac{\partial \zeta}{\partial \tau} = \frac{2}{\mathrm{Re}}\left(\frac{\partial^2 \zeta}{\partial \xi^2} + \frac{\partial^2 \zeta}{\partial \theta^2}\right) - \frac{\partial \psi}{\partial \theta}\frac{\partial \zeta}{\partial \xi} + \frac{\partial \psi}{\partial \zeta}\frac{\partial \theta}{\partial \xi} ; \quad (7.17a)$$

$$\frac{\partial^2 \psi}{\partial \xi^2} + \frac{\partial^2 \psi}{\partial \theta^2} = e^{2\xi}\zeta . \quad (7.17b)$$

Here, $\tau = U^0 t/a^0$ and $\mathrm{Re} = 2a^0 U^0/\nu^0$. Equations (7.17a,b) are those considered by Collins and Dennis (1973a,b) for sudden translation of a cylinder without rotation. Below, the rotation of the cylinder enters the problem through the parameter (see (7.23b))

$$\alpha = \frac{a^0 \omega^0}{U^0} . \quad (7.18)$$

Boundary conditions for (7.17a,b) are:

$$\psi = 0, \quad \frac{\partial \psi}{\partial \xi} = 0, \text{ when } \xi = 0 , \quad (7.19a)$$

$$e^{-\xi}\frac{\partial \psi}{\partial \xi} \to \sin\theta, \quad e^{-\xi}\frac{\partial \psi}{\partial \theta} \to \cos\theta, \text{ as } \xi \to \infty , \quad (7.19b)$$

and all of the dependent variables in the flow domain must be periodic functions of θ with period 2π. But for the moment, we do not have inital conditions at $\tau = 0$. Collins and Dennis (1973a,b) used Fourier-series substitution for ψ and ζ to reduce the governing equations (7.17a,b) to sets of partial differential equations in τ and ξ, and these were solved by exact analysis for small values of τ (1973a) and by numerical methods for larger values of τ (1973b), the same method was used by Badr and Dennis (1985).

7.2.2 Method of Solution

The equations for the Fourier components (F_n, f_n), of $\psi(\xi, \theta, \tau)$ are obtained from (7.17b):

$$\frac{\partial^2 F_n}{\partial \xi^2} - n^2 F_n = e^{2\xi} G_n \quad (n = 0, 1, 2, \ldots) , \quad (7.20a)$$

$$\frac{\partial^2 f_n}{\partial \xi^2} - n^2 f_n = e^{2\xi} g_n \quad (n = 1, 2, \ldots) , \quad (7.20b)$$

where (G_n, g_n) are the Fourier components of $\zeta(\xi, \theta, \tau)$ and boundary conditions for (7.20a,b) follow from (7.19a,b). Substituting of Fourier series in (7.17a) and integrating with respect to θ from 0 to 2π give an equation for $G_0(\xi, \tau)$,

$$e^{2\xi}\frac{\partial G_0}{\partial \tau} = \frac{2}{\mathrm{Re}}\frac{\partial^2 G_0}{\partial \xi^2} + \sum_{(n=1 \text{ to } \infty)} \frac{\partial}{\partial \xi}[n(F_n g_n - f_n G_n)] . \quad (7.21)$$

7.2 Unsteady-State Flow past a Rotating and Translating Cylinder

The sets of equations satisfied by $G_n(\xi,\tau)$ and $g_n(\xi,\tau)$ for general integer values of n are obtained by substituting Fourier series in (7.17a), multiplying by $\cos(n\xi)$ or $\sin(n\xi)$, respectively, and integrating from $\theta = 0$ to $\theta = 2\pi$. The sets of equations which result are

$$e^{2\xi}\frac{\partial G_n}{\partial \tau} = \frac{2}{\mathrm{Re}}\left(\frac{\partial^2 G_n}{\partial \xi^2} - n^2 G_n\right) - \frac{n}{2}f_n\frac{\partial G_0}{\partial \xi} + \frac{1}{2}S_n\ ; \quad (7.22a)$$

$$e^{2\xi}\frac{\partial g_n}{\partial \tau} = \frac{2}{\mathrm{Re}}\left(\frac{\partial^2 g_n}{\partial \xi^2} - n^2 g_n\right) - \frac{n}{2}F_n\frac{\partial G_0}{\partial \xi} + \frac{1}{2}T_n\ . \quad (7.22b)$$

The expressions for S_n and T_n are given in Badr and Dennis (1985, p. 453). In theory, it is necessary to solve the infinite sets of equations (7.20a,b), (7.21), and (7.22a,b), but in practice, one has to solve only a finite number of equations! The boundary conditions for (7.20a,b) are derived from (7.19a) and for (7.21) and (7.22a,b), they follow from (7.19b):

$$\text{as } \xi \to \infty\ :\ G_0 \to 0,\ G_n \to 0,\ g_n \to 0,\quad n = 1, 2, \ldots\ , \quad (7.23a)$$

and it is also necessary to assume that

$$\int_0^\infty e^{2\xi} G_0(\xi,\tau)\,\mathrm{d}\xi = 2\alpha\ . \quad (7.23b)$$

Conditions (7.23a,b) ensure that the external flow conditions, derived from (7.19b) for $\xi \to \infty$, are satisfied and that the circulation remains zero around a contour surrounding the cylinder and with all points of it at an infinite distance from the cylinder.

7.2.3 Determination of the Initial Flow

The problem of determining the initial expression for $\psi(\xi,\theta,0)$ and $\zeta(\xi,\theta,0)$ is based on the hypothesis that the initial flow is governed by the usual boundary-layer theory in which a layer of thickness $\delta(\tau) = (\tau/\mathrm{Re})^{1/2}$ surrounds the cylinder following a sudden start. Therefore, we introduce variables appropriate for this layer defined by

$$\eta = \frac{\xi}{\delta(\tau)}, \quad (7.24a)$$

and then transform all the appropriate equations using (7.24a) together with the scaling of variables

$$(F_n, f_n) = \delta(\tau)(F_n^*, f_n^*),\quad (G_n, g_n) = (1/\delta(\tau))(G_n^*, g_n^*)\ . \quad (7.24b)$$

If we set $\tau = 0 \Rightarrow \delta(0) = 0$ for the start of motion, then we obtain the following equations:

$$\frac{\partial^2 G_n^*}{\partial \eta^2} + 2\eta \frac{\partial G_n^*}{\partial \eta} + 2G_n^* = 0 ,\tag{7.25a}$$

$$\frac{\partial^2 g_n^*}{\partial \eta^2} + 2\eta \frac{\partial g_n^*}{\partial \eta} + 2g_n^* = 0 ,\tag{7.25b}$$

$$\frac{\partial^2 F_n^*}{\partial \eta^2} = G_n^* ,\tag{7.26a}$$

$$\frac{\partial^2 f_n^*}{\partial \eta^2} = g_n^* .\tag{7.26b}$$

The solutions of these equations that satisfy the boundary conditions which are obtained from the transformed conditions at $\eta = 0$, (7.23b) and also two integral conditions for G_n^* and g_n^*, analogous to (7.23b), [see, for instance, Badr and Dennis (1985, p. 452 and 454)], after (7.24a,b) have been applied, are

$$G_0(\xi, 0) = \frac{4\alpha}{\pi^{1/2}} \exp(-\eta^2), \ G_n(\xi, 0) = 0, \ n \neq 0 ,$$

$$F_0(\xi, 0) = -2\alpha \left[\eta(1 - \mathrm{erf}(\eta)) + \pi^{-1/2}(1 - \exp(-\eta^2)) \right] ,$$

$$F_n(\xi, 0) \equiv 0, \ f_n(\xi, 0) = 2 \left[\eta \, \mathrm{erf}(\eta) - \pi^{-1/2}(1 - \exp(-\eta^2)) \right] \lambda_n ,\tag{7.27}$$

where $\lambda_1 = 1$, $\lambda_n = 0$ for $n = 2, 3, \ldots$. From (7.27), we obtain the initial expression for $\psi(\eta, \theta, 0)$ and $\zeta(\eta, \theta, 0)$. In particular,

$$\zeta(\eta, \theta, 0) = 2\pi^{-1/2}(\alpha + 2\sin\theta) \exp(-\eta^2) .$$

7.2.4 Results of Calculations and Comparison with the Visualization of Coutanceau and Ménard (1985)

Badr and Dennis (1985) present results obtained for Re = 200 and Re = 500, at two rotational speeds $\alpha = 1/2$ and $\alpha = 1$. In Fig. 7.5, we present the results of calculations by Badr and Dennis (1985, p. 479 and p. 482) for Re = 500 and $\alpha = 1$, for two times $\tau = 6.0$ and $\tau = 10.00$. Figure 7.6 shows a comparison of calculated [Badr and Dennis (1985, p. 479 and p. 482)] and experimental instantaneous streamlines for Re = 200, $\alpha = 0.5$ and times $\tau = 3.0$, $\tau = 6.0$ [Badr and Dennis (1985, p. 468)]. We note that the corresponding flow visualizations of Coutanceau and Ménard (1985) are exhibited in their Figs. 6 and 9, and for continuity of their study, Badr and Dennis give some selected comparisons for one case. It can be seen that the calculations are virtually identical to the flow visualizations.

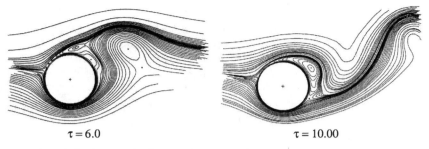

Fig. 7.5. Calculations for Re = 500, $\alpha = 1$, for $\tau = 6.0, 10.00$

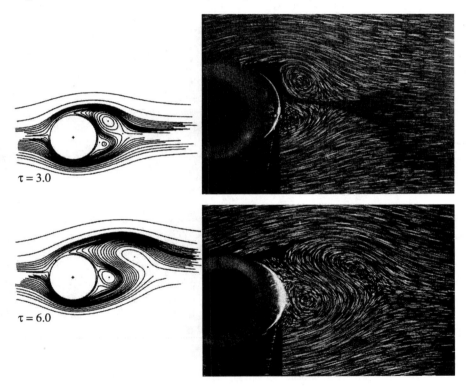

Fig. 7.6. Comparison with the visualizations Re = 200, $\alpha = 0.5$, for $\tau = 3.0, 6.0$

7.2.5 A Short Comment

This short comment concerns the determination of the initial expression for $\psi(\eta, \theta, 0)$ and $\zeta(\eta, \theta, 0)$. First, we stress again that Badr and Dennis (1985) assume that the initial flow is governed by the usual boundary-layer theory in which a layer of thickness $\delta(\tau) = (\tau/\text{Re})^{1/2}$ surrounds the cylinder following a sudden start! But, according to our asymptotic structure of unsteady-state

NSF equations at Re $\gg 1$ (see Sect. 4.3, Chap. 4), it seems, rather, that the initial flow (close to $\tau = 0$) is governed by a viscous Rayleigh type layer in which a layer of thickness $\Delta(\tau^{1/2}) = \tau^{1/2}/\text{Re}$ surrounds the cylinder following a sudden start, and then, it is necessary to introduce the new cordinate $\rho = \xi/\Delta(\tau^{1/2})$ in place of η. Obviously, in such a case, the initial expressions for $\psi(\eta, \theta, 0)$ and $\zeta(\eta, \theta, 0)$ are different.

The question is, what is the influence of this change in the initial expression for $\psi(\eta, \theta, 0)$ and $\zeta(\eta, \theta, 0)$ on the instantaneous streamlines of the flow for *high* Re and *small time* τ?

The above problem solved by Badr and Dennis (1985) is an initial boundary-value problem for unsteady-state Navier equations, when initial data are dependent on η rather than on ξ, at least for large Re $\gg 1$! But, in this case, the problem considered has two different scales ξ and η, and when Re $\gg 1$, it seems that our theory in Sect. 4.3.4 (Chap. 4), concerning the adjustment from the inviscid, one-dimensional, unsteady-state gas dynamics problem, is realized. In this case, the unsteady-state boundary-layer theory is very significant close to the wall of the cylinder for the various values of time τ, when we take into account the associated nonviscous (inviscid outer) flow. Concerning the case of an impulively-started cylinder, see the paper by Bouard and Coutanceau (1980).

7.3 Ekman and Stewartson Layers

Below, in this investigation, as a typical problem, we consider a disk of radius a^0 which rotates at an angular velocity Ω^0 in an unbounded viscous, incompressible, medium which itself rotates coaxially at angular velocity $(1 - \text{Ro})\Omega^0$. Our problem will be linearized in the small Rossby number Ro. The configuration is clarified in Fig. 7.7, where the various regions of the flow field are indicated. Let a^0, $1/\Omega^0$, U^0 and ρ^0 characterize the typical length, time, relative velocity and density. Lengths have been made dimensionless with a^0, velocities with $a^0\Omega^0$, and pressure with $\rho^0 a^{0^2} \Omega^{0^2}$.

In Fig. 7.7, I = Ekman layer, II = Stewartson layer, III = inner region, IV = outer region, and V = upper region.

Two important dimensionless parameters appear, the Ekman number,

$$\text{Ek} = \frac{\nu^0}{\Omega^0 a^{0^2}}, \tag{7.28a}$$

where ν^0 is the constant kinematic viscosity, and the Rossby number is

$$\text{Ro} = \frac{U^0}{\Omega^0 a^0}. \tag{7.28b}$$

The former is a gross measure of how the typical viscous force compares to the Coriolis force, and it is, in essence, the inverse Reynolds number for the

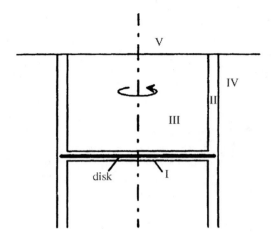

Fig. 7.7. The configuration

flow. Likewise, the Rossby number, a ratio of the convective acceleration to the Coriolis force, provides an overall estimate of the relative importance of nonlinear terms. The Ekman number is very small in most cases of interest where the primary effects of rotation are displayed. Practical values of 10^{-5} are usual and henceforth the assumption Ek $\ll$ 1 is made without further statement. The Rossby number Ro is of unit magnitude or less; linear theories presume an infinitesimal value.

In 1957, Keith Stewartson published a paper [Stewartson (1957)] in which he considered the shear layers between two coaxial rotating planes whose center disks rotate at a slightly different angular velocity. He found that if the deviations of the angular velocities of the disks from that of the planes are equal but opposite, a shear layer of thickness $\text{Ek}^{1/3}$ exists; if the deviations are equal in the same sense, an additional layer of thickness $\text{Ek}^{1/4}$ appears.

This last layer is necessary to fit the azimuthal velocity of the inner region to that of the outer region. Greenspan gave, in his monograph (1990), a clear and pertinent account of these so-called "Stewartson layers", and Moore and Saffman (1969) presented an analysis of different possibilities for a variety of situations. Hide and Titman (1967) performed an experimental investigation on a rotating disk of finite radius placed in a cylindrical tank which itself is rotating at a slightly different angular velocity – they showed the physical existence of the Stewartson layer.

Here, we consider the problem investigated by van de Vooren (1992) who showed that if a disk of finite radius and the surrounding medium rotate coaxially at slightly different angular velocities, an axial layer in the form of a cylindrical shell exists at the edge of the disk. This shell of thickness $O(\text{Ek}^{1/3})$ has length $O(1/\text{Ek})$ in the axial direction. The most characteristic element is the axial velocity of $O(\text{Ek}^{1/6})$ which is larger than everywhere else in the field. We observe that at the point where the Stewartson layer is joined to the Ekman layer, a logarithmic singularity exists in the pressure and this singularity is responsible for deflecting the boundary layer flow to axial flow in the Stewartson layer.

We observe [see, for instance, Gans (1983)], in general, that the time-dependent motion of a viscous fluid in a container rotating at Ω^0 is characterized by boundary layers on the container surfaces if Ek $\ll 1$. Let the frequency of the motion, measured in a corotating coordinate system, be $\omega^0 \Omega^0$. If $\omega^0 \approx 1$, then the length scale of the boundary layer is $(\nu^0/\Omega^0)^{1/2}$; unless $|\omega^0|$ equals twice the normal component of the unit rotational vector, scales of $a^0 \mathrm{Ek}^{1/3}$ and $a^0 \mathrm{Ek}^{1/4}$ are possible. If the normal vector and rotational vectors are parallel, the former scale vanishes. The Stewartson layers are a special case of these boundary layers (and free shear layers) on characteristic surfaces of finite area for the associated inviscid problem.

7.3.1 General Equations and Boundary Conditions

For an axially symmetrical configuration, the steady-state dimensionless equations of motion in an inertial system of reference are

$$u\frac{\partial u}{\partial r} + w\frac{\partial u}{\partial z} - \frac{1}{r}v^2 = -\frac{\partial p}{\partial r} + \mathrm{Ek}\left[\frac{\partial^2 u}{\partial r^2} + \frac{\partial}{\partial r}\left(\frac{u}{r}\right) + \frac{\partial^2 u}{\partial z^2}\right],$$

$$u\frac{\partial v}{\partial r} + w\frac{\partial v}{\partial z} + \frac{1}{r}uv = \mathrm{Ek}\left[\frac{\partial^2 v}{\partial r^2} + \frac{\partial}{\partial r}\left(\frac{v}{r}\right) + \frac{\partial^2 v}{\partial z^2}\right],$$

$$u\frac{\partial w}{\partial r} + w\frac{\partial w}{\partial z} = -\frac{\partial p}{\partial z} + \mathrm{Ek}\left[\frac{\partial^2 w}{\partial r^2} + \frac{1}{r}\frac{\partial w}{\partial r} + \frac{\partial^2 w}{\partial z^2}\right], \quad (7.29a)$$

and the equation of continuity is

$$\frac{1}{r}\frac{\partial(ru)}{\partial r} + \frac{\partial w}{\partial z} = 0, \quad (7.29b)$$

where u, v, and w are the radial (r), azimuthal, and axial (z) velocities, respectively; p is the pressure. To satisfy the equation of continuity (7.29b), a stream function ψ is introduced by

$$u = \frac{1}{r}\frac{\partial \psi}{\partial z}, \quad w = -\frac{1}{r}\frac{\partial \psi}{\partial r}. \quad (7.30)$$

The boundary conditions are

$$z = 0, r < 1 \;:\; u = 0, \; v = r, \; w = 0, \text{ at the disk},$$

$$(z^2 + r^2)^{1/2} \to \infty \;:\; u \to 0, \; v \to (1 - \mathrm{Ro})r, \; w \to 0, \; p \to \frac{1}{2}(1 - \mathrm{Ro})^2 r^2,$$

$$\text{at infinity.} \quad (7.31)$$

7.3.2 The Ekman Layer

Because Ro $\ll 1$, it is assumed that all deviations from the original (basic unperturbed) flow are proportional to Ro (linearized theory). Close to the disk, introducing the boundary layer coordinate $\zeta = z/\operatorname{Ek}^{1/2}$, we may write

$$\psi = \frac{\operatorname{Ro}}{2} \operatorname{Ek}^{1/2} r^2 h(\zeta), \quad u = \frac{\operatorname{Ro}}{2} r \frac{dh(\zeta)}{d\zeta},$$
$$v = r(1 - \operatorname{Ro}) g(\zeta),$$
$$w = -\operatorname{Ro} \operatorname{Ek}^{1/2} h(\zeta), \quad p = \frac{1}{2}(1 - 2\operatorname{Ro}) r^2, \qquad (7.32)$$

because the third equation of (7.29a) shows that $\partial p/\partial \zeta = O(\operatorname{Ek})$.

For $|\operatorname{Ro}| < O(1)$, with (7.32), we derive the following leading-order problem in place of (7.29a,b) and (7.31):

$$\frac{d^2 h}{d\zeta^2} = 4(g - 1), \quad \frac{dh}{d\zeta} + \frac{d^2 g}{d\zeta^2} = 0;$$
$$\zeta = 0 : h = 0, \quad \frac{dh}{d\zeta} = 0, \quad g = 0,$$
$$\zeta \to \infty \text{ (matching)} : \quad \frac{dh}{d\zeta} \to 0, \quad g \to 1. \qquad (7.33)$$

The solution of system (7.33) is obviously

$$g(\zeta) = 1 - \exp(-\zeta)\cos\zeta, \quad h(\zeta) = 1 - \exp(-\zeta)[\sin\zeta + \cos\zeta]. \qquad (7.34)$$

We observe that outside the disk (for $r > 1$), we can have only the original flow $h = 0$, $g = 1$ with $v = (1-\operatorname{Ro})r$. Hence, at $r = 1$, the Ekman layer, characterized by solution (7.34), *suddenly ends, which means that large changes occur in the radial direction and this gives rise to a Stewartson layer.*

7.3.3 The Stewartson Layer

Scaling of the quantities in the Stewartson layer (close to $r = 1$) can be taken most easily from Greenspan (1990, pp. 98 and 99). To comply with the rapid changes in the radial direction, a stretched coordinate r^* is introduced by

$$r^* = \frac{(r-1)}{\operatorname{Ek}^\alpha}, \quad 1 > \alpha > 0, \qquad (7.35)$$

and r^* and z are the independent variables in the Stewartson layer (when $\alpha = 1/3$). In this case, it is necessary to consider the following expansions:

$$\psi = \operatorname{Ro}[\operatorname{Ek}^{1/2} \psi_1 + \operatorname{Ek}^{5/6} \psi_2 + \ldots],$$
$$u = \operatorname{Ro}[\operatorname{Ek}^{1/2} u_1 + \operatorname{Ek}^{5/6} u_2 + \ldots],$$
$$v = (1 - \operatorname{Ro})r + \operatorname{Ro}[\operatorname{Ek}^{1/6} v_1 + \operatorname{Ek}^{1/2} v_2 + \ldots],$$
$$w = \operatorname{Ro}[\operatorname{Ek}^{1/6} w_1 + \operatorname{Ek}^{1/2} w_2 + \ldots],$$
$$p = \frac{1}{2}(1 - \operatorname{Ro})^2 r^2 + \operatorname{Ro}[\operatorname{Ek}^{1/2} p_1 + \operatorname{Ek}^{5/6} p_2 + \ldots]. \qquad (7.36)$$

212   7 Some Viscous Fluid Motions and Problems

This gives an axial flux $O(\mathrm{Ek}^{1/2})$ which is the deflected radius flux of $O(\mathrm{Ek}^{1/2})$ from the Ekman layer.

We observe that the second term in the expansion of w is required to match the term $w = -\mathrm{Ro}\,\mathrm{Ek}^{1/2}\,h(z/\mathrm{Ek}^{1/2})$ in the inner Ekman region, [see (7.32)], for $r^* \to \infty$.

Substituting (7.36) in (7.29a) and boundary conditions (7.29b) leads to the following set of equations for the first approximation [when $|\mathrm{Ro}| < O(Ek^{1/6})$]:

$$2v_1 = \frac{\partial p_1}{\partial r^*}, \quad 2u_1 = \frac{\partial^2 v_1}{\partial r^{*2}}, \quad \frac{\partial p_1}{\partial z} = \frac{\partial^2 w_1}{\partial r^{*2}}, \quad \frac{\partial u_1}{\partial r^*} + \frac{\partial w_1}{\partial z} = 0. \tag{7.37}$$

For the second approximation, we derive [valid only if $|\mathrm{Ro}| < O(\mathrm{Ek}^{1/2})$]

$$2v_2 = \frac{\partial p_2}{\partial r^*}, \quad 2u_2 = \frac{\partial^2 v_2}{\partial r^{*2}} + \frac{\partial v_1}{\partial r^*},$$

$$\frac{\partial p_2}{\partial z} = \frac{\partial^2 w_2}{\partial r^{*2}} + \frac{\partial w_1}{\partial r^*}, \quad \frac{\partial u_2}{\partial r^*} + \frac{\partial w_2}{\partial z} + u_1 = 0. \tag{7.38}$$

From relation (7.30) with (7.35) we find also that

$$u_1 = \frac{\partial \psi_1}{\partial z}, \quad u_2 = \frac{\partial \psi_2}{\partial z} - r^* u_1, \quad w_1 = -\frac{\partial \psi_1}{\partial r^*}, \quad w_2 = -\frac{\partial \psi_2}{\partial r^*} - r^* w_1, \tag{7.39}$$

and eliminating all functions except ψ_1 and ψ_2 yields, finally, as fundamental equations,

$$\frac{\partial^6 \psi_1}{\partial r^{*6}} + 4\frac{\partial^2 \psi_1}{\partial z^2} = 0; \tag{7.40a}$$

$$\frac{\partial^6 \psi_2}{\partial r^{*6}} + 4\frac{\partial^2 \psi_2}{\partial z^2} = 3\frac{\partial^5 \psi_1}{\partial r^{*5}}. \tag{7.40b}$$

For the boundary conditions at $z = 0$ for (7.40a,b), we observe that the Ekman layer which has thickness $O(Ek^{1/2})$ is reduced in the z coordinate to $z = 0$, and hence $z = 0$ must correspond to the outer edge of the Ekman layer.

The point $r^* = 0$, $z = 0$ is a singular point, and it is necessary to investigate the region connecting the Stewartson and Ekman layers! For $r^* < 0$, at the outer edge of the Ekman layer using (7.32), (7.34), and (7.35),

$$z = 0 : \psi = \frac{1}{2}\mathrm{Ro}\,\mathrm{Ek}^{1/2} + \mathrm{Ro}\,\mathrm{Ek}^{5/6}\,r^* + \ldots$$

Thus,

$$\psi_1 = 1/2, \quad \psi_2 = r^*, \quad w_1 = 0, \quad w_2 = -1. \tag{7.41a}$$

For $r^* > 0$, where there is no Ekman layer,

$$\psi_1 = 0, \quad \psi_2 = 0, \quad w_1 = 0, \quad w_2 = 0. \tag{7.41b}$$

On the other hand,

$$z \to \infty \;:\; \psi_1 \text{ is bounded},\tag{7.41c}$$

$$r^* \to -\infty \;:\; \psi_1 \text{ is bounded}, r^* \to \infty \;:\; \psi_1 = 0.\tag{7.41d}$$

The main solution in the Stewartson layer; determination of ψ_1.
The stream function ψ_1 is solved with the aid of Fourier transformation. The reader can find the main solution of the Stewartson layer in van de Vooren (1992, pp. 136–143). We observe, first, that ψ_1 depends only on the similarity variable $\kappa = z/r^{*3}$. Then, for any finite value of z and r^* varying from $-\infty$ to $+\infty$, ψ_1 varies from $1/2$ to 0. Thus, the axial mass flow in the Stewartson layer for any finite z is equal to $2\pi \operatorname{Ro} \operatorname{Ek}^{1/2}/2$. This is exactly equal but opposite to the axial mass flow in the inner region which is $-2\pi \operatorname{Ro} \operatorname{Ek}^{1/2} \int_0^1 h(\infty) r \, dr$, and $h(\infty) = 1$ is determined in the Ekman layer.

For finite z, there is no interchange between the two mass flows. That the axial velocity is constant in the inner region is due to the Taylor–Proudman theorem [for this theorem and a proof, see Meyer (1971, pp. 139–142)]. However, there is a small $\nu^0 = O(\operatorname{Ek})$ viscosity, and this causes the axial velocity in the inner region to diminish at a distance $z = O(1/\operatorname{Ek})$, where the upper region V (Fig. 7.7) begins. The axial velocity in the Stewartson region diminishes with increasing z as a result of the widening of the layer proportional to $z^{1/3}$.

In the upper region, the two axial mass flows begin to annihilate each other. Now, from $w_1 = -\partial \psi_1/\partial r^*$ and $\partial p_1/\partial z = \partial^2 w_1/\partial r^{*2}$, we also determine $p_1(r^*, z)$, and the infinitely large pressure at the singularity $r^* = 0$, $z = 0$, is the fundamental reason for the deviation of the flow from the Ekman boundary layer toward the Stewartson layer.

Second approximation in the Stewartson layer: determination of ψ_2.
The Stewartson layer gives rise to axial velocities $O(\operatorname{Ek}^{3/2})$ in the inner and outer regions. However, in the inner region, there is a more important axial velocity $w = -\operatorname{Ro} \operatorname{Ek}^{1/2}$, due to the Ekman layer, which is lacking in the outer region. Van de Vooren (1992, pp. 146–150) investigated the reduction of the axial velocity of $O(\operatorname{Ek}^{1/2})$ in the inner region to $O(\operatorname{Ek}^{3/2})$ in the outer region with the aid of the second approximation of the solution of (7.40b) for ψ_2. This second approximation, which is valid for $|\operatorname{Ro}| < O(\operatorname{Ek}^{1/2})$, gives contributions in the Stewartson layer of the following orders of magnitude: $O(\operatorname{Ek}^{5/6})$ for ψ, u, and p; $O(\operatorname{Ek}^{1/2})$ for v and w.

7.3.4 The Inner, Outer, and Upper Regions

First, the flow in the *inner* region for $r \uparrow 1$ should be matched to the flow in the Stewartson layer for $r^* \to -\infty$. The leading-order approximate equations in the inner (i) and *outer* (o) regions are the same and only ψ_i and w_i and ψ_o

and w_o are linear in z, the other functions are independent of z. The leading-order approximate equations are modified only when z becomes $O(1/\text{Ek})$. We observe that matching to the inner region and the outer region yields further terms in the expansions for $r \to 1$ of the solutions in the inner and outer regions.

In the *upper* (u) region, with $z_u = z/(1/\text{Ek})$, the orders of magnitude are the same as the inner region, which means that w is also $O(\text{Ek}^{1/2})$, but, in place of $\partial p_i/\partial r = 0$, we derive

$$\frac{\partial p_u}{\partial r} = \frac{\partial^2 w_u}{\partial r^2} + \frac{1}{r}\frac{\partial w_u}{\partial r},$$

and for $z_u \downarrow 0$, the limits of ψ_u and w_u are different when r is smaller or larger than 1:

$$r < 1,\ \psi_u \to (1/2)r^2,\ w_u \to -1;\quad r > 1,\ \psi_u \to 0,\ w_u \to 0.$$

It follows that the solution in the upper region contains a singularity at the point $r = 1$, $z_u = 0$.

7.3.5 Comments

In the above problem, we have two small parameters, Ro and Ek, and it seems possible to construct a consistent asymptotic theory when these two parameters are related by a similarity relation:

$$|\text{Ro}| = \lambda^0 \text{Ek}^\beta, \text{ with } \lambda^0 = O(1) \text{ and } \beta > 0,$$

with β a scalar, which must be determined! This determination is performed such that the asymptotic theory constructed would be more significative!

On the other hand, according to the van de Vooren (1992) theory, a rotating disk placed in a fluid rotating coaxially at a slightly different angular velocity, where the ratio of the angular velocities is

$$(1 - \text{Ro}) \approx 1 \text{ with } \text{Ro} \ll 1,$$

shows a Stewartson layer at its edge of width $O(\text{Ek}^{1/3})$ and height $O(1/\text{Ek})$ provided

$$|\text{Ro}| < O(\text{Ek}^{1/6}).$$

Consequently, if, for example we assume a similarity relation

$$\beta = 1/6 + \alpha, \text{ with } 0 < \alpha \ll 1,$$

and in this case,

$$|\text{Ro}|/\text{Ek}^{1/6} = \lambda^0 \text{Ek}^\alpha \ll 1,$$

then the situation corresponding to the existence of a Stewartson layer is characterized by the following limiting process:

Ek fixed, Ro $\downarrow$ 0, and then Ek $\downarrow$ 0 .

On the contrary, if

$$\beta = 1/6 - \alpha, \text{ then } |\text{Ro}| / \text{Ek}^{1/6} = \lambda^0 / \text{Ek}^\alpha \gg 1 ,$$

and the situation corresponding to the existence of a Stewartson layer is rather characterized by the limiting process:

Ro fixed, Ek $\downarrow$ 0, and then Ro $\downarrow$ 0 .

The limiting process,

$$\text{Ek} \downarrow 0 \text{ and } \text{Ro} \downarrow 0, \text{ such that } |Ro| = \lambda^0 \text{Ek}^\beta ,$$

is the more significant and gives the possibility of taking into account *simultaneously two small effects* which (via coupling-similarity parameter λ^0) lead to an effect of the order $O(1)$.

We note also that the rotation of the medium can be realized by thinking of a cylindrical tank rotating at an angular velocity $(1 - \text{Ro})\Omega^0$, but in this case, the top and bottom of the tank must be at a distance from the disk larger than $O(1/\text{Ek})$ so that the Stewartson layer of the disk is not influenced by the top and bottom of the tank.

7.4 Low Reynolds Number Flows: Further Investigations

7.4.1 Unsteady-State Adjustment to the Stokes Model in a Bounded Deformable Cavity $\Omega(t)$

We consider the unsteady-state Navier system for the velocity vector u and the pseudo-pressure π:

$$\text{In } \Omega(t) : \boldsymbol{\nabla} \cdot \boldsymbol{u} = 0 , \tag{7.42a}$$

$$\frac{\partial \boldsymbol{u}}{\partial t} + (\boldsymbol{u} \cdot \boldsymbol{\nabla})\boldsymbol{u} + \boldsymbol{\nabla}\pi = \frac{1}{\text{Re}}\boldsymbol{\nabla}^2 \boldsymbol{u} , \tag{7.42b}$$

with the boundary condition,

$$\boldsymbol{u} = \boldsymbol{V}(t) \text{ on } \partial\Omega(t) , \tag{7.43a}$$

and initial condition

$$t = 0 : \boldsymbol{u} = \boldsymbol{u}^0(\boldsymbol{x}) , \tag{7.43b}$$

where both $\boldsymbol{V}(t)$ and $\boldsymbol{u}^0(\boldsymbol{x})$ are given data of problem (7.42a,b), (7.43a,b), at low Reynolds number Re $\ll$ 1.

First Limiting Process: Re $\downarrow$ 0 with t and x fixed. In this case, we assume the existence of the following expansions for u and π:

$$u = u_s + \ldots, \quad \pi \equiv \frac{1}{\mathrm{Re}} \Pi = \Pi_s + \ldots . \tag{7.44}$$

From (7.42a,b) with (7.44), we derive the classical Stokes linear system for $u_s(t, x)$ and $\Pi_s(t, x)$:

$$\nabla \cdot u_s = 0, \tag{7.45a}$$
$$\nabla^2 u_s = \nabla \Pi_s, \tag{7.45b}$$

and for the above Stokes system, the only possibility is to assume

$$u_s = V(t) \text{ on } \partial\Omega(t), t \text{ fixed}. \tag{7.45c}$$

Second Limiting Process: Re $\downarrow$ 0 with $\tau = t/\mathrm{Re}$ and x fixed. Such a limiting process local in time is necessary if we want to take into account the unsteady state! In this case, again from (7.42a,b) with (7.44), we derive the following unsteady-state equations for $u_L(\tau, x)$ and $\Pi_L(\tau, x)$ in place of (7.45a,b):

$$\nabla \cdot u_L = 0, \tag{7.46a}$$
$$\frac{\partial u_L}{\partial \tau} = \nabla^2 u_L - \nabla \Pi_L, \tag{7.46b}$$

when we assume, in place of (7.44), the existence of the following local expansions in time when Re $\downarrow$ 0:

$$u = u^*(\tau, x; \mathrm{Re}) = u_L + \ldots, \quad \Pi \equiv \Pi^*(\tau, x; \mathrm{Re}) = \Pi_L + \ldots . \tag{7.47}$$

For (7.46a,b) local in time, we can associate the conditions

$$u_L = V(0) \text{ on } \partial\Omega(0), \tag{7.46c}$$
$$\tau = 0 : u_L = u^0(x), \tag{7.46d}$$

if we assume that the deformation of the bounded cavity is performed only relative to time variable t (no structure local in time τ),

$$V(t) \equiv V(\mathrm{Re}\,\tau) = V(0) + \mathrm{Re}\,\tau V'(0) + \ldots, \quad \mathrm{Re} \downarrow 0.$$

Consequently, the problem local in time is governed by (7.46a,b) for $u_L(\tau, x)$ and $\Pi_L(\tau, x)$ with boundary and initial conditions (7.46c,d). $V(0) = 0$ and $\Omega(0)$ denote the considered cavity at rest, independent of time t.

7.4 Low Reynolds Number Flows: Further Investigations

The Adjustment Problem to the Stokes Steady Classical Solution.
Let $u_s(0, x)$ be assumed *different from initial given data* $u^0(x)$. We stress that $u_s(0, x)$ is the value at $t = 0$ of the Stokes (7.45a–c), "quasi-steady-state" problem [t is a parameter in this Stokes problem via $V(t)$], and if it is necessary, we can determine the value of $\Pi_s(0, x)$ by inverting the operator ∇ in Stokes equation (7.45b): $\Pi_s(0, x) = \nabla^{-1}[\nabla^2 u_s(0, x)]$.

To formulate the adjustment problem to the Stokes classical solution, we introduce

$$U(\tau, x) = u_L(\tau, x) - u_s(0, x), \quad P(\tau, x) = \Pi_L(\tau, x) - \Pi_s(0, x), \quad (7.48)$$

and in this case, we obtain the following unsteady-state adjustment problem,

$$\nabla \cdot U = 0, \quad \frac{\partial U}{\partial \tau} + \nabla P = \nabla^2 U, \quad (7.49a)$$

$$U = 0 \text{ on } \partial\Omega(0), \tau = 0 : U = u^0(x) - u_s(0, x), \quad (7.49b)$$

because $\nabla \cdot u_s(0, x) = 0$.

Solution of problem (7.49a,b). Obviously, in a undeformable cavity, $\Omega(t = 0) = \Omega^0$, we can write the following solution for $U(\tau, x)$ and $P(\tau, x)$:

$$U(\tau, x) = \sum \alpha_k \exp(-\lambda_k \tau) \Phi_k(x),$$
$$P(\tau, x) = \sum \alpha_k \exp(-\lambda_k \tau) \Psi_k(x), \quad k = 1 \text{ to } \infty. \quad (7.50)$$

With (7.50), from (7.49a,b), we derive the following classical spectral problem for $[\Phi_k(x), \Psi_k(x); \lambda_k]$:

$$\nabla^2 \Phi_k - \nabla \Psi_k = \lambda_k \Phi_k; \quad \nabla \cdot \Phi_k = 0, \Phi_k(x) = 0 \text{ on } \partial\Omega(0), \quad (7.51)$$

where the eigenvalues λ_k are positive and real, such that $\lambda_k \to \infty$ with $k \to \infty$. Then, the following relation is satisfied:

$$\alpha_k \int_{\Omega(0)} \Phi_k \, dv = \int_{\Omega(0)} \Phi_k \cdot [u^0(x) - u_s(0, x)] \, dv. \quad (7.52)$$

From solution (7.50), obviously, when $\tau \to +\infty$, $[U(\tau, x)$ and $P(\tau, x)]$ tend to zero exponentially, and the adjustment is assured.

Consequently, by matching,

$$u_L(+\infty, x) = u_s(0, x), \quad \Pi_L(+\infty, x) = \Pi_s(0, x). \quad (7.53)$$

Conclusion. We stress, again, that

$$u^0(x) \neq u_s(0, x), \quad (7.54)$$

because, for instance, $\nabla \cdot \boldsymbol{u}_s(0, \boldsymbol{x}) = 0$, but, in general, for the given data, $\boldsymbol{u}^0(\boldsymbol{x})$, obviously, $\nabla \cdot \boldsymbol{u}^0(\boldsymbol{x}) \neq 0$! Condition (7.54) gives the possibility of determining coefficient α_k according to (7.52), when the solution of the eigenvalue (spectral) problem is known.

We see that the initial layer close to $t = 0$ gives the possibility of solving the singular behavior $[\boldsymbol{u}_s(0, \boldsymbol{x}) \neq \boldsymbol{u}^0(\boldsymbol{x})]$ of the Stokes solution when t tends to zero!

7.4.2 On the Wake in Low Reynolds Number Flow

The term "wake" means an imprint cast by an obstacle placed in a stream. In low Reynolds number unsteady-state flow behind an impulsively started cylinder, the imprint that develops as time progresses is in form of a recognizable regime. In it, the flow is basically axial. The results obtained by Bentwich and Miloh (1984, p. 5) suggest that the ratio between the velocity components

$$\frac{-\partial \psi / \partial x}{\partial \psi / \partial y} \text{ is of } O[\Delta(\mathrm{Re})\,\mathrm{Re}]\,,$$

where $\Delta(\mathrm{Re})$ is a small gauge function $[\Delta(\mathrm{Re}) \to 0$, with $\mathrm{Re} \to 0]$. Note that, at every instant, the axial component $\partial \psi / \partial y$ is uniform across the wake, and its magnitude is obtained by setting $Y = 0$ over $0 < X < \infty$, where (X, Y) are the outer Cartesian coordinates in the later outer solution. Because of the fluids' incompressibility, axial variations in that component are compensated thanks to tranverse motion, but it is much smaller in magnitude.

We observe that a qualitatively similar situation prevails in the wake trailing a two-dimensional obstacle placed in a steady-state stream when the Reynolds number is high. In this case, again that region is distinct, and the flow in it is basically axial. But there are some quantitative differences. In particular, in the high Reynolds number wake, the ratio between the velocity components, as defined above, is of $O(\mathrm{Re}^{1/2})$, and this feature is associated with the scaling adopted when Re is high. It reveals that in such a case, the wake flow is governed by the boundary-layer equation. Thus, though the latter constitutes a balance between the effects of inertia and viscosity, in the case at hand, the wake is viscosity dominated.

For low Reynolds number *unsteady-state* flow, as reported (see Chap. 5), the solution consists of *three* expansions that represent the flow at $\mathrm{Re} \ll 1$ in *three* different space-time subdomains (see Fig. 5.1).

In Bentwich and Miloh (1984), it is shown that there a *fourth* subdomain also exists. And consequently, it is necessary to consider an *additional expansion and match this expansion with the other three*. The later fourth expansion represents the flow late in the process in a wake region, which extends all the way downstream to infinity, and its width is comparable to the diameter of the obstacle.

7.4 Low Reynolds Number Flows: Further Investigations

According to Bentwich and Miloh (1984) (see the Introduction), the late outer (Oseen) expansion is not valid everywhere beyond the vicinity of the moving cylinder for Re $\ll 1$ flow behind an impulsively started cylinder because the outer representation of the stream function has a discontinuous (second) derivative across the half-plane behind the moving cylinder! This discontinuity is significant. It is an inherent feature of the outer (Oseen) flow field because a cylinder acquires a uniform velocity instantaneously and consequently, it appears that one cannot construct a solution of the unsteady-state fourth-order Oseen leading equation

$$\left(\frac{\partial}{\partial T} + \frac{\partial}{\partial X} - \nabla^2\right) \nabla^2 \psi^{Os} = 0 , \qquad (7.55)$$

with $T = (\text{Re}^2)t$, $X = (\text{Re})x$, which prevails in the outer "Oseen" subdomain, that satisfies all appropriate conditions. By that we mean that it matches local in time and inner (Stokes) expansions, meets the regularity conditions at infinity, and has its zero to third derivative continuous everywhere in the outer field. Obviously, this finding raises the question whether the three-expansion solution, proposed in Bentwich and Miloh (1982), is valid. In Bentwich and Miloh (1984), the authors show that it is, in the sense that a potential (outer) solution is a valid representation of high Reynolds number flow past an obstacle. Thus, just as the irrotational solution holds throughout the exterior of the obstacle but not on its surface, the three-expansion solution proposed in Bentwich and Miloh (1982) holds almost but not quite throughout the late outer Oseen, space-time subdomain under discussion. Furthermore, just as the inadequacy of classical irrotational flow can be mended by appending a boundary layer, the discontinuity of the later outer Oseen unsteady-state expansion along the half-plane behind the cylinder can be mended by constructing an additional fourth expansion in a "late-wake" subdomain. It expresses the flow over this plane and its immediate vicinity late in the process.

We observe that the discontinuity in the second derivative, in the later outer Oseen unsteady-state solution of (7.55), is irreconcilable with the assumption that the unsteady-state fourth-order Oseen equation (7.55) holds everywhere beyond the immediate vicinity of the obstacle and there is *no way* to "erase" it! We note that the solution of (7.55) fails along the positive axis and also where the cylinder undergoes uniform acceleration (because of the irregular behavior of the corresponding component over the semi-infinite plane downstream of the cylinder's axis). Naturally, none of these difficulties would emerge had one assumed that this plane and its immediate vicinity are not part of the outer Oseen subdomain late in the process. Bentwich and Miloh (1984; §3, pp. 4 and 5) just assume that a solution for such a fourth subdomain for an impulsively started cylinder is developed, and it is observed that in late-wake expansion, the corresponding stream function is a function of (T, X, y).

7.4.3 Oscillatory Disturbances as Admissible Solutions and their Possible Relationship to the Von Karman Sheet Phenomenon

In Bentwich (1985), a pertinent paper, the author observes that, in unsteady-state flow at Re ≪ 1, when only one term is retained in the inner (Stokes) expansion, two terms in the outer (Oseen), and two in the wake expansion, the solution for the steady-state components are determinable and equal to those recorded [see, for instance, Proudman and Pearson (1957)].

However, the scheme also admits a large variety of nontrivial solutions for time-dependent components and, in Bentwich (1985), attention is focused on those representing *oscillatory modes* of disturbance flow of indeterminable frequency and amplitude. In the inner field, such a single mode has the form of a rotationally symmetrical pattern, and far downstream, it is in the form of a sequence of vorticity packets of alternate signs, equally spaced along the wake's center plane (see Figs. 7.8 and 7.9).

The flow field resulting from such a disturbance superposed on a uniform stream bears a remarkable resemblance to the well-kown Von Karman vortex sheet. We observe that in the preceding papers by Bentwich and Miloh (1978, 1982), the unsteady-state soon dies out. In the flow under consideration, the uniform steady-state streaming motion is similarly dominant, whence the

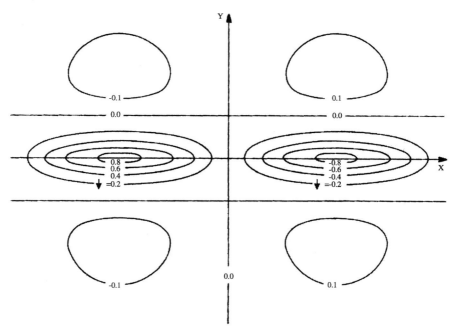

Fig. 7.8. Oscillatory disturbance flow pattern far downstream

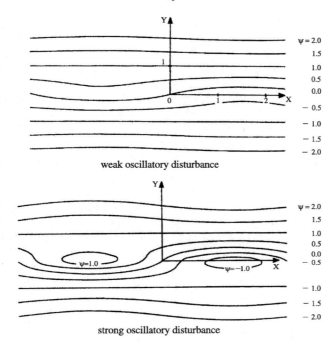

Fig. 7.9. Flow pattern in the wake

applicability of unsteady-state equations for fourth regions (local in time, Stokes, Oseen, and in wake) to the case at hand.

We stress again that attention is focused on disturbance-flow solutions that persist and have nonvanishing traces in all three domains, and it is found that timewise-oscillatory ones have these features. Unfortunately, within the framework of the approximation adopted, the amplitude and frequency characterizing these cannot be determined. The net result is that, in addition to the recorded steady-state solution for the low Reynolds number flow under discussion (flow past a cylinder), one gets persistent time-dependent indeterminable disturbances.

Finally, it follows that, within the context of Proudman and Pearson's (1957) scheme, *the solution for the flow past a cylinder is not unique, unless a steady-state is postulated* (at least when the solution satisfies the requirement that the disturbance flow at infinity be finite). The complete flow field downstream, that consisting of a single oscillatory disturbance mode superposed on a uniform stream [the steady-state disturbance, proportional to $\Delta(\mathrm{Re})/\mathrm{Re}$, with

$$\Delta(\mathrm{Re}) = \left[\ln \frac{1}{\mathrm{Re}} + k\right]^{-1}, \text{ with } k = 3.703 ,\tag{7.56}$$

according to Kaplun (1957), dies out there], is shown by Fig. 7.9 and in the strong disturbance bears a marked resemblance to the Von Karman vortex sheet. On the other hand, the persistence of the pattern shown in Fig. 7.8 is at first glance surprising. The closed-loop streamlines constitute evidence that vorticity packets of alternate signs exist in the flow, and these retain their strength as they move steadily dowstream. Yet, it follows from the Navier (incompressible!) equations and the nature of the medium, that vorticity decays in the presence of viscosity, particularly when the Reynolds number is small and that effect is dominant. However, the point made here is that the flow pattern shown in Fig. 7.8 does not represent either reality or an exact solution of the Navier equations. It is only an approximation and should so be viewed. It shows that most of the vorticity generated at the obstacle's surface decays in compliance with the theory of viscous (incompressible!) flows and the persistence pattern is induced by a residual, very small amount of vorticity, which decays at a rate that is too slow to be accounted for by the approximate solution.

Concerning here the problem of the determinacy and uniqueness, we note only that for the outer field, when $R = (\text{Re})r$, $T = (\text{Re}^2)t$ for the stream function ψ, we obtain the following equation

$$\frac{\partial}{\partial t}(\boldsymbol{\nabla}^{*2}\psi) + \text{Re}\left[\frac{\partial(\boldsymbol{\nabla}^{*2}\psi;\psi)}{\partial(X;Y)}\right] = \boldsymbol{\nabla}^{*4}\psi \ , \tag{7.57}$$

where (X,Y) are the outer Cartesian coordinates and $\boldsymbol{\nabla}^{*2}$ is the Laplace operator in terms of these coordinates, at infinity, the following condition must be satisfied:

$$\psi \sim Y/\text{Re} \ \text{as} \ \text{Re} \to \infty \ , \tag{7.58}$$

and condition (7.58), it seems, opens the door to *nonuniqueness*! All that it requires is that the disturbance created by the obstacle should be finite, and this condition is therefore rather loose compared with the no-slip condition on the cylinder $(r = 1)$, which fixes the value $(= 0)$ of

$$\psi \ \text{and} \ \frac{\partial \psi}{\partial r} \ \text{in a pointwise manner on} \ r = 1 \ . \tag{7.59}$$

But, and this is important, nothing as definite as conditions (7.59) can be prescribed at infinity, and for two reasons. First, if the stream is unbounded, then the drag is balanced by its momentum deficiency, and the second reason is that, when tunnel experiments are conducted, care is taken that the flow should be uniform and steady-state upstream. Downstream, the flow is left uncontrolled. Thus, when the Reynolds number is high (!) enough, vortex sheets and other deviations from uniformity occur there. Now, outside laboratory environments, in truly unbounded flow past an obstacle, flow downstream is even less controlled. Consequently, the looseness of condition (7.58) and the disturbances it allows are an inherent feature of the flow under discussion.

7.4.4 Some References

In the recent paper by Alassar and Badr (1999), the problem of uniform steady-state viscous flow over an oblate spheroid is solved in the low Reynolds number range $0.1 \leq \text{Re} \leq 1.0$. The full Navier equations are written in the stream function–vorticity form and are solved numerically by the series-truncation method. Spheroids that have axis ratios ranging from 0.245 to 0.905 are considered. The numerically computed drag coefficients are compared with previous analytical formulae which were based on the solution of the (linear) Stokes equations. As expected, the deviations between the numerical results and the analytical formulae are small only for low Reynolds number flow, and they increase when Re increases. The range of validity of the analytical solutions [for instance, with the analytical drag formulae given by Payne and Pell (1960) and Breach (1961)] has been estimated.

In a typical problem of this nature, the quantities sought are the drag, surface vorticity, surface pressure distribution, and streamlines. Figure 7.10 shows the streamlines and vorticity patterns for the two cases: (a) Re = 0.1 and (b) Re = 1.0 when the parameter $\xi_0 = \tanh^{-1}(b/a) = 0.25$. This parameter ξ_0 defines the surface of the spheroid and is related to the axis ratio (b/a).

For the Stokes problem, various rigorous mathematical results exist; the reader can consult the comprehensive and pertinent study by Ladyzhenskaya (1975). Some rigorous results are also in Babenko (1976, 1982) and Babenko and Vasiliev (1973). In Pogu and Tournemine (1982), the reader can find a study of 2-D biharmonic equation acting in unbounded domains by a variational approach. In particular, this study is applied to the prob-

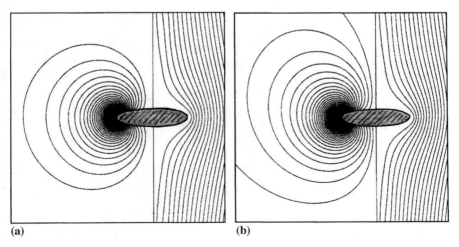

Fig. 7.10. Streamlines and vorticity patterns for: (a) Re = 0.1 and (b) Re = 1.0

lem of low Reynolds number flow, and the authors obtain, for an arbitrary profile, an *existence and uniqueness theorem* for the leading-order term of the *Stokes inner* (close to the wall of an obstacle) expansion. The paper by Tamada et al. (1983) considers the case when Re $\ll$ 1, but $|\log \text{Re}|^{-1}$ is not small, and in Miyagi and Kamei (1983), the standing vortex behind a disk normal to uniform flow is considered. The unsteady-state flow about a sphere at *low to moderate* Re was also considered by Chang and Maxey (1995). In the book edited by H.K. Kuiken (1996), pertinent informations are presented concerning *slow viscous flows*. Low Reynolds number, fully developed, 2-D *turbulent channel flow* with system rotation is considered in the paper by Nakabayashi and Kitoh (1996), and Nakanishi et al. (1997) give asymptotic solutions, again, for 2-D low Reynolds number flow around an impulsively started cylinder. The problem of displacement of fluid droplets from solid surfaces in low-Reynolds-number shear flows is considered by Dimitrakopoulos and Higdon (1997, 1998). Recently, the theory and an experiment on the low Reynolds number has been applied in the paper by Wong et al. (1998) to study the expansion or contraction of a bubble pinned at a submerged tube tip. The general solution of the particle momentum equation in unsteady-state Stokes flow is found in Coimbra and Rangel (1998).

An interesting aspect of low Reynolds number asymptotics is related to the Bénard–Marangoni thermocapillary-instability problem (for a deformable free surface; see Sect. 7.5 of this chapter). In this case, an asymptotic approach at Re $\ll$ 1, coupled with a long-wave approximation [see, for instance, Zeytounian (1998, pp. 260 and 261)] gives a so-called KS–KdV (Kuramoto, Sivashinski-Korteweg, de Vries) equation:

$$\frac{\partial h}{\partial t} + 2h\frac{\partial h}{\partial x} + \alpha\frac{\partial^2 h}{\partial x^2} + \beta\frac{\partial^3 h}{\partial x^3} + \gamma\frac{\partial^4 h}{\partial x^4} = 0 \,, \tag{7.60}$$

for the dimensionless 2-D film thickness $h(t,x)$, where the coefficients α, β, and γ take into account the influence of the (low) Reynolds and Marangoni numbers, Biot, and (high) Weber numbers, under the condition that Re / Fr^2 = 1.

Low Reynolds number flow around an oscillating circular cylinder was also considered recently [but at low Keulegan–Carpenter numbers; see review article by McNown (1991)] by Dütsh et al. (1998).

Finally, we mention the pioneer paper of Kaplun and Lagerstrom (1957).

7.5 The Bénard–Marangoni Problem: An Alternative

In so-called *Bénard thermal problem*, when an infinite horizontal viscous expansible liquid (of density ρ) layer (the one-layer simplified system) is heated from below, *two* physical effects are coupled: first, the convection effect linked

7.5 The Bénard–Marangoni Problem: An Alternative

with the gravity force and, second, the wavy motion of the free surface, open to ambient passive air. The motion induced by tangential gradients of variable (with temperature T) surface tension, $\sigma = \sigma(T)$, is a very important aspect of the modern theory of thin films and is customarily called the Marangoni effect (after one of the first scientists to explain this effect). More precisely, on the bounding surface at $x_3 = 0$, this horizontal liquid layer is in contact with a solid wall at constant temperature T_B, and at the level $x_3 = d$ is a free surface which separates the liquid layer from the atmosphere at constant temperature T_a and constant atmospheric pressure p_a, that has negligible viscosity and density. At the free surface, Newton's law of heat transfer is invoked [see, for instance, Davis (1987) and Joseph (1976)], and the surface tension $\sigma(T)$ decreases linearly with temperature T; thus,

$$\sigma(T) = \sigma(T_0) - \gamma(T - T_0), \tag{7.61}$$

where $\gamma = -\mathrm{d}\sigma/\mathrm{d}T$ is the constant rate of change of surface tension with temperature which is positive for most liquids. In (7.61), T_0 is the constant temperature of the free surface in the purely static basic state, when

$$T_s = T_s(x_3) = T_B - \beta x_3, \tag{7.62}$$

because a constant vertical temperature gradient, $-\mathrm{d}T_s/\mathrm{d}x_3 = \beta > 0$, is imposed on the expansible liquid layer. Obviously, $T_0 = T_B - \beta d$, and when we take into account Newton's law of heat transfer for $T_s(x_3)$, we can write $T_0 - T_a = \beta k(T_0)/q_s = \beta d/\mathrm{Bi}_s$, where

$$\mathrm{Bi}_s = \frac{dq_s}{k(T_0)}, \quad \text{and} \tag{7.63a}$$

$$\beta = \frac{\mathrm{Bi}_s}{1 + \mathrm{Bi}_s}\left[\frac{T_B - T_a}{d}\right], \tag{7.63b}$$

Bi_s is the *conduction* Biot number which accounts for heat transfer at the flat interface $x_3 = d$. The basic temperature state (7.62) describes pure conduction in a liquid at rest, when the undisturbed free surface is at $x_3 \equiv d$.

We assume that the perturbed deformable free surface is represented in a Cartesian coordinate system $(0; x_i)$, with $i = 1, 2, 3$, (in which $\boldsymbol{g} = -g\boldsymbol{k}$ acts in the negative x_3 direction) by the equation,

$$x_3 = d + a\eta(t, x_1, x_2) \equiv h(t, x_1, x_2). \tag{7.64}$$

The expansible viscous liquid, with the equation of state $\rho = \rho(T)$, is a Newtonian fluid with viscosities $\lambda(T)$ and $\mu(T)$, specific heat $C(T)$, and thermal conductivity $k(T)$; $\kappa = k/\rho C$ is the thermal diffusivity and $\nu = \mu/\rho$ is the kinematic viscosity. In (7.65), the constant q^0 is the unit thermal surface conductance between the deformable free surface and the air. At the deformable free surface, (7.64), Newton's law of heat transfer is $(q^0 \neq q_s)$

$$k(T)\frac{\partial T}{\partial n} + q^0(T - T_a) = 0 ,\tag{7.65}$$

where n is the distance in the direction of the outward normal $\boldsymbol{n}$ to the free surface. In recent book by Guyon et al. (2001) the reader can find a physical description of the Marangoni effect. In paper by Ruyer-Quil and Manneville (1998) the modeling of the film flows down inclined planes is considered.

7.5.1 Dimensionless Dominant Equations

For a weakly expansible liquid, when the dimensionless parameter,

$$\wp = \beta d \alpha_T, \text{ with } \alpha_T = -\left[\frac{\mathrm{d}(\log \rho)}{\mathrm{d}T}\right]_{T=T_0} ,$$

is *small*, it is possible to derive a simplified dimensionless dominant problem from the full NSF dimensionless problem [see, for instance, Zeytounian (2002b, Chap. 10)]. Let the coefficients with the subscript "$_0$" be reference values for these coefficients at $T = T_0$. First, we introduce the *dimensionless perturbations* for pressure and temperature by

$$\theta = \frac{(T - T_0)}{\beta d} ,\tag{7.66a}$$

$$\pi = \frac{1}{\mathrm{Fr}^2}\left(\frac{p - p_a}{g d \rho_0} + x_3' - 1\right) ,\tag{7.66b}$$

where Fr^2 is the square of the Froude number, $\mathrm{Fr}^2 = (\nu_0/d)^2/gd$. We observe that the introduction of pressure perturbation π, according to (7.66b), is deduced from a careful dimensionless analysis of the NSF full problem written with dimensionless variables (with ') $x_i' = x_i/d$ and $t' = t/(d^2/\nu_0)$ for the dimensionless functions $v_i = u_i/(\nu_0/d)$, θ, and π, which depend on t' and x_i'. The introduction of θ and π makes it possible to derive the Oberbeck-Boussinesq approximate equations asymptotically in a consistent way, when $\wp$ tends to zero (see Sect. 2.3.2 in Chap. 2). First, with (7.66a,b), from NS-F equations, we derive the following "dominant" dimensionless equations, when we neglect the terms proportional to the *square* of the small parameter $\wp$:

$$\frac{\partial v_k}{\partial x_k'} = \wp \frac{\mathrm{d}\theta}{\mathrm{d}t'} ,\tag{7.67a}$$

$$(1 - \wp\theta)\frac{\mathrm{d}v_i}{\mathrm{d}t'} + \frac{\partial \pi}{\partial x_i'} - \frac{\wp}{\mathrm{Fr}^2}\theta\boldsymbol{k} - \boldsymbol{\nabla}'^2 v_i = \wp L(v_i, \theta) ,\tag{7.67b}$$

$$[1 - \wp\theta(1 + \Gamma^0)]\frac{\mathrm{d}\theta}{\mathrm{d}t'} - \frac{1}{\mathrm{Pr}}\boldsymbol{\nabla}'^2\theta - 2\,\mathrm{Bq}\,\mathrm{Fr}^2(d_{ij}')^2 = \wp N(v_i, \theta) .\tag{7.67c}$$

7.5 The Bénard–Marangoni Problem: An Alternative

We note that in (7.67a–c),

$$d'_{ij} = \frac{1}{2}\left[\frac{\partial v_i}{\partial x'_j} + \frac{\partial v_j}{\partial x'_i}\right], \quad \frac{d}{dt'} = \frac{\partial}{\partial t'} + v_i \frac{\partial}{\partial x'_i},$$

$$\nabla'^2 = \frac{\partial^2}{\partial x'^2_1} + \frac{\partial^2}{\partial x'^2_2} + \frac{\partial^2}{\partial x'^2_3},$$

$$L(v_i, \theta) = \left[1 + \frac{\lambda_0}{\mu_0}\right] \frac{\partial}{\partial x'_i}\left(\frac{d\theta}{dt'}\right) - M^0 \frac{\partial}{\partial x'_j}(2\theta d'_{ij}),$$

$$N(v_i, \theta) = \mathrm{Bq}[p'_a + 1 - x'_3 + \mathrm{Fr}^2\,\pi]\frac{d\theta}{dt'} - 2\,\mathrm{Bq}\,\mathrm{Fr}^2\,M^0 \theta (d'_{ij})^2$$
$$- \frac{K^0}{\mathrm{Pr}}\frac{\partial}{\partial x'_i}\left[\theta\left(\frac{\partial \theta}{\partial x'_i}\right)\right],$$

where

$$p'_a = \frac{p_a}{gd\rho_0}, \quad \text{and:} \quad \Gamma^0 = \left[\frac{d\log C/dT}{d\log \rho/dT}\right]_0,$$

$$M^0 = \left[\frac{d\log \mu/dT}{d\log \rho/dT}\right]_0, \quad K^0 = \left[\frac{d\log k/dT}{d\log \rho/dT}\right]_0.$$

Of course, it is assumed that Γ^0, M^0, and K^0 are $O(1)$, when $\wp \to 0$.

For the derivation of the dominant equations (7.67a–c) we have also taken into account the following approximate relations, valid with an error of $O(\wp^2)$:

$$\rho(T) \cong \rho_0[1 - \wp\theta], \quad \mu(T) \cong \mu_0[1 - \wp M^0 \theta],$$
$$k(T) \cong k_0[1 - \wp K^0 \theta], \quad C(T) \cong C_0[1 - \wp \Gamma^0 \theta].$$

In the above dimensionless equations and relations, $\mathrm{Pr} = \nu_0/\kappa_0$ and $\mathrm{Bq} = g/\beta C_0$ are the Prandtl number and the analog of a Boussinesq number.

7.5.2 Dimensionless Dominant Boundary Conditions

For these (dominant) equations (7.67a–c), it is necessary to derive a set of dominant dimensionless boundary conditions, for v_i, θ, and π. First we note that, in the free surface boundary condition, it is necessary to introduce the operator $\nabla_\|$, which is the surface gradient and H, the mean curvature; we can write

$$\nabla_\| = \nabla - n(n \cdot \nabla) \quad \text{and} \quad H = -(1/2)(\nabla_\| \cdot n).$$

Now, we write the free surface equation (7.64) in dimensionless form, $x'_3 = 1 + \delta\eta(t', x'_1, x'_2) \equiv h'$, where $\delta = a/d$, is the dimensionless amplitude parameter for dimensionless free surface deformation $\eta(t', x'_1, x'_2)$ relative to the plane $x'_3 = 1$. Then, we derive the following five dominant dimensionless boundary conditions at the deformable free surface, $x'_3 = 1 + \delta\eta(t', x'_1, x'_2) \equiv h'$:

$$\pi = (\delta/\operatorname{Fr}^2)\eta + 2d'_{ij}n'_i n'_j + [\operatorname{We} - \operatorname{Ma}\theta](\boldsymbol{\nabla}'_\| \cdot \boldsymbol{n}') + \wp P(v_i,\theta) ; \qquad (7.68\mathrm{a})$$

$$d'_{ij}t'^{(s)}_i n'_j + \frac{1}{2}\operatorname{Ma} t'^{(s)}_i \frac{\partial\theta}{\partial x'_i} = \wp Q^{(s)}(v_i,\theta), \ s = 1,2 ; \qquad (7.68\mathrm{b})$$

$$\boldsymbol{\nabla}'\theta \cdot \boldsymbol{n}' + \frac{\operatorname{Bi}^0}{\operatorname{Bi}_s}[\operatorname{Bi}_s \theta + 1] = \wp K^0 \theta \boldsymbol{\nabla}'\theta \cdot \boldsymbol{n}' ; \qquad (7.68\mathrm{c})$$

$$v_3 = \delta\left[\frac{\partial\eta}{\partial t'} + v_1 \frac{\partial\eta}{\partial x'_1} + v_2 \frac{\partial\eta}{\partial x'_2}\right], \ \delta = \frac{a}{d} . \qquad (7.68\mathrm{d})$$

In the two ($s = 1$ and 2) dimensionless dominant boundary conditions (7.68b), $t'^{(s)}_i$ and $t'^{(s)}_i$ are the dimensionless components of two orthonormal tangent vectors to the free surface $x'_3 = h'$ ($\equiv 1 + \delta\eta$), and n'_i are the dimensionless components of the outward normal $\boldsymbol{n}'$ to the free surface:

$$\boldsymbol{t}'^{(1)} = \left[\frac{1}{\sqrt{N'_1}}\right]\left(1; 0; \frac{\partial h'}{\partial x'_1}\right) ;$$

$$\boldsymbol{t}'^{(2)} = \left[\frac{1}{\sqrt{N'_1 N'}}\right]\left(-\frac{\partial h'}{\partial x'_1}\frac{\partial h'}{\partial x'_2}; 1 + \left(\frac{\partial h'}{\partial x'_1}\right)^2; \frac{\partial h'}{\partial x'_2}\right) ;$$

$$\boldsymbol{n}' = \left[\frac{1}{\sqrt{N'}}\right]\left(-\frac{\partial h'}{\partial x'_1}; -\frac{\partial h'}{\partial x'_2}; 1\right) .$$

In the above relations according to Pavithran and Redeekopp (1994), the tangential and normal unit vectors to the deformed free surface, $x'_3 = h'(t', x'_1, x'_2)$, are written in terms of the (x'_1, x'_2, x'_3) Cartesian system of coordinates. In boundary condition (7.68a), we have the following explicit formula for $(\boldsymbol{\nabla}'_\| \cdot \boldsymbol{n}')$:

$$\boldsymbol{\nabla}'_\| \cdot \boldsymbol{n}' = -(N')^{-3/2}\left\{N'_2 \frac{\partial^2 h'}{\partial x'_1 \partial x'_1} + N'_1 \frac{\partial^2 h'}{\partial x'_2 \partial x'_2}\right.$$
$$\left. - 2\frac{\partial^2 h'}{\partial x'_1 \partial x'_2}\frac{\partial h'}{\partial x'_1}\frac{\partial h'}{\partial x'_2}\right\},$$

where

$$N'_1 = 1 + \left(\frac{\partial h'}{\partial x'_1}\right)^2 ; \ N'_2 = 1 + \left(\frac{\partial h'}{\partial x'_2}\right)^2 ; \ N' = 1 + \left(\frac{\partial h'}{\partial x'_1}\right)^2 + \left(\frac{\partial h'}{\partial x'_2}\right)^2 .$$

In (7.68a,b), for the terms $P(v_i,\theta)$ and $Q(v_i,\theta)$, we have:

$$P(v_i,\theta) = \frac{\lambda_0}{\mu_0}\frac{\mathrm{d}\theta}{\mathrm{d}t'} - 2M^0 \theta d'_{ij}n'_i n'_j ; \ Q^{(s)}(v_i,\theta) = M^0 \theta d'_{ij}t'^{(s)}_i n'_j , \ s = 1,2 .$$

The condition (7.68c) is derived from (7.65) when we take into account the relation $(T_0 - T_a) = \beta d/\operatorname{Bi}_s$, where Bi_s is given by (7.63a) and is always different from zero (see (7.63b)) in the Bènard thermal problem because $0 < \beta \neq 0$.

7.5 The Bénard–Marangoni Problem: An Alternative   229

Finally, we write the no-slip condition and the condition for θ,

$$v_i = 0 \text{ and } \theta = 1, \text{ at } x'_3 = 0, \tag{7.68e}$$

if we take into account (7.66a) and the definition of $T_0 = T_B - \beta d$. In the dimensionless boundary conditions, (7.68a,b), the dimensionless parameters

$$\text{We} = \sigma_0 \frac{d}{\rho_0(\nu_0)^2}, \quad \text{Ma} = \frac{\gamma d^2 \beta}{\rho_0(\nu_0)^2}, \quad \text{Bi}^0 = \frac{dq^0}{k(T_0)}, \tag{7.69}$$

are, respectively, the Weber, Marangoni and *convection* Biot numbers when we take (7.61) into account. We note also that $\text{Cr} = 1/\text{We}$ and $\text{Bn} = \text{Cr}/\text{Fr}^2$, are, respectively, the crispation (or capillary) and the Bond numbers. We observe that in the framework of a linear theory (when $\delta \ll 1$) it is necessary to take $\text{Bi}^0 \equiv \text{Bi}^0(\delta)$ and to reconsider from (7.68c) the linear problem of Sect. 3.5.

7.5.3 The Rayleigh–Bénard (RB) Thermal Shallow Convection Problem

In the above dimensionless formulation of the classical Bénard thermal-free surface problem via the full NSF dimensionless problem, there are two mechanisms responsible for driving the instability: the first is the *density variation* generated by the thermal expansion of the liquid, the second results from the *free-surface tension gradients* due to temperature fluctuations at the upper free surface of the liquid layer.

It is usual in the literature [see, for instance, Drazin and Reid (1981)], to denote as *Rayleigh–Bénard* (RB) *shallow* thermal convection, the instability problem produced mainly by *buoyancy* (including, possibly, the Marangoni and Biot effects in a *nondeformable free surface*).

Naturally, in the above full mathematical formulation of the dimensionless Bénard thermal-free surface problem governed by (7.67a–c) and boundary conditions, (7.68a–e), *both buoyancy and surface-gradient effects are operative* for a weakly ($\wp \ll 1$) expansible, viscous, and heat-conducting liquid, so it is important (from our point of view!) to ask

How are the two effects coupled when $\wp \to 0$?

To extract a tractable problem from the complicated dimensionless system of dominant equations, (7.67a–c), and dominant boundary conditions, (7.68a–e), which we have just presented above, we will now undertake an asymptotic analysis based mainly on the small parameter $\wp \to 0$!

Boussinesq limiting process. More precisely, from (7.67b), when $\wp \to 0$, it is obvious that if we want to take into account the buoyancy term, $(\wp/\text{Fr}^2)\theta \boldsymbol{k}$, then, according to Zeytounian (1989), it is necessary to impose the following "*Boussinesq limiting process*":

$$\wp \to 0 \text{ and } \mathrm{Fr}^2 \to 0, \text{ with } \wp/\mathrm{Fr}^2 \equiv \mathrm{Gr} = O(1), \tag{7.70a}$$

where

$$\mathrm{Gr} = \beta \alpha_T d^4 g / (\nu_0)^2 \tag{7.70b}$$

is the Grashof number and the associated Rayleigh (Ra) number is $\mathrm{Ra} = \mathrm{Pr}\,\mathrm{Gr} = \beta \alpha_T d^4 g / \nu_0 \kappa_0$. Now, if we consider the free-surface boundary condition (7.68a) for π, then, as a consequence of $\mathrm{Fr}^2 \ll 1$, the first term, $(\delta/F^2)\eta$, on the right-hand side of (7.68a) is *unbounded* when $\mathrm{Fr}^2 \to 0$, if we assume that $\delta = O(1)$. Thus, when $\mathrm{Fr}^2 \ll 1$,

the free-surface pressure condition (7.68a) is (asymptotically) consistent with the Boussinesq limiting process (7.70a), only for a small free-surface amplitude parameter δ, such that

$$\frac{\delta}{\mathrm{Fr}^2} = \delta^* = O(1), \text{ when } \delta \to 0 \text{ and } \mathrm{Fr}^2 \to 0. \tag{7.71}$$

As a consequence of the condition $Fr^2 \ll 1$, the Boussinesq limiting process, (7.70a), with (7.70b), is valid only if the *thickness d* of the liquid layer is such that

$$d \gg [(\nu_0)^2/g]^{1/3} \approx 1\,\mathrm{mm}, \tag{7.72a}$$

according to $\mathrm{Fr}^2 \ll 1$. On the other hand, when we assume that $\mathrm{Bq} = O(1)$, then the OB model equations [see, for instance, (2.75a–c)] are valid only when

$$\beta \cong g/C_0. \tag{7.72b}$$

Formulation of the RB rigid-free model problem with the buoyancy, Marangoni, and Biot effects. Therefore, if we take into account (7.68a) and consider large Weber numbers, $\mathrm{We} \gg 1$, such that

$$\delta\,\mathrm{We} = \mathrm{We}^* = O(1), \tag{7.73}$$

then, because $\delta \ll 1$, in conditions (7.68a) and (7.68b) the terms,

$$2d'_{ij} n'_i n'_j \text{ and } d'_{ij} t'^{(s)}_i n'_j, \text{ are at least } O(\delta),$$

we derive, in the limit, the following RB (rigid-free-leading-order) problem, which takes into account the buoyancy, Marangoni, and Biot effects at the *nondeformable* surface $x_3 = 1$:

$$\frac{\partial v_{k0}}{\partial x'_k} = 0,$$
$$\frac{dv_{i0}}{dt} + \frac{\partial \pi_0}{\partial x_i} - \mathrm{Gr}\,\theta_0 \boldsymbol{k} = \boldsymbol{\nabla}^2 v_{i0},\ i = 1,2,3,$$
$$\frac{d\theta_0}{dt} = \frac{1}{\mathrm{Pr}} \boldsymbol{\nabla}^2 \theta_0, \tag{7.74a}$$

7.5 The Bénard–Marangoni Problem: An Alternative 231

$v_{10} = v_{20} = v_{30} = 0$, and $\theta = 1$, at $x_3 = 0$,

$v_{30} = 0$, and $\dfrac{\partial^2 v_{30}}{\partial x_3^2} = \text{Ma} \left[\dfrac{\partial^2 \theta}{\partial x_1^2} + \dfrac{\partial^2 \theta}{\partial x_2^2} \right]$, at $x_3 = 1$,

$\dfrac{\partial \theta}{\partial x_3} + \text{Bi}^0 \theta + 1 = 0$, at $x_3 = 1$.

$$\dfrac{\partial^2 \eta}{\partial x_1^2} + \dfrac{\partial^2 \eta}{\partial x_2^2} - \dfrac{\delta^*}{We^*} \eta = -\dfrac{1}{We^*} \pi_0(t, x_1, x_2, 1) , \tag{7.74b}$$

when $\text{Bi}_s \equiv \text{Bi}^0$. The deformation of the free surface $\eta(t, x_1, x_2)$ is determined, when the perturbation of pressure $\pi_0(t, x_1, x_2, x_3)$ is known at $x_3 = 1$, after the resolution of the RB problem for v_{i0}, θ_0, π_0 by the last equation of the system (7.74b).

Note that RB problem (7.74a,b) is asymptotically derived from the dominant (relative to $\wp \ll 1$) problem formulated in Sects. 7.5.1 and 7.5.2, when $\wp \to 0$, $\text{Fr}^2 \to 0$, $\delta \to 0$, and $\text{Cr} \to 0$, with the following *three similarity* relations: $\wp / \text{Fr}^2 = \text{Gr} = O(1)$; $\delta / \text{Fr}^2 = \delta^* = O(1)$; $\delta / \text{Cr} = We^* = O(1)$, and in such a case $\text{Bn} = \text{Cr} / \text{Fr}^2 = O(1)$. For a particular case, when $\delta^* = We^* = \text{Bi}^0 = \text{Ma} \equiv 0$, we obtain the classical RB rigid-free shallow convection problem. The RB problem (7.74a,b) has been considered recently by Dauby and Lebon (1996), but unless the last equation in (7.74b) for $\eta(t, x_1, x_2)$. Concerning the equation of state $\rho = \rho(T)$, it is necessary to note that, from relation (7.66b),

$$\dfrac{(p - p_a)}{g d \rho_0} = 1 - x_3' + \text{Fr}^2 \pi ,$$

and it is clear that the presence of pressure p in a full baroclinic equation of state $\rho = \rho(T, p)$, in place of $\rho = \rho(T)$, does not change [in the Boussinesq (7.70a) limit] the form of the derived OB approximate equations (7.74a) and conditions (7.74b), because $\text{Fr}^2 \to 0$. In any case the isothermal compressibility of the considered weakly compressible liquid is small relative to α_T. Finally, it is necessary to note that [Zeytounian (1997), (1998)]

it is not consistent (from an asymptotic point of view) to take into account simultaneously the buoyancy effect and the deformation of the free surface in the RB thermal convection model problem (for a weakly expansible liquid).

7.5.4 The Bénard–Marangoni (BM) Problem

In reality, the well-known Bénard (1900) *convective cells* are primarily induced by free-surface tension gradients resulting from temperature variations across the free surface (the so-called Marangoni effect). Thus, it seems justified to use the term Bénard–Marangoni (BM) thermocapillary instability problem when, as in Bénard's experiments, the dominant, acting, driving force is the surface-tension gradient on the deformable free surface (without the influence of the buoyancy force). From (7.68a), to derive the full BM model thin film problem, when the free-surface deformation $\delta = O(1)$ in the free-surface conditions plays an essential role, it is necessary to assume also that the Froude number Fr is $O(1)$. Consequently, we must consider the following *incompressible limit process*:

$$\wp \to 0 \text{ and } \text{Fr}^2 = O(1) \,. \tag{7.75}$$

But in this case, Bq $\ll 1$ for most liquids beause

$$\frac{C_0 \Delta T}{g} \gg d \approx \left(\frac{\nu_0^2}{g}\right)^{1/3} . \tag{7.76}$$

As a consequence of (7.75), the buoyancy term in (7.67b) is negligible (in a leading-order approximation) for a weakly expansible liquid, and consequently, in place of the Boussinesq equations, we derive the classical "*incompressible* model equations" from the dominant dimensionless equations (7.67a–c):

$$\begin{aligned}
\frac{\partial v_{k0}}{\partial x_k} &= 0 \,, \\
\frac{dv_{i0}}{dt} + \frac{\partial \pi_0}{\partial x_i} &= \boldsymbol{\nabla}^2 v_{i0} \,, \\
\frac{d\theta_0}{dt} &= \frac{1}{Pr} \boldsymbol{\nabla}^2 \theta_0 \,,
\end{aligned} \tag{7.77a}$$

written without the primes and where $d/dt = \partial/\partial t + v_{i0}\partial/\partial x_i$.

Because $\text{Fr}^2 = O(1)$ and $\delta = O(1)$, we must write the *deformable* free-surface conditions for (7.77a):

$$\begin{aligned}
\pi_0 &= \frac{(h-1)}{\text{Fr}^2} + 2d_{ij0}n_in_j + [\text{We} - \text{Ma}\,\theta_0](\boldsymbol{\nabla}_{\|} \cdot \boldsymbol{n}) \,; \\
d_{ij0}t_i^{(s)}n_j &+ \frac{1}{2}\,\text{Ma}\,t_i^{(s)}\frac{\partial \theta_0}{\partial x_i} = 0, \quad s=1,2 \,; \\
\boldsymbol{\nabla}\vartheta_0 \cdot \boldsymbol{n} &+ \frac{\text{Bi}}{\text{Bi}_s}[\text{Bi}_s\,\vartheta_0 + 1] = 0 \,; \\
v_{30} &= \frac{\partial h}{\partial t} + v_{10}\frac{\partial h}{\partial x_1} + v_{20}\frac{\partial h}{\partial x_2},
\end{aligned} \tag{7.77b}$$

at the deformable free surface $x_3 = h(t, x_1, x_2)$.

Obviously, again,

$$v_{10} = v_{20} = v_{30} = 0, \quad \text{and} \quad \theta = 1, \text{ at } x_3 = 0 \,. \tag{7.77c}$$

These boundary conditions are written for the unknowns, v_{i0}, π_0, θ_0, and consequently, we obtain a coupled BM problem for the limiting functions,

$[v_{i0}, \pi_0, \theta_0] = \lim[v_i, \pi, \theta]$, when $\wp \to 0$, and $\mathrm{Fr}^2 = O(1)$, and also for the thickness $h(t, x_1, x_2)$, of the unknown free surface of the film.

The above full BM model problem (7.77a–c) is complicated, but for thin-film flow, we can apply the long-wave approximation.

Although Bénard was aware of the role of surface tension and surface tension gradients in his experiments, it took five decades to unambiguously assess, experimentally and theoretically [see, for instance, the papers by Block (1956) and Pearson (1958)], that surface tension gradients rather than buoyancy are the cause of Bénard cells in thin liquid films. Only in 1997 have I proved this almost evident physical fact through an asymptotic approach. Finally, we can formulate the following "alternative":

> *If the buoyancy is taken into account, then the free-surface deformation effect is negligible, and we can only partially take into account the Marangoni effect, or this free-surface deformation effect is taken into account and consequently the buoyancy does not play a significant effect in the Bénard–Marangoni full thermocapillary problem.*

The first author to explain the effect of surface tension gradients on Bénard convection was Pearson (1958). In Pearson (1958) the reader can find also a more judicious modeling of the heat transfer, in place of Newton's law of cooling (7.65).

7.6 Some Aspects of Nonadiabatic Viscous Atmospheric Flow

In Chap. I (*The rotating earth and its atmosphere*) and Chap. II (*Dynamical and thermodynamical equations for atmospheric motions*) of Zeytounian (1991, pp. 1–35) the reader can find a short introduction to meteorological fluid dynamics (MFD). In geophysical fluid dynamics (GFD) which considers the problems that arise primarily in meteorology and oceanography, the classical book is Pedlosky (1987); a more recent book is Salmon (1998) which gives a very readable description of major problems in this topic. We mention also the book by Monin (1988). In Zeytounian (1990), the reader can find a "theory" related to asymptotic modeling of atmospheric flow. Below, we present only some aspects of this theory which concerns mainly nonadiabatic viscous atmospheric flow.

A significant model system of equations for nonadiabatic viscous atmospheric motion is the so-called "large-synoptic scale, hydrostatic (nontangent and nonadiabatic) viscous equations," the L-SSHV equations.

7.6.1 The L-SSHV Equations

In a coordinate frame rotating with Earth, we consider the full NSF equations and take into account the Coriolis ($2\boldsymbol{\Omega} \wedge \boldsymbol{u}$) force, the gravitational

234   7 Some Viscous Fluid Motions and Problems

acceleration (modified by the centrifugal force) g, and the effect of thermal radiation Q, where $\boldsymbol{u} = (\boldsymbol{v}, w)$ is the (relative) velocity vector as observed in Earth's frame rotating at the angular velocity $\boldsymbol{\Omega}$, ρ is the atmospheric density, p is the atmospheric pressure, and T the absolute temperature. We assume that

$$\frac{d\mathcal{R}}{dz} = \rho_s Q, \text{ with } \mathcal{R} = \mathcal{R}(T_s), \tag{7.78}$$

and we note that ρ_s in relation (7.78) is the so-called standard density function of only the altitude z, and Q, which is a heat source (thermal radiation), is assumed to depend only on the "mean" standard atmosphere. Doing this, we consider only a mean, standard distribution for Q, and also for the eddy viscosity μ and thermal eddy conductivity k, and ignore variations therefrom for the perturbed atmosphere in motion. The thermodynamic functions, p_s, ρ_s, and T_s for the standard atmosphere satisfy the following equations:

$$\frac{dp_s}{dz} + g\rho_s = 0, \ p_s = R\rho_s T_s, \tag{7.79a}$$

$$k\frac{dT_s}{dz} + \mathcal{R}(T_s) = 0. \tag{7.79b}$$

Now, as the vector of rotation of earth $\boldsymbol{\Omega}$ is directed from south to north according to the axis of the poles, it can be expressed as follows:

$$\boldsymbol{\Omega} = \Omega^0 \boldsymbol{e}, \text{ with } \boldsymbol{e} = (\boldsymbol{k}\sin\phi + \boldsymbol{j}\cos\phi), \tag{7.80}$$

where ϕ is the algebraic latitude of point P^0 of the observation on Earth's surface, around which atmospheric flow is analyzed ($\phi > 0$ in the Northern Hemisphere). The unit vectors directed to the east, north, and zenith, in the opposite direction from $\boldsymbol{g} = -g\boldsymbol{k}$ (the force of gravity), are denoted by $\boldsymbol{i}$, $\boldsymbol{j}$, and $\boldsymbol{k}$, respectively.

The dimensionless dominant equations. It is helpful to employ spherical coordinates λ, ϕ, r, and in this case u, v, w again denote the corresponding relative velocity components in these directions, respectively, increasing azimuth (λ), latitude (ϕ), and radius (r). But it is very convenient to introduce the following transformations:

$$x = a_0 \cos\phi^0 \lambda; \ y = a_0(\phi - \phi^0); \ z = r - a_0, \tag{7.81}$$

where ϕ^0 is a reference latitude and for $\phi^0 \approx 45°$, $a_0 \approx 6300$ km. The origin of this right-handed curvilinear coordinate system lies on Earth's surface (for flat ground, where $r = a_0$) at latitude ϕ^0 and longitude $\lambda = 0$. We assume therefore that the atmospheric motion occurs in a mid-latitude region, distant from the equator, around some central latitude ϕ^0 and therefore, $\sin\phi^0$, $\cos\phi^0$ and $\tan\phi^0$ are all of order unity. Although x and y are, in principle, new longitude and latitude coordinates in terms of which the basic NSF atmospheric equations may be rewritten without approximation, they

7.6 Some Aspects of Nonadiabatic Viscous Atmospheric Flow

are obviously introduced in the expectation that for small $\delta = L^0/a_0$, they will be the Cartesian coordinates of the so-called "f^0-plane approximation". For this, it is necessary to introduce nondimensional variables and functions.

Finally, after careful dimensional analysis, we derive the following set of dimensionless dominant equations for the horizontal velocity $\boldsymbol{v} = (u, v)$, vertical velocity w, and thermodynamic functions p, ρ, T:

$$\rho\left\{\frac{d\boldsymbol{v}}{dt} + \left[\frac{1}{\mathrm{Ro}}\frac{\sin\phi}{\sin\phi^0} + \frac{\delta\tan\phi}{1+\varepsilon\delta z}\right](\boldsymbol{k}\wedge\boldsymbol{v})\right\} + \frac{1}{1+\varepsilon\delta z}\frac{1}{\gamma M^2}\mathbf{D}p$$
$$= \frac{1}{\varepsilon^2\,\mathrm{Re}}\frac{\partial}{\partial z}\left(\mu\frac{\partial\boldsymbol{v}}{\partial z}\right) + O(\varepsilon)\,; \qquad (7.82\mathrm{a})$$

$$\frac{\partial p}{\partial z} + \rho = \gamma M^2 O(\varepsilon^2)\,; \qquad (7.82\mathrm{b})$$

$$\frac{d\rho}{dt} + \rho\left\{\frac{\partial w}{\partial z} + \left[\frac{1}{1+\varepsilon\delta z}\right](\mathbf{D}\cdot\boldsymbol{v} - \delta\tan\phi v + \varepsilon\delta w)\right\} = 0\,; \qquad (7.82\mathrm{c})$$

$$\rho\frac{dT}{dt} - \frac{\gamma-1}{\gamma}\frac{Dp}{Dt} = \frac{1}{\varepsilon^2\,\mathrm{Re}\,Pr}\left\{\frac{\partial}{\partial z}\left(k\frac{\partial T}{\partial z}\right)\right.$$
$$\left. + \Pr\mu(\gamma-1)M^2\left[\frac{1}{1+\varepsilon\delta z}\left|\frac{\partial}{\partial z}(\frac{\boldsymbol{v}}{1+\varepsilon\delta z})\right|^2\right] + \sigma\frac{d\mathcal{R}}{dz}\right\} + O(\varepsilon^2)\,, \qquad (7.82\mathrm{d})$$

with $p = \rho T$. In the above equations,

$$\frac{d}{dt} = \frac{\partial}{\partial t} + \frac{1}{1+\varepsilon\delta z}\boldsymbol{v}\cdot\mathbf{D} + w\frac{\partial}{\partial z}\,,$$

$$\mathbf{D} = \frac{\cos\phi^0}{\cos\phi}\frac{\partial}{\partial x}\boldsymbol{i} + \frac{\partial}{\partial y}\boldsymbol{j},\ \boldsymbol{k}\cdot\mathbf{D} = 0\,.$$

The Rossby number is $\mathrm{Ro} = U^0/f^0 L^0$, with $f^0 = 2\Omega^0\sin\phi^0$, the Mach number is $M = U^0/(\gamma R T^0)^{1/2}$, $\varepsilon = H^0/L^0$ is the ratio of vertical and horizontal characteristic atmospheric lengths, such that $\mathrm{Bo} \equiv 1$, and the Prandtl number is $\Pr = C_p\mu^0/k^0$, where k^0 is the value of k on the ground. In (7.82d), the parameter $\sigma = (R/gk^0)\mathcal{R}^0$, with $\mathcal{R}^0$ a characteristic value of the radiative (standard) flux $\mathcal{R}(T_s)$. The Strouhal number is equal to unity, and the characteristic time $t^0 = L^0/U^0$. Finally, we observe that $\phi = \phi^0 + \delta y$.

For the large spectrum of synoptic atmospheric motions close to Earth,

$$\varepsilon \ll 1 \text{ but } \mathrm{Re} \gg 1\,.$$

Large-synoptic scale, hydrostatic (nontangent and nonadiabatic) viscous equations. Obviously, the set of dimensionless dominant equations (7.82a–d) is complicated; so below we consider an approximate form of this set, which is derived when we consider the following limiting process:

$$\varepsilon \to 0 \text{ and } \mathrm{Re} \to \infty, \text{ such that } \varepsilon^2\,\mathrm{Re} \equiv \mathrm{Re}_\perp = O(1)\,, \qquad (7.83)$$

with *time t and position vector* $\boldsymbol{x} = (x, y, z)$ *fixed*.

With (7.83), the set of dimensionless dominant equations (7.82a–d) is significantly simplified. We obtain (written with the same notations by simplicity) the following L-SSHV equations for the limiting values of $\boldsymbol{v} = (u, v)$, w and p, ρ, T, dependent on the time-space variables (t, x, y, z):

$$\rho \left\{ \frac{d\boldsymbol{v}}{dt} + \left(\frac{1}{Ro} \frac{\sin \phi}{\sin \phi^0} + \delta \tan \phi \right) (\boldsymbol{k} \wedge \boldsymbol{v}) \right\}$$
$$+ \frac{1}{\gamma M^2} \mathbf{D} p = \frac{1}{\mathrm{Re}_\perp} \frac{\partial}{\partial z} \left(\mu \frac{\partial \boldsymbol{v}}{\partial z} \right) ; \tag{7.84a}$$

$$\frac{\partial p}{\partial z} + \rho = 0 ; \tag{7.84b}$$

$$\frac{d\rho}{dt} + \rho \left\{ \frac{\partial w}{\partial z} + \mathbf{D} \cdot \boldsymbol{v} - \delta \tan \phi \right\} = 0 ; \tag{7.84c}$$

$$\rho \frac{dT}{dt} - \frac{\gamma - 1}{\gamma} \frac{Dp}{Dt} = \frac{1}{\mathrm{Re}_\perp} \frac{1}{\mathrm{Pr}} \left\{ \frac{\partial}{\partial z} \left(k \frac{\partial T}{\partial z} \right) \right.$$
$$\left. + \mathrm{Pr} \, \mu (\gamma - 1) M^2 \left| \frac{\partial \boldsymbol{v}}{\partial z} \right|^2 + \sigma \frac{d\mathcal{R}}{dz} \right\} ; \tag{7.84d}$$

$$p = \rho T , \tag{7.84e}$$

with

$$\frac{d}{dt} = \frac{\partial}{\partial t} + \boldsymbol{v} \cdot \mathbf{D} + w \frac{\partial}{\partial z} ,$$
$$\mathbf{D} = \frac{\cos \phi^0}{\cos \phi} \frac{\partial}{\partial x} \boldsymbol{i} + \frac{\partial}{\partial y} \boldsymbol{j} .$$

These L-SSHV, nontangent and nonadiabatic equations (7.84a–e) constitute a significant system for large-scale atmospheric motion in a thin layer, such as the troposphere, around Earth. For these L-SSHV equations, we write the following boundary conditions *on flat ground* $z = 0$:

$$\boldsymbol{v} = 0, \; w = 0, \text{ and } k \frac{dT}{dz} + \sigma \mathcal{R} = 0, \text{ on } z = 0 . \tag{7.85}$$

Here, we leave unspecified the behavioral conditions at high altitude when $z \uparrow \infty$, and in the horizontal directions for x and y tending to infinity.

Concerning the initial conditions for L-SSHV equations (7.84), we must give [see, below, (7.87)] only the initial values for $\boldsymbol{v}$ and T (or p), and they (in general) have nothing to do with the corresponding initial conditions for the full NSF atmospheric equations.

Consequently, for the boundary-value problem [(7.84a–e), (7.85)], it is necessary to formulate an unsteady-state adjustment problem analogous to that considered by Guiraud and Zeytounian (1982) for the *primitive Kibel equations*.

The primitive Kibel (nonviscous, tangent, and adiabatic) equations are derived from the above L-SSHV equations, when $1/\mathrm{Re}_\perp \equiv 0$ and $\delta = 0$.

7.6 Some Aspects of Nonadiabatic Viscous Atmospheric Flow

Actually, the derivation and the analysis of this unsteady-state, local in time, adjustment problem to the L-SSHV main model is an open problem!

In spite of this fact, the L-SSHV set of equations can be used as a theoretical basis for the various investigations of features of atmospheric motions depending on parameters Ro, $Re_\perp$, δ, Pr, *and* M^2.

The L-SSHV equations in pressure coordinates: HV model equations. Hydrostatic equation (7.84b) makes it possible to write the relation,

$$\frac{\partial}{\partial z} = -\rho \frac{\partial}{\partial p},$$

and this allows us to develop a standard procedure for changing variables from t, x, y and z, to new (pressure) variables, τ, ξ, η and p, with

$$\tau \equiv t, \ \xi \equiv x, \ \eta \equiv y, \ z = \mathcal{H}(\tau, \xi, \eta, p), \tag{7.86a}$$

where $\mathcal{H}$ has to be considered as one of the unknown functions.

We keep the notation $\mathbf{D}$ for the horizontal gradient on the isobaric surface, $p = \text{const}$, with the components $[(\cos\phi^0/\cos\phi)(\partial/\partial\xi), (\partial/\partial\eta)]$ and we set $\mathrm{d}/\mathrm{d}\tau = \partial/\partial\tau + \mathbf{v}\cdot\mathbf{D} + \omega\partial/\partial p$, where

$$\omega = \frac{\mathrm{d}p}{\mathrm{d}\tau} = \rho\left[\frac{\partial \mathcal{H}}{\partial \tau} + \mathbf{v}\cdot\mathbf{D}\mathcal{H} - w\right], \tag{7.86b}$$

is the vertical pseudo velocity that is the rate of variation of pressure following air particles.

Finally, with (7.86a,b), as a result, for $\mathbf{v}$, ω, $\mathcal{H}$, T, and ρ, functions of variables τ, ξ, η, p, we obtain the following *"hydrostatic viscous,"* HV model equations, in place of L-SSHV equations:

$$\frac{\mathrm{d}\mathbf{v}}{\mathrm{d}\tau}\left[\frac{1}{\mathrm{Ro}}\frac{\sin\phi}{\sin\phi^0} + \delta\tan\phi\right](\mathbf{k}\wedge\mathbf{v}) + \frac{1}{\gamma M^2}\mathbf{D}\mathcal{H}$$

$$= \frac{1}{\mathrm{Re}_\perp}\frac{\partial}{\partial p}\left(\rho\mu\frac{\partial \mathbf{v}}{\partial p}\right);$$

$$\frac{\partial \omega}{\partial p} + \mathbf{D}\cdot\mathbf{v} - \delta\tan\phi = 0;$$

$$T = -p\frac{\partial \mathcal{H}}{\partial p}$$

$$\frac{\mathrm{d}T}{\mathrm{d}\tau} - \frac{\gamma-1}{\gamma}\frac{T}{p}\omega = \frac{1}{\mathrm{Re}_\perp}\frac{1}{\mathrm{Pr}}\left\{\frac{\partial}{\partial p}\left(\rho k\frac{\partial T}{\partial p}\right)\right.$$

$$\left. + \mathrm{Pr}\,\mu(\gamma-1)M^2\left|\rho\frac{\partial \mathbf{v}}{\partial p}\right|^2 - \sigma\rho\frac{\mathrm{d}\mathcal{R}}{\mathrm{d}p}\right\}, \tag{7.87}$$

where $\rho = p/T$ and we assume that μ, k, and $\mathcal{R}$ are known functions only of p. For the HV system of equations (7.87), it is necessary to impose the following boundary conditions on flat ground in place of (7.85):

$$v = 0, \quad \omega = \rho \frac{\partial \mathcal{H}}{\partial \tau}, \quad \text{and} \quad \rho k \frac{dT}{dp} = \sigma \mathcal{R}, \quad \text{on } \mathcal{H} = 0. \tag{7.88a}$$

The boundary conditions that must be applied at the upper end of the troposphere, $p = 0$, and at infinity in the horizontal planes $\mathcal{H} = \text{const}$, may be, for instance, that the operator

$$\mathcal{L}(\mathcal{H}) \equiv \left[\frac{p^2}{K_s(p)}\right] \left\{ \left|\frac{\partial \mathcal{H}}{\partial p}\right|^2 + |\mathbf{D}\mathcal{H}|^2 \right\}, \tag{7.88b}$$

acting on $\mathcal{H}$, decays sufficiently rapidly at infinity, where, by definition,

$$K_s(p) = T_s(p) \left\{ \frac{\gamma - 1}{\gamma} - \left(\frac{p}{T_s(p)}\right) \frac{dT_s(p)}{dp} \right\} > 0. \tag{7.88c}$$

7.6.2 The Tangent HV (THV) Equations

The β effect. The tangent hydrostatic (nonadiabatic) viscous (THV) equations are derived from HV model equations (7.87) when we take into account the β effect. For this, first, we observe that

$$\frac{\sin \phi}{\sin \phi^0} = 1 + \frac{\delta}{\tan \phi^0} y + O(\delta^2), \quad \text{because } \phi = \phi^0 + \delta y,$$

and consequently,

$$\frac{1}{\text{Ro}} \frac{\sin \phi}{\sin \phi^0} = \frac{1}{\text{Ro}} + \beta y,$$

with an error of $O(\delta^2)$, where

$$\beta = \frac{\delta}{\text{Ro} \tan \phi^0}.$$

On the other hand, $\tan \phi = \tan \phi^0 [1 + O(\delta)]$, with $\tan \phi^0 \approx 1$ for $\phi^0 \approx 45°$.

The THV equations. Now, if we consider the following limiting process:

$$\delta \to 0, \quad \text{with } \beta = O(1), \tag{7.89}$$

then, we derive the THV equations with the β effect, which generalize the classical Kibel primitive tangent equations written in pressure coordinates. These THV equations take the following form:

7.6 Some Aspects of Nonadiabatic Viscous Atmospheric Flow   239

$$\frac{d\boldsymbol{v}}{d\tau} + \left[\frac{1}{\text{Ro}} + \beta y\right](\boldsymbol{k} \wedge \boldsymbol{v}) + \frac{1}{\gamma M^2}\boldsymbol{D}\mathcal{H} = \frac{1}{\text{Re}_\perp}\frac{\partial}{\partial p}\left(\rho\mu\frac{\partial \boldsymbol{v}}{\partial p}\right) ;$$

$$\frac{\partial \omega}{\partial p} + \boldsymbol{D}\cdot\boldsymbol{v} = 0; \quad T = -p\frac{\partial \mathcal{H}}{\partial p};$$

$$\frac{dT}{d\tau} - \frac{\gamma-1}{\gamma}\frac{T}{p}\omega = \frac{1}{\text{Re}_\perp}\frac{1}{\text{Pr}}\left\{\frac{\partial}{\partial p}\left(\rho k\frac{\partial T}{\partial p}\right)\right.$$

$$\left.+ \text{Pr}\,\mu(\gamma-1)M^2\left|\rho\frac{\partial \boldsymbol{v}}{\partial p}\right|^2 - \sigma\rho\frac{d\mathcal{R}}{dp}\right\} . \qquad (7.90)$$

Equations (7.90) constitute a consistent model for atmospheric *tangent* motions in a thin layer as the troposphere.

THV equations without the β effect. In Sect. 7.6.3, we start from (7.90) but, as a simplified case *without* the β effect, and we assume the existence of a similarity relation between $\text{Re}_\perp$ and the Kibel number

$$\text{Ki} = \frac{(1/f^0)}{t^0} , \qquad (7.91a)$$

which is the Rossby number Ro, when the Strouhal number is equal to unity ($S \equiv 1 \Rightarrow t^0 = L^0/U^0$). A judicious and consistent choice is

$$\frac{1}{\text{Re}_\perp} = \kappa_0\,\text{Ki}, \text{ with } \kappa_0 = O(1) , \qquad (7.91b)$$

and we observe that

$$\frac{\text{Ki}}{\text{Re}_\perp} = \text{Ek}_\perp = \frac{\mu\rho}{f^0(H^0)^2} \qquad (7.91c)$$

is the Ekman number associated with $\text{Re}_\perp$ which, according to (7.91b), is a small parameter of the order $O(\text{Ki}^2)$.

With (7.91b), we obtain the following set of equations from (7.90), when the β effect is *neglected*:

$$\text{Ki}\left\{\frac{\partial \boldsymbol{v}}{\partial \tau} + (\boldsymbol{v}\cdot\boldsymbol{D})\boldsymbol{v} + \omega\frac{\partial \boldsymbol{v}}{\partial p}\right\} + (\boldsymbol{k}\wedge\boldsymbol{v}) + \frac{\lambda_0}{\text{Ki}}\boldsymbol{D}\mathcal{H}$$

$$= \kappa_0\,\text{Ki}^2\frac{\partial}{\partial p}\left(\rho\mu\frac{\partial \boldsymbol{v}}{\partial p}\right) ; \qquad (7.92a)$$

$$\frac{\partial \omega}{\partial p} + \boldsymbol{D}\cdot\boldsymbol{v} = 0; \quad T = -p\frac{\partial \mathcal{H}}{\partial p}; \quad \rho = \frac{p}{T} ; \qquad (7.92b)$$

$$\text{Ki}\left\{\frac{\partial T}{\partial \tau} + \boldsymbol{v}\cdot\boldsymbol{D}T + \omega\left[\frac{\partial T}{\partial p} - \frac{\gamma-1}{\gamma}\frac{T}{p}\right]\right\}$$

$$= \frac{1}{\text{Pr}}\kappa_0\,\text{Ki}^2\left\{\frac{\partial}{\partial p}\left(\rho k\frac{\partial T}{\partial p}\right) + \text{Pr}\,\mu(\gamma-1)\frac{1}{\gamma\lambda_0}\text{Ki}^2\left|\rho\frac{\partial \boldsymbol{v}}{\partial p}\right|^2 - \sigma\rho\frac{d\mathcal{R}}{dp}\right\} ,$$

$$(7.92c)$$

where

$$\lambda_0 = \frac{1}{\gamma} (\text{Ki}/M)^2 .$$

The similarity relation (7.91b) is motivated by the fact that it corresponds to the least degeneracy of (7.90) when Ki $\downarrow 0$.

7.6.3 The Quasi-Geostrophic Model

Below, we discuss only briefly the asymptotic derivation of the quasi-geostrophic model. The reader can find a complete asymptotic theory of small Kibel numbers in Guiraud and Zeytounian (1980) paper.

The main limit: Ki tends to zero. Our main small parameter is the Kibel number Ki, and we expand $\mathcal{H}$, $\boldsymbol{v}$, ω, ρ, and T according to the following scheme:

$$(\mathcal{H}, \rho, T) = (\mathcal{H}_0, \rho_0, T_0) + \text{Ki}(\mathcal{H}_1, \rho_1, T_1) + \text{Ki}^2(\mathcal{H}_2, \rho_2, T_2) + O(\text{Ki}^3) , \tag{7.93a}$$

$$(\boldsymbol{v}, \omega) = (\boldsymbol{v}_0, \omega_0) + \text{Ki}(\boldsymbol{v}_1, \omega_1) + \text{Ki}^2(\boldsymbol{v}_2, \omega_2) + O(\text{Ki}^3) . \tag{7.93b}$$

From (7.92a–c), with (7.93a,b), we find, first, that $(\mathcal{H}_0, \rho_0, T_0)$ do not depend on horizontal variables, and then we assume that they do not depend on time τ either. Although this does not follow directly from the above equations, it will be found consistent with the constancy of $\mathrm{d}\mathcal{R}/\mathrm{d}p \equiv Q(p)$, which has been assumed previously. Of course, $T_0 = -p\,\mathrm{d}\mathcal{H}_0/\mathrm{d}p$, and $\rho_0 = p/T_0$, but we don't know yet how T_0 depends on p. Then, assuming

$$-p\left[\frac{\mathrm{d}T_0}{\mathrm{d}p} - \frac{\gamma-1}{\gamma}\frac{T_0}{p}\right] \equiv K_0(p) \neq 0 , \tag{7.94}$$

we find from (7.92c), $\omega_0 = 0$, and the same (7.92c) leads, for T_1, to

$$\frac{\partial T_1}{\partial \tau} + \boldsymbol{v}_0 \cdot \mathbf{D}T_1 - \frac{K_0(p)}{p}\omega_1$$
$$= \frac{1}{\Pr}\kappa_0 \left\{ \frac{\partial}{\partial p}\left(\rho_0 k_0 \frac{\partial T_0}{\partial p}\right) - \sigma\rho_0 Q(p), \right\} . \tag{7.95a}$$

Of course, from (7.92a), we derive the geostrophic balance,

$$(\boldsymbol{k} \wedge \boldsymbol{v}_0) + \lambda_0 \mathbf{D}\mathcal{H}_1 = 0 , \tag{7.95b}$$

and the continuity equation (7.92b) leads to (because $\omega_0 \equiv 0$)

$$\mathbf{D} \cdot \boldsymbol{v}_0 = 0 . \tag{7.95c}$$

7.6 Some Aspects of Nonadiabatic Viscous Atmospheric Flow

The quasi-geostrophic equation. Going to higher order, we find that

$$\frac{\partial \boldsymbol{v}_1}{\partial \tau} + (\boldsymbol{v}_0 \cdot \mathbf{D})\boldsymbol{v}_1 + \boldsymbol{k} \wedge \boldsymbol{v}_1 + \lambda_0 \mathbf{D}\mathcal{H}_2 = 0,$$
$$\frac{\partial \omega_1}{\partial p} + \mathbf{D} \cdot \boldsymbol{v}_1 = 0; \quad T_1 = -p\frac{\partial \mathcal{H}_1}{\partial p},$$
$$\boldsymbol{v}_0 = \lambda_0(\boldsymbol{k} \wedge \mathbf{D}\mathcal{H}_1), \tag{7.96}$$

and, by eliminating $\boldsymbol{v}_1$, ω_1, and T_1, from (7.95a) and (7.96), we get the following *quasi-geostrophic potential vorticity equation*:

$$\frac{\mathrm{d}_{qg}}{\mathrm{d}\tau}(\Lambda \mathcal{H}_1) = 0, \tag{7.97}$$

where

$$\frac{\mathrm{d}_{qg}}{\mathrm{d}\tau} = \frac{\partial}{\partial \tau} + \lambda_0(\mathbf{D}\mathcal{H}_1 \wedge \mathbf{D}),$$

with

$$\Lambda = \lambda_0 \mathbf{D}^2 + \frac{\partial}{\partial p}\left\{\left[\frac{p^2}{K_0(p)}\right]\frac{\partial}{\partial p}\right\}.$$

We observe that the quasi-geostrophic model equation (7.97) contains one derivation with respect to time τ and, consequently, only one initial condition must be supplied for $\mathcal{H}_1$, via an unsteady-state adjustment problem. We observe also that when Ki tends to zero, the flat ground with the equation $\mathcal{H} = 0$, leads to equation $\mathcal{H}_0(p) = 0$, and below, it is assumed that $p = 1$ (in dimensionless form) is the solution of this equation. At $p = 1$, a boundary condition must be derived via the Ekman boundary-layer problem.

Adjustment to geostrophic balance (7.95b). It is not difficult to verify that the unsteady-state adjustment is satisfied by setting $t^* = \tau/\mathrm{Ki}$ and applying the initial limiting process Ki $\downarrow 0$, with t^* fixed. Let us set f^* for any quantity f considered as a function of t^* instead of τ. We expand the functions according to

$$(\boldsymbol{v}, \omega, \mathcal{H}, \rho, T) = (\boldsymbol{v}_0^*, \omega_0^*, \mathcal{H}_0^*, \rho_0^*, T_0^*) + \mathrm{Ki}(\boldsymbol{v}_1^*, \omega_1^*, \mathcal{H}_1^*, \rho_1^*, T_1^*) + \dots,$$
$$\tag{7.98a}$$

and substituting in (7.92a–c), one finds first,

$$(\mathcal{H}_0^*, \rho_0^*, T_0^*) = (\mathcal{H}_0(p), \rho_0(p), T_0(p)),$$

with

$$T_0^* = -p\frac{\partial \mathcal{H}_0}{\partial p}, \text{ and } \rho_0^* = \frac{p}{T_0(p)}. \tag{7.98b}$$

To find equations for $(v_0^*, \omega_0^*, \mathcal{H}_1^*, T_1^*)$, we have to go to higher order:

$$\frac{\partial v_0^*}{\partial t^*} + k \wedge v_0^* + \lambda_0 \mathbf{D}\mathcal{H}_1^* = 0 \,,$$

$$\frac{\partial \omega_0^*}{\partial p} + \mathbf{D} \cdot v_0^* = 0; \; T_1^* = -p\frac{\partial \mathcal{H}_1^*}{\partial p} \,,$$

$$\frac{\partial T_1^*}{\partial t^*} - \frac{K_0(p)}{p}\omega_0^* = 0 \,. \tag{7.98c}$$

The system of equations (7.98c) governs the unsteady-state process of adjustment to geostrophy. To solve (7.98c), we may, without loss of generality in the analysis, set

$$v_0^* = \mathbf{D}\phi_0^* + k \wedge \mathbf{D}\psi_0^* \,,$$

and a single equation for ϕ_0^* is derived:

$$\frac{\partial^2}{\partial t^{*2}}\left\{\frac{\partial}{\partial p}\left[\frac{p^2}{K_0(p)}\left(\frac{\partial \phi_0^*}{\partial p}\right)\right]\right\} + \lambda_0 \mathbf{D}^2 \phi_0^*$$
$$+ \frac{\partial}{\partial p}\left[\frac{p^2}{K_0(p)}\left(\frac{\partial \phi_0^*}{\partial p}\right)\right] = 0 \,. \tag{7.98d}$$

But it is necessary to give initial conditions for v_0^* and $\mathcal{H}_1^*$. For v_0^*, we may use the initial value v^0 of the horizontal velocity v imposed to equation (7.92a). For the initial value of $\mathcal{H}_1^*$ we may use the initial value for $\mathcal{H}$, which is related to the initial value of T, imposed on (7.92c) by the relationship $T = -p\partial \mathcal{H}/\partial p$, but this works only if the initial values $(T^0, \mathcal{H}^0)$, for T and $\mathcal{H}$, may be set in the form $(\mathcal{H}^0, T^0) = (\mathcal{H}_0(p), T_0(p)) + \text{Ki}(\mathcal{H}_1^0, T_1^0)$.

Then,

$$t^* = 0 \; : \; v_0^* = v^0, \; \mathcal{H}_1^* = \mathcal{H}_1^0 \,. \tag{7.98e}$$

Whenever the initial value appropriate for (7.92c) cannot be put into such a form, we must expect that another adjustment process holds!

When $K_0(p) = 1$, (7.98d) is the one derived by I.A. Kibel [see §4.2, (1957/1963)]. Ilya Afanasievitch was able (in 1955) to settle the main issue of the adjustment problem which is to know whether or not v_0^* and $\mathcal{H}_1^*$ evolve toward the geostrophic balance (7.95b) when t^* tends to infinity; concerning this problem see, also, the book by Monin (1972). As a matter of fact,

$$\lim_{t^* \uparrow \infty}(v_0^*, \mathcal{H}_1^*) = [v_0(\tau = 0, \xi, \eta, p), \mathcal{H}_1(\tau = 0, \xi, \eta, p)] \,, \tag{7.98f}$$

and

$$\text{at } \tau = 0 \; : \; (k \wedge v_0) + \lambda_0 \mathbf{D}\mathcal{H}_1 = 0 \,. \tag{7.98g}$$

7.6 Some Aspects of Nonadiabatic Viscous Atmospheric Flow

There is an important observation which was known to Ilya Afanasievitch and which concerns the way in which,

$$\lim_{t^* \uparrow \infty} \mathcal{H}_1^* \text{ is related to the initial values } \boldsymbol{v}^0, \mathcal{H}_1^0 .$$

In relation with the above question we can obtain the following equation between $\boldsymbol{v}_0^*$ and $\mathcal{H}_1^*$:

$$\frac{\partial}{\partial t^*} \{\boldsymbol{k} \cdot (\mathbf{D} \wedge \boldsymbol{v}_0^*)\} + \frac{\partial}{\partial p} \left\{ \left[\frac{p^2}{K_0(p)} \right] \frac{\partial \mathcal{H}_1^*}{\partial p} \right\} = 0 .$$

Now, if we integrate this equation between $t^* = 0$ and $t^* = \infty$ and if we use the geostrophic balance for the limiting values of $\boldsymbol{v}_0^*$ and H_1^* when $t^* \uparrow \infty$, then

$$(A\mathcal{H}_1)_{\tau=0} = \boldsymbol{k} \cdot \{\mathbf{D} \wedge \boldsymbol{v}^0\} + \frac{\partial}{\partial p} \left\{ \left[\frac{p^2}{K_0(p)} \right] \frac{\partial \mathcal{H}_1^0}{\partial p} \right\} , \quad (7.98\text{h})$$

where Λ is the operator which appears in the quasi-geostrophic main model equation (7.97). We observe that

> the initial condition for the quasi-geostrophic operator $\Lambda\mathcal{H}_1$ in (7.97) is related, according to (7.98h), to the initial values $\boldsymbol{v}^0$ and $\mathcal{H}_1^*$ for the tangent hydrostatic, viscous (THV) nonadiabatic and tangent, model equations.

These (hydrostatic) initial values, $\boldsymbol{v}^0$ and H_1^*, must be derived from an unsteady-state adjustment problem to hydrostatic balance (7.84b) [which is a consequence of the hydrostatic (long-wave) limiting process (7.83)]. This adjustment problem has been considered for Kibel primitive equations by Guiraud and Zeytounian (1982).

The Ekman layer and the boundary condition at $p = 1$ for quasi-geostrophic main model equation (7.97). Below, from investigation of the steady-state Ekman layer, which leads to the classical Ackerblom problem, we derive, via matching, the boundary condition at $p = 1$ for the quasi-geostrophic main model equation (7.97). For this, we consider a local inner region near flat ground $p = 1$ which corresponds to

$$|p - 1| = O(\text{Ki}) \Rightarrow p^* = \frac{1-p}{\text{Ki}} , \quad (7.99)$$

and we call it the steady-state Ekman region. Within this region, the independent variables are τ, ξ, η, and p^*, and we consider the following inner limiting process:

$$\text{Ki} \to 0 \text{ with } \tau, \xi, \eta \text{ and } p^* \text{ fixed} . \quad (7.100\text{a})$$

With (7.100a), we consider the following inner expansion:

$$(\boldsymbol{v}, \omega, \mathcal{H}, T, \rho)$$
$$= (\boldsymbol{v}_{\text{Ek}}, 0, 0, T_0(1), \rho_0(1)) + \text{Ki}(\boldsymbol{v}^1_{\text{Ek}}, \omega_{\text{Ek}}, \mathcal{H}_{\text{Ek}}, T_{\text{Ek}}, \rho_{\text{Ek}}) + \ldots, \quad (7.100b)$$

and we obtain the following set of equations, for $\boldsymbol{v}_{\text{Ek}}$, ω_{Ek}, $\mathcal{H}_{\text{Ek}}$ and T_{Ek} from starting equations (7.92a–c):

$$(\boldsymbol{k} \wedge \boldsymbol{v}_{\text{Ek}}) + \lambda_0 \mathbf{D}\mathcal{H}_{\text{Ek}} - \kappa_0 \frac{\partial}{\partial p^*}\left(\frac{\partial \boldsymbol{v}_{\text{Ek}}}{\partial p^*}\right) = 0 \, ;$$

$$\frac{\partial \omega_{\text{Ek}}}{\partial p^*} = \mathbf{D} \cdot \boldsymbol{v}_{\text{Ek}} \, ;$$

$$T_0(1) = \frac{\partial \mathcal{H}_{\text{Ek}}}{\partial p^*} \, ;$$

$$\frac{1}{Pr}\kappa_0 \frac{\partial}{\partial p^*}\left(\frac{\partial T_{\text{Ek}}}{\partial p^*}\right) = 0 \, , \quad (7.100c)$$

when we assume that (with dimensionless quantities) $\rho_0(1) = 1$, $k(1) = 1$ and $\mu(1) = 1$. From the boundary conditions on flat ground (7.88a) and (7.99),

$$\text{on } \mathcal{H}_{\text{Ek}} = 0 \, : \, \boldsymbol{v}_{\text{Ek}} = 0, \, \omega_{\text{Ek}} = \frac{\partial \mathcal{H}_{\text{Ek}}}{\partial \tau}, \, \frac{\partial T_{\text{Ek}}}{\partial p^*} = \sigma \mathcal{R}(1) \, , \quad (7.100d)$$

if we also assume that $T_0(1) = 1$ and that the main radiative transfer does not have an Ekman structure. The flat ground in the Ekman layer theory is characterized by $p^* = p^*_{w0} + \text{Ki}\, p^*_{w1} + \ldots$, and from matching with the main quasi-geostrophic region, we can write

$$T_{\text{Ek}} = T_1(\tau, \xi, \eta, 1) + \left(\frac{dT_0}{dp}\right)_{p=1} p^*, \text{ with } T_1|_{p=1} = -\left(\frac{\partial \mathcal{H}_1}{\partial p}\right)_{p=1},$$

but from the third equation of (7.100c), with $T_0(1) = 1$,

$$\frac{\partial \mathcal{H}_{\text{Ek}}}{\partial p^*} = 1 \Rightarrow \mathcal{H}_{\text{Ek}} = \mathcal{H}_1(\tau, \xi, \eta, 1) + p^* \, ,$$

and $H_{\text{Ek}} = 0$ implies

$$p^*_{w0} = -\mathcal{H}_1(\tau, \xi, \eta, 1) \, . \quad (7.100e)$$

The Ackerblom problem. Now, in the first of the equations (7.100c) for $\boldsymbol{v}_{\text{Ek}}$, we set $\boldsymbol{v}_{\text{Ek}} = \boldsymbol{v}_0(\tau, \xi, \eta, 1) + \boldsymbol{v}'_{\text{Ek}}$, and from matching with the main quasi-geostrophic region, when p^* tends to infinity, we obtain

$$\lim_{p^* \uparrow \infty} \boldsymbol{v}'_{\text{Ek}} = 0 \text{ and } [\boldsymbol{k} \wedge \boldsymbol{v}_0 + \lambda_0 \mathbf{D}\mathcal{H}_1]_{p=1} = 0 \, ,$$

7.6 Some Aspects of Nonadiabatic Viscous Atmospheric Flow

and we derive the following relationship: $\lambda_0 \mathbf{D}\mathcal{H}_{\mathrm{Ek}} + \mathbf{k}\wedge \mathbf{v}_{\mathrm{Ek}} = \mathbf{k}\wedge \mathbf{v}'_{\mathrm{Ek}}$. Finally, for $\mathbf{v}'_{\mathrm{Ek}}$, we recover the following classical *Ackerblom problem*:

$$\kappa_0 \frac{\partial}{\partial p^*}\left(\frac{\partial \mathbf{v}'_{\mathrm{Ek}}}{\partial p^*}\right) - \mathbf{k}\wedge \mathbf{v}'_{\mathrm{Ek}} = 0,$$

$$\mathbf{v}'_{\mathrm{Ek}} = -\mathbf{v}_0|_{p=1}, \text{ on } p^* = -\mathcal{H}_1(\tau,\xi,\eta,1),$$

$$\mathbf{v}'_{\mathrm{Ek}} \to 0 \text{ when } p^* \to \infty. \tag{7.100f}$$

The solution of (7.100f) is obtained by a standard method:

$$\mathbf{v}'_{\mathrm{Ek}} - \mathrm{i}\mathbf{k}\wedge \mathbf{v}'_{\mathrm{Ek}} = -\left[\mathbf{v}_0|_{p=1} - \mathrm{i}\,(\mathbf{k}\wedge \mathbf{v}_0)|_{p=1}\right] E^*, \tag{7.101a}$$

where

$$E^* = \exp\left\{-\frac{1+\mathrm{i}}{(2\kappa_0)^{1/2}}\left[p^* + \mathcal{H}_1|_{p=1}\right]\right\}. \tag{7.101b}$$

Now, from the second equation of (7.100c), we obtain, according to condition (7.100d) for ω_{Ek},

$$\omega_{\mathrm{Ek}} = \int_{p^*_{w0}}^{p^*} (\mathbf{D}\cdot \mathbf{v}_{\mathrm{Ek}})\,\mathrm{d}p^* + \left(\frac{\partial \mathcal{H}_{\mathrm{Ek}}}{\partial \tau}\right)_{p^*=p^*_{w0}}, \tag{7.101c}$$

where p^*_{w0} is given by (7.100e). Therefore, when $p^* \uparrow \infty$, after a straigtforward calculation, we obtain the following relation for $\omega^\infty_{\mathrm{Ek}}$:

$$\omega^\infty_{\mathrm{Ek}} = \frac{\partial\left(\mathcal{H}_1|_{p=1}\right)}{\partial \tau} - \lambda_0 \kappa_0^{1/2} \mathbf{D}^2 \mathcal{H}_1\Big|_{p=1}. \tag{7.101d}$$

Finally, by matching between the Ekman inner region ($p^* \uparrow \infty$) and the main quasi-geostrophic outer region ($p \downarrow 0$), we derive the following boundary condition at $p = 1$ for the main quasi-geostrophic model equation (7.97):

$$\left\{\frac{\partial}{\partial \tau} + \frac{1}{K_0(1)}\left(\frac{\partial}{\partial \tau} + \mathbf{v}_0\cdot \mathbf{D}\right)\frac{\partial}{\partial p} - \lambda_0 \kappa_0^{1/2}\mathbf{D}^2\right\}\mathcal{H}_1 = 0,$$

on $p = 1$, \hfill (7.101e)

where $\mathbf{v}_0 = \lambda_0 \mathbf{k}\wedge \mathbf{D}\mathcal{H}_1$.

This last boundary condition, on flat ground $p = 1$, *takes into account the influence of the viscous Ekman steady-state boundary layer on the main quasi-geostrophic flow* governed by (7.97).

246   7 Some Viscous Fluid Motions and Problems

7.7 Miscellaneous Topics

Obviously, it is not possible in the framework of this single volume to give (even more or less) a complete account of the multiple aspects of viscous motions. But, I think that the reader, reading the various account presented till now has a good general view of the complexity of viscous flows. Below, I want to comment very briefly on some particular problems where the viscosity also plays an important role.

7.7.1 The Entrainment of a Viscous Fluid in a Two-Dimensional Cavity

An interesting property of steady-state solutions of the Navier equations for an incompressible and viscous fluid in 2-D flow was studied by Prandtl (1904 and 1961) and Batchelor (1956). The PB theory asserts that vorticity should, in the limit of infinite Re, become constant within a region of nested closed streamlines. The PB theory has never been rigorously established mathematically due to the nonlinear and nonlocal complexity of the inviscid limit of Navier flow. For proper a priori conditions and an endeavor toward a rigorous mathematical approach to PB theory, see the two papers by Kim, cited in Kim (1998), where proof of the existence and uniqueness of the PB-type boundary-layer solution for a circular domain was presented. In the proof, he assumed that the velocity at the wall differs only slightly from the Euler limit velocity. This was examined by Wood (1957) and by Feynman and Lagerstrom (1956). According to Kim (1998), we observe that an another important issue, where there are again very few results, concerns the constants which arise in the PB theory, determining the vorticities of the domains of closed streamlines as well as the global topology of eddies (e.g., their number), compatible with boundary data. The "bare" PB theory establishes the constancy of vorticity in any such domain without determining the value of the constant. Assuming that a relevant Navier incompressible viscous solution exists, the value or values of the constant must be determined by viscous effects near the wall, which should, in principle, be accessible within the classical boundary-layer theory. This was discussed by Batchelor (1956) for the circular domain, where the boundary-layer structure is particularly simple; a more general class (for a domain close to circular) was considered by Wood (1957). Feynman and Lagerstrom (1956) extended their approach to eddies of arbitrary shape, and Lagerstrom (1975) summarized these results and have discussed a number of related issues.

Prandtl's theorem and the Batchelor–Wood formula. For steady-state 2-D incompressible flow, when $\boldsymbol{u} = (u, v)$ and $u^2 + v^2 = q$, from (s is the arc length)

$$\frac{1}{\text{Re}} \int_{\psi=c^*} \frac{\partial \omega}{\partial n} \, ds = 0 \Rightarrow \gamma(c^*) \omega'(c^*) = 0 \, , \tag{7.102}$$

with $\gamma(c^*) = \int_{\psi=c^*} q \, ds$, because c^* determines an arbitrary streamline within the eddy and because $\gamma > 0$, $0 < \psi < c^*$ for the eddy, the conclusion is that

$$\omega = \omega_0 = \text{const for } 0 \le \psi < c^* \, ,$$

and this is the result stated by Prandtl:

> a steady-state region of closed streamlines, over which, in the limit of large Reynolds number Re, viscous stresses are negligible, will in the limit be a region of constant vorticity (Prandtl's theorem).

But, Prandtl's theorem *does not by itself determine the constant* ω_0. Its value, reasoning from the physical picture just indicated, *must be a result of the action of viscous stresses*. If $\boldsymbol{u}_e(x,y)$ is the Euler limit of velocity vector $\boldsymbol{u}$ over a simply connected bounded domain D, we can write

$$\boldsymbol{u}_e = \omega_0 \left(\frac{\partial \Psi}{\partial y}; -\frac{\partial \Psi}{\partial x} \right), \ \boldsymbol{\nabla}^2 \Psi = -1, \ (x,y) \in D, \ \Psi = 0 \text{ on } \partial D \, .$$

Setting $q_e(s) = \omega_0 \, (\partial \Psi / \partial n)|_{\partial D}$, conventional boundary-layer matching requires that q_e be the velocity that must be attained at the outer edge of the boundary layer. In von Mises variables, (s, ψ^*), where

$$\psi^* = \text{Re}^{1/2} \, \psi, \ u_{\text{BL}} = q(s, \psi^*) \, ,$$

the boundary-layer problem is

$$\frac{\partial q}{\partial s} - \frac{dq_e}{ds} = q \frac{\partial^2 q}{\partial \psi^{*2}}, \ q(s,0) = q_w(s), \ q(s, \infty) = q_e(s) \, , \tag{7.103}$$

where $q(s+L, \psi^*) = q(s, \psi^*)$, where L is the perimeter of D. When q^2 is bounded for large $\psi^* = \text{Re}^{1/2} \, \psi$, we derive the *Batchelor–Wood formula*:

$$\int_0^{2\pi} q_w^2(\theta) \, d\theta = \int_0^{2\pi} q_e^2 \, d\theta = \frac{\pi}{2} \omega_0^2 \, . \tag{7.104}$$

This determines ω_0 in terms of the wall data, assuming that the orientation may be inferred independently from the data [for this last point, see Kim (1998)].

Kim's (1998) results. The analytical setting by Kim (1998) offers a controlled environment for discussing the birth of an eddy and multiple eddy vorticity in the context of PB theory. According to Kim (1998), the bifurcation structure of steady-state flow might also be studied in this setting, allowing

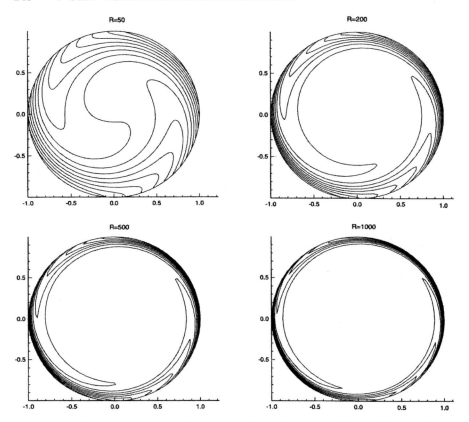

Fig. 7.11. Equivorticity lines for various Re and $\varepsilon = 0.2$ [from SIAM J. Appl. Math. **58**(5), 1394 (1998)]

a basic discussion of the link between stable flow topology and boundary data. To study the effect of finite Re on (7.104), a matched asymptotic expansion of the velocity in two parameters ε [Kim perturbs the rigid body rotation in a unit disk with the wall velocity $u(1,\theta) = 1 + \varepsilon f(\theta)$] and Re, is calculated up to the order of $O(\varepsilon^2/\mathrm{Re}^{1/2})$ and compared with the numerical results. For the resulting vorticity, including the correction for finite Re, ω_{full}, Kim (1998) obtained values in reasonably good agreement with the numerical results. In Fig. 7.11, we observe that by Re = 200, a distinct BL structure has developed periodically, and it becomes well established by Re = 500, thus indicating the rate of approach to the PB state when $\varepsilon = 0.2$.

Unsteady-state case of Guiraud and Zeytounian (1984a). In a 2-D cavity D with boundary $\partial D = \Sigma$, where the unit normal, denoted by $\boldsymbol{n}$ is directed into the cavity, the following unsteady-state Navier (for an incompressible, viscous fluid) problem is considered for velocity vector $\boldsymbol{u}(t,s,n) = (u,v)$

and pressure $p(t, s, n)$, where s is the abscissa along Σ and n the distance to the wall:

$$\frac{\partial \boldsymbol{u}}{\partial t} + (\boldsymbol{u} \cdot \boldsymbol{\nabla})\boldsymbol{u} + \boldsymbol{\nabla} p = \varepsilon^2 \boldsymbol{\nabla}^2 \boldsymbol{u}, \ \boldsymbol{\nabla} \cdot \boldsymbol{u} = 0, \ \varepsilon^2 = \frac{1}{\mathrm{Re}} \ ; \qquad (7.105\mathrm{a})$$

$$\boldsymbol{u}|_{\Sigma} = \boldsymbol{u}_w, \ \boldsymbol{u}|_{t=0} = 0 \ . \qquad (7.105\mathrm{b})$$

Of course $\boldsymbol{u}_w$ is L^0-periodic with respect to abscissa s, where L^0 is the length of the boundary and $\mathrm{Re} = U^0 L^0 / \nu^0$, where U^0 is a velocity scale of the order of $|\boldsymbol{u}_w|$ and ν^0 is the kinematic viscosity. When $\boldsymbol{u}_w = \boldsymbol{u}_w^\infty(s)$ exists, for $t \to \infty$, it is expected that $\boldsymbol{u}^\infty$ and p^∞ also exist when $t \to \infty$ and when ε is small (Re is large) $\boldsymbol{u}^\infty$ and p^∞ are ruled by nonviscous flow with a constant vorticity according to the PB state. In GZ (1984a), a study is presented for the unsteady-state process of entrainment that occurs on a time-scale $t = O(\mathrm{Re})$, and the constant vorticity PB state is established for $t \gg \mathrm{Re}$. In GZ (1984a), a functional equation is derived which models the distribution of vorticity during the main phase of entrainment. For a circular cavity, it shows that vorticity decays exponentially to its steady-state value.

In the initial stage of motion, as long as the time is $O(1)$, we may try an *outer, inviscid* expansion for $\boldsymbol{u} = (u, v)$ and p

$$(\boldsymbol{u}, p) = (\boldsymbol{u}_0, p_0) + \varepsilon(\boldsymbol{u}_1, p_1) + \ldots \qquad (7.106)$$

and $\boldsymbol{u}_0 = 0$, $p_0 = 1$. This outer expansion (7.106) should be matched with an inner boundary layer, valid near wall Σ. Using the *local coordinates* s and $n^* = n/\varepsilon$ and velocity components u and $v^* = v/\varepsilon$, with

$$(u, v^*, p) = (u_0^*, v_0^*, p_0^*) + \varepsilon(u_1^*, v_1^*, p_1^*) + \ldots \ , \qquad (7.107)$$

we find that $p_0^* = p_0 = 1$ and that the *initial phase* of the boundary-layer motion starts as a Rayleigh flow:

$$u_0^* = u_w \, \mathrm{Erfc}\left(\frac{n^*}{t^{1/2}}\right) , \qquad (7.108)$$

whenever u_w starts *impulsively*! For other laws of motion, see, for example, our ONERA Technical Note Zeytounian (1968a). Such solutions are valid for $t \ll 1$.

When $t = O(1)$, we expect a periodic boundary-layer solution to be formed, and when $t \gg 1$, we expect this periodic boundary layer solution to be steady-state, but it would be valuable to provide a numerical solution [see Kim (1998)]. Here, we are interested only in the output of this solution:

$$v_0^{*\infty} = v_0(t, s, \infty) \ .$$

Going back to the outer expansion (7.106), we find that

$$\boldsymbol{u}_1 = \boldsymbol{\nabla}\phi, \ p_1 = -\frac{\partial \phi}{\partial t}, \ \boldsymbol{\nabla}^2 \phi = 0, \ \left.\frac{\mathrm{d}\phi}{\mathrm{d}n}\right|_\Sigma = v_0^{*\infty} \ . \qquad (7.109)$$

Thus, during the whole period $t = O(1)$, even the boundary velocity has attained its limiting value; the entrainment of the interior (outer) flow is rather weak (of the order ε). As expected, the interior flow is ultimately entrained at a velocity $O(1)$, so we must investigate the flow for a very long time. To investigate *the long-term behavior*, we set

$$T = \varepsilon^2 t, \tag{7.110}$$

because the main phase of entrainment occurs on a time-scale $O(\varepsilon^{-2})$.

Now, for $\boldsymbol{u}(T/\varepsilon^2, s, n) = \boldsymbol{U}(T, s, n; \varepsilon^2)$, $p(T/\varepsilon^2, s, n) = P(T, s, n; \varepsilon^2)$, we try a long-term expansion, when T, s, n are fixed, for $\varepsilon \downarrow 0$:

$$(\boldsymbol{U}, P) = (\boldsymbol{U}_0, P_0) + \ldots + \varepsilon^2 (\boldsymbol{U}_1^*, P_1^*) + \ldots, \tag{7.111a}$$

and from the starting viscous problem (7.105a,b), rewritten with T in place of t, we find, first,

$$\boldsymbol{U}_0 = -(\boldsymbol{k} \wedge \boldsymbol{\nabla}\Psi_0), \tag{7.111b}$$

where $\boldsymbol{k}$ is the unit normal to the plane of flow, and then for Ψ_0, we derive the following problem:

$$-\boldsymbol{\nabla}^2 \Psi_0 = \Omega(T, \Psi_0), \; \Psi_0|_\Sigma = 0. \tag{7.111c}$$

Now, the main problem is to determine the function $\Omega(T, \Psi_0)$. When this is done, (7.111c) may be solved numerically. We observe that the flow ruled by (7.111c) is quasi-steady-state, and we expect it to be formed with a closed streamline. Let C_0 designate one of them. From the linear equations derived for $(\boldsymbol{U}_1^*, P_1^*)$, as a consequence of (7.105a,b) with (7.110) and (7.111a), we can prove that the following constraint must be satisfied:

$$\int_{C_0} \boldsymbol{\tau} \cdot \left[\frac{\partial \boldsymbol{U}_0}{\partial T} - \boldsymbol{\nabla}^2 \boldsymbol{U}_0 \right] \mathrm{d}s = 0, \tag{7.112}$$

for each closed streamline of the leading interior flow, and (7.112) is a *compatibility relation* for the solvability of linear equations for $(\boldsymbol{U}_1^*, P_1^*)$. But, with (7.111b,c), we find that

$$\boldsymbol{\tau} \cdot \boldsymbol{\nabla}^2 \boldsymbol{U}_0 = \boldsymbol{\tau} \cdot (\boldsymbol{k} \wedge \Psi_0) \frac{\partial \Omega}{\partial \Psi_0}.$$

Finally, we derive the following *Guiraud–Zeytounian formula*

$$\int_{C_0} \boldsymbol{\tau} \cdot \frac{\partial \boldsymbol{U}_0}{\partial T} \mathrm{d}s + \frac{\partial \Omega}{\partial \Psi_0} \int_{C_0} (\boldsymbol{\tau} \cdot \boldsymbol{U}_0) \mathrm{d}s = 0, \tag{7.113}$$

and this is the main result of our (1984) analysis with J.P. Guiraud.

U_0 being a functional of $\Omega(T, \Psi_0)$, so (7.113) appears as a rather intricate *functional equation* for $\Omega(T, \Psi_0)$, and it seems difficult in any specific configuration to determine the function $\Omega(T, \Psi_0)$ concretely!

Nevertheless, we observe from the analysis that time $t = O(\varepsilon^{-2})$ appears to be the proper time for the entrainment process, and at the end of this process, when $T \ll 1$, we expect that $\partial U_0/\partial T \ll 1$, such that we recover the Prandtl–Batchelor formula $\partial \Omega/\partial \Psi_0 = 0$.

This result is obtained, as suspected, from a double limiting process $t \uparrow \infty$, first, and, then, $\varepsilon = 1/\mathrm{Re}^{1/2} \to 0$. We may be more precise by stating that it occurs under limit $(\varepsilon^2 t) \to \infty$.

In a *circular cavity* of nondimensional radius $r = 1$, for the streamfunction $\psi(T, r)$ related to the *long-term behavior*, we write

$$\psi(T, r) = \psi_0(r) + \psi^*(T, r), \tag{7.114}$$

where $\psi_0(r) = (V_0/2)[r^2 - 1]$, and V_0 is given by a boundary-layer analysis. For instance, if the velocity of the wall is steady-state when $T = O(1)$ and is $v_w(\theta)$, then

$$2\pi V_0^2 = \int_0^{2\pi} v_w(\theta) \, \mathrm{d}\theta = \mathrm{const}. \tag{7.115}$$

In a circular cavity, in place of (7.111b,c) with (7.113), and (7.114), we must consider the following problem for the function $\psi^*(T, r)$:

$$\frac{\partial^2 \psi^*}{\partial r^2} + \frac{1}{r}\frac{\partial \psi^*}{\partial r} + \omega^* = 0,$$
$$\frac{\partial^2 \psi^*}{\partial T \partial r} + \left(\frac{\partial \omega^*}{\partial \psi_0}\right) \frac{\mathrm{d}\psi_0}{\mathrm{d}r} = 0,$$
$$r = 1 : \frac{\partial \psi^*}{\partial r} = \psi^* = 0, \ r = 0 : \frac{\partial \psi^*}{\partial r} = 0, \tag{7.116}$$

if we take into acount that $\Omega = 2V_0 + \omega^*(T, \psi_0(r))$. The second equation of (7.116) may be integrated in r and then substituted in the first. In this way we obtain a single equation for $\psi^*(T, r)$,

$$\frac{\partial \psi^*}{\partial T} = \frac{\partial^2 \psi^*}{\partial r^2} + \frac{1}{r}\frac{\partial \psi^*}{\partial r} + \frac{\mathrm{d}F(T)}{\mathrm{d}T}, \tag{7.117}$$

in place of two equations of (7.116), where the arbitrary function $F(T)$ comes from the integration of the second equation of (7.116). Now, if we consider the classical eigenvalue problem,

$$\frac{\partial \chi}{\partial T} = \frac{\partial^2 \chi}{\partial r^2} + \frac{1}{r}\frac{\partial \chi}{\partial r}, \ r = 0, 1 : \frac{\partial \chi}{\partial r} = 0, \tag{7.118}$$

then we have the following eigenfunctions:

$$\chi(T,r) = \exp\left(-k_1^2 T\right) J_0(k_n r), \quad (7.119)$$

when $\psi^*(T,r) = \chi(T,r) + F(T)$, where J_0 is the Bessel zeroth-order function with $\mathrm{d}J_0(k_n r)/\mathrm{d}r|_{r=1} = 0$, and, of course, due to linearization, it is the first zero which is relevant, $k_1 = 3.85$. The arbitrary function $F(T)$ must be chosen such that $F(T) = -\chi(T,1)$, and consequently, the solution for $\psi^*(T,r)$ is given by

$$\psi^*(T,r) = A^0 \exp\left(-k_1^2 T\right) [J_0(k_1 r) - J_0(k_1)], \quad (7.120)$$

and A^0 is an arbitrary constant. This analysis gives strong support to the conjecture that (7.113) is the proper equation that rules the behavior of $\Omega(T, \Psi_0)$ and that this, coupled with (7.111b,c) and with corresponding matching with the boundary layer, will fully determine the (unique) solution.

Finally, we note that, recently, a very similar "uniqueness" problem was investigated by Buldakov et al. (2000), which considers the uniqueness of steady-state flow past a rotating cylinder with suction. It is known that the Euler nonviscous equations admit a family of solutions where circulation is an arbitrary parameter. For modeling non-separated flow around a body with a sharp trailing edge, say an airfoil, one may use the Kutta–Joukowski–Villat condition [see, for instance, Zeytounian (2002a, Chap. 4, Sect. 4.3)] for determining the unique magnitude of the circulation; this condition has a viscous nature and eliminates the singular pressure gradient leading to separation from the sharp edge; however, it cannot be used for smooth bodies! In Buldakov et al. (2000), the unique solution is determined by the requirement that it satisfies a certain invariant condition which can be derived from the

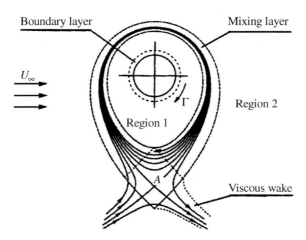

Fig. 7.12. The structure of the flow past a rapidly rotating cylinder with weak suction

full Navier equations (vorticity flux through any closed contour equals zero) for arbitrary Re, and which must be valid in the asymptotic limit.

We observe that the periodicity condition [Glauert (1957)] does not give a unique solution for suction velocity much greater than $1/\mathrm{Re}$.

When suction is of the order $(1/\mathrm{Re})^{1/2}$, the circulation in the potential flow (slightly perturbed by the displacement effect of the boundary layer) remains constant, and using suitable scaling, it may be chosen as 2π. The streamline pattern of such a flow with weak suction is shown in the above Fig. 7.12.

7.7.2 Unsteady-State Boundary Layers

Unsteady-state boundary-layer growth. First, we consider, again, the temporal evolution of a laminar incompressible boundary layer from rest at $t = 0$ (see Sect. 4.5.2 in Chap. 4). The functions $\mathcal{L}_s(y)$, the solution of the problem (4.130), are defined in Zeytounian (1968a) as

$$\mathcal{L}_s(y) = \frac{A_s}{s!} \int_\infty^y (y - y')^s \exp(-y'^2) \, dy' , \qquad (7.121)$$

and with $\mathcal{L}_s(0) = 1$: $A_0 = -(2/\pi^{1/2})$, $A_1 = 2$, $A_s = 2sA_{s-2}$, $A_{-1} \equiv 1$.

Functions (7.121) have the following main properties:

$$\frac{d^k \mathcal{L}_s(y)}{dy^k} = \frac{A_s}{A_{s-k}} \mathcal{L}_{s-k}(y) ,$$

$$\mathcal{L}_s(y) = \mathcal{L}_{s-2}(y) + \frac{A_s}{sA_{s-1}} y \mathcal{L}_{s-1}(y) ,$$

$$\mathcal{L}_0(y) = -\frac{2}{\pi^{1/2}} \int_\infty^y \exp(-y'^2) \, dy', \quad \mathcal{L}_{-1}(y) = \exp(-y^2) ,$$

$$\mathcal{L}_1(y) = \exp(-y^2) - \pi^{1/2} y \mathcal{L}_0(y) . \qquad (7.122)$$

The solution of the equation

$$\Lambda_m(g(y)) \equiv g'' + 2yg' - 2mg = \sum b_k \mathcal{L}_k(y) \text{ when } k = 1 \text{ to } r ,$$

with the conditions

$$g(0) = Y^0 \text{ and } g(\infty) = 0 ,$$

is, for $k \neq m$,

$$g(y) = Y^0 \mathcal{L}_m(y) + \frac{1}{2} \sum_{(k=1 \text{ to } r)} \left[\frac{b_k}{k - m} \right] [\mathcal{L}_k(y) - \mathcal{L}_m(y)] . \qquad (7.123a)$$

For the equation,

$$g'' + 2yg' - 2mg = \sum b_{kj} \mathcal{L}_k(y) \mathcal{L}_j(y),$$
with $g(0) = Y^0$ and $g(\infty) = 0$,

when $m - (k+j) = r = 1$, a solution is

$$g(y) = Y^0 \mathcal{L}_m(y) + \frac{1}{2} \sum b_{kj} \left[\frac{A_k A_j}{A_{k+1} A_{j+1}}\right] [\mathcal{L}_{k+1}(y) \mathcal{L}_{j+1}(y) - \mathcal{L}_m(y)]. \tag{7.123b}$$

In Zeytounian (1968a, Appendix), the case of $r > 1$ is also considered, and a more general solution is considered when the right-hand side in equation $\Lambda_m[g(y)] = \phi(y)$ has the following complicated form:

$$\phi(y) = \mathcal{L}_i(y)\mathcal{L}_j(y)\ldots\mathcal{L}_k(y)$$
$$\text{``}p\text{-terms''}$$
$$= \sum_{r=1 \text{ to } p} \Pi_{q-r}^{(r)}(y) \exp\left(-ry^2\right) \mathcal{L}_0^{p-r}(y), \tag{7.123c}$$

where $i + j + \ldots + k = q$.

In (7.123c), $\Pi_{q-r}^{(r)}(y)$ is a polynomial of order $(q-r)$ in y. We observe that we obtain the following formula:

$$\mathcal{L}_m(y) = \frac{A_m}{A_0} P_m(y) \mathcal{L}_0(y) + Q_{m-1}(y) \exp\left(-y^2\right), \tag{7.123d}$$

where the $P_m(y)$ are related to the classical Hermite–Tchebicheff polynomial, $H_m(y)$, such that

$$P_m(y) = \frac{1}{i^m 2^m m!} H_m(iy), \ i = (-1)^{1/2}, \tag{7.124a}$$

and $Q_{m-1}(y)$ are polynomials of order $(m-1)$ such that

$$Q_{m-1}(y) = Q_{m-3}(y) + \frac{1}{m}\frac{A_m}{A_{m-1}} Q_{m-2}(y), \tag{7.124b}$$

with $m = 2, 3, \ldots$, and $Q_{-1}(y) = 0$ and $Q_0(y) = 1$.

For example, for $m = 3$,

$$\mathcal{L}_3(y) = -\frac{\pi}{2} y \left(3 + 2y^2\right) \mathcal{L}_0(y) + \left(1 + y^2\right) \exp\left(-y^2\right).$$

In Zeytounian's (1968a) ONERA Technical Note, the reader can find a simple application of the above relations for a cylinder. The growth of the incompressible boundary layer on a body of revolution in *spinning motion* whose outer flow and angular velocity are also expressed in power series of $t^{1/2}$ was studied by Pop (1971), according to our theory; the separation characteristics for a sphere are presented. Some aspects of unsteady-state laminar compressible flow near a flat plate (which is infinite-Rayleigh problem in boundary-layer approximation) is investigated in Zeytounian (1970b), valid for almost all values of time. When we take into account the derived solution in a power series of $t^{1/2}$, the numerical results show the influence of compressibility on the laminar boundary-layer characteristics, as a function of the reference (upstream) Mach number of the outer flow far from the plate. In particular, this solution tends toward the asymptotic solution (for large values of time) obtained by Van Dyke (1952) for a flat plate impulsively set in motion. In Zeytounian's second ONERA Technical Note (1970a) devoted to the theory of the unsteady-state compressible boundary layers, the reader can find, in particular, an application to a disk, initially at rest, starting uniform rotation. The Appendix of Zeytounian (1970a) contains a detailed study of the problem of unsteady-state convection near an infinite flat plate placed in a gravity field and set at an angle to the horizontal (in the framework of the boundary-layer approximation), a group of general solutions for the unsteady-state Navier equations and also some results relative to the Rayleigh problem, using complete (unsteady-state one-dimensional) NSF equations. As regards the choice of method to be applied to expansions in power series of $t^{1/2}$, it should be pointed out that Oleinik (1969) succeeded in substantiating its merits by furnishing conclusive proof (for an incompressible flat-plate boundary layer) that series so constructed are true representations of the solution; the incident error is of the order of magnitude of the neglected term. Obviously enough, it would be futile to look to the method for something beyond its scope, using it to obtain the steady-state limit solution. Still, the example of the plate exponentially set in translation motion shows that our calculations tie in reasonably well with the standard steady-state solution of the Rayleigh problem. In Zeytounian (1968a, 1970a), the reader can find various references to some aspects of unsteady-state laminar boundary-layer flow up to 1970. In *Recent Research on Unsteady Boundary Layers* by Eichelbrenner (1972), and in Riley (1975), the reader can also find information and references.

See also Stuart (1971b), Telionis (1975), Riley and Vasantha (1989) and Telionis book (1981).

Boundary layer separation. We mention first two review papers, J.C. Williams III (1977) and S.-F. Shen (1978), where the reader can find classical results. More recent references are available in recent papers by Degani, Li, and Walker (1998), Degani, Walker, and Smith (1998) and also Cassel (2001). In particular, the boundary layer that develops in the leading-edge region of an airfoil impulsively started from rest at an angle of attack has been studied. It is well-known that the flow in the vicinity of the leading nose can be represented by the problem of flow past a parabola at an angle of attack, and this is the generic problem considered by Degani, Li, and Walker (1998). In this case, when the mainstream flow is at an angle of attack to the airfoil, a portion of the boundary layer will be exposed to an *adverse pressure gradient*. Once the angle of attack exceeds a certain *critical value*, it is demonstrated that unsteady-state boundary-layer separation will occur in the leading-edge region in the form of an *abrupt focused* boundary-layer eruption and this process is likely to initiate the formation of the *dynamic stall vortex*. We observe that the separation in the leading-edge region develops in a *zone of relatively limited streamwise extent* over a *wide range of angles of attack* and this suggests that

> *localized control measures (such as suction!) may possibly be effective in inhibiting separation.*

On the other hand, as the critical angle is approached from below, the wall shear stress ultimately goes to zero at a critical streamwise location s_c, and when the angle of attack is within an asymptotically small range $O(\mathrm{Re}^{-2/5})$ of the critical angle, an *interactive region* develops near s_c has a small streamwise extent $O(\mathrm{Re}^{-1/5})$.

Within this interactive region, steady-state solutions may be found which contain a *bubble of recirculating flow*; this phenomenon is known as *marginal separation* [see, for instance, Stewartson et al. (1982), Ruban (1982), and more recently Elliott and Smith (1987)]. According to Degani, Li, and Walker (1998, pp. 710, 711), the temporal development of the displacement thickness, defined by

$$\delta^* = \int_0^\infty \frac{1-u}{U_\infty}\, \mathrm{d}n \;,$$

is shown in Fig. 7.13, for $a = 2.0$ (plotted curves are at $t = 1, 2, 3, 4, 5, 5.817$). $U_\infty(\xi) = (\xi + a)/(\xi^2 + 1)^{1/2}$ is the inviscid slip velocity along the surface

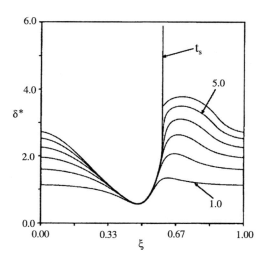

Fig. 7.13. Temporal development of the displacement thickness δ^*

of the parabola for the constant a. The parabolic coordinate ξ is related to the Cartesian coordinate x by $\xi = \pm x^{1/2}$ and in the boundary layer $n = \eta \operatorname{Re}^{1/2} \left(\xi^2 + 1\right)^{1/2}$, where $\eta = 0$ is the surface of the parabola.

Unsteady-state boundary-layer development across moving walls in the limit of an infinite Re was investigated by Degani, Walker, and F.T. Smith (1998), using both Eulerian and Lagrangian formulations. In the paper, two model problems are considered, a translating and rotating cylinder and a vortex convected in a uniform flow above an infinite flat plate; the calculated results show that

> *unsteady-state separation is delayed with increasing wall speed and is eventually suppressed, when the speed of the separation singularity approaches that of the local mainstream velocity.*

For the use of Lagrangian variables in describing of unsteady-state boundary-layer separation, see the paper by Cowley et al. (1990), Van Dommelen and Cowley (1990), and Van Dommelen (1991).

We observe also that a *moving surface* [see Ece et al. (1984)] acts to *inhibit the separation process*, but, in general, the minimum wall speed required to achieve suppression is not known!

Supplemental references. First, for the *unsteady-state compressible* boundary layers, we mention the book by Stewartson (1964). It seems that Moore (1951) first considered the unsteady-state laminar compressible boundary layer over an insulated surface. Ostrach (1955) and later Moore and Ostrach (1957) extended the theory to include the effects of heat transfer but confined their attention to flat plate flow. Flows with nonvanishing pressure gradients (flows about wedges) were studied first by Sarma (1965), and the unsteady-state laminar compressible boundary layer flow in the immediate vicinity of a 2-D stagation point due to an incident stream whose velocity varies arbitrarily with time was considered by Vimala and Nath (1975). The small amplitude oscillation theory for compressible oscillating boundary layers is considered in Telionis and Gupta (1977) paper already mentioned.

The higher-order unsteady-state boundary-layer solution for an impulsively started circular cylinder at uniform velocity and for an exponentially accelerating circular cylinder in incompressible, relatively high Reynolds number flow of short duration were considered by Nam (1990). Nam observes that the compounding singularity in the higher-order solutions (because the Navier equations should not yield any singularity) reveals the *limitation* of the boundary-layer approximation and in this sense, the appearance of a singularity could be a strong indication of *imminent separation*!

We believe that, after Van Dommelen and Shen (1980) clearly showed that a singularity exists at $\tau \approx 1.5$, where τ is a dimensionless time normalized by a/U_∞, i.e., radius/velocity of the cylinder, the *controversy* over the existence of a singularity in the boundary-layer solution of unsteady-state flow was put to rest.

Cowley (1983) and Ingham (1984) both confirmed this result and Henkes and Veldman (1987) suggested some *connection between the singularity and separation*!

As a complement to the above discussion, we mention again the paper by Riley and Vasantha (1989) who developed a method for calculating the unsteady-state high Reynolds number flow, which includes the limiting boundary-layer case. Boundary-layer calculations have been carried out for the impulsive flow past a circular cylinder and the flow induced when a line vortex is introduced into the neighborhood of a circular cylinder.

These calculations terminate in a singular behavior at a finite time, with a structure that is in accord with the analysis of Van Dommelen and Shen (1980). The solutions were extended to high, but finite, values of the Reynolds number using a viscous–inviscid interactive model which is based on a thin-layer approximation.

Calculations using this model indicate that the boundary-layer singularity is removed – no singularity appeared in the framework of the interactive calculations. The flow structure at large finite Reynolds number is dominated by the appearance of a single eddy close to the cylinder and slightly ahead of the vortex, as predicted by boundary-layer calculations [see Riley and Vasantha (1989, p. 252, Fig. 4a), reproduced from Ta Phuoc Loc and Bouard (1985)].

According to F.T. Smith (1988), finite time breakup can occur in any unsteady-state interacting boundary layer!

Finally, we mention again the very recent paper by Duck and Dry (2001) relative to a class of unsteady-state, nonparallel, 3-D disturbances to boundary-layer flows, where the reader can find recent references relative to unsteady-state boundary-layers.

7.7.3 Various Topics Related to Boundary-Layer Equations

First, a survey of higher order boundary-layer theory is given by Van Dyke (1962a,b,c, 1969, 1970). In particular, the Van Dyke (1962c) paper, concerning second-order compressible boundary-layer theory with applications to *blunt bodies in hypersonic flow*, has been an important step in the development of high-speed aerodynamics and its applications in *astronautics*. The following effects are important in second-order boundary-layer theory: (1) *longitudinal* and *transverse body curvature*, (2) boundary-layer *displacement*, (3) *vorticity interaction* in a *curved shock*, (4) *entropy gradient*, and (5) *slip-flow* and *temperature jump*. Thus, it is possible to determine the wall shear stresses and the heat transfer at moderate Reynolds numbers without solving the full NSF equations. On the other hand, because flow of this type occurs with *rentry bodies* (space shuttles), the work devoted, in particular, to 3-D boundary-layer flow at the stagnation point of a general body [see, for instance, Papenfuss (1977)] contributes to *optimizing the geometry at the stagnation point to obtain minimal heat transfer*.

We note that recently, Brazier, Aupoit, and Cousteix (1991, 1992) considered the second-order effects in hypersonic laminar boundary layers via the so-called "*defect approach*," coupled with asymptotic expansions, which ensures smooth merging of the viscous flow into inviscid flow, even when the inviscid profiles vary significantly through the boundary layer. In particular, when the *wall temperature is low* and thus the displacement effect is negligible, first-order defect calculations can give *good results* and reproduce NSF solutions with *reasonable accuracy at a lower cost*. It could then be a valuable tool for design task. In a more recent paper by Séror et al. (1997), an asymptotic defect boundary-layer theory is applied to thermochemical non-equilibrium hypersonic flow problem. The influence of *wall cooling on hypersonic boundary-layer separation and stability* was investigated by Cassel, Ruban, and Walker (1996).

In Nickel (1979), the 3-D fluid flow of a steady- or unsteady-state moving incompressible continuous medium is considered and several *Crocco-type transformations* are applied that reduce the system of three equations of the Prandtl boundary layer to a system of two strongly coupled parabolic equations; recently, in the paper by Van Oudheusden (1997), a *complete Crocco integral* is derived for 2-D laminar boundary-layer flow across an adiabatic wall for Pr = 1.

The review paper by Cousteix (1986) is devoted to *3-D and unsteady-state boundary-layer computations*, and F.T. Smith (1982) gives a pertinent survey relative to the *high Reynolds number theory of laminar flow*. A 3-D boundary-layer calculation on a *yawed body* of revolution by Allen and Riley (1994) provided a fundamental insight into boundary-layer separation of the *"open" type* (which is characterized by a *line of singularities* in the boundary-layer solution). In particular, it has been shown that the solution for a cylinder of circular section, at small angles of yaw, develops a singularity at a point that we interpret as the point at which the flow first *erupts* from the boundary layer to form a streamwise vortex configuration. In Fig. 7.14, the distortion of the cross-flow streamlines in the neighborhood of singular line $\theta = \theta_1$ reveals

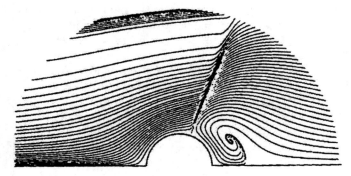

Fig. 7.14. A cross section of the stream surface in the boundary layer

260   7 Some Viscous Fluid Motions and Problems

erupting singular behavior. In Fig. 7.14 itself, the boundary layer has been radially magnified by a factor of $O(Re^{1/2})$.

In Timoshin (1996), the *marginal singularities* in the boundary-layer flow on a downstream-moving surface are considered.

More recently, Yan and Riley (1996) investigated the boundary-layer flow around a *submerged* circular cylinder *induced by free-surface traveling waves*. *Sound radiation* during local laminar *breakdown in a low-Mach* number slightly compressible boundary layer is considered by Wang et al. (1996), and in Asmolov (1995), the duty-gas flow in a laminar boundary layer *over a blunt body* is studied. For a flat plate, the stability of such a *duty-gas* laminar boundary layer is investigated by Asmolov and Manuilovich (1998), and the boundary-layer separation due to *gas thermal expansion* is studied in Higuera (1997). Breuer et al. (1997) considered the late stages of transition induced by a low-amplitude wave packet in a laminar boundary layer and in Brooker et al. (1997), a nonparallel linear analysis of a *vertical boundary layer* in a differentially heated cavity is performed. *Non-similarity* solutions to corner boundary-layer equations were considered by Duck et al. (1999). A *phase-equation approach* to boundary-layer instability in *Tollmien–Schlichting waves* is considered in Hall (1995). Steady-state boundary-layer solutions for a *swirling stratified fluid in a rotating cone* are obtained in Hewit et al. (1999). Simple cases of the *streamline-curvature instability* in 3-D boundary layers are considered by Itoh (1996), and the linear stability of flat-plate boundary-layer flow of fluids with *temperature-dependent viscosity* is considered in Wall and Wilson (1997). A generalized critical-layer analysis of fully coupled *resonant-triad interaction* in boundary layers was performed by Lee (1997). Weidman (1997) considered the *Blasius* boundary-layer flow over an *irregular leading edge*. Finally, the *transient* boundary-layer *heat transfer* from a flat plate subjected to a *sudden change in heat flux* was considered recently by Harris et al. (2001).

7.7.4 More on the Triple-Deck Theory

The survey paper, A view of the triple deck, by Meyer (1983) is a pertinent point of view on the various facets of the triple-deck theory. Meyer illuminates of the main features of the rational structure of the theory to a degree that they had yet (in 1983!) emerged; for the more complete and detailed triple-deck Stewartson–Neiland–Sychev theory, the reader can consult Chap. 12 in Zeytounian (2002b) devoted to asymptotic modeling of fluid flow phenomena. In Meyer (1983), some fundamental concepts are introduced, which characterize this new strong interactive Stewartson–Neiland triple-deck theory: first, the limit concept underlying Prandtl's boundary-layer theory for weak interaction, then a triplet of notions (mass-flow bound, penetration, localization) and also the "upstream condition". Meyer shows how their applications to Navier equations results in a need for more than two simultaneous boundary-layer limits. A third needed limit is also developed that completes a definite triple deck of limit concepts. But it leaves us far short of a definite system of differentials for describing local flow equations. Finally, Meyer gives a comprehensive description of the strong interaction between three decks, explores the successes and failures of this strong interactive triple-deck theory and sketches briefly Sychev's ideas (proposal!) about separation. The Meyer "enterprise" may be described as an attempt to enable the "nonspecialist" (!) reader to benefit thoroughly from such surveys as Brown and Stewartson (1969), Stewartson (1974, 1981) and Messister (1978). A numerical technique for the triple-deck problem was considered, in particular, by Napolitano et al. (1979) and see also Ragab and Nayfeh (1980a) who consider the second-order triple-deck equations.

Triple-deck solutions for subsonic flow past humps, steps, concave or convex corners, and wedged trailing edges were considered by F.T. Smith and Merkin (1982), and in Korolev (1987), the problems of the interaction of a boundary layer with external flow are solved. In Mauss (1994), the reader can find a pertinent new approach to asymptotic modeling for separating boundary layers, and Saintlos and Mauss (1996) consider separating boundary layers in a channel. More recently, Lagrée (1999) uses the triple-deck theory to investigate the thermal mixed convection problem and gives references related to this problem. We observe that the problem of the thermal response of an incompressible Blasius boundary layer has been posed by Zeytounian (1987, pp. 223–228). In Zeytounian (1991a, pp. 176–185), the reader can find a steady-state model of a triple-deck problem for local thermal prediction in an atmospheric boundary layer; in such a case, the basic undisturbed upstream infinity flow is characterized by an Ekman layer profile, and we derive a coupled system of local equations for the velocity components, u, w, and perturbation of the temperature θ, valid in a lower viscous layer close to heated flat ground:

$$u\frac{\partial u}{\partial x} + w\frac{\partial u}{\partial z} + \frac{B^*}{\gamma}\int_\infty^z \frac{\partial \theta}{\partial x}dz' + \frac{dP(x)}{dx} = \frac{\partial^2 u}{\partial z^2},$$

$$\frac{\partial u}{\partial x} + \frac{\partial w}{\partial z} = 0,$$

$$u\frac{\partial \theta}{\partial x} + w\frac{\partial \theta}{\partial z} = \frac{1}{\Pr}\frac{\partial^2 \theta}{\partial z^2}, \qquad (7.125a)$$

with the following conditions:

$$z = 0 : u = w = 0, \text{ and } \theta = \lambda^*\Theta(x) \text{ when } 0 \leq x \leq 1,$$

$$x \to -\infty : u \to z, w \to 0, \theta \to 0, P(x) \to 0, A(x) = \frac{dA}{dx} \to 0,$$

$$z \to +\infty : u \to z + A(x), w \to -z\frac{dA}{dx}, \theta \to 0. \qquad (7.125b)$$

We observe that this triple-deck problem is derived when in the similarity relation (between local Rossby (Ro) and Reynolds (Re) numbers based on local scale length l^0),

$$2\,\mathrm{Ro} = \mathrm{Re}^{-1/a}, \quad \text{when } 2 < m \equiv \frac{2a}{a-1}, \tag{7.126a}$$

we choose $m = 5$ which is the value used by F.T. Smith (1973) and Smith et al. (1977). The local length l^0 is linked with the thermal condition on flat ground in dimensionless form,

$$T = 1 + \tau^0 \Theta\left(\frac{x^*}{l^0}\right), \quad \text{where } \Theta\left(\frac{x^*}{l^0}\right) \neq 0 \text{ only if } \left|\frac{x^*}{l^0}\right| \leq 1.$$

For $m = 5$, we have the possibility of simplifying the starting problem via the Boussinesq approximation with two similarity relations (between the Boussinesq number (Bo), the Mach number (M), and the thermal parameter τ^0):

$$\frac{\mathrm{Bo}}{M} = B^* \quad \text{and} \quad \frac{\tau^0}{M} = \lambda^*. \tag{7.126b}$$

The specification of the triple-deck problem in (7.125a,b) is completed by the relation between the perturbation of pressure π and the function $A(x)$:

$$\frac{\partial}{\partial x}\left[\left(\frac{\partial \pi}{\partial z}\right)_{z=0}\right] = B^{*2}\mu^0 \frac{\mathrm{d}A}{\mathrm{d}x} + \gamma \frac{\mathrm{d}^3 A}{\mathrm{d}x^3}, \tag{7.126c}$$

where $\mu^0 > 0$ is the stratification parameter. Finally, we find that

$$\pi(x, z=0) \equiv P(x),$$

and from (7.126c), the pressure $P(x)$ driving the flow in the lower deck is itself induced in the main stream (i.e., the upper deck) by the displacement thickness $A(x)$ of the lower deck transmitted through the middle deck by the passive effect of displacement of the streamlines. We stress again that

> *strong singular self-induced coupling arises because the lower deck problem (7.125a,b) does not accept $P(x)$ as data known prior to resolution,*

as in classical Prandtl boundary-layer problems.

In the recent paper by Exner and Kluwik (2001), inspired by Lagrée (1999) and El Hafi (1994), the authors considered thermally induced separation of laminar free-convection flow and provided some pertinent references.

Viscous–inviscid interactive problem. Finally, a last comment concerning the very recent paper by Veldman (2001) which presents his *personal view* of the history of viscous–inviscid interactive methods – a history closely related to the evolution of MMAE. In Fig. 7.15, according to Veldman (2001,

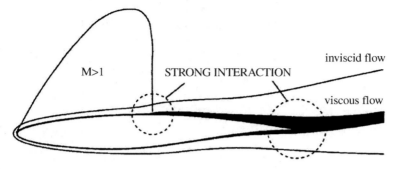

Fig. 7.15. Subdivision of the flow field around a airfoil

p. 190) paper, the subdivision of the flow field around a airfoil in an inviscid-flow region and in a viscous-shear layer is sketched (exaggerated in thickness).

It is now well known that the main challenge in solving Prandtl's boundary-layer equations has been to overcome the singularity at a point of steady-state flow separation and that the Stewartson–Neiland–Sychev triple-deck theory has inspired a solution to this challenge. The reader can find in Veldman's paper, a short, but pertinent, description of the Prandtl boundary-layer concept and flow separation which gives the breakdown of the boundary-layer equations. The triple-deck (three-layered) structure is summarized. According to Veldman's description, matching of middle and upper decks yields two kinds of pressure terms in an expansion for pressure (in the steady-state 2-D flow for the Navier equations):

$$p(x,y) = \text{Re}^{-2\alpha/3} p^{(p)}(x_\alpha, y_\alpha) + \text{Re}^{2\alpha/3 - 1/2} p^{(\delta)}(x_\alpha, y_\alpha) + \ldots . \quad (7.127)$$

When we observe that the flow in the lower deck is still governed by Prandtl's boundary-layer equations, with a pressure that is again constant in the vertical direction, $p(x,y) = \text{Re}^{-2\alpha/3} P(x_\alpha)$, we see that the term $p^{(p)}$ matches the pressure in the middle deck, so that it satifies $p^{(p)}(x_\alpha, 0) = P(x_\alpha)$. The term $p^{(\delta)}$ arises due to displacement effects and is related to the horizontal velocity perturbation in the upper deck. Figure 7.16 shows the relative order of the two terms in (7.127), and it reveals the essential character of the triple-deck structure; only for $\alpha = 3/8$ are both pressure terms in (7.127) equally important (the *principle of least degeneracy*), and this constitutes the essence of the triple deck.

Finally, the triple-deck equations [a narrow region around a singular point S of extent $x_\alpha = x - x_s = O(\text{Re}^{-\alpha})$, $\alpha < 1/2$] consist of Prandtl's boundary-layer equations, with boundary condition (from the lower deck analysis)

$$u\left(x_\alpha, \text{Re}^{-\beta} y\right) = \text{Re}^{-\alpha/3} \left[a \, \text{Re}^{-\beta} y + aA(x_\alpha) + \ldots\right], \quad (7.128)$$

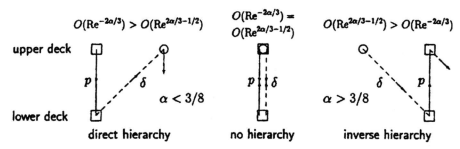

Fig. 7.16. Hierarchy between pressure contributions in lower and upper decks at a streamwise length scale $x = O(\mathrm{Re}^{-\alpha})$

with $\beta = (1/2) + \alpha/3$ and $a = 0.332$ (from the classical Blasius problem), where the function A is related to the displacement thickness as follows from the expansions in the middle deck and a second relation between pressure and displacement given by a Cauchy–Hilbert integral (as a consequence of the resolution of the Laplace equation in the upper deck in incompressible flow):

$$P(x_\alpha) = \frac{1}{\pi} \int_{-\infty}^{\infty} \frac{(\mathrm{d}A/\mathrm{d}\xi)}{x_\alpha - \xi} \mathrm{d}\xi . \tag{7.129}$$

Veldman discusses thoroughly the main consequences of the triple-deck structure for the numerical treatment and comments on some aspects of the non-asymptotic point of view (in particular, the reflections of LeBalleur (1977) at ONERA in France). Viscous–inviscid interactive methods are also discussed by Veldman and, here, we mention again the paper by LeBalleur (1978). Finally, an application to transonic airfoil flow is demonstrated. Veldman notes also that "in 1975, Lagerstrom described his view of the triple deck, and today we know that his paper contains the essential message required to overcome the singularity at a point of flow separation – *"boundary-layer and inviscid flow have to be solved simultaneously."*

It is through this type of insight that the use of viscous–inviscid interactive methods in engineering applications can flourish; in Veldman (2001), the reader can find a list of pertinent papers in this direction. Readers might also wish to consult Chap. 14 of the enlarged edition of Schlichting's well-known *Boundary-Layer Theory* [Schlichting and Gersten (2000)] for a more detailed discussion.

In conclusion, we mention, first, the recent paper by F.T. Smith (1996) where starting governing equations are (in the framework of a "composite approach"),

$$\frac{\partial u}{\partial x} + \frac{\partial v}{\partial y} = 0,$$

$$\frac{\partial u}{\partial t} + u\frac{\partial u}{\partial x} + v\frac{\partial u}{\partial y} = -\frac{\partial p}{\partial x} + \frac{1}{\text{Re}}\frac{\partial^2 u}{\partial y^2}, \qquad (7.130a)$$

$$\frac{\partial v}{\partial t} + u\frac{\partial v}{\partial x} + \Gamma v\frac{\partial v}{\partial y} = -\frac{\partial p}{\partial y},$$

$$u \to u_e(t), \quad \frac{\partial p}{\partial x} \to Q(t) \text{ as } y \to \infty,$$

$$u = v = 0 \text{ at } y = 0, \qquad (7.130b)$$

upstream and downstream in x constraints, initial conditions at $t = 0$.

In (7.130b), $u_e(t)$ is the given unsteady-state free-stream velocity, independent of x, with streamwise pressure gradient $Q(t)$ such thaty

$$\frac{\mathrm{d}u_e(t)}{\mathrm{d}t} = -Q(t).$$

The "unsteady-state composite scheme" is then modified, by Smith, to accommodate, iteratively, the Navier form and then the Euler form for comparison. The agreement between these sets of results tends to be very close in the parameter range studied. It seems that the way is now open, in principle, for many useful applications in dynamic stall and boundary-layer transitions.

In recent investigations by Mauss and Cousteix (2002a,b), a new approach is proposed for constructing a uniformly valid approximation: the successive complementary expansion method (SCEM). Contrary to the MMAE, the starting point is an assumed form of the approximation, and the matching principle is a by-product of the method not at all necessary to construct a uniformly valid approximation. With the SCEM, consistency is ensured by the possibility of finding the different components of the assumed expansion, and if a mistake is introduced in the assumed expansion, contradictions appear in the calculations. In particular, via SCEM, it is possible to concentrate more information in the first term of the expansion, and this is certainly an interesting feature because in the asymptotic methods the series are very often divergent and applicable when the small parameter goes to zero. In practice, however, it is desirable to apply the results when the small parameter is not really small. In this respect, it is valuable to include terms in the first elements of the expansion that are negligible when the small parameter goes to zero but are no longer negligible when this small parameter is not really small. In the same way, it is interesting to respect the boundary conditions exactly and not asymptotically.

A preliminary application at high Reynolds number led to the *notion of the strong interactive method in a very natural manner*. If a term that is negligible when the Reynolds number goes to infinity is included in the boundary-layer equations, it appears that the inviscid equations cannot be solved independently from the boundary-layer equations; the "extended boundary-layer equations" and the inviscid flow equations muts be solved simultaneously. These results must be confirmed by a deeper analysis. See also Cousteix and Mauss (2002).

7.7.5 Some Problems Related to Navier Equations for an Incompressible Viscous Fluid

In the paper by Perry and Chong (1986), an algorithm was developed that enables generating local Taylor-series-expansion solutions of the Navier and continuity equations to arbitrary order.

The initial aim of this work was to explore 3-D steady-state and unsteady-state flow-pattern topology. Using the algorithm in conjunction with *critical-point theory*, the authors discovered how to generate separation patterns with great control over their scale and topological properties. A critical point in a flow field is the point where the streamline slope is indeterminate. All possible patterns close to the critical point can be derived and classified, and sectional streamline patterns form saddles, nodes, or foci [see, for instance, Oswatitsch (1958), Lighthill (1963a), Perry and Fairlie (1974)]. Critical points are the salient features of a flow pattern and if their position and type are known, the rest of the pattern can be deduced qualitatively because the number of ways that the streamlines can be joined between the points is limited. Consequently, the *basic topology* and *qualitative transport properties* of the pattern can be *understood by using the critical-point concept*. First, flow patterns synthesized in 2-D are presented which are solutions of the Navier system; then a more general procedure was developed for synthesizing 3-D flow separation. In Perry and Chong (1986), examples of various symmetrical separation bubbles are shown in their Figs. 6, 7 and 8 (pp. 218–220). Note in Fig. 7.17a how fluid shown shaded on one side of the center plane finds its way to the focus on the other side of the center plane. This pattern has undergone a major change in topology because the original symmetrical pattern was structurally unstable, i.e., it has a saddle-to-saddle connection by a separatrix streamline [see Tobak and Peake (1982) and Perry and Hornung (1984) regarding structural stability].

Finally, we note that in the book edited by Moffatt and Tsinober (1989), a number of applications of topology to the physics of fluids can be found, where a classical aspect of "topological fluid dynamics," is considered, that is, the study of local qualitative properties of steady-state flows based on Taylor expansions of the governing fields. This point of view is mainly used to describe the flow in the vicinity of a point of separation of a solid boundary [see, for instance, Legendre (1956) paper working in ONERA]. In Brøns (1994), the topological fluid dynamics of interfacial flow is considered on the basis of a Navier incompressible and viscous system.

For interfacial flows, we mention the recent review paper by Scardovelli and Zaleski (1999) which is devoted to "direct simulation of free-surface and interfacial flow." The reader can find physical modeling of interfaces with

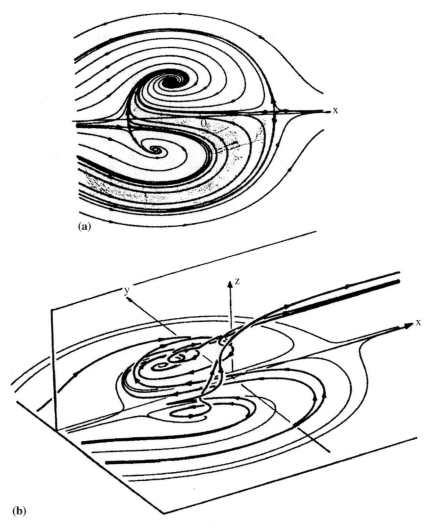

Fig. 7.17. (a) Surface flow pattern (limiting streamlines on the (x, y) plane). (b) Oblique view with some out-of-plane trajectories added (shown as heavy lines)

some relevant references. Figure 7.18 shows an example of droplet formation for a capillary number Ca = 1, Weber number We = 2000 and Re = 1000; the ratio of densities between the liquid and air is $\beta = 20$. The intermediate steps (between time = 4.0 and time = 24) are not shown, and the level lines of the vortices are omitted in (b) to avoid overcrowding the plot. The very small structures are formed because of the systematic stretching of a thin filament of liquid.

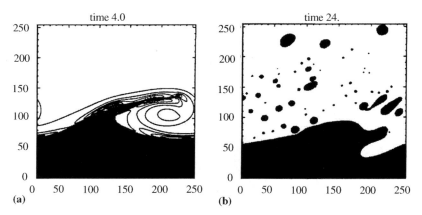

Fig. 7.18. Droplet formation at higher capillary (Ca = 1) and Weber (2000) numbers for Reynolds number Re = 1000 at $t = 4.0$ and $t = 24$

The steady-state problem of flow along a horizontal semi-infinite flat plate moving in its own plane through an incompressible viscous liquid just below the free surface was considered by Hofman (1993). The fact that this problem is a natural generalization of the classical Blasius boundary-layer problem gives more significance to the problem studied. Due to viscosity, small-amplitude gravity waves can be formed on the free surface. The problem is investigated when the Reynolds number is large (u_0 is a reference scale, h^0 the undisturbed thickness of the free-surface, ν^0 is the kinematic viscosity and g the gravity):

$$\text{Re} \gg 1 \text{ and } \frac{\text{Fr}}{\text{Re}} \ll 1 \Rightarrow h^{0\,2} \gg \frac{u_0}{g}\nu^0 \,, \tag{7.131}$$

where Fr is the Froude number; this makes the viscous terms in the free-surface conditions small enough to be neglected in the asymptotic procedure (expansions in terms of the $1/\text{Re}^{1/2}$), leaving the outer flow (to the order considered) entirely inviscid.

In the first-order approximation (terms proportional to $\text{Re}^{1/2}$), the solution obtained – the undisturbed stream for the outer flow (as if the liquid were inviscid), and Blasius flow for the boundary layer (as if the free surface were not present) – should be understood as proof that the usual separate analysis of viscosity and waves is correct, as a first approximation of the problem. The second-order approximation for the outer flow gives the inviscid flow past a thin parabolic cylinder of nose radius β^2/Re, as in Sect. 4.2.5, but, the free surface is involved. From the point of view of water wave theory, we are dealing with the waves caused by an obstacle in a steady-state stream, and in this case it is necessary to consider three regions in the outer flow (Fig. 7.19).

In region I, we consider infinitely deep flow with a free surface, in region II, we have a flow of slowly varying depth with a free surface, and in region III,

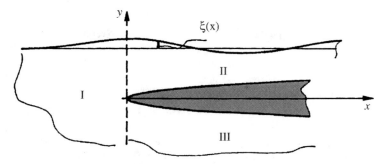

Fig. 7.19. Characteristic three regions of the outer flow

we consider again infinitely deep flow, but no free surface. When the Froude number approaches its critical value (Fr = 1) because of the strong interaction between viscosity and the free surface, even slight viscosity causes abrupt changes in the flow. The second-order approximation for the boundary-layer flow gives the free surface-boundary layer interaction and, in the outer flow, free-surface disturbances caused entirely by the effects of viscosity were obtained. These disturbances have different forms for subcritical (Fr < 1) and supercritical (Fr > 1) regimes of flow: for the subcritical regime, only the effects of viscosity induce a gravity wave of small amplitude that speads far downstream. The amplitude of this wave increases as the Froude number rises to its critical value, and there (in the transcritical flow regime) only slight viscosity causes abrupt changes in the flow field. On the other hand, the free surface also changes the flow in the boundary layer (but only on the upper side of the flate plate), increasing the *resistance when* Fr < 1, and *decreasing it when* Fr > 1, all compared with the values for the unbounded stream. And that seems to be the *most significant effect of boundary layer to free surface interaction*.

In the paper by Ranger (1994), explicit solutions are found for the stream function satisfying the steady-state Navier equations representing the steady-state 2-D motion of a viscous incompressible liquid. The solutions contain two arbitrary analytic functions and in general are confined to certain region of the (x,y) plane. In a complex variable formulation of the flow, it is necessary to consider the following system of two equations for the (real) stream function $\psi = \psi(x,y)$ and function $\phi = \phi(x,y)$ related to Bernoulli function $B = (p/\rho^0) + (1/2)|\boldsymbol{u}|^2$:

$$\frac{\partial^2 \phi}{\partial z \partial z^*} + i\frac{\partial^2 \psi}{\partial z \partial z^*} + \frac{1}{2\nu^0}\frac{\partial^2 \psi}{\partial z^{*2}} = h(z^*), \qquad (7.132a)$$

$$-\nu^0 \boldsymbol{\nabla}^2 \phi = -4\nu^0 \frac{\partial^2 \psi}{\partial z \partial z^*} = \frac{p}{\rho^0} + \left(\frac{\partial \psi}{\partial z}\right)\frac{\partial \psi}{\partial z^*}, \qquad (7.132b)$$

where $h(z^*)$ is arbitrarily an analytic in a suitable region of the fluid: $z = x+iy$ and $z^* = x-iy$. It is of interest to point out that the solutions for ψ are closely related to spatial nonlinear wave propagation in a conformally mapped region of the (x, y) plane because, by applying integrability conditions, an explicit general expression is constructed for ψ in terms of space variables.

Bunyakin et al. (1998) consider steady-state incompressible flow past an aerofoil with a cavity in its upper surface. The aerofoil and cavity shape is constructed to ensure that the high Reynolds number asymptotics of the flow is described by the Batchelor model discussed in Sect. 7.7.1. The result is a nontrivial example of steady-state flow, where the high Re asymptotics is described by the Batchelor model, and they show in detail the mechanisms and existence of such flows, but demonstrate that such flows are quite rare! We observe that the results obtained with asymptotic methods on the basis of the steady-state Navier equations have the advantage of assured self-consistency and objectivity. In the references, the reader can find a selection of very pertinent Russian works.

The so-called Hagen–Poiseuille flow problem consists of an incompressible viscous fluid flowing in a circular pipe of constant radius r^0 under the action of a constant applied pressure gradient

$$p^* = -4\rho^0 \nu^0 \left(\frac{U^0}{r^{0\,2}}\right) z^* .$$

At low enough values of $\mathrm{Re} = r^0 U^0 / \nu^0$, the flow realized is uniquely

$$u^* = U^0 \left[1 - \left(\frac{r}{r^0}\right)^2\right] z^* ,$$

in the usual cylindrical coordinate system.

Rotating Hagen–Poiseuille flow is considered by Barnes and Kerswell (2000), and in this case the pipe is further considered to be rotating at angular velocity $\Omega^0 z^*$ about its axis, which itself is rotating at $\omega^0 x^*$ about a diameter (see Fig. 7.20).

The addition of perpendicular rotation (ω^0) at low axial rotational rates is found only to stabilize the system and in the absence of axial rotation

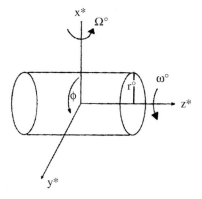

Fig. 7.20. A diagram of the geometry

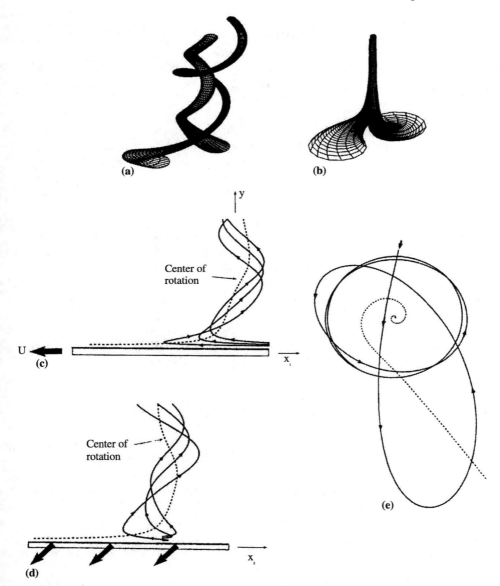

Fig. 7.21. (a), (b) Stream surfaces and vortex surfaces tubes viewed relative to the vortex core as it propagates at constant velocity over a stationary plane; (c), (d), (e) Projections of the streamlines together with that of the center of rotation in the three coordinate planes [from Studies in Appl. Math. **107**, 1 (2001)]

($\Omega^0 = 0$), the 2-D steady-state flow solution in *rotating Hagen–Poiseuille flow* which connects smoothly to Hagen–Poiseuille flow as $\omega^0 \to 0$ is found to be stable at all Re below 10^4. At high axial rotational rates, the superposition of perpendicular rotation produces "precessional" instability which is a supercritical Hopf bifurcation (see, for instance, Chap. 10) that leads directly to 3-D traveling waves.

Exact solutions of the Navier equations are used to describe pump flow that involves the interaction of several different components in the recent work of Öztekin et al. (2001). In particular, a detailed description is given of flow that involves both a swirl and a shear component. Steady-state pump flow and unsteady-state self-similar flow are also investigated. The *pump flows* (plane material surfaces that are orthogonal to the vertical y axis at any time remain orthogonal to the y axis at subsequent times) is local flow that is used to model complex global flows in the vicinity of the y axis when local, rather than global, conditions control the flow. Examples include Couette and Poiseuille flows, Blasius–Hiemenz stagnation point flow, von Karman axisymmetric swirling flow, Rayleigh shear flow, Ekman spiraling flow, Burgers vortex flow, and Kelvin flow.

In Fig. 7.21, (a) shows a typical stream-surface formed by streamlines crossing the perimeters of two circles that lie in some plane that is perpendicular to the y axis; (b) shows the vortex tube that is formed by vortex lines crossing the same circles; (c)–(e) show the projections of the streamlines, together with the projections of the trajectory of the center of symmetry in the three coordinates planes.

7.7.6 Low and Large Prandtl Number Flow

Large Prandtl numbers. In early work, the problem of natural convection from a vertical flat plate at large values of the Prandtl (Pr) number was discussed by Stewartson and Jones (1957) and independently by Kuiken (1968). Experimental studies on heat transfer at very high Pr numbers are reported in a review article by Soehngen (1969). In Crane (1976), the natural convection from a vertical cylinder is also determined in the limiting case of very large Pr number, where the Grashof number remains finite. When Pr $\to \infty$ with Gr (for the Grashof number Gr, see Sect. 7.5) finite, the temperature (*inner*) layer becomes vanishingly thin, and the momentum (*outer*) region becomes infinitely thick. In natural convection from a vertical cylinder, the solution in the inner region is determined in terms of a parameter which is roughly equal to the ratio of the thickness of this layer to the radius of the cylinder – it is valid up to a vertical height at which this parameter is about unity. In the inner layer whose thickness varies as $\text{Pr}^{-1/2}$, the solution was obtained by Stewartson and Jones (1957) and Kuiken (1968). Crane (1976) gives the solution for outer momentum flow and also a corrected approximation for inner flow.

For Earth's mantle, the Prandtl number is at least 10^{22}, essentially infinite, but it is necessary to *relax* the classical *Boussinesq approximation* and use the so-called *anelastic-gas equations* [see, for instance, Ogura and Phillips (1962)] valid in a *deep convecting layer*. In Jarvis and McKenzie (1980), these "deep convection" equations are used for 2-D time-dependent (which is a common feature of convection in deep layers) flow to investigate the effects of compressibility, viscous dissipation, adiabatic heating, and nonhydrostatic pressure gradients on the flow, energetics, and stability of a deep convecting layer. Of the non-Boussinesq features included in the model considered by Jarvis and McKenzie (1980), the variation of the adiabatic gradient has dominant effects on the character of the flow, and viscous dissipation and density stratification appear to affect only details of the 2-D flow structure, whereas decrease in the magnitude of the adiabatic gradient with height acts as an internal source of buoyancy.

In a more recent paper by Thess and Orszag (1995), surface-tension driven (Bénard–Marangoni) convection in a planar fluid layer is studied by numerical simulation of the 3-D time-dependent governing equations in the limit of infinite Prandtl number; the limit model equations are

$$0 = \frac{\partial p}{\partial x} + \nu^0 \boldsymbol{\nabla}^2 \boldsymbol{u}, \ \boldsymbol{\nabla} \cdot \boldsymbol{u} = 0, \ \frac{\partial T}{\partial t} + (u \cdot \boldsymbol{\nabla})T = k\boldsymbol{\nabla}^2 T \,, \tag{7.133}$$

and the emphasis is placed on the spatial scale of weakly supercritical flows and on generating small-scale structures in strongly supercritical flows. The case of high Marangoni number is also considered, and in this case, discontinuities of the temperature gradient are formed between convection cells, producing a universal spectrum $E = k^{-3}$ of the 2-D surface temperature field – but the mathematical problem of understanding the asymptotic regime for Ma $\to \infty$ remains unsolved! The properties of this limiting regime will certainly be different from classical Kolmogorov turbulence because only the Peclet number tends to infinity, whereas the Reynolds number remains small by virtue of the definition of the infinite Prandtl number model. This case strongly differs from "hard" Rayleigh–Bénard turbulence in which both Re $\to \infty$ and Pe $\to \infty$. In the present case, the flow is still dominated by viscosity, whereas the temperature isolines are strongly deformed and the thermal diffusivity can be neglected except in thin thermal layers of thickness $\delta \approx 1/\text{Pe}$.

It is known both from experiments and from numerical simulations [see, for instance, Travis et al. (1990)] of Rayleigh–Bénard convection as Pr $\to \infty$, that the velocity field and the temperature field exhibit nontrivial small-scale behavior in the limit of high Rayleigh number (and thus $Pe \gg 1$). In Thess and Orszag (1995), the analogous question is considered for thermocapillary driven Bénard convection at high Marangoni number.

Low Prandtl numbers. The study of thermal convection at very low Prandtl nmber has long been motivated by geophysical and astrophysical interest – for instance, Earth's liquid-core convection, it is thought, is governed by a Prandtl number of order $\Pr \approx 0.1$, owing to the *metallic nature* of its materials. In particular, the results of Thual (1992) show that the zero Prandtl number equations are a proper limit for small Prandtl number regimes. In the framework of Boussinesq model equations, for $\Pr = 0$, we derive the following system of limit equations for the velocity vector $\boldsymbol{u} = (u, v, w)$, pseudopressure π and temperature deviation θ:

$$\frac{\partial \boldsymbol{u}}{\partial t} + (\boldsymbol{u} \cdot \boldsymbol{\nabla}) \boldsymbol{u} = \boldsymbol{\nabla}\pi + \boldsymbol{\nabla}^2 \boldsymbol{u} + \mathrm{Ra}\, \theta \boldsymbol{k} ,$$
$$\boldsymbol{\nabla} \cdot \boldsymbol{u} = 0 ,$$
$$0 = \boldsymbol{\nabla}^2 \theta + w . \tag{7.134}$$

We observe that the inversion $\theta = -\boldsymbol{\nabla}^{-2} w$, where $w = \boldsymbol{u} \cdot \boldsymbol{k}$, with the boundary conditions $\theta = 0$ at $z = 0$, and $z = 1$, and the horizontal periodicity of the temperature, is a well-posed problem. In the list of references of the paper by Thual (1990), various pertinent papers are mentioned relative to low Prandtl number flow. A rich variety of regimes is observed in the case of $\Pr = 0$ convection, governed by the above limit equations as the Rayleigh number (Ra) is increased, and this multiplicity of regimes can be attributed to the close interaction between stationary and oscillatory instabilities (see, for instance, Chap. 9).

Bénard–Marangoni convection at a low Prandtl number was considered recently by Boeck and Thess (1999); the numerical simulations by these authors demonstrated that BM convection at low Prandtl number quickly becomes time-dependent. BM convection in $Pr \ll 1$ flow, i.e., liquid metals or semiconductor melts, plays an important role in industrial processes such as crystal growth [Davis (1987)] and electron beam evaporation [Schiller, Heisig and Panzer (1982); Pumir and Blumenfeld (1996)]. Due to the strong thermal forcing and the low kinematic viscosity of liquid metals, these flows usually have high Reynolds numbers, i.e., they are usually time-dependent or even turbulent. We observe that in the case of BM convection in the first equation of (7.134), the term proportional to Ra is neglected (see our "alternative" at the end of the Sect. 7.5) and when the Biot number is zero as boundary conditions, we can write:

$$\frac{\partial u}{\partial z} + \mathrm{Ma}\frac{\partial \theta}{\partial x} = 0,\ \frac{\partial v}{\partial z} + \mathrm{Ma}\frac{\partial \theta}{\partial y} = 0,\ w = 0,\ \frac{\partial \theta}{\partial z} = 0,\ \text{at } z = 1 ; \tag{7.135a}$$
$$\frac{\partial u}{\partial z} = 0,\ \frac{\partial v}{\partial z} = 0,\ w = 0,\ \theta = 0,\ \text{at } z = 0 \text{ (free-slip)} ; \tag{7.135b}$$
$$u = 0,\ v = 0,\ w = 0,\ \theta = 0,\ \text{at } z = 0 \text{ (no-slip)} . \tag{7.135c}$$

Finally, we mention the recent papers by Dawes (2001) where the rapidly rotating thermal convection at low Prandtl number is considered, and also Delgado-Buscalioni et al. (2001) who consider the flow transition of a low Prandtl number in an inclined 3-D cavity.

7.7.7 A final comment

Unfortunately, I will have to leave out of Chap. 7 some interesting topics directly related to viscous effects. For example, among others I mention *viscous transonic* and *hypersonic* flows, *Kolmogorov* flow, *sound* wave propagation in an inhomogeneous *thermoviscous* fluid, *vorticity dynamics* theory of 3-D flow *separation*, the influence of (vanishing) *viscous effects* in *turbomachinery* fluid flow, *vortex sheets*, and shock layer (*internal structure of shocks*) phenomena, *amplitude equations* in *dissipatives* systems (see Chaps 9 and 10, for instance), *free viscous* layers, viscous *thermocapillary* (BM) convection at *high Marangoni* number, *viscous separation of a vortex sheet* from a body, *wave damping* by a (thin) layer of viscous fluid, and the influence of *viscosity and wall thermal effect* on the *lee wave* phenomena in a stratified atmosphere.

8 Some Aspects
of a Mathematically Rigorous Theory

In this chapter the reader can find some basic tools for the *rigorous analysis* of the viscous (mainly Navier) equations.

In particular, for definiteness and simplicity, we consider the *Galerkin approximation* and *weak* solutions for the Navier equations, with periodic boundary conditions on $\Omega = [0, L]^d$, in the presence of a body force which is constant in time.

Unfortunately, the weak solutions of the Navier equations are not unique in any sense! Their construction as the limit of a subsequence of the Galerkin approximation leaves open the possibility that there is more than one *distinct limit*, even for the same sequence of approximations. *Nonunique* evolution would violate the basic tenets of classical Newtonian determinism and would render the Navier equations *worthless* as a predictive model. Naturally, if the weak solutions were smooth enough that all of the terms in the Navier equations made sense as "normal" functions, then we would say that the weak solutions were *"strong"* solutions.

As noted pertinently by Doering and Gibbon (1995),
"To many readers this kind of distinction may seem little more than a mathemaical formality of no real consequence or practical importance. The issues involved, however, go straight to the heart of the question of the validity and *self-consistency* of the Navier equations as a hydrodynamical model, and the mathematical difficulties have their source in precisely the same physical phenomena that the equations are meant to describe"!

In this chapter we also give some information concerning recent rigorous mathematical results (*existence, regularity*, and *uniqueness*) for incompressible and compressible viscous fluid flow problems.

In the two recent review papers by Zeytounian (1999, 2001), the reader can find a fluid-dynamic point of view on some mathematical aspects of fluid flows.

For the mathematically rigorous theory of the Navier incompressible viscous system, we mention the books by Ladhyzhenskaya (1969), Shinbrot (1973), Temam (1984), Constantin and Foias (1988), Galdi (1994/1998a,b), Doering and Gibbon (1995), P.-L. Lions (1996), and the recent survey by Temam (2000) which describes the Navier equations from the work of Leray in 1933 up to the recent research on attractors and turbulence. It is also necessary to mention the recent, very pertinent, short paper of Jean-Yves Chemin (2000), where the author gives an illuminating analysis of the Leray (1934) fundamental paper: "Sur le mouvement d'un liquide visqueux emplissant l'espace", published in Acta Math., 63 (1934), 193–248, and also the survey papers by Heywood (1976, 1980a,b,c, 1989). Concerning the mathematical legacy of Jean Leray see also Temam (1999).

For mathematically rigorous results of compressible viscous equations, we mention the book of P.-L. Lions (1998) and the review papers by Solonnikov and Kazhikov (1981) and Valli (1992). In the recent book by Lemarie-Rieusset (2002) the reader can find the recent developments in the Navier–Stokes (in fact, Navier) problem.

8.1 Classical, Weak, and Strong Solutions of the Navier Equations

Below, we consider the initial boundary-value problem for viscous, incompressible, and homogeneous fluid flow in a smoothly bounded two or three-dimensional domain Ω for $\boldsymbol{u}(t,\boldsymbol{x})$ and $p^*(t,\boldsymbol{x}) = \pi/\rho^0$:

$$\frac{D\boldsymbol{u}}{Dt} + \nabla p^* = \nu_0 \nabla^2 \boldsymbol{u} \text{ , and } \nabla \cdot \boldsymbol{u} = 0 \text{ , for } \boldsymbol{x} \in \Omega \text{ , } t > 0 \text{ ,} \quad (8.1a)$$

$$\boldsymbol{u}(0,\boldsymbol{x}) = \boldsymbol{u}^0(\boldsymbol{x}) \text{ , for } \boldsymbol{x} \in \Omega \text{ , and } \boldsymbol{u}(t,\boldsymbol{x}) = 0 \text{ , for } \boldsymbol{x} \in \partial\Omega \text{ , } t > 0 \text{ ,} \quad (8.1b)$$

where D/Dt is the material derivatve. Roughly, a weak solution is a function $\boldsymbol{u}(t,\boldsymbol{x})$ that satisfies an integral equation [(8.2) below], derived from the Navier equations (8.1a), and which would be a solution only if it had enough derivatives for all of the terms in the Navier equations to make sense. The concept is due to Leray (1933, 1934), but it was Hopf (1951) who first proved that weak solutions exist in the general domain Ω. Precisely, for any (smooth) divergence-free vector field test function $\phi(\boldsymbol{x})$ and times $t_2 > t_1 \geq 0$, weak solutions $\boldsymbol{u}(t,\boldsymbol{x})$ obey

$$\int_\Omega \phi(x) \cdot \boldsymbol{u}(t_2,\boldsymbol{x}) \, \mathrm{d}^d\boldsymbol{x} - \int_\Omega \phi(\boldsymbol{x}) \cdot \boldsymbol{u}(t_1,\boldsymbol{x}) \, \mathrm{d}^d\boldsymbol{x}$$

$$= \int_{t_1}^{t_2} \int_\Omega [\nu_0 \nabla^2 \phi(\boldsymbol{x})] \cdot \boldsymbol{u}(t,\boldsymbol{x}) \, \mathrm{d}^d\boldsymbol{x} \, \mathrm{d}t$$

$$+ \int_{t_1}^{t_2} \int_\Omega \boldsymbol{u}(t,\boldsymbol{x}) \cdot [\nabla \phi(\boldsymbol{x})] \cdot \boldsymbol{u}(t,\boldsymbol{x}) \, \mathrm{d}^d\boldsymbol{x} \, \mathrm{d}t \text{ .} \quad (8.2)$$

Relation (8.2) follows easily from the Navier equations if $\boldsymbol{u}(t,\boldsymbol{x})$ is a solution in the usual sense – simply multiply $\phi(\boldsymbol{x})$ into the Navier equation and

8.1 Classical, Weak, and Strong Solutions of the Navier Equations

integrate over space and time, integrating by parts where necessary. This does not imply, however, that $u(t,x)$ actually satisfies the Navier equations.

By "the usual sense", used above, we mean that each term in the Navier equations is well defined and smooth. However, just because $u(t,x)$ is, for instance, a square integrable function of space, it may not be that $\nu_0 \nabla^2 u$ is square integrable as well.

What follows from the construction of weak solutions is that they are square integrable at each instant and that the L^2 norms of their first derivatives are square integrable in bounded intervals of time. In fact, the weak solutions satisfy a form of the integrated energy equation known as Leray's inequality:

$$\frac{1}{2} \|u(t,\cdot)\|_2^2 + \nu_0 \int_0^t \|\nabla u(t',\cdot)\|_2^2 \mathrm{d}t' \leq \frac{1}{2} \|u(0,\cdot)\|_2^2 . \qquad (8.3)$$

We know that the kinetic energy in a fluid in volume Ω bounded by $\partial \Omega$ is

$$\frac{1}{2} \int_\Omega |u|^2 \mathrm{d}^d x = \frac{1}{2} \|u\|^2 ,$$

where we have introduced the usual notation $\| \cdot \|_2$ for the norm in $L^2(\Omega)$, the Hilbert space of square integrable functions.

Note that the inequality in (8.3), rather than the equality, follows from direct manipulation of the Navier equations (8.1a). This says that the energy at time t, plus total energy dissipated up to time t, is not larger than the initial kinetic energy (when the total work performed by the body force is neglected), although it need not be equal to it. Hence the classical energy equation,

$$\frac{\mathrm{d}}{\mathrm{d}t} \left[\frac{1}{2} \|u\|_2^2 \right] = -\nu_0 \|\nabla u\|_2^2 , \qquad (8.4)$$

does not automatically follow from the construction of the weak solutions. In conclusion, the weak solutions satisfy only energy inequality (8.3) rather than energy equality (8.4) that is derived for classical solutions of the Navier equations, which is, more or less, a function $u(t,x)$, satisfying (8.1a) and (8.1b), for which all the terms appearing in the equations (8.1a) are continuous functions of their arguments.

The classical solutions of Navier equations are unique (more precisely, if the solutions are classical, then it is easy to prove their uniqueness), smooth, infinitely differentiable functions. We note, however, that there are no known examples of weak solutions [according to Doering and Gibbon (1995, p. 104)] that actually violate energy equality!

On the other hand, it has not been shown that the weak solutions are unique in any sense – their construction as the limit of a subsequence of the Galerkin approximations (see the Sect. 8.2) leaves open the possibility that there is more than one distinct limit, even for the same sequence of approximations, and non-unique evolution would violate the basic tenets of classical Newtonian determinism and would render the Navier equations worthless as a predictive model!

Naturally, any classical solution of the initial boundary-value problem, (8.1a), (8.1b), for the Navier equations satisfies the integral equation (8.2) derived from the Navier equations, but on the other hand, it is easy to imagine functions $u(t, x)$ that satisfy this integral equation (8.2), but are not sufficiently differentiable for the corresponding Navier equations (8.1a) to make any sense at all.

It is not too hard to show that a weak solution that is smooth enough is actually a classical solution if $\partial \Omega$ is also smooth.

The weak solution is in some sense a solution of the Navier equations (8.1a), although it is not clear that the derivatives $\partial u/\partial t$ and $\nabla^2 u$ in these equations ever have any meaning. If the weak solutions were smooth enough that all of the terms in the Navier equations made sense as normal functions, then we would say that the weak solutions are "strong" solutions.

For *strong solutions*, each term in the Navier equations exists as an element of Hilbert space, at least. Strong solutions are infinitely differentiable in all variables, and all of the terms in the equations are continuous [see, for instance, Kaniel and Shinbrot (1967) and also the book by Shinbrot (1973, Chap. 13)]. Strong solutions were first proved to exist by Kiselev and Ladyzhenskaya (1957). A more accessible reference for those who do not read Russian is the book by Ladyzhenskaya (1969) and the review paper of Ladyzhenskaya (1975).

Using (as in Sect. 8.2) the integrated energy equation for the Navier equations and Galerkin's method, it is easy to prove the existence of a weak solution for the Navier problem (8.1a), (8.1b). The set of values of time t at which a weak solution can fail to be strong (and, therefore, unique) is empty in 2-D.

It would be nice to prove this in 3-D, but all efforts to do so have so far failed, and the preponderance of evidence seems to indicate that the corresponding set is not empty in 3-D! But the set of values of t on which a weak solution is not strong must be rather small, and on the other hand, to say that a weak solution is strong is to say that it is exponentially small for large t [this is a kind of (asymptotic) stability theorem].

For uniqueness, it is necessary to consider two solutions, $u(t, x)$ and $u'(t, x)$, both of which satisfy the Navier problem (8.1a), (8.1b), and then to try to show that they must be equal. Toward this end, let us write down the Navier problem that governs the difference between two solutions, $v = u' - u$.

Now, if the energy evolutionary equation for v is derived from this Navier equation for v,

8.1 Classical, Weak, and Strong Solutions of the Navier Equations

$$\frac{\mathrm{d}}{\mathrm{d}t}\left[\frac{1}{2}\|v\|_2^2\right] + \int v \cdot [\nabla u] \cdot v \,\mathrm{d}^d x = -\nu_0 \|\nabla v\|_2^2, \tag{8.5}$$

and Poincaré's inequality and Gronwall's lemma are used [concerning these notions, see Doering and Gibbon (1995, pp. 32 and 36)], we deduce the following estimate:

$$\|v(t,\cdot)\|_2^2 \le \|v(0,\cdot)\|_2^2 \exp\left\{-\left(\frac{4\pi^2 \nu_0 t}{L^2}\right) + \int_0^t \|\nabla u(t',\cdot)\|_\infty \,\mathrm{d}t'\right\}, \tag{8.6}$$

where L is an appropriate length scale and $\|\nabla u(t',\cdot)\|_\infty$ is the maximum component of the supremum over the space of $|\partial u_i/\partial u_j|$ at time t.

So if we knew that the argument of the exponent on the right-hand side of (8.6) was finite – in particular, if we knew that $\|\nabla u\|_\infty$ had at most integrable singularities – then we could conclude that if v were initially equal to zero, then it would remain so.

This would yield uniqueness. However, for the weak solutions, we do not (yet) know that $\|\nabla u\|_\infty$ emains so well behaved, and this illustrates the quandary that we face!

Global weak solutions are square integrable at each instant, and their derivatives are square integrable in space and time, but these facts do not preclude singularities in the flow field across which $\|\nabla u\|_\infty$ is not integrable in time. Roughly speaking, as regards uniqueness, an integrability condition on $\|\nabla u\|_\infty$ for the Navier equations is the analog of a Lipschitz condition for an ODE.

To infer the uniqueness of solutions of the Navier equations, we must have more regularity than that provided directly by the process of constructing weak solutions. It is important to note [according, again, to Doering and Gibbon (1995, pp. 106, 107)], that the question of uniqueness of solutions is not the only motivation for inquiring into the regularity properties of solutions. If the smoothness of solutions is lost on some level, then this implies the presence of a small-scale structure in the flow. On the other hand, a discontinuity implies a macroscopic change over an infinitesimally microscopic interval.

On its own, this may not lead to any difficulties, but the incompressible, viscous, Navier equations are derived from a microscopic model of interacting particles (atoms or molecules) precisely in a limit of infinite separation of length and timescales between microscopic and macroscopic phenomena [see, for instance, the two papers by Bardos et al. (1991, 1993) cited in Chap. 1 of Zeytounian (2002a)].

If macroscopic hydrodynamic equations were then to generate singularities or structure on infinitesimal scales, this would signal a breakdown of the very ansatz that leads to their derivation. Clearly, this is a fundamental philosophical problem, but it is also a mathematical problem.

Rigorous derivation of the hydrodynamical equations relies on the smoothness of their solutions to show convergence from the microscopic scales [considered as "fluid dynamic limits of kinetic equations" – see, again, Chap. 1 in Zeytounian (2002a)]. So along with the question of uniqueness, it is important to monitor the length (and time) scales generated by the Navier equations.

As developed in Chaps. 6 and 7 of the book by Doering and Gibbon (1995), the relevant technical issues are closely related [in Chap. 6 of Doering and Gibbon (1995), the authors obtain the so-called "ladder theorem" for the Navier equations] – to derive the ladder theorem, it is necessary to introduce the idea of seminorms that contain derivatives of the velocity field higher than unity.

In Chap. 7 of their book, these authors analyze the regularity and length scales for 2-D and 3-D Navier equations.

Note that the length of time during which we are assured of a reasonably smooth solution t^* depends sensibly on the parameters of the starting Navier problem (8.1a), (8.1b):

> Increasing the viscosity lengthens t^*, as does smoothing the initial flow field [decreasing $\|\nabla u^0\|_2$].

In a periodic domain, global weak solutions exist in both 2-D and 3-D. A crucial distinction arises at the level $\|\nabla u\|_2^2$, which is proportional to the instantaneous rate of energy dissipation and, for the periodic boundary conditions, to the enstrophy (a global measure of the vorticity content of a flow, is the square of its L^2 norm, called the enstrophy). In 2-D, these quantities are forever finite for solutions of the Navier equations. In 3-D, they are bounded for short times, but even in the absence of boundaries and/or external forcing, we do not know if they remain well behaved for arbitrarily long times.

It is necessary to note that the above discussion relative to uniqueness is also related directly to the question of continuous dependence on the data.

In Heywood (1994), the reader can find various aspects of the Navier theory and their significance in the problem of the global existence of smooth solutions. Here we note only [according to Heywood (1994)] that

> "In 1934, Leray wrote a landmark paper concerning the existence, uniqueness, and regularity of solutions of the time dependent 3D Navier equations. For a fluid that fills all space, so that there are no boundaries to deal with, he proved two fundamental existence theorems. The first gave the local existence of a smooth solution – by 'local' it is meant that the solution is obtained on only a finite interval of time $[0, T)$, unless the data are restricted to be small, in which case Leray's smooth solution does exist for all time. The estimates for the smoothness of the solution, that underlie this theorem, 'blow up' as $t \to T$. This does not mean that a singularity necessarily develops at time T, but only that the possibility is not excluded. Leray's second

existence theorem provides the global existence of what we now call a 'weak solution.' Leray called it a 'turbulent solution.' This theorem is based on the classical energy estimate, which is the only global estimate that we know for the Navier equations. It provides an estimate for the L^2 norm over space-time of the first spatial derivatives of the velocity, as well as an estimate for the spatial L^2 norm of the velocity, pointwise in time. But it seems that, in the 3D case, this is not enough to go on to prove that the solution is smooth, or even that it is unique and depends continuously on the data. Leray did provide a partial justification for his introduction of weak solutions by proving a uniqueness theorem that assures that so long as a smooth solution exists, a weak solution cannot differ from it. He also went on to prove an upper bound for the one-half dimensional Hausdorff measure of the set of times t at which singularities occur."

Today, 68 years later, despite tremendous progress in many aspects of the mathematically rigorous Navier theory, we have not yet answered the fundamental questions raised by this work of Leray – *we have not determined whether a solution that is initially smooth can develop a singularity at some later time, or whether singularities are an important feature of turbulence.* But are singularities really necessary to explain turbulence? 2-D flows are smooth, but can be highly unstable and apparently chaotic! The modern theory of bifurcations and attractors [as presented in Chap. 10, but from a simplified phenomenological point of view] seems to provide a perfectly satisfactory setting for a theory of turbulence (chaos – not strongly developed!) within the framework of smooth solutions!

8.2 Galerkin Approximations and Weak Solutions of the Navier Equations

For simplicity, now, as in Doering and Gibbon (1995), and definiteness, we consider the Navier equations (8.1a), with periodic boundary conditions on $\Omega = [0, l]^d$. We assume also that the initial data $\boldsymbol{u}^0(\boldsymbol{x})$ [in (8.1b)] describe a periodic, square integrable, divergence-free vector field. Moreover, the case of a mean zero $\boldsymbol{u}^0(\boldsymbol{x})$ is considered; evolution preserves this mean zero property for $\boldsymbol{u}(t, \boldsymbol{x})$ (these assumptions are necessary if we are going to use Poincaré's inequality with periodic boundary conditions – the problem in the presence of rigid boundaries is significantly more involved).

The Navier equations can be considered an infinite system of ODEs (a so-called "infinite-dimensional dynamical system"), and to manufacture global solutions, the idea is to write down explicit successive approximations consisting of solutions of finite systems of ODEs (the Galerkin approximations) and then to show that the approximations converge, in a certain sense, to solutions of the Navier PDEs. The solutions of the Navier PDEs produced in

284   8 Some Aspects of a Mathematically Rigorous Theory

this way are again the "weak" solutions of the Navier equations because the original Navier equations can only be shown to be satisfied in a weak sense.

As discussed in classical analysis, a velocity field may be Fourier transformed according to

$$\hat{u}(t, k) = \int_\Omega \exp[-i k \cdot x] u(t, x) \, d^d x \,, \tag{8.7a}$$

with inverse transform

$$u(t, x) = \left(\frac{1}{L^d}\right) \sum_{(k)} \exp[i k \cdot x] \hat{u}(t, k) \,, \tag{8.7b}$$

where the discrete wave numbers k are

$$k = \sum \left(\frac{2\pi n_j}{L}\right) e_j \,; \; n_j = 0, \pm 1, \pm 2, \ldots \quad \text{and} \quad j = 1 \text{ to } d \,, \tag{8.7c}$$

with e_j the unit vector in the jth direction.

Obviously, because $u(t, x)$ is real, the complex Fourier coefficients and their complex conjugates are related by

$$\hat{u}^*(t, k) = \hat{u}(t, -k) \,. \tag{8.7d}$$

The initial conditions for the Fourier transformed variables are

$$\hat{u}^0(k) = \int_\Omega \exp[-i k \cdot x] u^0(x) \, d^d x \,. \tag{8.8}$$

The temporal evolution of the wave number components $\hat{u}(t, k)$ are an infinite set of ODEs computed as the Fourier transform of the Navier equations (8.1a).

The Galerkin approximations are truncated Fourier expansions and, in particular, for a given integer $N > 0$, we consider the finite collection of complex variables $\hat{u}^N(t, k)$ for wave numbers k, according to (8.7c), but with $n_j = \pm 1$ to $\pm N$, and where the values of $\hat{u}^N(t, k)$ satisfy the reality condition (8.7d). If we assume (without loss of generality) that

$$\hat{u}^N(t, 0) = 0 \text{ for all } t > 0 \,,$$

then the finite system of coupled ODEs for $\hat{u}^N(t, k)$'s is

$$\frac{d}{dt}\left[\hat{u}^N(t, k)\right] = -\nu_0 k^2 \hat{u}^N(t, k)$$

$$+ \left(\frac{i}{L^3}\right) \left[I - \left(\frac{k k}{k^2}\right)\right] \cdot \sum \left[\hat{u}^N(t, k') \cdot k'' \hat{u}^N(t, k'')\right] \,; \tag{8.9a}$$

$$k \cdot \hat{u}^N(t, k) = 0 \,, \tag{8.9b}$$

8.2 Galerkin Approximations and Weak Solutions of the Navier Equations 285

with $\boldsymbol{k}' + \boldsymbol{k}'' = \boldsymbol{k}$ and where the sum over $\boldsymbol{k}'$ extends over the range where both $\boldsymbol{k}'$ and $\boldsymbol{k}''$ have components between $-2\pi N/L$ and $+2\pi N/L$, as in (8.7c), but with $n_j = \pm 1$ to $\pm N$.

The initial conditions for system (8.9a) are truncation of the initial condition for the full problem,

$$\hat{\boldsymbol{u}}^N(0, \boldsymbol{k}) = \hat{\boldsymbol{u}}^0(\boldsymbol{k}) ;$$

$$\text{for } \boldsymbol{k} = \sum \left(\frac{2\pi n_j}{L}\right) \boldsymbol{e}_j ; \quad n_j = \pm 1 \text{ to } \pm N ,$$

and $j = 1$ to d. (8.9c)

In (8.9a), $\mathbf{I}$ is the unit tensor and $[\mathbf{I} - (\boldsymbol{kk}/k^2)]$ is the projector onto divergence-free vector fields in wave number space. Hence, the Galerkin approximation's evolution looks like the transformed Navier equations, but modes with components larger than $2\pi N/L$ are set to zero.

Note that both the divergence-free condition in (8.9b) and the reality relations in (8.7d), for $\hat{\boldsymbol{u}}^N(t, \boldsymbol{k})$, are preserved by the evolution in (8.9a), so that as long as the initial condition satisfies the condition (8.9b) and reality conditions, then (8.9b) and reality conditions for $\hat{\boldsymbol{u}}^N(t, \boldsymbol{k})$ are preserved for all $t > 0$.

Without, the constraints in (8.9b), the Galerkin system (8.9a), are really M coupled ODEs for the complex components of $\hat{\boldsymbol{u}}^N(t, \boldsymbol{k})$ with

$$(d-1)(N^d + dN) = M .$$

Now, the local existence of continuous solutions for $\hat{\boldsymbol{u}}^N(t, \boldsymbol{k})$ is established by verifying the local Lipschitz condition satisfied by the right-hand side of (8.9a). In this case, a time $T > 0$ exists, depending on the magnitude of ν_0, L, N, as well as the magnitude of the initial condition, such that there is a unique continuous solution $\hat{\boldsymbol{u}}^N(t, \boldsymbol{k})$ on $[0, T]$. On the other hand, the global existence for $\hat{\boldsymbol{u}}^N(t, \boldsymbol{k})$ then follows from the observation that $\hat{\boldsymbol{u}}^N(t, \boldsymbol{k})$ is bounded uniformly in T.

This fact follows from the energy equation for the Galerkin approximations obtained from (8.9a) by multiplying by $\hat{\boldsymbol{u}}^{N*}(t, \boldsymbol{k})$, summing over $\boldsymbol{k}$, and adding the complex conjugate,

$$\frac{1}{2}\frac{d}{dt}\left[\sum_{\boldsymbol{k}} |\hat{\boldsymbol{u}}^N(t,\boldsymbol{k})|^2\right] = -\nu_0 \sum_{\boldsymbol{k}} k^2 |\hat{\boldsymbol{u}}^N(t,\boldsymbol{k})|^2 . \tag{8.10}$$

This is analogous to the energy evolutionary equation derived in the continuum case [see (8.4)]. The crucial point here is that the contributions from the nonlinear (quasi-linear) terms all cancel for the Galerkin approximations just as they do for the full exact Navier equation. The nonlinear terms transfer energy (in the form of Fourier coefficient amplitude) between the modes, but they neither create nor destroy energy.

Next, noting that $k^2 \geq (2\pi/L)^2$ in the sum on the right-hand side of (8.10), we easily obtain the following estimate (by applying Gronwall's lemma):

$$\sum_{\boldsymbol{k}} |\hat{\boldsymbol{u}}^N(T,\boldsymbol{k})|^2 \leq \max\left[\sum_{\boldsymbol{k}} |\hat{\boldsymbol{u}}^N(0,\boldsymbol{k})|^2\right]. \tag{8.11}$$

Therefore, a unique continuous solution can be extended over another interval of time $T' > 0$, which itself depends only on

$$\sum_{\boldsymbol{k}} |\hat{\boldsymbol{u}}^N(0,\boldsymbol{k})|^2,$$

at which time the same argument can be applied to the bound

$$\sum_{\boldsymbol{k}} |\hat{\boldsymbol{u}}^N(T+T',\boldsymbol{k})|^2$$

again by the same quantity.

Obviously, this process can be repeated indefinitely, and the global existence and uniqueness of continuous solutions are established [see, Doering and Gibbon (1995, pp. 97–100) for a detailed proof].

Finally, the continuum velocity vector fields reconstructed from the solutions of ODEs (8.9a) with (8.9b),

$$\boldsymbol{u}^N(t,\boldsymbol{x}) := \left(\frac{1}{L^d}\right) \sum \exp[i\boldsymbol{k} \cdot \boldsymbol{x}] \hat{\boldsymbol{u}}^N(t,\boldsymbol{k}), \tag{8.12}$$

for $|k_j| \leq 2\pi N/L$, satisfy the following PDEs:

$$\frac{\partial \boldsymbol{u}^N}{\partial t} + P^N[\boldsymbol{u}^N \cdot \nabla \boldsymbol{u}^N + \nabla p^{*N}] = \nu_0 \nabla^2 \boldsymbol{u}^N, \tag{8.13a}$$

$$\nabla \cdot \boldsymbol{u}^N = 0, \tag{8.13b}$$

with initial conditions

$$\boldsymbol{u}^N(0,\boldsymbol{x}) = P^N[\boldsymbol{u}^0(\boldsymbol{x})], \tag{8.13c}$$

where $P^N[\cdot]$ is the projection operator onto the first N Fourier modes. This projection operator is equivalent to the identity operator when acting on the Galerkin approximation $P^N[\boldsymbol{u}^N(t,\boldsymbol{x})] = \boldsymbol{u}^N(t,\boldsymbol{x})$, and for any projection operator,

$$(P^N)^2 = P^N, \quad \text{and} \quad P^N\left[\frac{\partial f}{\partial x_j}\right] = \frac{\partial}{\partial x_j}\left(P^N[f]\right). \tag{8.14}$$

8.2 Galerkin Approximations and Weak Solutions of the Navier Equations

It is not very hard to show that the total energy in the approximations, $(1/2)\|\boldsymbol{u}^N\|_2^2$, is bounded uniformly in time. It is also uniformly bounded in the order of the approximation N, because we can take the $N \to \infty$ limit on the right-hand side of (8.11),

$$\sum_{\boldsymbol{k}} |\hat{\boldsymbol{u}}^N(t,\boldsymbol{k})|^2 \leq \max \left[\sum_{(\text{all } \boldsymbol{k})} |\hat{\boldsymbol{u}}(0,\boldsymbol{k})|^2 \right] < \infty . \tag{8.15}$$

This observation is the basis of the concept of weak solutions of the Navier equations through Galerkin approximations.

The "weak" solutions of the Navier equations are obtained as a limit of the Galerkin approximations, as $N \to \infty$. These solutions are appropriately called weak because the sense of convergence of the approximations is in a weak topology.

We observe that, in the usual norm topology on a function space, a sequence of functions f_n converges to a limit (denoted f) iff:

$$\|f_n - f\| \to 0 , \quad \text{as } n \to \infty .$$

A weak topology on a functional space is one where a sequence converges iff for every continuous linear scalar-valued function on the space, which we denote by $(\hat{f}, f)$ with $\hat{f}$ a member of the dual space; the sequence obeys $(\hat{f}, f_n) \to (\hat{f}, f)$. Convergence in the norm topology implies convergence in the weak topology, but not the reverse. The central feature of the weak topology used in establishing the existence of limiting weak solutions of the Navier equations is that it can be shown that the Galerkin approximations are in a compact set in an appropriate space of functions with an appropriate weak topology. A compact set enjoys the property that any sequence contained in it has a convergent subsequence.

Hence there is a subsequence, $\boldsymbol{u}^{N_j}(t,\boldsymbol{x})$, whose limit, $\boldsymbol{u}(t,\boldsymbol{x})$, we identify as a solution of the Navier equations. At each instant, the limit is a square integrable vector field. The existence of global weak solutions was first obtained by Leray (1934), where the notion of weak solutions was first introduced. For the details of the functional settings and issues involved in establishing these limits, we refer the reader to the bibliographical notes below.

8.2.1 Some Comments and Bibliographical Notes

The level of the previous analysis does not exceed that in Rudin (1976). Functional space topologies are discussed in Rudin (1973) and Reed and Simon (1980).

We emphasize that we have focused on the periodic boundary conditions in Sect. 8.2; other situations are very different, for example, the relation [see (8.14)], $P^N[\partial f/\partial x_j] = \partial P^N[f]/\partial x_j$, does not generally hold for the projection operator $P^N[\cdot]$!

Modern analyses of Navier (incompressible and viscous) equations from the point of view of applied mathematicians may be found in books by Doering and Gibbon (1995), and P.-L. Lions (1996).

For relatively recent mathematical results on compressible fluid flows, we note, again, the survey paper by Valli (1992) and also the recent book by P.-L. Lions (1998). The two books by P.-L. Lions (1996 and 1998) form a unique and rigorous treatise on various mathematical aspects of fluid mechanics models. The main emphasis in Volume 1 is on the mathematical analysis of incompressible models (Navier equations, including the inhomogeneous case, and Euler equations). Known results and many new results about the existence and regularity of solutions are presented with complete proofs. The discussion contains many interesting insights and remarks. The text highlights, in particular, the use of modern tools and methods and also indicates many open problems. Volume 2 is devoted to new results for compressible models [in fact, isentropic (barotropic) NS equations]. The book by Doering and Gibbon (1995) is an introductory presentation, both physical and mathematical, of Navier incompressible and viscous equations, with an emphasis on unresolved problems on the regularity of solutions in space dimension three. The authors give a complete presentation of both the successes and failures of the methods used for treating these equations.

A final comment concerning the proof of rigorous mathematical results for fluid flows (presented in Sects. 8.3–8.4).

Among the existence and uniqueness theorems, those are obviously preferable that, in the strategy of their proof, suggest methodologies and algorithms for the numerical computation of the solutions and, when possible, give information on the qualitative behavior of the solutions; very good examples of this strategy are the results derived via the Galerkin approximations for viscous fluid flows in the books by Temam (1984) and Doering and Gibbon (1995). Unfortunately, this is not the case for the rigorous results obtained recently by P.-L. Lions through the so-called "compactness" method!

The main mathematical problem is to discover and specify the circumstances that give rise to solutions that persist forever. Only after having done that can we expect to construct proofs that such solutions exist, are unique, and are regular. This problem touches upon the theories of stability, bifurcations, turbulence, singularity, and blowup. Indeed, proofs of global (for all time) nonexistence in the literature do not, in general, imply finite-time pointwise blowup of the solution itself [see the paper by Ball (1997), for instance].

But, if one can couple the global nonexistence argument with a local continuation argument based on the assumption of an appropriate a priori bound for the solution, then in such a case, global nonexistence will imply finite-time blowup.

Concerning, more precisely, the existence and uniqueness problem, one of the fundamental questions that should be answered about any problem of mathematical physics is whether it is well set, that is, whether solutions actually exist and whether they are unique. This question is usually answered by existence and uniqueness theorems, which are results of a rigorous mathematical investigation of fluid dynamics problems (equations with the initial and boundary conditions – so-called "IBVP").

The main (local) existence theorem for NSF equations is the theorem asserting that the Cauchy problem and IBVP are solvable locally with respect to time, i.e., they have a solution in a certain time interval $(0, t^0)$, depending on the data.

The question of the solvability of the above problem on an arbitrary time interval $(0, T^0)$ independent of the data (global existence) is still open, but some new results were recently obtained by P.-L. Lions (1996 and 1998).

However, if we consider one-dimensional viscous compressible flow, then global existence theorems are proved [see, for instance, the review paper by Solonnikov and Kazhikov (1981) and also the book by Antontsev et al. (1983/1990)].

It is important to note that the difference in the recent results of P.-L. Lions, [for example, compared with the Valli (1992) results] is that he has been able to prove an existence theorem for compressible barotropic fluid flows (NS equations), which is global in time, without assuming smallness of the data. On the other hand, uniqueness is an open problem (the solution is "weak"). In some sense, P.-L. Lions has proven [see, for instance, P.-L. Lions (1993a,b)] the analog of the global existence result of Leray (1933, 1934), for the weak solution of the Navier incompressible viscous equation, and of Hopf (1951), who first proved that weak solutions exist in general domains V. The results of Valli (1992) – in the book edited by Rodrigues and Sequeira (1992) – yield the global existence of a "strong" and "regular" solution, but assuming that the data are small enough and in this case the uniqueness holds [see also Valli (1986), concerning global estimates and periodic solutions for barotropic, $p = p(\rho)$, compressible viscous fluid flows]. Concerning the compressible case we mention also the paper by Valli and Zajaczkowski (1986).

The next section is devoted to rigorous mathematical results obtained for Navier incompressible and viscous fluid flows. Sect. 8.4 is devoted to compressible and viscous fluid flows, related to NS and NSF equations. Finally, in Sect. 8.5, we present some concluding remarks.

8.3 Rigorous Mathematical Results for Navier Incompressible and Viscous Fluid Flows

We now want to touch on the existence of solutions of the Navier incompressible, viscous system (8.1a,b). We observe, first, that the viscosity, in principle, makes the problem more regular. For the Navier fluid flow, we can derive an energy inequality [see (8.3)] which gives us H_1 control on the solution $\boldsymbol{u}$. Thus we have enough compactness to obtain a global solution even in 3-D. However, the uniqueness of such a solution is an open problem. See the books by Ladhyzhenskaya (1969), Shinbrot (1973), Temam (1983/1995, 1984, 1988a, 1999, 2000), Wahl (1985), Doering and Gibbon (1995, especially Chaps. 5 to 7) for a classical existence, uniqueness, and regularity theory of Navier fluid flow. In the recent book by Temam and Miranville (2001) the reader can also find a mathematical modeling approach of the continuum mechanics.

In Shinbrot (1973), the reader can find a complete detailed mathematical theory (available at the beginning of the 1970s) for Navier incompressible viscous equations (weak and strong solutions, the existence and uniqueness of a weak solution, the existence and uniqueness of a strong solution, and a reproductive property of the Navier equations).

In particular, if the external force $\boldsymbol{f}$ in the Navier equation has period $\tau = t_1 - t_0$, where t_1 and t_0 are any two values of time t, then the reproductive property obviously implies that the solution is periodic.

For a precise mathematically rigorous definition of classical, weak, and strong solutions of Navier equations, see Shinbrot (1973, pp. 155 and 193) and Temam (1984, §3.6). More recent results are available in the book by Temam (1984) for stationary and nonstationary Navier equations and also for the so-called stationary Stokes equations (see §13), with a given external force $\boldsymbol{f}$ governed by the following equations:

$$-\nu_0 \Delta \boldsymbol{u} + \nabla p = \boldsymbol{f}, \quad \text{and} \quad \operatorname{div} \boldsymbol{u} = 0, \text{ in } \Omega, \tag{8.16a}$$

$$\boldsymbol{u} = \boldsymbol{b}, \quad \text{on } \Gamma = \partial \Omega, \tag{8.16b}$$

where $\boldsymbol{b}$ is a given vector function.

Girault and Sequeira (1992), show that the weighted Sobolev space of Hanouzet (1971) in 3-D and those of Giroire (1987) in 2-D are a good framework for obtaining well-posed variational formulations of steady-state Stokes flow (8.16a,b) in exterior domains, either in primitive variables or in terms of the stream function and vorticity. Naturally, it is necessary to assume a certain decay at infinity by means of weights obtained from Hardy's inequalities. See also Girault and Sequeira (1991). On the other hand, the aim of Pileckas (1996) is to review known results on the Stokes and Navier equations in domains with noncompact boundaries in an aperture domain Ω. This author considers (8.16a) in weighted functional spaces that reflect the decay properties of the solutions. Beirão da Veiga (1997b) gives a simplified version of the L^2-regularity theorem for solutions of a nonhomogeneous system of the Stokes type, where the following equation is considered in place of div $\boldsymbol{u} = 0$:

$$\lambda p + \operatorname{div} \boldsymbol{u} = h, \; \lambda \geq 0 \text{ is a real parameter.} \tag{8.17}$$

For $\lambda = 0$, we obtain the Stokes equation with div $\boldsymbol{u} = h$, where h is a given scalar field. The author also gives a corresponding approximation theorem.

8.3 Navier Incompressible and Viscous Fluid Flows

It is important to note that the common approaches to the Stokes problem require an additional set of nontrivial results, connected to particular functional spaces that are specific to that problem. This fact leaves the Stokes system outside the elementary theory of elliptical differential equations. The proof given by Beirão da Veiga (1997b) does not require any of these particular results (if $h \neq 0$). In this same regard, note that the nonhomogeneous Stokes system (with f and h) can be reduced, as well, to the homogeneous one ($h = 0$). This reduction also requires further specific technical devices due to constraint div $u = 0$.

On the other hand, note that certain inequalities [for example, Leray's inequality (8.3)] supplied the existence of a weak solution and such inequalities, valid if a solution exists and implying its existence, are called a priori inequalities. The strong solution is no different from the weak solution, in that it satisfies an a priori inequality. But, instead of these inequalities, we can use, as in Sect. 8.2, Galerkin's method/approximation, considered as a truncated Fourier expansion, to provide a new proof of the existence of a weak solution of the initial-value problem for the Navier equation [see also, for example, Temam (1984, Chap. II, §1)].

Navier fluid flow is very different when dealing in domains with or without a boundary. In the absence of a boundary, the solution for the 2D Navier initial-value problem can be constructed along the same lines as those for 2-D Eulerian nonviscous fluid flow. The Navier equation for 2-D flow in terms of *vorticity* ω reads

$$\frac{D\omega}{Dt} = \nu_0 \Delta_2 \omega \,, \tag{8.18}$$

and it is necessary to construct the solutions of Navier fluid flow by simply replacing the current lines by the stochastic analog [the above equation for ω can be interpreted as a Fokker–Planck nonlinear type of equation associated with a stochastic equation – for more details, see Marchioro and Pulvirenti (1994)]. Another possibility for constructing Navier fluid flow is to use the splitting method (with L^∞ norm of the vorticity). But, when boundaries are present, the problem is more difficult, and we no longer control the L^∞ norm of the vorticity. Actually, the boundary can produce vorticity, which is a consequence of the interaction of a fluid with the wall. Then, other methods, based on energy inequality, can be used. Heywood (1980a,c) constructed classical solutions of the Navier equations for both stationary and nonstationary boundary-value problems in arbitrary three-dimensional domains with smooth boundaries. For the proof of the principal existence theorem for nonstationary problems, the initial velocity must have merely a finite Dirichlet integral. The solution is constructed, using Galerkin's method with eigenfunctions of the Stokes operator as basis functions.

In a recent paper, Simon (1989/1990) considered the flow of a viscous, incompressible and nonhomogeneous fluid that has a variable density [viscous isopycnic (isochoric!) model]. This flow (which is not Navier flow!) is considered in a domain $\Omega \subset \mathbf{R}^3$ with boundary Γ during a time interval $[0,T]$. The velocity $\boldsymbol{u}$, the pressure p, and the density $\rho > 0$ satisfy the following equations in $\Omega \times]0,T[$:

$$\frac{\partial(\rho\boldsymbol{u})}{\partial t} + \nabla \cdot (\boldsymbol{u}\rho\boldsymbol{u}) - \mu\Delta\boldsymbol{u} = \rho\boldsymbol{f} - \nabla p, \qquad (8.19a)$$

$$\frac{\partial \rho}{\partial t} + \nabla \cdot (\rho\boldsymbol{u}) = 0, \quad \nabla \cdot \boldsymbol{u} = 0, \qquad (8.19b)$$

coupled with the following boundary and initial conditions:

$$\boldsymbol{u} = 0, \quad \text{on } \Gamma \times]0,T[, \qquad (8.20a)$$

$$\rho = \rho^0 \text{ and } \rho\boldsymbol{u} = \rho^0\boldsymbol{u}^0, \quad \text{at } t = 0 \text{ in } \Omega. \qquad (8.20b)$$

This author proves the existence of a global (in time) solution $(\boldsymbol{u},p,\rho)$ for which $\rho\boldsymbol{u}$ satisfies a weak initial condition. For this solution, $\boldsymbol{u}$ and $\rho\boldsymbol{u}$ are not necessarily t-continuous, and $\boldsymbol{u}(t=0)$ and $(\rho\boldsymbol{u})(t=0)$ are not defined. The initial density $\rho^0 > 0$ is not required to have a positive lower bound. But, if in addition ρ^0 is not too small, $\boldsymbol{u}$ is t-continuous at $t=0$ and satisfies the strong initial conditions

$$(\rho\boldsymbol{u})(t=0) = \rho^0\boldsymbol{u}^0 \quad \text{and} \quad \boldsymbol{u}(t=0) = \boldsymbol{u}^0, \qquad (8.20c)$$

then $(\boldsymbol{u},p,\rho)$ is a global strong solution.

Chapter 2 of the book by P.-L. Lions (1996) is also devoted to system (8.19a,b), but with a variable coefficient μ, a continuous function of the density ρ. In this case, in place of the term $\mu\Delta\boldsymbol{u}$ in (8.19a), we have the term, $\nabla \cdot [2\mu\mathbf{D}(\boldsymbol{u})]$, where the components of $\mathbf{D}(\boldsymbol{u})$ are the d_{ij}. Pertinently, P.-L. Lions notices that (8.19a,b) are also adapted to the description of some multiphase flows and in particular to fluid initially surrounded by vacuum and observes that, in these problems, the regularity of the free boundary is a widely open problem.

Unfortunately, this viscous isochoric system of three equations (8.19a,b), when $\mu \neq 0$, is not a consistent degeneracy of the NSF full system of equations (1.30)–(1.32), with (1.33) – at least from the point of view of asymptotic modeling. The change, $DS/Dt = 0$ to $D\rho/Dt = 0$, where S is the specific entropy (and as consequence $\nabla \cdot \boldsymbol{u} = 0$), is asymptotically valid only in the framework of the compressible Euler equations [see, for instance, Sect. 4.6, in Chap. 4, of Zeytounian (2002a)].

In Amann (1994), the motion of a viscous incompressible fluid occupying Ω is considered, and the no-slip boundary condition on $\partial\Omega$ and the initial velocity field $\boldsymbol{u}^0(\boldsymbol{x})$ are assumed to be known. The stress tensor for an incompressible and viscous fluid is specified by a constitutive law as a smooth function of the linear rate-of-deformation tensor. In general, a fluid is said to be Newtonian if the stress tensor is a linear function of the rate-of-deformation tensor [as in (2); see the Introduction], and non-Newtonian

8.3 Navier Incompressible and Viscous Fluid Flows

otherwise. If the given vector field of the body-force density $f = 0$ and initial velocity $u^0 = 0$, the system of equations governing the motion has a trivial zero solution, corresponding to the rest state of the fluid within the Ω, a bounded smooth domain in $\mathbf{R}^3$. According to Amann (1994), the solution is stable under suitably small perturbations of $f = 0$ and $u^0 = 0$, and the perturbed solution decays exponentially to zero, provided this is true for the perturbing body force. It seems that the exponential decay of the perturbed solution to zero in the presence of an outer force field is a new result, even for Navier (Newtonian fluid) classical equations.

For the regularity of stationary Navier equations in bounded domains, we mention, first, Frehse and Rùzicka (1994) and also Frehse and Rùzicka (1995) and Struwe (1995). The series of results obtained in these papers and the techniques used have added considerably to our understanding of Navier equations and represent an important development in mathematical analysis.

Tani and Tanaka (1995) paper is devoted to large-time existence of surface waves in incompressible viscous fluids with or without tension. There are several papers treating the problem considered in this paper – see, for instance, the references in Tani and Tanaka (1995, pp. 305–306). In a more recent paper, Tani (1996) established the small-time existence for the three-dimensional Navier equation of an incompressible fluid with a free surface.

The results obtained by Biagioni and Gramchev (1996) are applied to obtain global strong solutions of the Navier system with appropriate (singular) initial data.

In the absence of a uniqueness theorem for the Navier equations, the properties proved for solutions obtained by different methods could not coincide! Hence, it is interesting to construct solutions with useful "additional" properties, solutions which are (very likely) those that have a physical meaning. In this direction, Beirão da Veiga (1985) studied methods of constructing suitable weak solutions to the Navier equations – basically, a suitable weak solution verifies the local energy inequality (estimate) up to the boundary – other properties required by the definition [see Caffarelli et al. (1982)] follow directly from the equations, if the data are smooth enough. Beirão da Veiga (1995), extended, in a very natural way, the classical Prodi (1959, 1960)–Serrin (1963) sufficient condition concerning the regularity of solutions of nonstationary equations.

The analog of the quasi-hydrodynamic system of Elizarova and Chetverushkin (1986) for a viscous incompressible fluid is derived in Sheretov (1994) by a formal generalization. The kinetic energy balance equation for this mathematical model is obtained, and a theorem for the dissipation of the total kinetic energy of a fluid over time is proved. It is shown that solutions of the Navier system, which describe Couette and Poiseuille flows, are also exact solutions of the new system. The quasi-hydrodynamic equations can be interpreted as a regularization of the Navier equations which is suitable for constructing dissipatively stable numerical algorithms.

It is also known that the actual existence of solutions has been proved for appropriately regularized hydrodynamic equations.

The asymptotic behavior of the Cauchy problem, as $t \to \infty$, for the equations of the dynamics of a heat-conducting fluid is considered in A.V. Glushko and Ye.G. Glushko (1995), and Bernardini et al. (1995) deal with a problem where the Navier equations are coupled with the heat equation. The stability of weak rarefaction waves of systems of conservation laws describing one-dimensional viscous media with strictly hyperbolic flux functions is considered by Szepessy and Zumbrun (1996). Ladyzhenskaya and Solonnikov (1980) considered the problem of determinating solutions of boundary-value problems for steady-state Stokes and Navier equations that have an infinite Dirichlet integral. The paper by Solonnikov (1993a) is devoted to the boundary-value problem with discontinuous boundary data for the Navier equations in 3-D, and the three papers by Solonnikov (1993b, 1994, 1995a) are devoted to the problem with free boundaries.

Heywood (1994) gives some remarks on the possible global regularity of solutions of the 3-D Navier equation and Cottet (1992) paper, which is devoted to 2-D incompressible fluid flow with singular initial data, gives some essential features and details of the concept of concentration which is the key step in the proof of global existence for a positive vortex sheet. In Cottet (1992), the reader can find a result which can be obtained by classical compactness arguments; namely the existence of solutions of the Navier equations for any measure of initial vorticity.

Chapter 6 of Doering and Gibbon (1995) is devoted to the so-called "ladder theorem" for the Navier equations for $\boldsymbol{u}$ and p. For this problem, see also papers by Bartucelli et al. (1991) and with Malham (1993). For a history and further references, see Constantin and Foias (1988) and Temam (1983/1995) book. To derive the ladder theorem for the Navier equations, it is necessary to introduce the idea of seminorms H_N which contain derivatives of the velocity field higher than unity ($N \geq 1$). The function (velocity $\boldsymbol{u}$) bounded in L^2 can still display spatial singularities, and the only general information we have about weak solutions is that the time integral of H_1 is bounded according to Leray's inequality [see, Doering and Gibbon (1995), §5.3].

Constantin (1995b) treats some fundamental problems related to turbulence. The author presents, first, various estimates (in particular, for the rate of dissipation of energy), discusses the validity of the Kolmogorov–Obukhov law, and gives some results on the formation of singularities in solutions of the 3-D Euler equations. It is also interesting to note here the recent paper by Caflisch et al. (1996) concerning the geometry of singularities for steady-state Boussinesq equations. For a pertinent review of this paper, see the Featured Review by Michael Siegel in Math. Rev. 98f: **35**, 121–122. This paper presents the first rigorous analysis (supported by numerical computations) of generic singularity types for the steady-state Boussinesq equations describing 2-D stratified flow. This is an important development in a widely

8.3 Navier Incompressible and Viscous Fluid Flows

followed area, and this paper is pioneering in its use of geometric methods to analyze singularities of partial differential equations arising in problems from fluid dynamics. The phrase "singularity type" refers to the equivalence class of singularities that are preserved under an analytic change of variables. "Generic" means that a singularity is stable with respect to perturbations and that a solution with a singularity type that is not stable may be perturbed into a solution having only stable singularity types. In Caflisch et al. (1996), complex-valued solutions of the steady-state Boussinesq equations are considered, and the singularities are allowed to have complex spatial positions. That is, if $\boldsymbol{x}$ is a suitably chosen spatial coordinate, then the singularity may occur at complex values of x and for complex values of the velocity $\boldsymbol{u} = (u, v)$, density ρ, and vorticity ω.

Finally, most of the results given in the book by P.-L. Lions (1996) are new, and the proof of the new results is presented with full details. The analysis of the new results is not easy. The adaptation of the classical ideas of Leray (1933) to nonhomogeneous flow has been the object of a series of contributions ranging over 30 years, and optimal results for weak solutions in any space dimensions and with density-dependent viscosity were just recently (1993) obtained. As described in the book, the approach relies heavily on the analysis of the weak solution of "conservative" transport equations with singular coefficients considered by DiPerna and P.-L. Lions (1989a,b).

8.3.1 Navier Equations in an Unbounded Domain

There is extensive literature on the regularity theory for the Navier equations [see the bibliography in Heywood (1980a,b,c)]. For the decay of solutions in an exterior domain, as $t \to \infty$, Heywood obtained decay of order $(1/t)^{1/4}$ for the nonhomogeneous problem governing the perturbations that are studied in proving the stability of solutions of the exterior stationary problem. In homogeneous boundary values and forces, decay is of order $(1/t)^{1/2}$ for the nonstationary problem. Finally, in posing problems for the Navier equations in general unbounded domains, Heywood took into account the necessity of setting "auxiliary conditions" appropriate for the domain.

For the existence of the steady-state Navier flow past a body in a plane, we note the paper by Amick (1986). This author considered the problem of finding a solution $(p, \boldsymbol{w})$ of the steady-state Navier equations in an exterior plane domain P:

$$-\nu \Delta \boldsymbol{w} + (\boldsymbol{w} \cdot \nabla)\boldsymbol{w} = -\nabla p \quad \text{and} \quad \nabla \cdot \boldsymbol{w} = 0 \text{ in } P, \quad (8.21\text{a})$$

$$\boldsymbol{w} = 0 \text{ on } \Gamma, \quad \boldsymbol{w} \to \boldsymbol{w}_\infty \text{ as } |\boldsymbol{x}| \to \infty. \quad (8.21\text{b})$$

In 1933, Leray (1933) approached problem (8.21a,b) by solving a sequence of approximate problems in P_R:

an annular bounded region $(|\boldsymbol{x}|^2 < R^2)$,

and proved the existence of a solution $(p_R, \boldsymbol{w}_R)$ with the additional property that

$$\int_{P_R} |\nabla w_R|^2 \, d\Omega_R < \text{const.} \tag{8.21c}$$

Existence and asymptotic behavior for strong solutions of the Navier equations in whole space was obtained in Beirão da Veiga (1987a) – the author was mainly interested in finite energy solutions. The paper by Beirão da Veiga and Secchi (1987) continues the previous study of Beirão da Veiga (1987a).

Borchers and Miyakawa (1992) studied the existence of a weak solution in an arbitrary unbounded domain D, which goes to zero in L^2, as $t \to \infty$, with explicit rates. The L^2-decay problem for Navier flows was first posed by Leray (1933) forn case $D = \boldsymbol{R}^3$. The first (affirmative) answer was given by Kato (1984a) for $D = \boldsymbol{R}^n$, $n = 3, 4$, through his study of strong solutions in general L^p spaces. A different approach was then taken by Schonbek (1985), which is based on the Fourier decomposition of the fluid velocity $\boldsymbol{u}$.

The existence and uniqueness of the stationary plane flow of a viscous and incompressible fluid at a low Reynolds number in exterior domains is considered in the paper by Galdi (1993), and Ben-Artzi (1994) who establishes an interesting existence and uniqueness result for 2-D Navier equations with initial vorticity in L^1. It is shown also in that paper that, as the viscosity vanishes, there is a subsequence of solutions to Navier that converge to a solution of Euler!

The purpose of Brezis (1994) is to show that one technical assumption in Ben-Artzi (1994) [the behavior condition (1.19)] is not necessary for uniqueness. Note that Ben-Artzi (1994) presents a new approach to the existence and uniqueness theory for solutions of the system of unsteady-state Navier equations in $\boldsymbol{R}^2$ that describe the motion of a viscous incompressible fluid in a whole plane with initial condition $\boldsymbol{u}(\boldsymbol{x}, 0) = \boldsymbol{u}^0(\boldsymbol{x})$ and the behavior condition

$$\boldsymbol{u}(\boldsymbol{x}, t) \to |\boldsymbol{x}| \to \infty, \quad \text{for } 0 < t < \infty. \tag{8.21d}$$

This approach is based entirely on elementary comparison principles for linear parabolic equations. First, smooth classical solutions for a very restricted class of initial data are obtained and for this, an iterative method, successively solving a linear equation of a "convection-diffusion" type is used. The results are then extended by continuity, allowing rather general initial functions $\boldsymbol{u}^0(\boldsymbol{x})$. It is also possible to extend the results to measure-valued initial data.

However, the "well-posedness" (in the sense of strong continuous dependence on initial data) is *clearly lost in this extension*!

Ben-Artzi (1994) (see pp. 334 and 335) also presents a short review of existing literature concerning 2-D Navier equations and corresponding limiting (nonviscous) Euler equations.

8.3 Navier Incompressible and Viscous Fluid Flows

The main objective of the paper by Galdi and Sohr (1995) is to show the following result. Let $[\boldsymbol{u} = (u,v), p]$ be a (smooth) solution of steady-state 2-D Navier equations corresponding to $\boldsymbol{f}$ (the negative of the body force), a prescribed vector field of bounded support ($\boldsymbol{f}$ decays sufficiently fast at large distances), that satisfies the condition

$$\boldsymbol{u}(\boldsymbol{x}) \to \boldsymbol{u}_\infty \quad \text{as} \quad |\boldsymbol{x}| \to \infty, \tag{8.22}$$

where $\boldsymbol{u}_\infty = (1,0)$ is a given constant vector, and also has a finite Dirichlet integral

$$\int_\Omega (\nabla \boldsymbol{u} : \nabla \boldsymbol{u}) \, \mathrm{d}\Omega < M, \tag{8.23}$$

where M depends only on the data. If for some $s \in (1, \infty)$ and $r > 0$,

$$\int_{|\boldsymbol{x}|>r} |v(\boldsymbol{x})|^s \, \mathrm{d}x < \infty, \tag{8.24}$$

then $[\boldsymbol{u} = (u,v), p]$ is a physically reasonable (Ph.R) solution. By definition, a Ph.R solution satisfies the condition,

$$\boldsymbol{u}(\boldsymbol{x}) - \boldsymbol{u}_\infty = O\left(|\boldsymbol{x}|^{-1/4-\varepsilon}\right), \tag{8.25}$$

for large $|\boldsymbol{x}|$ and some $\varepsilon > 0$. Existence in the class Ph.R was proved in a general context by Galdi (1993, pp. 1–33). Ω (the region of flow) is a 2-D domain lying in the complement of a compact region (an obstacle problem of steady-state plane flow around an obstacle).

It is clear that (8.23) alone cannot control the convergence of $\boldsymbol{u}(\boldsymbol{x})$ to a constant vector $\boldsymbol{u}_0$ as $|\boldsymbol{x}|$ tends to infinity, and it is easy to find examples of solenoidal vector functions that satisfy (8.23) and grow at large distances. The problem of the coincidence of $\boldsymbol{u}_0$ and $\boldsymbol{u}_\infty$ remains open.

Amick (1988) showed that any (sufficiently smooth) $\boldsymbol{u}(\boldsymbol{x})$ that satisfies the steady-state Navier equations and the condition at the boundary $\partial\Omega$ [on the obstacle $\boldsymbol{u}(\boldsymbol{x}) = 0$ for $\boldsymbol{x} \in \partial\Omega$], with $\boldsymbol{f} = 0$ and satisfying (8.23), is necessarily bounded. Even if these solutions also satisfy (8.22), do they exhibit the basic features expected from the physical point of view? For instance, we expect that they satisfy the energy equation and that, for $\boldsymbol{u}_\infty \neq 0$, the associated flow presents an infinite wake in the direction of $\boldsymbol{u}_\infty$. These properties are related to the asymptotic structure of solutions at large distances. Motivated by this asymptotic structure, various authors introduced the class of Ph.R solutions that satisfy (8.25)! Of course, in 3-D, the velocity field of a Ph.R solution has behavior different from that stated in (8.25). We note here, again, the two recent books of Galdi (1994/1998a,b) treating the mathematical theory of the Navier equation (linearized steady-state problems and nonlinear steady-state problems).

For the decay properties of strong solutions of unsteady-state Navier equations in 2-D unbounded domains, see Kozono and Ogawa (1993). More recently, Padula and Petunin (1994) considered a viscous incompressible heat-conductive, diffusive nonhomogeneous fluid in an exterior domain, in the Oberbeck–Boussinesq approximation. These authors construct a global regular solution with finite kinetic energy and provide a decay rate for the L^2 norm of such a solution. Steady-state Navier equations in a plane aperture domain are considered in the paper by Galdi, Padula and Solonnikov (1996), and it is shown that for a given small aperture flux, the problem has at least one solution that decays at large distances as $|x|^{-1}$. The symmetrical structure is essential for the method used. The uniqueness of weak solutions is also tackled, and an asymptotic expression for the solution at $|\boldsymbol{x}| \to \infty$ is obtained. Galdi et al. (1997) consider a new approach to the existence theory for Navier (incompressible and viscous) equations, using a technique of Kato (1984a). They obtain global existence and convergence to the steady state of a Navier unsteady-state incompressible and viscous fluid flow past an obstacle that is started from rest (the so-called "Finn starting problem"). For the exterior initial boundary problem for Navier equations, see also Shibata (1995). A uniqueness theorem related to nonstationary Navier flow past an obstacle is proved in Heywood (1979). In Heywood (1989), the reader can find a pertinent discussion of the decay of the Navier solution, as $t \to \infty$; the qualitative properties of steady-state Navier solutions in unbounded domains, problems with infinite energy and also proper posing of problems in unbounded domains.

In Bruneau and Fabrie (1996), new efficient boundary conditions for Navier equations and a well-posed result are obtained. They first establish a family of "natural" boundary conditions on the open boundary that lead to well-posed problems for Navier equations and also prove the uniqueness of weak solutions in 2-D. Finally, we mention a second paper by Kozono and Ogawa (1994).

8.3.2 Some Recent Rigorous Results

First, relative to the regularity of solutions of the Navier equations via the truncation method, we mention the two papers by Beirão da Veiga (1997a, 1998). The author is interested in studying sufficient conditions for regularity of solutions – because bounded solutions are regular, the author is concerned with sufficient conditions for boundedness on $\Omega \times (0,T)$, where Ω is an open connected subset of $\boldsymbol{R}^n$, $n \geq 3$. The proof relies on the so-called "truncation method," introduced in DeGiorgi (1957) for studying scalar elliptical equations and developed further by many authors [see, in particular, Ladyzhenskaya et al. (1968)].

In Shananin (1996), some theorems on the "microlocal" regularity of solutions of stationary and nonstationary Navier systems in a domain Ω,

8.3 Navier Incompressible and Viscous Fluid Flows

such that $R^n \supset \Omega$ are presented without detailed proofs. In Sell (1996), the following result is obtained:

The weak solutions of the Navier equations on any bounded, smooth 3-D domain have a global attractor for any positive value of the viscosity, and under added assumptions, this global attractor consists entirely of strong solutions.

The Boussinesq equations for heat transfer in a viscous incompressible fluid are considered in a 3-D bounded domain by Guo and Yuan (1996), and a global weak solution is constructed so that u^2 satisfies a differential inequality in the sense of distributions, where u stands for the velocity vector. The inequality becomes a trivial equality if the solution is smooth enough. In 2-D, it is proved that the weak solution is unique. In a very interesting paper by Necas et al. (1996), the authors give a sketch of the proof that any weak solution to Navier type problem,

$$-\nu_0 \nabla^2 U + aU + a(x \cdot \nabla)U + (U \cdot \nabla)U + \nabla P = 0 , \tag{8.26a}$$

$$\text{div}\, U = 0 , \tag{8.26b}$$

in R^3, that belongs to $[L^3(R^3)]^3$ is trivial; consequently, there are no singular solutions to the nonstationary Navier equations of the Leray form in R^3:

$$u = \frac{1}{(2a(T-t))^{1/2}} U\left(\frac{x}{(2a(T-t))^{1/2}}\right) . \tag{8.27}$$

The investigations of Zhi-Min Chen and Price (1996) show that time-dependent, periodic, 2-D Navier flow on a 2-D torus, excited by a prescribed spatially sinusoidal external force $(4k^2 \sin 2k, 0)$, (the so-called "Kolmogorov problem"), exhibits $(3k)^{1/2}$ Hopf bifurcation[1] and when the Reynolds number increases through each of these bifurcation values, a periodic flow arises from the steady-state solution. Furthermore, no second step bifurcation was observed in the flows as Reynolds number varies.

He Cheng (1997) constructs a class of weak solutions to the Navier equations, which are second-order spatial derivatives and first-order time derivatives, of pth-power summability for $1 < p \leq 5/4$.

Among the more recent rigorous investigations devoted to Navier flow, we mention the following: Ross (1997) concerning Navier approximations in exterior domains; Miyakawa (1997) on L^1 stability of stationary Navier flows in

---

[1] The Hopf bifurcation plays an important role in describing the behavior of nonlinear phenomena that occurs in the theory of dynamical systems [see, for example, Marsden and McCracken (1976), Guckenheimer and Holmes (1983)]. In Chap. 10 devoted to a phenomenological approach to finite-dimensional dynamical systems, the reader can find a discussion about the various types of bifurcations in dissipative dynamical systems and associated scenarios (routes) to chaos.

R^n with $n \geq 3$; H. Kato (1997) devoted to the existence of periodic solutions of the Navier equations; Guo LiHui (1997) concerning a finite dimensionality property for planar Navier equations; Cho Chung Ki and Choe Hi Jun (1997) related to the L^∞ bound of weak solutions of Navier systems; Chae Dongo and Nam Hee-Seok (1997) concerning the local existence and blowup criterion for Boussinesq equations; Germond (1997) related to a second-order convergence result for the approximation of Navier equations by an incremental projection method; and Beirão da Veiga (1997b) remarks on the smoothness of the $L^\infty(0, T; L^3)$ solutions of 3-D Navier equations.

Finally, we mention two papers published in 1998. The first is by A.V. Glushko (1998), where existence and uniqueness theorems are established for the solution of the initial boundary-value problem in the half-space for a system of linearized dynamic equations for a viscous exponentially stratified fluid. It is shown that the problem is well-posed in special functional spaces. The second is by Planchon (1998), where the author constructs global solutions to the Navier equations with initial data small in a Besov space [see, for instance, Peetre (1976), Triebel (1983) and Cannone (2000)]. Under additional assumptions, the author shows that they behave asymptotically (for large time) like self-similar solutions. Of course, the limiting solutions are invariant under the scaling, so

$$v(x, t) = \frac{1}{t^{1/2}} V\left(\frac{x}{t^{1/2}}\right). \tag{8.28}$$

Such self-similar solutions were studied previously by Cannone and Planchon (1996) and also by Giga and Myakawa (1989). In Cannone and Planchon (1996) a rigorous proof of the previous heuristic approach is given.

8.4 Rigorous Mathematical Results for Compressible and Viscous Fluid Flows

For barotropic, compressible, viscous fluid flows (governed by the NS equations), when

$$p = p(\rho) \quad \text{and} \quad \mu > 0 \quad \text{and} \quad \mu_v > 0, \tag{8.29a}$$

we have a global in time existence theorem that has been proven by Valli (1983), extending some ideas of Matsumura (1981) and Matsumura and Nishida (1981, 1982). If we assume for simplicity in this barotropic/isentropic case that

$$\mu = 1, \; \mu_v > \frac{2}{d}, \; \rho^0(\boldsymbol{x}) = \rho_0 = 1 \quad \text{and} \quad p(\rho) \equiv \rho, \tag{8.29b}$$

and consider the new unknown,

$\sigma(t, \boldsymbol{x}) = \rho(t, \boldsymbol{x}) - 1$, which satisfies $\int_\Omega \sigma \, \mathrm{d}\Omega = 0$, (8.29c)

then we can rewrite the starting NS equations in the following form:

(a) $\dfrac{\partial \boldsymbol{u}}{\partial t} - \Delta \boldsymbol{u} - \nabla \operatorname{div} \boldsymbol{u} + \nabla \sigma = \boldsymbol{f} + \boldsymbol{H}$ in $Q_\infty :=]0, +\infty[\times \Omega$,

(b) $\dfrac{\partial \sigma}{\partial t} + \operatorname{div} \boldsymbol{u} = L$ in Q_∞,

(c) $\mathrm{H} := -(\boldsymbol{u} \cdot \nabla)\boldsymbol{u} - \dfrac{\sigma}{(1+\sigma)}[-\nabla \sigma + \Delta \boldsymbol{u} + \nabla \operatorname{div} \boldsymbol{u}]$,

(d) $L := -\sigma \operatorname{div} \boldsymbol{u} - \boldsymbol{u} \cdot \nabla \sigma$.

It is possible to eliminate the contributions of terms $\nabla \sigma$ and $\operatorname{div} \boldsymbol{u}$ by multiplying (a) by $\boldsymbol{u}$ and (b) by σ and integrating in Ω. Adding the two equations, from

$$\int_\Omega \sigma \operatorname{div} \boldsymbol{u} \, \mathrm{d}\Omega = -\int_\Omega \nabla \sigma \cdot \boldsymbol{u} \, \mathrm{d}\Omega,$$

one obtains [here we will indicate each equivalent norm in $H^k(\Omega)$ by $\|\cdot\|_k$]:

$$\frac{\mathrm{d}}{\mathrm{d}t}[(\|\boldsymbol{u}\|_0)^2 + (\|\sigma\|_0)^2] + (\|\nabla \boldsymbol{u}\|_0)^2 < c(\|\boldsymbol{f}\|_{-1})^2 + \mathrm{NL},\quad (8.30)$$

where "NL" indicates some norms related to nonlinear terms (which we expect to be "good"; they will be controlled by the linear terms because we are assuming small initial data). This procedure can be repeated for $\partial \boldsymbol{u}/\partial t$ (and for "tangential" and "interior" derivatives of $\boldsymbol{u}$ up to order two) in all of these cases, the boundary conditions permit integration by parts. The reader can find the details of the derivation of "good" estimates for the nonlinear terms in Valli (1983). But nonbarotropic (baroclinic) compressible fluid flows require some additional estimates which permit "enriching" the energy and dissipative terms. In a more recent paper by Zajaczkowski (1994), the motion of a viscous compressible fluid in $\boldsymbol{R}^3$, bounded by a free surface that is under surface tension and constant exterior pressure, is considered. Assuming that the initial density is sufficiently close to constant, the initial domain is sufficiently close to a ball, the initial velocity is sufficiently small, and the external force vanishes, the existence of a global, in time, solution is proven, which satisfies, at any moment, the properties prescribed at the initial moment.

Here, we mention, again, the review paper by Solonnikov and Kazhikov (1981), where, for viscous fluids, the existence theory available at the beginning of the 1980s is presented together with a detailed bibliography, and also Valli (1992) concerning the existence of global solutions of the equations that describe the motion of barotropic, compressible, viscous fluid flow. The existence of stationary solutions of NS equations is considered in Valli (1987). We note also that, in particular, the uniqueness theorem has been proven by Valli (1981) and Itaya (1976), and in Valli (1992), the reader can find a detailed bibliography of mathematical results for compressible flows. In the short Note of P.-L. Lions (1993b), the author proves some compactness properties of sequences of solutions of compressible, isentropic NS equations. These results are one of the key ingredients of the existence proof of global solutions by the author [see, for instance, P.-L. Lions (1993a)]. The paper by Kazhikov (1994) concerns "some new statements for IBVPs for NS equations of viscous gas." In this survey paper, the author considers three different topics for compressible viscous fluids: the free boundary problem, viewed as a unilateral boundary value problem with nonlocal restrictions; a control problem for the one-dimensional Burgers equations; and a linearized problem obtained by means of the usual Stokes approximation of the momentum equation. For the first problem, the existence and uniqueness of the solution are obtained via a priori estimates and a penalty argument. The existence of a solution of the second problem is shown, when the cost functional is connected with the final state at $t = T$, and the control is given by the initial datum. Finally, the uniqueness of the solution of the third problem is proven if the density ρ is bounded and strictly positive, and in that case, the existence of strong solutions is also shown. Padula (1994) surveys the mathematical theory of compressible NS equations for 3-D steady-state and unsteady-state motion mainly for barotropic fluid flows. The author discusses the existence both in bounded domains and exterior domains with asymptotics at space infinity. Although some of the results discussed are known, their presentation demonstrates that the decomposition method has several advantages. This paper also studies the existence problem of nonstationary problems as well as the stability of thermally conducting fluids between horizontal layers – again the author applies the decomposition method. As more recent results, we mention, first, a recent, complete paper by Novotny and Padula (1994); in this paper, 3-D steady-state flow, uniform at infinity, of a viscous compressible fluid past a fixed obstacle is considered, and the authors solve the fundamental questions of existence, uniqueness, and asymptotics for permanent exterior flow of an ideal viscous gas in thermal equilibrium (the authors consider only isothermal motion, prove the existence and uniqeness of solutions in L^p spaces, $p > 3$, and study their regularity as well as their decay at infinity). Such a basic problem has attracted the attention of many authors, and more than a hundred papers have been written on approximate numerical computation, but the well-posedness questions from a rigorous mathematical point of view have been considered only by a few authors. In particular, for the potential flow of inviscid fluids, these questions were studied in the middle of the twentieth century [see, for instance, the review paper by Serrin (1959a), for references cited therein]. Such severe assumptions on the flow may be required because the vorticity of the flow far from the obstacle is expected to be almost zero (an exponential decay rate can be proved for incompressible viscous fluids). Note that, for motions in bounded domains, a series of important advances in the mathematical theory of viscous steady-state flow of compressible fluids has been made during the past few years.

For unbounded domains, we quote [according to Novotny and Padula (1994)] the existence result obtained in an exterior domain when the value of the velocity at infinity $\boldsymbol{u}_\infty$ is zero and that of Padula (1993), in the whole of $\boldsymbol{R}^3$ for $\boldsymbol{u}_\infty$ sufficiently small. We note that systems considered in Novotny and Padula (1994), which are important for compressible fluids, are also suitable for the Stokes and Oseen equations and Navier incompressible viscous fluids. From the physical point of view, such equations are a suitable linear approximation of the full system when the whole (Stokes-like) or a part (Oseen-like) of the convective term is disregarded in the momentum equation only. This means that the inertial term is much smaller than that due to friction [slow motion/large viscosity]. However, it is possible to apply the process of proof to heat-conducting fluids as well as to consider the small nonzero boundary condition on $\partial \Omega$ (Ω is an domain exterior to some compact region). In a more recent paper by Novotny and Padula (1997), compressible isothermal flows past a moving 3-D body were considered. The existence and uniqueness of solutions satisfying certain decay conditions at infinity (Ph.R. solutions) are proved. The results are physically meaningful. In computational fluid dynamics, there has been a great deal of discussion about the boundary conditions to be applied at so-called far-field boundaries. These boundaries are those introduced by truncating an unbounded domain to a bounded domain. For (compressible) NS equations, the boundary conditions that have frequently been employed have been derived from the Euler equations, that is, the viscous terms in the NS equations are ignored! This creates the puzzling situation that the NS equations with these boundary conditions are apparently an ill-posed problem, but the computations frequently yield good results. In Nordström (1995), the author derives a set of boundary conditions that are well-posed with the NS equations and are numerically useful. Nordström (1995) also contains a good discussion of the theory of well-posed boundary conditions, and numerical results are presented indicating the usefulness of the proposed boundary conditions. Hoff (1997) proves the global existence of a weak solution of the NSF equations (for compressible, viscous, and heat-conducting fluid flow) in 2-D and 3-D, when the initial density is close to constant in functional space, $L^2 \cap L^\infty$, the initial temperature is close to constant in L^2, and initial velocity is small in $H^s \cap L^4$, where $s = 0$, when $n = 2$, and $s > 1/3$, when $n = 3$ (the L^p norms must be weighted slightly when $n = 2$), and see also Hoff (1995). In Hoff (1996), the reader can find a discussion about some recent global existence results for a weak solution of the NS isentropic flow in 2-D and 3-D when the Cauchy data are small but possibly discontinuous. The results discussed in this last paper extend the earlier work [Hoff (1995)] of the author. For new regularity results for 2-D flow of multiphase isochoric (density-dependent but incompressible) viscous fluids, we mention the paper by Desjardin (1997a). In Desjardin (1997b), there are two main results. First, better regularity of spatially periodic solutions of isentropic compressible NS equations have shown as long as the density is bounded, and second, a partial uniqueness result for the same problem is proved, namely, any weak solution with bounded density must be equal to a strong solution if it exists.

In Serre (1986, 1991), the reader can find some rigorous results on viscous and compressible fluid flows. We note that the Valli (1983) results have been extended to the complete NSF system of equations by Valli and Zajaczkowski (1986), where some results on inflow–outflow problems – when we have the condition $(\boldsymbol{u}\cdot\boldsymbol{n})|_{\partial\Omega} \neq 0$, can also be found. In Valli (1986), the reader can find some results concerning global estimates and periodic solutions for barotropic $[p = p(\rho)]$ compressible viscous fluid flows. The mathematical theory of NS equations is also considered in Padula (1993, 2000). The recent paper by Njamkepo (1996) concerns a global existence result for a one-dimensional model in gas dynamics, a viscous polytropic gas with internal capillarity and higher viscosity (bipolar fluids) under periodic conditions. By linearization, the author constructs a weak solution on a possibly short time interval. For zero external force, the proof of the global in time existence and uniqueness of the solution is given. In Vaigant and Kazhikov (1997), the reader can find an existence result of global solutions of 2-D NS equations for barotropic plane-parallel nonstationary motion, which is periodic in space variables. A new element of this work is the additional assumption that the coefficient of bulk viscosity depends on pressure (or on density) and grows polynomially at infinity. The recent paper by Kobayashi Takayuki (1997) is related to the local energy decay of higher derivatives of solutions for the equations of motion of compressible viscous and heat-conductive gases in an exterior domain in $\boldsymbol{R}^3$. The 2-D steady-state problem for a viscous, isothermal, compressible fluid in an exterior domain was investigated in the recent paper by Galdi, Novotny and Padula (1997). Nonzero velocity is assumed at infinity, and existence with uniqueness are proved for small data. Proving existence and uniqueness for a corresponding linear problem, the authors decompose this last problem into three: the Neumann problem for the Poisson equation, the nonhomogeneous Oseen problem, and a boundary-value problem for the transport equation; sharp estimates are proved for these problems.

Finally, in the more recent paper by Zhouping Xin (1998), a sufficient condition on the blowup of smooth solutions of the compressible NS equations in arbitrary space dimensions with initial density of compact support is presented.

As an immediate application, it is shown that *any smooth solutions will blow up in finite time as long as the initial densities have compact support and an upper bound, which depends only on the initial data.*

Another implication is that *there are no global small (decay in time) or even bounded (when all viscosity coefficients are positive) smooth solutions of the NS equations for compressible polytropic fluids, no matter how small the initial data are, as long as the initial density has compact support.*

This is in *contrast* to the classical theory of the global existence of small solutions of the same system, where initial data are a small perturbation of a constant state that is not a vacuum! The blowup of smooth solutions of the compressible Euler system with initial density and velocity of compact support is a simple consequence of the above result. The key idea in the above analysis is the faster decay of the total pressure in time in the presence of vacuum [for compressible Euler equations, the faster decay of total pressure was first studied by Chemin (1990)].

8.4.1 The Incompressible Limit

Here, we consider only the incompressible limit from the point of view of applied mathematicians, who are mainly interested in a rigorous justification of the formal derivation of the Navier equations for an incompressible and viscous fluid from the NS (barotropic!) compressible fluid flows equations.

Viscous stationary compressible fluids were studied in Beirão da Veiga (1987b, 1994b), and viscous nonstationary compressible fluids were studied in Klainerman and Majda (1982), Kreiss et al. (1991), Bessaïh (1992), P.-L. Lions (1993c), and in recent works by Hagstrom and Lorenz (1998) and P.-L. Lions and Masmoudi (1998). For viscous compressible fluids, these authors considered only equations for isentropic (barotropic) flow of a polytropic gas.

For example, Beirão da Veiga (1994b) studied the behavior of the solution $(\rho, \boldsymbol{v})$ of the following compressible barotropic Cauchy problem (8.31):

$$\frac{\partial \rho}{\partial t} + \boldsymbol{v} \cdot \nabla \rho + \rho \nabla \cdot \boldsymbol{v} = 0 \,,$$

$$\rho \left[\frac{\partial \boldsymbol{v}}{\partial t} + (\boldsymbol{v} \cdot \nabla) \boldsymbol{v} \right] + \lambda^2 \left(\frac{\mathrm{d}p}{\mathrm{d}\rho} \right) \nabla \rho = \nu \Delta \boldsymbol{v} + \mu \nabla (\nabla \cdot \boldsymbol{v}) \,,$$

$$p(\lambda, \rho) = \lambda^2 p(\rho) \,,$$

$$\rho(0, \boldsymbol{x}) = 1 + \rho^0(\boldsymbol{x}) \,, \quad \boldsymbol{v}(0, \boldsymbol{x}) = \boldsymbol{v}^0(\boldsymbol{x}) \,; \tag{8.31}$$

as (simultaneously) the (Mach) number $(1/\lambda)$ goes to zero, the constant viscosity ν converges to a value $\nu^* \geq 0$, μ stays bounded, and $(\rho^0, \boldsymbol{v}^0)$ converges to $(0, \boldsymbol{u}^0)$. The limit Navier equations are the equations of motion of an incompressible viscous fluid for $(\boldsymbol{u}, \pi)$ with density $\rho^* \equiv 1$ and viscosity $\nu^* \geq 0$, i.e., (8.32):

$$\nabla \cdot \boldsymbol{u} = 0 \,, \quad \frac{\partial \boldsymbol{u}}{\partial t} + (\boldsymbol{u} \cdot \nabla)\boldsymbol{u} + \nabla \pi = \nu^* \Delta \boldsymbol{u} \,,$$

$$\boldsymbol{u}(0, \boldsymbol{x}) = \boldsymbol{u}^0(\boldsymbol{x}) \,, \quad \nabla \cdot \boldsymbol{u}^0 = 0 \,, \tag{8.32}$$

for some pseudo-pressure $\pi(t, \boldsymbol{x})$!

According to Beirão da Veiga (1994b, pp. 314 and 315), it is important to note that the Cauchy problems (8.31) and (8.32) can be considered as a dynamical system with the solutions,

$$[\rho, \boldsymbol{v}](t) \quad \text{and} \quad [1, \boldsymbol{u}](t) \,,$$

which describe continuous trajectories in the Hilbert space H^k, the data space. Hence, the natural and optimal result is to prove that trajectories converge to the limit trajectory in the H^k norm, uniformly with respect to time! This is a significant accomplishment in the theory of the "incompressible limit of compressible fluid motion."

In a more recent paper Hagstrom and Lorenz (1998) consider the system describing unsteady-state, isentropic, compressible flow of a viscous polytropic gas [analogous to (8.31)] in two space dimensions (x, y) and under 2π-periodic solutions in x and y, and show that the solution of the limit (Navier incompressible, viscous) equations (8.32) remain smooth for all time, if the *Mach number is sufficiently small* and the initial data are *almost incompressible*, but is not assumed that the initial data are small. To the leading-order, the solution consists of the corresponding incompressible flow plus a highly oscillatory part describing sound waves. But it is not clear if the above result generalizes to Euler flow ($\nu = \mu = 0$). The limit incompressible Euler solution ($\boldsymbol{u}_\mathrm{E}, \pi_\mathrm{E}$) is

C^∞ for $0 \leq t < \infty$, but generally does not tend to zero!

A generalization of this result to 3-D flow is straightforward, however, if we assume that the initial incompressible velocity field $\boldsymbol{u}^0(\boldsymbol{x})$ leads to a classical solution $(\boldsymbol{u}, \pi)$ for $0 \leq t < \infty$, which – together with all derivatives – tends to zero as $t \to \infty$ [exponentially; see Kreiss and Lorenz (1989, Chap. 9)].

Finally, concerning the paper by P.-L. Lions and Masmoudi (1998), which extends (somewhat unprecisely) the results announced in P.-L. Lions (1993c), we note first, that P.-L. Lions developed a multidimensional theory of weak solutions for an isentropic system and established connections with the theory of weak solutions for incompressible flow.

More of these results are global in time and without size restriction on the initial data. In P.-L. Lions and Masmoudi (1998), the starting compressible insentropic viscous system of equations is the same as in Hagstrom and Lorenz (1998). The more recent paper is Desjardins et al. (1999), where the incompressible limit is considered for solutions of isentropic (NS) equations with Dirichlet boundary conditions.

The goal of the P.-L. Lions and Masmoudi (1998) work is to justify completely(?) the formal derivation (from a fluid mechanician's viewpoint) of incompressible viscous (or inviscid) models and, more precisely, to pass to the limit in the global weak solutions of compressible isentropic Navier–Stokes equations, whose existence was recently proven by P.-L. Lions (1993b, 1998). An important step (in the three papers mentioned) in the proof of the convergence to incompressible model (8.32), is the fact that ρ goes to 1 when the Mach number tends to zero! Unfortunately, this is not the case when we consider, as in Zeytounian and Guiraud (1980a,b), unsteady-state fluid flow within a cavity that has a time-deformable wall (see, for instance, Sects. 6.1 and 6.5 in Chap. 6 devoted to low Mach asymptotics of NSF equations).

The dependence of solutions of the equations of motion of compressible, viscous fluids on the Mach number (M) and on the viscosity coefficients, when $M \to 0$ [slightly compressible fluids or low Mach number (hyposonic) flows], is a very interesting but difficult problem and is strongly dependent on the considered fluid motion (internal or external, atmospheric) via boundary and initial conditions and on Coriolis and gravitational forces! The reader can find various references in Zeytounian and Guiraud (1984) and Zeytounian (1994, 2000a, 2002a), where a fluid dynamics approach to low Mach number flow is considered from the point of view of asymptotic methods, as in Chap. 6.

8.5 Some Concluding Remarks

Before all, we note, again, that the review paper by Gresho (1992) related to numerical simulations of Navier incompressible fluid flows and the recent *Handbook* of computational fluid dynamics, edited by R. Peyret (1996), give a complete view of so-called numerical fluid dynamics. Now, if we restrict ourselves to a Galerkin approximation of 2-D Navier equations based on the spaces V_m, i.e., the family w_j of eigenfunctions of the associated Stokes (linear) problem, then, if m is sufficiently large, the behavior of the Galerkin approximation u_m is completely determined by the behavior, as $t \to \infty$, of a certain number m^* of its mode. For the Galerkin approximations, it is also interesting to note that important properties of these approximations can be established with ease when they satisfy the energy balance equations. This is one more reason for looking for such approximations in fluid flow problems, and they can be constructed in many cases. For the classical Bénard convection problem, Treve (1983) proves that there is a unique way of constructing Galerkin approximations of any order such that the energy balance equations are satisfied exactly and gives a simple proof of the boundedness of the approximations so constructed, indicates how bounds can be obtained for quantities of physical interest such as the Nusselt number, and demonstrates the global asymptotic stability of the purely conductive solution for subcritical Rayleigh numbers. We mention also that Galerkin (spectral) methods are explored for the numerical simulation of incompressible flow within simple boundaries in Orzag (1971).

Foias and Saut (1986) show that, for the evolutionary Navier equations, the ratio of enstrophy to energy has a limit, as $t \to \infty$, which is an eigenvalue of the associated Stokes operator. This fact has direct consequences for the global structure of the Navier equations. This allows us to construct a flag of nonlinear spectral manifolds of the Navier operator in the space R of initial data. These are analytic manifolds of finite dimension that are invariant for the Navier equations and completely determine the energy decay of the solution. On the other hand, Ghidaglia (1986) was motivated by the study of attractors and of the long-time behavior for infinite-dimensional dynamical systems arising from fluid mechanics [namely, Navier equations, magnetohydrodynamic equations, and thermohydraulic equations], from combustion [Kuramoto–Sivashinsky (KS) equations], and from optics [the nonlinear Schrödinger (NLS) equation].

In conclusion, note that the point of view of a fluid dynamician, to well-posedness of fluid dynamics problems, is often very different from the point of view of applied mathematicians. Most (if not all) numerically computable solutions of fluid dynamics IBVPs should be well-posed – naturally, the "numerical" definition of a well-posed problem is too strict and that we should permit weaker, generalized solutions that do not exact so much smoothness – especially (for the Navier equation) with regard to the constraints that result from $\nabla \cdot \boldsymbol{u} = 0$ in $\Omega + \partial\Omega$ for $t > 0$. If the starting formulated fluid flow problem, after computation through an effective and stable numerical code, leads to a good realistic fluid flow configuration, which is coherent with the experiences and results of measurement techniques, then this starting fluid flow problem is (almost always?) well-posed from a fluid dynamician's/numerician's point of view!

We recognize that the so-called "mathematical topics in fluid dynamics" have remained closed to the mainstream of applied mathematics and mathematical physics due in large part to the technical nature of the "rigorous" investigations, often phrased in the unfamiliar language of abstract functional analysis. Doering and Gibbon (1995), on the contrary, present some rigorous mathematical techniques and results of these solvability studies in a more familiar context, explaining and developing the tools progressively.

Nonlinear phenomena are so numerous and varied that one cannot expect any one subject, such as fluid dynamics, to exemplify them all! But, because of the complexity and variety of fluid dynamic phenomena and the (relative!) simplicity and exactitude of the governing equations, special depth and beauty are expected in the mathematical theory. Thus, it is a source of pleasure and fascination that many of the most important questions in the theory remain yet to be answered and seem certain to stimulate contributions of depth, originality, and influence, far into the future – in Heywood (1989), the reader can find a discussion of some of these questions. Finally,

8.5 Some Concluding Remarks

we mention the short (but pertinent) relatively recent (and last) paper by Jean Leray (1994), concerning some aspects of theoretical (mathematical) fluid mechanics.

In selecting, citations and references in Sects. 8.2 to 8.4, we have included those that we judged provide useful additional information, but these are not exhaustive and cannot provide a balanced assessment of the scholarship of the many scientists and mathematicians who have contributed to the subject. After perusing this chapter, the reader might be disappointed by the absence of any details related to the rigorous mathematical results cited! I think that several pertinent books exist where the reader can find these details [see, for instance, books cited before by Shinbrot (1973), Temam (1984), Doering and Gibbon (1995), and P.-L. Lions (1996)]. In the recent book by P.-L. Lions (1998), the reader can find new results concerning the existence and uniqueness of compressible viscous fluid flows and many references. We mention, here, also some more recent references: Feireisl and Petzeltova [(1998) who consider the steady-state solutions to the NS compressible equations]; Masmoudi [(1998) who considers the Euler, zero-viscosity limit of Navier incompressible equations with Coriolis force, and takes into account the Ekman layer – unfortunately, the formulation of the starting problem is much too simplified!]; Nazarov and Pileckas [(1998) who study the asymptotics of the solutions of steady-state Stokes and Navier problems in domains with paraboloid outlets (which are a case intermediate between cylindrical and conical outlets) to infinity]; Hishida [(1999) who proves an existence theorem for 3-D unsteady-state Navier flow in the exterior of a rotating obstacle]; Feireisl et al. [(1999) who prove the existence of weak time-periodic solutions of the NS equations for isentropic flows in 3-D when the external force is time-periodic with a positive period]; and Feireisl and Petzeltova [(1999) who investigate the large-time dynamics of weak solutions of the NS equations for isentropic compressible flow confined to a domain $\Omega \subset R^3$ and $t \in (0, \infty)$]; Duchon and Robert [(1999) who study the local equation for weak solutions of 3-D incompressible Euler and Navier equations and give a simple proof of Onsager's conjecture)]; Socolescu [(2000) who shows that the asymptotic behavior of a smooth function (velocity solution) with a bounded Dirichlet integral in an exterior domain (for steady-state 2-D Navier equations) is controlled by the asymptotic behavior of its first Fourier coefficient (asymptotic mean value)]; Gallagher et al. [(2000) who study the existence and uniqueness of axisymmetric solutions of 3-D unsteady-state Navier equations with no-slip and axisymmetric initial conditions]; Chacon Rebollon and Guillén Gonzalez [(2000) who prove, by intrinsic analysis, an existence result for the hydrostatic approximation of NS equations, but written in a very simplified form(!), according to two papers by J.-L. Lions et al. (1992, 1994)]; Feireisl [(2000a) who proves the existence of a compact global attractor for NS isentropic equations in 3-D by the approach of Sell (1996)]; Golse and Levermore [(2001) who show that renormalized solutions of the Boltzmann equation in some appropriate scaling converge to a unique limit governed by the Stokes and heat equations, as the Knudsen number vanishes [see, for instance, Chap. 1 in Zeytounian (2002a)]; and recently Golse and Saint-Raymond (2001) have considered again the NS limit for the Boltzmann equations; Caltagirone and Vincent [(2001) who consider a tensorial penalization method for solving the full NSF equations, when the viscous stress is decomposed in viscosities of compression, elongation, shearing, and rotation, and a numerical solver is used – several numerical applications being proposed]; and Gallagher and Planchon [(2001) who consider the infinite energy solutions to the NS equations]. Finally, in the book edited by Málek et al. (2000), the reader can find six survey contributions that focus on several open problems of theoretical fluid mechanics for both incompressible and compressible fluids; in particular, we mention the paper by Feireisl (2000b, pp. 35–66) concerning the dynamical systems approach to NS isentropic equations of compressible fluids and the paper by Masmoudi (2000, pp. 119–158) relative to asymptotic problems and compressible–incompressible limit. See also Málek et al. (1999).

The discussion, comments and references cited in this chapter are an "invitation" to further reading! Today, despite tremendous (70 years later, the fundamental Leray paper) progress in many mathematical aspects of fluid flows theory, it is necessary to be just a little more modest, if we have in mind the (large) number of problems that still remain open! For instance, we have not determined whether a solution that is initially smooth can develop a singularity at some (finite!) later time or whether singularities are a fundamental feature of turbulence!

On the other hand, unfortunately, the fluid dynamics problem considered by the Applied Mathematicians, in most cases, at the time of their rigorous analysis are very simplified and do not take into account the initial and boundary conditions which correspond to physics of the problem itself. Indeed, we have the impression that often the fluid dynamics problems are considered as a formal application for the various abstract mathematical analysis, and the relation between the physics of the problem and the mathematical assumptions (inequalities, space, norme, ...), which are necessary for the obtention of rigorous results, is do not clear!

9 Linear and Nonlinear Stability of Fluid Motion

In this chapter, we investigate some qualitative and quantitative properties of solutions of fluid dynamics equations (mainly Navier equations), give some results for the stability of various flows, and discuss some instabilities. First, Sect. 9.1 is devoted to a general discussion of different (qualitative) aspects of the linear and nonlinear stability theory of fluid motion, and in Sect. 9.2, the fundamental ideas concerning the linear and nonlinear stability of fluid motion are briefly presented. Using a technique of the Lyapunov–Schmidt type derived from bifurcation theory and perturbation methods, coupled with multiple scales technique, we present in Sect. 9.3, a unified approach to nonlinear hydrodynamic stability. In Sect. 9.4, we consider some facets of the Rayleigh–Bénard (RB) convective instability (buoyancy effect) problem and the Bénard–Marangoni (BM) thermocapillary instability (deformable free-surface effect) problem. In Sect. 9.5, we discuss some features of nonlinear Couette–Taylor viscous flow between two rotating concentric (infinite) cylinders. Finally, in Sect. 9.6, some concluding remarks are given relative to the stability of viscous fluid flows.

9.1 Some Aspects of the Theory of the Stability of Fluid Motion

The theory of stability poses, first, quite a natural question:

Given an evolutionary fluid dynamic equation, we want to know whether a small perturbation of the initial conditions produces effects on the solution that are uniformly small in time!

This question is obviously directly related to the third (3) condition concerning the continuous dependence on initial conditions in the (Hadamard) definition of a well-posed fluid flow problem considered in Chap. 9 of Zeytounian (2002a).

Unfortunately, in general, the perturbations grow exponentially in time, and the main problem is to seek the conditions for which this does not happen! But, it is necessary to understand, first,

what is stability?

Consider, as a simple example, a fluid in a basic state of steady-state motion. Now, suppose that the fluid is disturbed, possibly controlled, for example, by vibrating part of the boundary of the fluid or possibly uncontrolled because the fluid picks up any random alteration in its environment, e.g., by someone walking past the apparatus. What happens to the fluid after this disturbance? There are two main possibilities: (1) The disturbance may generate waves in the fluid that propagate through it but do not pick up energy from the basic state (as the waves on the surface of water). Nevertheless, there will always be dissipative processes present that damp out the disturbance, and the final state is one in which the fluid has essentially returned to its basic state, possibly a little hotter than it was! (2) The disturbance grows by extracting energy from the basic state. Often this process is weakened by dissipation and even reversed if the dissipation is sufficiently strong. The final outcome of the disturbance may be that the fluid switches to another steady or quasi-steady state, or that turbulent motion ensues.

Because random disturbances can never be prevented, it is clear that only if possibility (1) occurs is the basic state realizable, and two questions to be answered, then, are (a) what are the parameters that will determine whether (1) or (2) occurs? (b) if (2) occurs, what is the final state of the fluid? But note that the second question has meaning only if it is supposed that a parameter, say Reynolds number (Re), is varied until the switch from (1) to (2) is made. Otherwise the problem as formulated has no possible realization because the basic state can never be achieved.

9.1.1 Linear, Weakly Nonlinear, Nonlinear, and Hydrodynamic Stability

Linear stability. When we consider a *linear theory*, we assume that the amplitude of the disturbance is infinitesimally small. Thus, we can answer the first question partially by saying that the flow is stable [possibility (1)] or unstable [possibility (2)] for appropriate values of the parameters for disturbances that are sufficiently small. Obviously, it is quite possible for a flow to be stable to infinitesimal disturbances but unstable if the amplitude of the disturbance is finite and sufficiently large (even if a fluid flow is stable to small-amplitude disturbances, one may wish to know whether this stability survives finite-amplitude perturbations); the reverse, of course, is impossible.

The second question (if unstable, what is the final state?) cannot be answered by linear theory. Paradoxically, even in the absence of stability, the linearized problem is still a good approximation of the full (nonlinear) problem under certain closeness assumptions at time zero, even on an arbitrary (but fixed a priori) interval of time. On the other hand, it is interesting to note, again, that the problem of stability must be treated as an initial-value problem because the classical normal-mode approach implicitly ignores the initial growth phase of the disturbance or perturbation of the basic state or flow. This normal-mode procedure is not even formally permissible because

9.1 Some Aspects of the Theory of the Stability of Fluid Motion

the spectrum of the disturbances is continuous, not discrete, and therefore the transient behavior cannot be described by a linear superposition of an infinite number of discrete modes. Because this normal-mode approximation is valid only asymptotically (as $t \to \infty$), it may not correctly predict the breakup time.

There are at least two main methods at our disposal to study the stability of a steady-state flow of incompressible viscous (Navier) fluid flow. The first are the various *energy methods* that are the main methods in so-called "global theory." These methods lead to a variational problem for the first critical viscosity for Navier incompressible and viscous fluid flows and a definite criterion that is sufficient for the global stability of the basic flow – see, for instance, below, the Reynolds–Orr, energy, sufficient stability criterion. On the other hand, energy analysis leads to global statements about stability that take form as criteria sufficient for stability. Two typical examples are (a) steady-state flow is stable to arbitrary disturbances when the viscosity is larger than a critical value, and (b) statistically stationary turbulence of a given intensity cannot exist when the viscosity is larger than another critical value (which depends on the intensity of the turbulence).

The second is the method of *linearized stability,* based on the study of the spectrum of the linear spatial part in the equations for perturbation. Whereas the first method yields a sufficient criterion for energy stability, no matter how large the initial values are, the second method furnishes only a condition for the stability of the steady-state flow to small perturbations. But, in concrete cases, the required smallness of the perturbations is difficult to control. The admissible perturbations may be so small that, from the point of view of physics, no perturbations of finite amplitude are admitted. Therefore, these perturbations are called *infinitesimal.* The importance of the second method lies in the fact that it gives a sufficient condition for instability. Von Wahl (1994) is interested in the coincidence of the sufficiency criterion for stability via the energy method, which is thus a necessary criterion for instability, and the sufficiency condition for instability via the method of linearized stability.

We emphasize here that there is a third, more recent method for approaching the stability problem, in particular, if the motionless state is perturbed. For this [see, again, Wahl (1994)], we refer to the comprehensive work of Galdi and Padula (1990). The examples considered in Wahl (1994) are plane parallel shear flow with a non-symmetrical profile in an infinite rotating layer and the effect of rotation on convection.

The *linear stability* of a stationary solution is, by definition, determined by the evolution of *infinitesimal perturbations.* For infinitesimal perturbations, the quadratic term is negligible (of higher order) and so may be dropped in the evolutionary equations for the perturbations – then the system's evolution may be linearized, and the temporal evolution of this linear (albeit generically nonconstant coefficient) system of equations may be reduced to an *eigenvalue problem* by imposing an $\exp(-i\sigma_n t)$ time dependence on solutions, and superposing (relative to n) to obtain the general solution.

The operator whose spectrum is of interest in the eigenvalue problem (with the associated boundary conditions) is not generally self-adjoint, so the eigenvalues typically have both real and imaginary parts. If the *real parts are all positive*, then the amplitude of every infinitesimal perturbation *decays exponentially with time*. The underlying stationary solution is then said to be *"linearly stable."* If the larger *real part of any eigenvalue is exactly zero*, then the stationary solution is said to be *"marginal."* Note that linear stability does not guarantee that any finite perturbation decays. Linear stability is a relatively weak notion, quite distinct from some more robust stability conditions relevant to finite amplitude perturbations. If at least one eigenvalue has a *negative real part*, however, then some infinitesimal perturbation *grows exponentially*. The associated stationary solution (fixed point in phase space) is then *"linearly unstable."* This is a *strong notion of instability*:

> *It guarantees that there is some deviation which will grow and also that it will be amplified exponentially in time.*

But a small perturbation growing exponentially will eventually grow large enough for the neglected "nonlinear terms" to come into play, and then the linearized evolutionary system no longer applies.

> *Linear instability is a sufficient condition for instability, but linear or marginal stability is a necessary condition only for stability.*

Weakly nonlinear stability. The survival of the stability of a fluid flow to finite-amplitude perturbations is a natural question to ask for slightly subcritical flow, for example, where if bifurcation is subcritical (for bifurcations, see Chap. 10, Sect. 10.2), then the stability will apply only up to some threshold disturbance amplitude. This sort of question has traditionally been addressed using *weakly nonlinear* perturbation expansions, and in this case, the well-known Landau equation for the amplitude function $A(t)$ characterizing the disturbance:

$$\frac{\mathrm{d}A}{\mathrm{d}t} = \lambda A + kA|A|^2, \qquad (9.1)$$

shows how the growth rate is modified by nonlinear effects.

A more realistic (weakly) nonlinear situation is a localized disturbance that is a combination of spatial modulation with nonlinear terms. In this case, we can derive the following nonlinear Schrödinger (NLS) type equation:

$$\frac{\partial A}{\partial t} = \alpha \frac{\partial^2 A}{\partial \xi^2} + \beta A + kA|A|^2, \qquad (9.2)$$

for the amplitude function $A(t,x)$, characterizing the disturbance.

9.1 Some Aspects of the Theory of the Stability of Fluid Motion

In (9.1) and (9.2), t and x are slowly varying time-space variables. But weakly nonlinear theory is possible only when the relative growth rate of disturbances is small, e.g., near a neutral curve. An intensive application of weakly nonlinear asymptotic theory is a problem relative to *transition from laminar to turbulent flow* which has been a subject of study for many decades, and has been frequently reviewed. Recent reviews include those by Stuart (1986), Bayly, Orszag and Herbert (1988), Huerre and Monkewitz (1990), and Kleiser and Zang (1991). In Cowley and Wu (1994), the reader can find asymptotic approaches to transition modeling.

Nonlinear stability. A rigorous *nonlinear criterion* is often based on the existence of a particular function, called the *Lyapunov function*, which plays the role of controlling motion and gives the possibility of rigorously proving the stability of some stationary fluid flows. It is remarkable how a direct nonlinear approach is easier and more powerful than any procedure based on linearization. In addition, we want to stress that the linearization method cannot give, in some cases, conclusive results for the stability of the solution. For mathematically rigorous results, we mention the books by Straughan (1992), Galdi and Rionero (1985), Iooss and Joseph (1980), and Joseph (1976, 1983).

The phenomenon of nonlinear instability is due mainly to a permanent energy cascade toward short wavelengths. There are numerous methods for avoiding "nonlinear instability," but a rather specific way of suppressing this nonlinear instability is based on optimization of some functionals. In Alekseev (1994), an accuracy functional for finite difference schemes is proposed for some problems in hydrodynamics.

Hydrodynamic stability. The *hydrodynamic stability theory* has many applications. We refer the reader to the classical book by Drazin and Reid (1981) and to a pertinent review paper by Guiraud (1980). In Guiraud's paper, the reader can find a brief account of the energy method and of the basis of the linear theory, a heuristic scheme for a nonlinear theory based on techniques derived from bifurcation theory, a method akin to Lyapunov–Schmidt's combined with a multiple-scale technique. This method is also used in De Conink et al. (1983), and see, for instance, Sect. 9.3. The review paper by Debnath (1987) gives information concerning nonlinear instability in applied mathematics (related to Navier and Euler fluid flows, nonlinear wave trains, thermal instability, etc., with 226 references). In the theory of hydrodynamic stability, Poiseuille flow, as well as the Taylor–Couette problem, play the role of a benchmark for analytical and numerical approaches. In both cases, neutral stability curves can be obtained only numerically, in contrast to the Bénard thermal convection problem. Afendikov and Mielke (1995) considered the bifurcations of Poiseuille flow between parallel plates for three-dimensional solutions with large spanwise wavelength. By using analytical tools of bifurcation theory, such as center manifold techniques, many otherwise difficult, accessible solution types can be derived together with their stability properties. But, in the end, the relevant coefficients of the reduced problem on the center manifold are again calculated numerically!

In stability theory, some authors use the *method of Lyapunov–Schmidt* to decompose the space of solutions and equations into finite-dimensional and infinite-dimensional parts. The infinite part can be solved, and the resulting finite-dimensional problem has all of the information about bifurcation. Other authors use the *center manifold* to reduce the problems to finite dimensions – this method uses the fact that solutions in various problems are attracted to the center manifold, which is *finite-dimensional*. Both methods are good for proving existence theorems. Though they can also be used to construct solutions, but they involve redundant computations.

Iooss and Joseph (1980) apply the implicit function theorem to justify the direct, sequential computation of power series solutions in an amplitude, using the Fredholm alternative, as the most (!) economical way to determine qualitative properties of the bifurcating solutions and to compute them. Full nonlinear equations must be used to establish stability to any finite-amplitude perturbations, no matter how small.

In Joseph (1976) (see the Notes for Chapter II, pp. 54–63), the reader can find information concerning theories and methods for studying stability problems. Finally, see also Benjamin (1976) concerning the applications of Leray–Schauder degree theory to problems of hydrodynamic stability.

9.1.2 Reynolds–Orr, Energy, Sufficient Stability Criterion

A sufficient condition for stability is that the kinetic energy in the deviations should eventually vanish, and for Navier incompressible and viscous fluid flow, the L^2 norm of a perturbation $\delta \boldsymbol{u}$ evolves according to [see (8.5)]

$$\frac{\mathrm{d}}{\mathrm{d}t}\left[\frac{1}{2}\left\|\delta\boldsymbol{u}\right\|_2^2\right] + \int [\delta\boldsymbol{u}\cdot\nabla]\boldsymbol{U}\cdot\delta\boldsymbol{u}\,\mathrm{d}^d\boldsymbol{x} = -\nu_0 \left\|\nabla\delta\boldsymbol{u}\right\|_2^2 , \qquad (9.3)$$

where $\boldsymbol{U}(\boldsymbol{x})$ is the stationary base solution and the arbitrary velocity fields $\boldsymbol{u}(t,\boldsymbol{x})$ solution of the Navier equation is decomposed according to

$$\boldsymbol{u}(t,\boldsymbol{x}) = \boldsymbol{U}(\boldsymbol{x}) + \delta\boldsymbol{u}(t,\boldsymbol{x}) . \qquad (9.4)$$

Clearly, the viscosity works to *inhibit perturbation*, but the influence of the base solution is more subtle. To establish the "nonlinear stability" of $\boldsymbol{U}(\boldsymbol{x})$, it is sufficient to show that the L^2 norm of any *relevant* perturbation *vanishes* as $t \to \infty$.

To show this, it is sufficient to show that the L^2 norm of any relevant perturbation *vanishes exponentially*. In particular, this is the case if the right-hand side of (9.3) is negative and less than or equal to

$$-C^0 \left\|\delta\boldsymbol{u}\right\|_2^2 \quad \text{for some positive constant } C^0 \text{ uniformly}$$
$$\text{for all relevant flow fields } \delta\boldsymbol{u} .$$

9.1 Some Aspects of the Theory of the Stability of Fluid Motion 317

This last assertion follows from Gronwall's inequality, and a proof is presented in Doering and Gibbon (1995, p. 32). In such a case, any finite-amplitude, finite-energy perturbation would decay, and the underlying base solution would be *nonlinearly stable*.

The above method, by which the study of hydrodynamic stability for Navier flow is reduced to the study of the increase of kinetic energy in the deviations, is called the *energy method* introduced by Reynolds (1894) and Orr (1907).

9.1.3 An Evolution Equation for Studying the Stability of a Basic Solution of Fluid Flow

To test for stability or instability, we consider a hydrodynamic configuration, whose motion is ruled by the following evolutionary equation (see, for instance, Sect. 2.2.6, Chap. 2):

$$S \frac{du}{dt} = F(u), \qquad (9.5)$$

where $u(t)$ is a function of time t with values in some function space that must be specified in each concrete situation and S is the Strouhal number. In closed terms, this mean that, for each fixed value of time, u is a function of the space variables that have differentiability properties defined precisely by the function space to which it belongs. By F is meant a nonlinear differential operator in whose definition boundary conditions are taken care of. For example, if one is to study the motion of a *Navier*, incompressible and viscous fluid and if u stands for velocity $\boldsymbol{u}$ and Re is the appropriate Reynolds number, according to the Navier equation,

$$\boldsymbol{F}(\boldsymbol{u}) = -[\boldsymbol{u} \cdot \nabla \boldsymbol{u} - \nabla p] + \frac{1}{\mathrm{Re}} \nabla^2 \boldsymbol{u}, \qquad (9.6a)$$

with the constraint: $\nabla \cdot \boldsymbol{u} = 0$. $\qquad (9.6b)$

As is well known, p should be interpreted, from a thermodynamic point of view, as the part of the sress associated with this constraint (9.6b). Correspondingly, ∇p may be considered as adjusting itself so that $\nabla \cdot \boldsymbol{F} = 0$.

If one is interested in a *Rayleigh–Bénard* (RB) convection problem, u stands for the couple $(\boldsymbol{u}, \theta)$, where $\boldsymbol{u}$ is the velocity and θ is the perturbed temperature. In this case of thermal convection, F stands for the couple $(\boldsymbol{F}, \phi)$ with

$$\boldsymbol{F}(\boldsymbol{u}) = -[\boldsymbol{u} \cdot \nabla \boldsymbol{u} - \nabla p] + \nabla^2 \boldsymbol{u} + \mathrm{Gr}\,\theta \boldsymbol{e}_z, \qquad (9.7a)$$

$$\phi = -\boldsymbol{u} \cdot \nabla \theta + \frac{1}{\mathrm{Pr}} \nabla^2 \theta, \qquad (9.7b)$$

where $\boldsymbol{e}_z$ is the unit vector of the vertical, drawn upward.

Here (9.6b) again holds, and in the same way as above, ∇p adjusts itself such that $\nabla \cdot \boldsymbol{F} = 0$. Two nondimensional parameters arise in (9.7a,b): the Prandtl number Pr and the Grashof number Gr; their product, Pr Gr = Ra, is the Rayleigh number.

The problem of hydrodynamic stability may be posed in the following terms:

Let u_0 be a particular solution of (9.5), for example, a solution independent of time, a so-called stationary solution, which should accordingly satisfy

$$F(u_0) = 0 \ . \tag{9.8}$$

Any other solution may be written as

$$u = u_0 + v \ , \tag{9.9}$$

and the problem of the stability of the solution u_0 amounts to studying whether or not v *remains small*, in some sense, that is, according to the norm in some function space, as long as v is *sufficiently small, in the same sense, at the initial time*.

In what follows, we shall assume that u_0 is independent of time and that the equation for v may be written as (see, for instance, the derivation of (2.61) in Sect. 2.2.6)

$$\frac{\mathrm{d}v}{\mathrm{d}t} = L(u_0)v + Q(v,v) \ . \tag{9.10}$$

In (9.10), $L(u_0)$ is a linear operator, depending on stationary solution u_0, and $Q(v,v)$ is a quadratic operator, such that

$$Q(av, av) = a^2 Q(v,v) \ , \quad \text{if } a = \text{const.} \tag{9.11a}$$

Moreover, we assume that (for any v and w):

$$Q(v+w, v+w) = Q(v,v) + M(v,w) + Q(w,w) \ , \tag{9.11b}$$

where $M(v,w)$ is bilinear and is a function of two variables – this occurs because Q is quadratic (a function of one variable) and the left-hand side of (9.11b) has been straightforwardly expanded.

It is interesting to note that, if we take the inner product $(\cdot \, ; \cdot)$ of (9.10) with v we obtain the equation

$$\frac{\partial}{\partial t}\left[\frac{|v|^2}{2}\right] = (L(u_0)v; v) \ , \tag{9.12a}$$

because

$$(Q(v,v); v) = -(v \cdot \nabla v; v) = \int_D u_j \left(\frac{\partial u_i}{\partial x_j}\right) u_i \, \mathrm{d}x = 0 \ . \tag{9.12b}$$

Now let

$$\gamma = \inf \left\{ \frac{(L(u_0)v; v)}{[|v|^2/2]} \right\}, \qquad (9.13)$$

subject to the constraint div $u = 0$ and $u = 0$ on ∂D. This variational problem is similar to the classical variational characterization of the first eigenvalue of the Laplacian. Consequently, from (9.12a,b), we obtain

$$\frac{\partial}{\partial t}\left[\frac{|v|^2}{2}\right] - \gamma |v|^2 \leq 0, \qquad (9.14)$$

and integrating this differential inequality, we get

$$|v(t)| < |v(0)| \exp(\gamma t). \qquad (9.15)$$

If $\gamma = 0$, all perturbations decay exponentially to zero in the L^2 norm, regardless of the initial size of the disturbance, and the flow u_0 is *unconditionally stable*.

9.2 Fundamental Ideas on the Theory of the Stability of Fluid Motion

More precisely, let F be a phenomenon associated with a fluid motion that occurs in a domain $\Omega \subset \mathbf{R}^3$, bounded at least in one direction. The state vector of F is denoted by $u(t,x)$ and $\mathcal{H}$ denotes the space of the states of F. We assume that $\mathcal{H}$ is a Hilbert space and denote again the scalar (inner) product by $(\cdot;\cdot)$ and the induced norm by $\|\cdot\|$. Further, we assume that, on $\mathcal{H}$, F is modeled by initial-value problem (9.16a,b):

$$\frac{\partial u}{\partial t} + Lu + N(u) = 0, \quad u(0) = u^0, \qquad (9.16a)$$

where

$L = $ an autonomous linear operator,

$N = $ an autonomous nonlinear operator such that $N(0) = 0$, and

$u^0 = $ a prescribed initial state. $\qquad (9.16b)$

The Cauchy problem (9.16a,b) admits the trivial solution $u = 0$, and we will study the stability of this solution recalling that the stability of any solution can be reduced to the study of the stability of the zero solution.

9.2.1 Linear Case

Any solution $u(t,x)$ of the linearized form of (9.16a),

$$\frac{\partial u}{\partial t} + Lu = 0 \;, \quad u(0) = u^0 \;, \tag{9.17}$$

can be written as a linear combination of eigensolutions,

$$u_n = v_n(x) \exp(-\sigma_n t) \;, \tag{9.18a}$$

where

$$v_n(x) = a_n(x) + i b_n(x) \;, \tag{9.18b}$$

is an eigenvector associated with $\sigma_n = \lambda_n + i\mu_n$, and when L is *compact*, it follows that the eigenvalues can be ordered as follows:

$$\lambda_1 < \lambda_2 < \ldots < \lambda_n \ldots$$

The zero solution is said to be linearly stable if $\lambda_1 > 0$, and this implies $u \to 0$ exponentially when $t \to \infty$. From, $\sigma_n v_n = L v_n$, it turns out that (v_n^* is the complex conjugate of v_n)

$$\sigma_n(v_n; v_n^*) = (L v_n; v_n^*) \;,$$

and hence [see (9.18b)],

$$\lambda_n = \frac{(La_n; a_n) + (Lb_n; b_n)}{\|a_n\|^2 + \|b_n\|^2} \;. \tag{9.19}$$

When L is *symmetrical* (Hermitian, and in this case the eigenvalues are real numbers $\sigma_n \equiv \sigma_n^*$), it can be shown that

$$\lambda_1 = \min_{u \in \mathcal{H}} \left\{ \frac{(Lu; u)}{\|u\|^2} \right\} \;. \tag{9.20}$$

Now, we assume that L depends on two nondimensional parameters, α and β. The locus $\lambda_1(\alpha, \beta) = 0$ is called a locus of *neutral stability* because it separates the states of linear stability from those of instability.

Exchange of stability and a point of convective bifurcation. Let β' be a fixed value. The lowest value $\alpha^l(\beta')$ of α such that

$$\lambda_1(\alpha^l(\beta'), \beta') = 0 \;, \quad \alpha < \alpha^l(\beta') \to \lambda_1(\alpha, \beta') > 0 \;, \tag{9.21}$$

is the *critical value* of α of linear stability associated with β'.

Let $\alpha < \alpha^l(\beta')$, and let α grow. When α reaches $\alpha^l(\beta')$, one passes from a state of linear stability to a state of instability. The state associated with

9.2 Fundamental Ideas on the Theory of the Stability of Fluid Motion

$[\alpha^l(\beta'), \beta']$ is called a bifurcation point according to the following definitions. If (μ_1 is the imaginary part of σ_1)

$$\lambda_1(\alpha^l(\beta'), \beta') = 0 \to \mu_1(\alpha^l(\beta'), \beta') = 0 , \qquad (9.22)$$

then one says (by definition) that

the principle of exchange of stability holds,

and in the state $[\alpha^l(\beta'), \beta']$, the equation $\partial u/\partial t + Lu = 0$ admits, other than the steady-state solution $u = 0$, the steady-state solution

$$u_1 = v_1 . \qquad (9.23)$$

The later denotes the *secondary steady-state motion* arising when α increases through $\alpha^l(\beta')$, and then the state $[\alpha^l(\beta'), \beta']$ is said to be a point of convective bifurcation.

A point of Hopf bifurcation (overstability). If

$$\lambda_1(\alpha^l(\beta'), \beta') = 0 \quad \text{but} \quad \mu_1(\alpha^l(\beta'), \beta') \neq 0 , \qquad (9.24)$$

then in the state $[\alpha^l(\beta'), \beta']$, the equation $\partial u/\partial t + Lu = 0$ admits, other than the solution $u = 0$, the *oscillatory motions*

$$U_1 = a_1 \sin \mu_1 t - b_1 \cos \mu_1 t , \qquad (9.25a)$$

$$U_2 = a_1 \cos \mu_1 t + b_1 \sin \mu_1 t , \qquad (9.25b)$$

where $v_1 = a_1 + ib_1$, is a nontrivial eigenfunction associated with $\sigma_1 = i\mu_1$. Then the state $[\alpha^l(\beta'), \beta']$ is said to be a point of *Hopf bifurcation* (or of *overstability*). Obviously, when the eigenvalues are *real numbers*, then (9.22) is automatically satisfied, and it is said that the *principle of exchange of stability holds in the strong sense.*

When domain Ω is unbounded. When Ω is unbounded, the eigensolutions are spatially periodic in the direction in which Ω is unbounded. Then, the eigenvalues σ_n depend on the *wave number* k, and therefore,

$$\lambda_1 = f(\alpha, \beta, k) . \qquad (9.26)$$

Then, to obtain the critical value $\alpha^l(\beta')$ associated with β', one has to solve the equation (a so-called "eigenvalue relation"),

$$f(\alpha, \beta', k) = 0 . \qquad (9.27)$$

Let $\alpha = \phi(\beta', k)$ be the solution of (9.27). Then,

$$\alpha^l(\beta') = \min_{k + \mathbf{R}^+} \phi(\beta', k) \qquad (9.28)$$

denotes the critical value of α under the second of conditions (9.21). In the plane (k,α), solution $\alpha = \phi(\beta',k)$ separates the (linearly) stable states from the unstable ones. When (9.22) does not hold, then the (linearly) stable states are *separated* from the unstable ones by different curves (or manifolds) of *neutral stability*. These curves are associated with the *convective and Hopf bifurcation and need not intersect*.

9.2.2 Nonlinear Case

The zero solution of (9.16a) is said to be nonlinearly stable in the norm of $\mathcal{H}$ if

$$\forall \varepsilon > 0 \,, \exists \delta(\varepsilon) > 0 : \|u^0\| < \delta \to \|u(t)\| < \varepsilon \,, \forall t > 0 \,. \tag{9.29}$$

The zero solution of (9.16a) is said to be (nonlinearly)*attractive* if

$$\exists \gamma > 0 : \|u^0\| < \gamma \to \lim \|u(t)\| = 0 \,, \quad \text{when } t \to \infty \,. \tag{9.30}$$

If $\gamma = \infty$, $u = 0$ is said to be *unconditionally attractive*. Referring to Galdi and Rionero (1985) for an in-depth analysis, we summarize here two essential steps of nonlinear stability in fluid motion. The *first step* consists of obtaining an inequality like the following:

$$\frac{\mathrm{d}\|u\|}{\mathrm{d}t} \leq R^0 I - D + \beta \|u\|^\alpha D \,, \tag{9.31}$$

where R^0 is a nondimensional number, α and β are positive constants, and I and D are quadratic functions (i.e., integrals) of u and ∇u, with ∇D positive definite.

The *second step* consists of resolving the variational problem,

$$\frac{1}{R_N^0} = \max_{\mathcal{H}} \left(\frac{I}{D}\right) \,, \tag{9.32a}$$

that is well set if a Poincaré inequality holds, i.e.,

$$\exists s = \text{const.} > 0 : \|u\| < sD \,, \forall u \in \mathcal{H} \,. \tag{9.32b}$$

Then,

$$R < R_N$$
$$\|u^0\|^\alpha < \frac{R_N - R}{\beta R_N} \to \|u\| \leq \|u^0\| \exp\left[-\left(\frac{1}{s}\right) At\right] \,, \tag{9.32c}$$

with

$$A = \frac{R_N - R}{R_N} - \beta \|u^0\|^\alpha \,, \tag{9.32d}$$

9.2 Fundamental Ideas on the Theory of the Stability of Fluid Motion

i.e., the *conditional asymptotic exponential stability*, and we remark that (i) when L is symmetrical, it can be shown that $R_l = R_N$; (ii) when L is not symmetrical, generally, $R_N \leq R_l$; and (iii) when (9.31) holds with $\beta = 0$, by (9.32b), it follows that

$$R < R_N \to \|u^0\|^\alpha < \|u\| \leq \|u^0\| \exp[-\left(\frac{1}{s}\right) A^0 t], \tag{9.32e}$$

with $A^0 = 1 - (R/R_N)$, i.e., $R < R_N$ ensures *unconditional asymptotic stability*. (iv) When (9.31) holds with $\beta > 0$, the condition on the initial data

$$\|u^0\|^\alpha < \frac{(R_N - R)}{\beta R_N} \tag{9.32f}$$

can be very severe. Further, $R \to R_N$ implies $\|u^0\| \to 0$.

On taking (iv) into account, it follows that any choice of the topology of $\mathcal{H}$ that allows $R_N = R_l$, without restriction on the initial data appears of the highest interest. Apart from the classical Bénard problem for thermal convection, it is known that this happens only for few physically interesting fluid motions, and here as example we consider the problem of thermal convection in a layer of dielectric fluid subject to an alternating current.

More precisely, for this problem, the choice of the Lyapunov function

$$L^*(t) = \frac{1}{2}\left[\frac{1}{\Pr}\|\boldsymbol{u}\|^2 + m\|\theta\|^2 + L\|\nabla\phi\|^2\right], \tag{9.33a}$$

plays an important role; $\boldsymbol{u}$ is the perturbed velocity vector, θ the perturbed temperature, ϕ the perturbed electrical potential, m a positive constant (the Lyapunov parameter) that will be chosen later, and L is the so-called Roberts number (electric Rayleigh number).

More precisely the linear stability of the basic motionless state with the Lyapunov second method is considered by introducing a Lyapunov function

$$L(t) = \frac{1}{2}\left[\frac{1}{\Pr}\|\boldsymbol{u}\|^2 + (\mathrm{Ra} + L)\|\theta\|^2 - L\|\nabla\phi\|^2\right], \tag{9.33b}$$

and it is proved that the linear operator associated with the norm $L(t)$ is symmetrical; then the necessary and sufficient conditions of linear stability are that

> the critical values L_c found by Roberts coincide with those obtained with $L(t)$ – the Lyapunov method.

It is easy to verify that the Lyapunov function $L^*(t)$ is equivalent to $L(t)$ in $\mathcal{H}$, and $\mathcal{H}$ denotes here the closure of the space of admissible functions with respect to the norm $[\|\boldsymbol{u}\|^2 + \|\theta\|^2 + \|\nabla\phi\|^2]^{1/2}$, in $L^2(V)$ – the starting perturbation equations hold in the 3-D planar region $z \in (0,1)$, and it is assumed that the unknown functions are sufficiently smooth and satisfy a plane tilting pattern in the x, y directions, so they define a perturbation cell V over the lateral boundaries of which their contributions are equal and cancel out in the ensuing integration by parts. As result,

for any $\text{Ra} \in [0, 657.511]$, the condition $L < L_c$ implies unconditional nonlinear exponential stability according to the inequality
$$L^*(t) \leq L^*(0) \exp[-2\pi^2 (1 - m^*)t] ,$$

where $m^* = m^*(\text{Ra}, L, m) = 1$ at criticality and $m^* < 1$ is a stability condition.

It is interesting to note that the energy method imposes a severe restriction on the initial data and then on conditional nonlinear stability. On the contrary, the role of the choice of the Lyapunov function $L^*(t)$ is to give the opportunity of overcoming this hard restriction and to obtain global nonlinear stability. Of course, in view of the equivalence of $L(t)$ and $L^*(t)$, the condition $L < L_c$ ensures also the unconditional nonlinear stability of the basic motionless state with respect to $L(t)$. Concerning the choice of the Lyapunov function in the stability of fluid motions, see also Rionero (1988).

9.3 The Guiraud–Zeytounian Asymptotic Approach to Nonlinear Hydrodynamic Stability

In the theory of hydrodynamic stability, much progress has been made through the use of asymptotic techniques [see, for instance, Stuart (1971a), Bouthier (1972, 1973), Stewartson (1976), Guiraud and Zeytounian (1978), Guiraud (1980), F.T. Smith (1981), and de Conink et al. (1983)].

In this asymptotic approach, the main tool is decomposition in mormal modes and expansion techniques, and this way was opened by Stuart (1958, 1960a,b) and Watson (1960b). However, when the modes are not discrete but spread along a continuum, the "Stuart–Watson technique" failed.

In a pioneering work, DiPrima et al. (1971) tried to deal with the continuous spectrum by considering many discrete normal modes. Their main results were recovered by Stewartson and Stuart (1971) who succeeded in dealing directly with the continuous spectrum by clever use of the multiple-scales technique, but restricted to 2-D perturbations.

Their analysis was extended to 3-D by Davey et al. (1974). The above three last works were not oriented toward dealing with the whole temporal evolution of perturbations, from an initial one, up to the stage when damped modes have been brought to a very low level and the growth of amplified ones has been limited by nonlinearities.

However, this task may easily be completed for pure discrete modes by an extension of the Stuart (1960a) and Watson (1960b) techniques using tools borrowed from bifurcation theory and applied formally. This is explained in Sect. 9.3.2.

When dealing with modes spread over a continuum, the extension of the Stewartson and Stuart (1971) technique to the whole of temporal evolution is not at all obvious, and the difficulty has its root in the fact that, to our knowledge, no standard mathematically rigorous technique exists for dealing with bifurcation from a continuous spectrum. In 1978, Guiraud and Zeytounian (1978) succeeded in filling this gap, in a purely formal way, for continuously spread Tollmien–Schlichting waves that exhibit close similarity to the technique used for discrete modes.

In this section, we present a kind of unified theory [according to de Conink et al. (1983)] in which the purely discrete and the continuous cases are treated by two facets of a somewhat unique technique. We are not interested in any specific application but rather in presenting the technique for deriving economically the equation or set of equations that rule the evolution of the amplitude of the most rapidly amplified modes of linear theory.

In Sect. 9.3.5, as a particular case, we derive a system of equations for Rayleigh–Bénard thermal convection between two infinite parallel planes. We show that nonlinearity occurs as quadratic interactions between modes that have wave-number vectors of equal modules but rotated through $\pi/3$. This kind of quadratic interaction has nothing to do with that due, for example, to variation of viscosity with temperature, as described by Palm (1960) and thoroughly discussed in Palm (1975). The interaction we consider seems to occur each time a continuum of modes, with wave-number vectors spread over a thin circular ring, is allowed to interact with itself, quadratically. The result is that six modes that have the ends of their wave-number vectors located at the six vertices of a regular hexagon centered at the origin, each interacts with its two neighbors, a configuration described in a short note by de Coninck (1979). We show also that, due to the slight thickness of the ring, diffusion occurs similarly to that described in the pioneering work of Stewartson and Stuart (1971).

This quadratic interaction is expressed through a coefficient which includes integration of products of functions of the vertical coordinate z which give the vertical structure of normal modes. A careful examination (performed by J.P. Guiraud) of the statement of Schüter et al. (1965), concerning the vanishing of the quadratic interaction in the nonlinear interaction of a finite number of discrete modes (in their work on stability of steady-state finite-amplitude convection), shows that this is true only for boundaries that are both rigid or both free!

9.3.1 Linear Theory

According to linear theory, we simply replace (9.10) by

$$\frac{dv}{dt} = L(u_0)v \ . \tag{9.34}$$

If v remains small all the time, provided that it is small initially, one says that the hydrodynamic configuration characterized by u_0 is linearly stable or that it is stable to infinitesimal perturbations. One says also that the linearization principle holds if linear stability implies stability for the nonlinear problem. A wealth of results are known concerning linear stability, and it is not the purpose of Sect. 9.3 to attempt any review. We refer, for example, to Chandrasekhar (1961/1981), Lin (1955), and Joseph (1976). Here, we shall limit ourselves to mentioning the points which will be useful for the rest of Sect. 9.3, making a distinction between two very different situations.

Confined hydrodynamic configurations. By this, we mean situations where u is defined on a bounded spatial domain. Under such conditions, one may prove that $L(u_0)$ has a discrete spectrum. There exists an infinite sequence of eigenvalues, in fact complex, namely $\sigma_n, n = 1, 2, \ldots$ with associated eigenfunctions v_n, such that

$$L(u_0)v_n = \sigma_n \ , \quad \text{and the equation } L(u_0)v - \sigma = f \ , \tag{9.35}$$

with a given f, is uniquely solvable provided σ is distinct from all σ_n. Of course, the proof of this property must be provided in each particular case [see Iooss and Joseph (1980)], and this has much to do with the kind of function space used. Although this is a very important question, we shall not consider it here.

Let $\sigma_n = \lambda_n + i\mu_n$, and we are interested here in the unstable case but restrict ourselves to "weak instability"; for this, we partition the indexes n into two parts N and A:

$$n \in N \Rightarrow \lambda_n = O(\varepsilon^q) \quad \text{and} \quad n \in A \Rightarrow \lambda_n < 0 \ , \tag{9.36}$$

where ε is a small parameter and q an exponent that is introduced here for later convenience. We must comment on the partition (9.36), which we have introduced as a *working hypothesis* to be checked in each particular case.

The flow u_0 depends on parameters (for example, a Reynolds number Re), and when Re < Re$_c$, the configuration is stable, whereas it is unstable for Re > Re$_c$. By continuity, this corresponds to the fact that, when Re crosses Re$_c$ increasing, some eigenvalues σ_n, those corresponding to $n \in N$, cross the imaginary axis from left to right. When Re is close to Re$_c$ on either side, the λ_n corresponding to indexes $n \in N$ are small, and this is expressed through the first relation of (9.36). The most frequent are those when N has *just one element* with σ_n *real* or *two elements* with *a complex conjugate of* σ_n, σ_n^*.

9.3 The Guiraud–Zeytounian Asymptotic Approach

Unconfined perturbations. By this, we mean any situation where u is defined on an unbounded spatial domain. Here we shall be much more restrictive. Let us use Cartesian coordinates $\boldsymbol{x} = (x_k; k = 1, 2, 3)$ and $\boldsymbol{x} = (\boldsymbol{x}_\|, \boldsymbol{x}_\perp)$, where $\boldsymbol{x}_\|$ may have one or two components and $\boldsymbol{x}_\perp$ has the others. We assume that two conditions are fulfilled: first, u_0 does not depend on $\boldsymbol{x}_\|$; second, the spatial domain is cylindrical according to the decomposition $\boldsymbol{x} = (\boldsymbol{x}_\|, \boldsymbol{x}_\perp)$, in the sense that it is defined by

$$\boldsymbol{x}_\perp \in V \quad \text{and} \quad \boldsymbol{x}_\| \in E_\| , \tag{9.37}$$

where V is a bounded domain and $E_\|$, is the whole space corresponding to $\boldsymbol{x}_\|$. As examples, we mention *Poiseuille* and *Couette* plane flows and *Bénard thermal* convection for which $\boldsymbol{x}_\|$ has two components, and *Couette-Taylor* circular flow for which $\boldsymbol{x}_\|$ has only one component (the coordinate along the axis of the concentric cylinders).

To treat the linear stability problem, one uses a Fourier transform, which is an extension of that defined by

$$v_F = Fv = \int_{E_\|} \exp(-i\boldsymbol{k} \cdot \boldsymbol{x}_\|) v(\boldsymbol{x}_\|, \boldsymbol{x}_\perp) \, d\boldsymbol{x}_\| , \tag{9.38}$$

for functions "well behaved" at infinity.

Thanks to the independency of u_0 from $\boldsymbol{x}_\perp$, the linear evolution equation (9.34) may be transformed into

$$\frac{dv_F}{dt} = L_F(u_0, \boldsymbol{k}) v_F , \tag{9.39}$$

where $L_F(u_0, \boldsymbol{k})$ is an operator that is differential with respect to the variables of $\boldsymbol{x}_\perp$ and algebraic with respect to $\boldsymbol{k}$.

Under general conditions, L_F has a discrete spectrum of eigenvalues $[\sigma_n(\boldsymbol{k})]$ with eigenfunctions $v_{F,n}$ and $L_F v_{F,n} = \sigma_n v_{F,n}$. If for any $\boldsymbol{k}$ and n, $\lambda_n = \text{Real}(\sigma_n) < 0$, then the solution u_0 is stable.

Here, we propose to study a situation analogous to that defined by (9.36), when for some n (in fact for $n = 0$), there exists a domain in the $\boldsymbol{k}$ plane where $\lambda_n > 0$ (in this case the solution u_0 is unstable) and is characterized by

$$\boldsymbol{k} \in \Delta , \; n = 0 \quad \Rightarrow \quad \lambda_0 = O(\varepsilon^2) , \tag{9.40a}$$

$$\boldsymbol{k} \notin \Delta , \; n = 0 \text{ or } n > 0 \quad \Rightarrow \quad \lambda_n < 0 . \tag{9.40b}$$

On the major part of Δ, $\lambda_n > 0$, and if $\boldsymbol{x}_\|$ has only one component, Δ is an interval of the real line. In many other cases, $\boldsymbol{x}_\|$ has two components as well as $\boldsymbol{k}$. We observe that the choice of $q = 2$ in (9.40a) is a direct consequence of the choice of (9.45) below [see, for instance, Guiraud (1980, p. 40)].

Let us set $\boldsymbol{k} = (a, b)$. Then the domain Δ is an area in the (a, b) plane with small dimensions. In this case, among all the situations that may occur, two are prominent. The first, which corresponds to plane *Poiseuille* or *Couette flow* (and by extension to plane boundary-layer flow[1]) has the following property:

$$\sigma_0(a, -b) = \sigma_0(a, b) \quad \text{and} \quad \sigma_0(-a, b) = \sigma_0^*(a, b) , \tag{9.41a}$$

where the star denotes complex conjugation. In the second situation,

Δ has two connected components each of which
is a small area of $O(\varepsilon)$ in its two dimensions. (9.41b)

For historical reasons one says that, in this first situation, v is compounded of *Tollmien–Schlichting waves*. In the second situation, which corresponds to *Rayleigh–Bénard convection*, the following holds:

$$\sigma_0 = \sigma_0(|\boldsymbol{k}|) , \tag{9.42}$$

and Δ is a *small circular ring of thickness* $O(\varepsilon)$, centered at the origin in the (a, b) plane.

9.3.2 Nonlinear Theory – Confined Perturbations. Landau and Stuart Equations

We start from a normal-mode decomposition, and according to (9.36),

$$v = \sum_{n \in N} C_n v_n + \sum_{n \in A} C_n v_n = X + Y , \tag{9.43a}$$

and we set

$$X = P^1 v , \quad Y = P^2 v , \tag{9.43b}$$

thus defining P^1 and P^2 as two projection operators.

From (9.36), we see that

$$n \in N \Rightarrow \sigma_n = \varepsilon^q \lambda_n^0 + i\eta_n , \tag{9.43c}$$

---

[1] This extension is not at all trivial because of the occurrence of a continuous spectrum, even after Fourier transforming as in (9.38) – see, for instance, Salwen and Grosch (1981). But the whole of the continuous spectrum corresponds to damped modes, and it is expected that the Lyapunov–Schmidt decomposition is preserved, at least sufficiently close to the critical Reynolds number. Of course, when dealing with a boundary layer, one should also take into account the nonparallel-flow effects [see, Bouthier (1972, 1973), Gaster (1974), and F.T. Smith (1981)].

9.3 The Guiraud–Zeytounian Asymptotic Approach

and taking (9.43a) into account, we obtain the following system of two "bifurcation type" equations for X and Y in place of nonlinear evolution equation (9.10):

$$\frac{dX}{dt} = \varepsilon^q L_1 X + L_2 X + P^1 Q(X+Y, X+Y), \tag{9.44a}$$

$$\frac{dY}{dt} = L_A Y + P^2 Q(X+Y, X+Y). \tag{9.44b}$$

It may be stated that $\varepsilon^q L_1$ corresponds to $\varepsilon^q \lambda_n^0$, L_2 to $i\eta_n$, both for $n \in N$, and L_A corresponds to σ_n for $n \in A$.

Below, we shall concentrate only on the case when

$$P^1 Q(X, X) = 0. \tag{9.45}$$

In many applications, X is periodic with respect to $\boldsymbol{x}_\parallel$ and contains only the fundamental, and $Q(X, Y)$ contains only harmonics that are annihilated by the projection P^1. Assuming that $\varepsilon \ll 1$, we want to solve the system (9.44a,b) by asymptotic expansions, but we need to use a multiple-scale technique by setting

$$\tau = \varepsilon^q t, \quad X = \varepsilon^r X^*, \quad Y = \varepsilon^{2r} Y^*, \tag{9.46}$$

and, due to hypothesis (9.45), one can see that $r = 1$ and $q = 2$ are the relevant choices. In what follows, we ignore the transient unsteady-state phase $t = O(1)$ to concentrate on the nonlinear phase, when "long" time $\tau = O(1)$. In (9.46), X^* and Y^* are functions of both time t and τ. We substitute

$$(X^*, Y^*) = (X_0^*, Y_0^*) + \varepsilon(X_1^*, Y_1^*) + \varepsilon^2(X_2^*, Y_2^*) + \ldots, \tag{9.47}$$

With (9.46) and the condition (9.45) from (9.44a,b), with $r = 1$ and $q = 2$, we derive two equations for the functions X^* and Y^*; namely:

$$\frac{dX^*}{dt} - L_2 X^* + \varepsilon^2 \left[\frac{dX^*}{d\tau} - L_1 X^* - P^1 M(X^*, Y^*)\right] + O(\varepsilon^3) = 0, \tag{9.48a}$$

$$\frac{dY^*}{dt} - L_A Y^* - P^2 Q(X^*, X^*) + O(\varepsilon) = 0, \tag{9.48b}$$

and in (9.48a), we have taken into account the decomposition (9.11b).

Next, thanks to the fact that $\tau = O(1)$, we obtain, from (9.48a,b) for the leading-order term of the expansion (9.47), X_0^*, Y_0^*, the following solution:

$$X_0^* = \sum_{n \in N} X_{0,n}^* \exp(i\eta_n t), \tag{9.49a}$$

$$Y_0^* = \sum_{p \in N} \sum_{q \in N} Y_{0,pq}^* \exp[i(\eta_p + \eta_q)t], \tag{9.49b}$$

with
$$Y^*_{0,pq} = [i(\eta_p + \eta_q)\boldsymbol{I} - L_A]^{-1} P^2 Q(X^*_{0,p}, X^*_{0,q}) , \qquad (9.50)$$
where $\boldsymbol{I}$ is the unit operator.

Now, we consider the equation for the term X^*_2, seeing that we want to determine the dependence of $X^*_{0,n}$ relative to τ. For X^*_2, we obtain as equation:
$$\frac{dX^*_2}{dt} - L_2 X^*_2 + \sum_{n \in N} \left[\frac{dX^*_{0,n}}{d\tau} - L_1 X^*_{0,n} \right] \exp(i\eta_n t)$$
$$- \sum_{p \in N} \sum_{q \in N} \sum_{r \in N} P^1 M(X^*_{0,p}, Y^*_{0,qr}) \exp[i(\eta_p + \eta_q + \eta_r)t] = 0 . \qquad (9.51)$$
Next eliminating (secular) terms in the solution for X^*_2, we get:
$$\frac{dX^*_{0,n}}{d\tau} - L_1 X^*_{0,n} - \sum_{R_n} P^1 M(X^*_{0,p}, Y^*_{0,qr}) = 0 , \qquad (9.52)$$
and $\sum_{R_n}$ means the sum over all resonant (p, q, r) triplets, where R_n is the *resonance condition*,
$$R_n : \eta_p + \eta_q + \eta_r = \eta_n . \qquad (9.53)$$
As a particular case, when N contains only one element, we have the exchange of stabilities, and X^*_0 is well defined by one scalar $C(\tau)$ which must be a solution of the classical *Landau–Stuart equation*,
$$\frac{dC}{d\tau} = \lambda^0 C + KC^3 , \qquad (9.54)$$
where λ^0 is the unique λ^0_n of (9.43c) corresponding to the fact that N has but one element and the *amplitude $C(\tau)$ is real*.

We recover the result of the analysis of Stuart (1960a), Watson (1960a), and many others. We emphasize that the absence of a quadratic term in (9.54) is merely a consequence of (9.45). Whether or not this relation holds depends on the special problem at hand. Here, we mention that when the domain is bounded through periodicity conditions, (9.45) happens to hold true through cancelation of the spatial dependence of modes due to the integration implied by the projection operator P^1. When the domain is bounded completely round the walls, (9.45) does not hold in general, as stressed by Benjamin (1978b).

When N contains two elements, the corresponding σ_n are complex conjugates, and X^*_0 is well defined by a complex scalar $S(\tau)$ which must be a solution of the so-called Stuart equation,
$$\frac{dS}{d\tau} = \lambda^0 S + KS|S|^2 , \qquad (9.55)$$
and we recover again the result of Stuart (1960a) and Watson (1960b). In mathematically oriented literature, this situation is referred to as the Hopf bifurcation.

9.3.3 Nonlinear Theory – Unconfined Perturbations. General Setting

This case has generally been treated by artificially confining the perturbation via a periodicity condition imposed on the dependency with respect to x. This was done for parallel flows by Watson (1960b) and by Reynolds and Potter (1967). The analysis in Sect. 9.3.2 holds in this situation, and it is even the prototype of situations where (9.45) holds, leading to so-called sub- or supercritical bifurcations. Confined flow is, rather, transcritical as pointed out by Benjamin (1978b). The problem now is to generalize the technique of Sect. 9.3.2 to deal with truly unconfined perturbations while retaining the principle of the Lyapunov–Schmidt procedure. To our knowledge, the kind of treatment we explain below was presented for the first time by Guiraud and Zeytounian (1978) for the special case of parallel flow. We do not claim that our treatment is superior to that of Stewartson and Stuart, as far as the modeling of the physical situation is concerned.

We merely stress that our treatment with J.P. Guiraud of unconfined flow is quite parallel to the treatment of confined flow. If one recalls that this last treatment may be justified by a rigorous mathematical argument, it may be enlightening to know that, formally at least, a quite analogous treatment also exists for unconfined flow.

We start from the obvious observation that, after applying the Fourier transform (9.38), we may use the analog of (9.43a).

Let $w_{F,n}$ stand for the eigenfunctions of the adjoint to $L_F(u_0, \bm{k})$, and let F^{-1} be the inverse of the Fourier transform. We set

$$v = F^{-1}\left[\sum_n \langle Fv; w_{F,n}\rangle v_{F,n}\right] = X + Y , \qquad (9.56)$$

where

$$X = F^{-1}[\bm{1}_\Delta \langle Fv, w_{F,0}\rangle v_{F,0}] = P^1 v , \qquad (9.57a)$$

$$Y = F^{-1}[\bm{1}c_\Delta \langle Fv, w_{F,0}\rangle v_{F,0} + \sum_n \langle Fv, w_{F,n}\rangle v_{F,n}] = P^2 v , \qquad (9.57b)$$

and $\langle \cdot; \cdot \rangle$ is the scalar product with respect to which the adjoint of L_F is defined. On the other hand, we set

$$\bm{1}_\Delta = 1 \text{ if } \bm{k} \in \Delta \quad \text{and} \quad \bm{1}_\Delta = 0 \text{ if } \bm{k} \notin \Delta , \qquad (9.58)$$

with an analogous definition for $\bm{1}c_\Delta$ where c_Δ is the complement of Δ in the space of $\bm{k}$.

With these notations, we find, for X and Y, the following system of two equations, first derived in Guiraud and Zeytounian (1978):

$$\frac{dX}{dt} = L_N X + P^1 Q(X+Y, X+Y) , \tag{9.59a}$$

$$\frac{dY}{dt} = L_A Y + P^2 Q(X+Y, X+Y) . \tag{9.59b}$$

The main characteristic of the problem is that X belongs to a space of infinite dimension, in contrast to the confined case when it is of finite dimension. This is a source of very great mathematical difficulties. On the other hand, if we write

$$X = F^{-1}[\mathbf{1}_\Delta(\boldsymbol{k})\Psi(\boldsymbol{k},\boldsymbol{x}_\perp)v_{F,0}(\boldsymbol{k},\boldsymbol{x}_\perp)] , \tag{9.60}$$

we see that the dimension of the space in which X varies is "small" in a somewhat loosely defined sense, due to the fact that the measure of Δ is small. The main problem is how to take care of this observation!

Two points deserve attention. The first point is that $P^1 Q$ and $P^2 Q$ are rather awkwardly defined, and this is a heavy constraint when one wants to bring the calculations to completion in any specific situation. The second point is that ∇p comes into L_N, L_A, and also into $P^1 Q$ and $P^2 Q$, and this, again, is a source of complication in any specific situation. This was observed by Davey et al. (1974). Before, we considered the structure of a typical quasi-linear term in $Q(v,v)$:

$$\boldsymbol{u} \cdot \nabla \boldsymbol{u} = F^{-1}\left[J(F\boldsymbol{u}; F\boldsymbol{u}) \right] , \tag{9.61a}$$

with

$$J(\boldsymbol{f};\boldsymbol{f}) = \boldsymbol{f}_\| \div i\boldsymbol{k}\boldsymbol{f} + \boldsymbol{f}_\perp \div \nabla \boldsymbol{f} , \tag{9.61b}$$

where we have defined the convolution of $\boldsymbol{f}$ with $\boldsymbol{g}$

$$\boldsymbol{f} \div \boldsymbol{g}(\boldsymbol{k}) \equiv \int_{E_{F\|}} \boldsymbol{f}(\boldsymbol{k}') \cdot \boldsymbol{g}(\boldsymbol{k}-\boldsymbol{k}')\,d\boldsymbol{k}' . \tag{9.61c}$$

In (9.61c), $E_{F\|}$ stands for the space dual to $E_\|$ (see (9.37)). If $\boldsymbol{f}$ and $\boldsymbol{g}$ are both restricted to belong to X, we must restrict (9.61c) according to $\boldsymbol{k}' \in \Delta$ and $\boldsymbol{k} - \boldsymbol{k}' \in \Delta$, and if, furthermore, we want to compute $P^1 Q(X,X)$, we must restrict $\boldsymbol{k}$ to Δ. This observation is crucial for the analysis that follows.

9.3.4 Nonlinear Theory – Unconfined Perturbations. Tollmien–Schlichting Waves

The domain Δ is centered on $(a_0, 0)$, and we set

$$a = a_0 + \varepsilon a' , \quad b = \varepsilon b' , \tag{9.62a}$$

9.3 The Guiraud–Zeytounian Asymptotic Approach

with

$$\text{Real}\left[\frac{\partial \sigma_0(a_0,0)}{\partial a}\right] = 0 . \tag{9.62b}$$

a' and b' are $O(1)$ on Δ. We note also that $(a_0, 0)$ is the point where $\text{Real } \sigma_0 = \lambda_0$ has a maximum. Within Δ, we may set

$$\sigma_0 = -i\lambda_0 - \varepsilon i l^0 a' + \varepsilon^2 \left[\gamma_0 - \frac{1}{2}(\alpha a'^2 + \beta b'^2)\right] , \tag{9.63}$$

where λ_0, l^0, γ_0 are real and α and β are complex. As we consider an unstable configuration, we may assume that $\gamma_0 > 0$. Taking care of the fact that, in (9.59a), L_N corresponds to σ_0, we may set

$$L_N = A + \varepsilon B + \varepsilon^2 C + O(\varepsilon^3) . \tag{9.64}$$

Now, if we neglect the nonlinear terms, we find after decay of the transients for $t \gg 1$, that

$$v = \varepsilon^2 \, \text{Real} \left\{ \left[\exp \frac{(-\varepsilon^2 \gamma_0 t)}{t}\right] A(Fv^0; w_{F,0,0}) v_{F,0,0} E \right\} , \tag{9.65a}$$

where

$$A = A\left(\frac{x - l^0 t}{t^{1/2}}, \frac{y}{t^{1/2}}\right) . \tag{9.65b}$$

In (9.65a), we have set $E = \exp[i(a_0 x - \lambda_0 t)]$, the space-time modulation of the most amplified Tollmien–Schlichting wave according to the linear theory [for details, see Guiraud and Zeytounian (1978)]. The initial value of v has been taken as v^0, and $w_{F,0,0}$ and $v_{F,0,0}$ are the values assumed by $w_{F,0}$ and $v_{F,0}$ for $a = a_0$ and $b = 0$.

Finally, A is a function that depends on α and β. When passing to the nonlinear part of the theory, we make the fundamental observation that (9.45) holds in this case. The reason is that $P^1Q(X,X)$ is the sum of terms which are zero, thanks to the presence of the three factors $\mathbf{1}_\Delta$.

Now for the resolution of nonlinear equations (9.59a,b), we use the perturbation technique

$$X = \varepsilon(X_0' + \varepsilon X_1' + \varepsilon^2 X_2' + \ldots) , \tag{9.66a}$$

$$Y = \varepsilon^2(Y_0' + \ldots) , \tag{9.66b}$$

with various time and space scales suggested by (9.63) and (9.65a,b):

$$t_0 = t , \quad t_1 = \varepsilon t , \quad t_2 = \varepsilon^2 t , \tag{9.67a}$$

$$\tau = \varepsilon^4 \frac{\exp(\gamma_0 t_2)}{t_2} , (\xi, \eta) = \varepsilon \boldsymbol{x}_\| . \tag{9.67b}$$

Substituting (9.66a,b) and (9.67a,b) into (9.59a,b), we obtain the following system of equations for X_0' and Y_0':

$$\frac{dX_0'}{dt_0} - A X_0'$$
$$+ \varepsilon \left[\frac{dX_1'}{dt_0} - AX_1' + \frac{dX_0'}{dt_1} - BX_0'\right]$$
$$+ \varepsilon^2 \left[\frac{dX_2'}{dt_0} - A X_2' + \frac{dX_1'}{dt_1} - BX_1'\right.$$
$$\left. + \left(\gamma_0\tau - \frac{\tau}{t_2}\right)\frac{dX_0'}{d\tau} - CX_0' - P^1 M(X_0', Y_0')\right] + \ldots = 0, \quad (9.68)$$

$$\frac{dY_0'}{dt_0} = L_A Y_0' + P^2 Q(X_0', X_0'). \quad (9.69)$$

Solving for X_0' and then eliminating the secular terms in X_1', we get the following solution:

$$X_0' = \text{Real}[q_0(\tau, \xi - l^0 t_1, \eta) v_{F,0,0} E]. \quad (9.70a)$$

Then solving for Y_0' and noting that for $\tau = O(1)$, the transients have decayed, we obtain

$$Y_0' = \text{Real}[q_0^2 Y^* E^2)] + |q_0|^2 Y^{**}, \quad (9.70b)$$

where the terms Y^* and Y^{**} are defined in Guiraud and Zeytounian (1978, p. 395).

Finally, from (9.68), eliminating the secular terms in the equation for X_2', we find the following evolutionary equation:

$$\left(\gamma_0\tau - \frac{\tau}{t_2}\right)\frac{dX_0'}{d\tau} - CX_0' - \text{Res}[P^1 M(X_0', Y_0')] = 0, \quad (9.71)$$

where $\text{Res}[f]$ means that we use only the resonant part of f, that is the one oscillates according to the modulation E of the Tollmien–Schlichting wave.

Now, we know from the above that whenever $t_2 = O(1)$, the nonlinear term in (9.71) is of a higher order than the linear ones. We have to wait till $\tau = O(1)$ for the nonlinear term in (9.71) to be of the same order as the others, and then $\tau/t_2 \ll 1$.

Accordingly, taking into account the form of the operator C and the work of Davey et al. (1974), who computed the nonlinear term, we find that

$$\gamma_0\tau\frac{dq_0}{d\tau} - \gamma_0 q_0 - \frac{1}{2}\left[\alpha\frac{\partial^2 q_0}{\partial \xi^2} + \beta\frac{\partial^2 q_0}{\partial \eta^2}\right] = F(q_0), \quad (9.72a)$$

with

$$F(q_0) = K_1 q_0 |q_0|^2 + K_2 q_0 \Pi , \qquad (9.72b)$$

and

$$\frac{\partial^2 \Pi}{\partial \xi^2} + \frac{\partial^2 \Pi}{\partial \eta^2} = \frac{\partial^2 (|q_0|^2)}{\partial \eta^2} , \qquad (9.72c)$$

where K_1 and K_2 are constants.

As this is elucidated in Guiraud and Zeytounian (1978), *the process by which, from a fairly arbitrary initial perturbation, the wave packet is first organized and then evolves is related to four times scales of evolution – only during the last one do the nonlinear terms come into play, and the envelope equation (9.72a), with (9.72b,c), is relevant.*

9.3.5 Nonlinear Theory – Unconfined Perturbations. Rayleigh–Bénard Convection

These were treated in an incomplete way by De Coninck (1979). Here, we complete the analysis, but we give only the main points of this analysis; for the details, see De Conink et al. (1983).

We restate that we must deal with the velocity $\boldsymbol{u} = \boldsymbol{u}_T + w\boldsymbol{e}_z$ and the perturbed temperature θ. We also use the vertical component of vorticity $\omega = \boldsymbol{e}_z \cdot (\nabla \wedge \boldsymbol{u})$. Consequently, as v we use $(\omega, w, \theta)^T$ in the notation of a column matrix, and in such case we have for $Q(v,v)$:

$$Q(v,v) = \begin{pmatrix} -\boldsymbol{e}_z \cdot \{\nabla \wedge [(\boldsymbol{u} \cdot \nabla)\boldsymbol{u}]\} \\ \Delta^{-1}\{-\Delta(\boldsymbol{u} \cdot \nabla w) + D\nabla \cdot [(\boldsymbol{u} \cdot \nabla)\boldsymbol{u}]\} \\ -\boldsymbol{u} \cdot \nabla \theta \end{pmatrix} , \qquad (9.73)$$

with $D = \partial/\partial z$, and it might seem from (9.73) that Q does not depend only on v, but we readily find that

$$\boldsymbol{u}_T = F^{-1}\{|\boldsymbol{k}|^{-2}[i\boldsymbol{k}Dw_F - (\boldsymbol{e}_z \wedge i\boldsymbol{k})w_F]\} , \qquad (9.74)$$

and we see that $Q(v,v)$ is a quadratic function of v.

Now, we use (9.57a,b), (9.58), and (9.59a,b) again, but we must change from Cartesian coordinates (9.62) to polar defined by

$$\boldsymbol{k} = [k_c + \varepsilon k^*]\boldsymbol{e}(\phi) , \qquad (9.75)$$

where $\boldsymbol{e}(\phi)$ is a unit vector making the angle ϕ with a reference unit vector. Then on small area Δ, $k^* = O(1)$. Substituting (9.75) in

$$X = \frac{1}{(2\pi)^2} \iint_\Delta \exp(i\boldsymbol{k} \cdot \boldsymbol{x}_\parallel) H(\boldsymbol{k}, t) v_{F,0}(|\boldsymbol{k}|, z) \, d\boldsymbol{k} , \qquad (9.76)$$

we get

$$X = \frac{\varepsilon}{2\pi} \int_0^{2\pi} k_c \exp[ik_c \boldsymbol{x}_\| \cdot \boldsymbol{e}(\phi)] \Psi(\varepsilon \boldsymbol{x}_\| \cdot \boldsymbol{e}(\phi), \phi, t) v_{F,0,c}(z) \, d\phi \,, \tag{9.77}$$

where $v_{F,0,c}$ stands for $v_{F,0}$ evaluated at $|\boldsymbol{k}| = k_c$ and where the definition of Ψ may be found according to (9.76) and (9.77) from

$$\Psi(\psi, \phi, t) = \frac{1}{2\pi} \int_I \exp(ik^*\psi) H^*(k^*, \phi, t) \, dk^* \,, \tag{9.78}$$

where $\psi = \varepsilon \boldsymbol{x}_\| \cdot \boldsymbol{e}(\phi)$, I is a small segment (k^* sweeps this small segment), and in (9.78), we have written H^* for H expressed as a function of k^* and ϕ. Provided that k_c is chosen conveniently, we find that the operator $(d/dt) - L_N$ applied to X is equivalent to the operator,

$$\Lambda = \frac{\partial}{\partial t} - \varepsilon^2 \left[\delta + \gamma \frac{\partial^2}{\partial \psi^2} \right] \,, \tag{9.79}$$

applied to Ψ, and our main task, now, is to evaluate $P^1 Q(X, X)$ which is *not zero*, a point whose *importance* seems to have been noticed for the first time by De Conink (1979). With $|I|$ standing for the length of the interval I, we find the following expression:

$$P^1 Q(X, X) = \left\{ \left(\frac{3^{1/2}}{2} \right) \varepsilon^3 |I| \int_0^{2\pi} k_c F^0 \exp\left[\frac{ik\psi}{\varepsilon} \right] \Psi^+ \Psi^- \, d\phi \right\} v_{F,0,c}, \tag{9.80}$$

where

$$\Psi^+ = \Psi\left(\varepsilon \boldsymbol{x}_\| \cdot \boldsymbol{e}(\phi + \pi/3), \, \phi + \pi/3, t\right) \,, \tag{9.81a}$$

and

$$\Psi^- = \Psi\left(\varepsilon \boldsymbol{x}_\| \cdot \boldsymbol{e}(\phi - \pi/3), \, \phi - \pi/3, t\right) \,. \tag{9.81b}$$

For a detailed derivation of (9.80), see Guiraud and Zeytounian (1978, pp. 14 and 15).

Finally, as we have seen that

$$\frac{dX}{dt} - L_N X = \frac{\varepsilon}{2\pi} \int_0^{2\pi} k_c \exp\left[\frac{ik_c \psi}{\varepsilon} \right] \Lambda \Psi v_{F,0,c}(z) \, d\phi \,, \tag{9.82}$$

we obtain, to leading-order, the following equation for Ψ from (9.59a), (9.80), and (9.82), with (9.79):

$$\frac{\partial \Psi}{\partial \tau} - \delta \Psi + \gamma \frac{\partial^2 \Psi}{\partial \psi^2} = \kappa \Psi^+ \Psi^- \,, \tag{9.83}$$

where

$$\tau = \varepsilon^2 t, \quad \kappa = 3^{1/2}\pi|I|F^0,$$
$$\Psi^\pm = \Psi(\psi^\pm, \phi^\pm, t), \quad \text{with } \phi^\pm = \phi \pm \pi/3,$$
$$\psi^\pm = \varepsilon \boldsymbol{x}_\| \cdot \boldsymbol{e}(\phi \pm \pi/3). \tag{9.84}$$

In (9.80) and (9.84), the coefficient F^0 is as defined in Guiraud and Zeytounian [1978, p. 14, formula (5.36)].

Equation (9.83) is the main result of our analysis of RB waves, and γ and δ are real constants that are known from linear theory. The next task, a rather cumbersome one, would be to compute the constant κ.

The result should be unaffected by a change of axis, that is, by the change $\phi \to \phi + \chi$, and consequently, κ must be independent of ϕ. The case when Δ is a small circular ring in a wave-number space of dimension two, which occurs for RB convection, is highly special.

Each of the finite-dimensional systems, referred to above, occurs as a superposition of six modes with wave-number representatives forming a hexagonal pattern. The six-dimensional system may be described by prescribing an amplitude for each vertex of the hexagon. The so-called hexagonal pattern of cellular convection is recovered when all six amplitudes are equal.

The rolls correspond to the vanishing of all of the amplitudes but one. However, the quadratic interaction would then vanish identically.

Of course, the identification of one particular system with hexagonal cellular convection neglects the small spreading over the transversal dimension of Δ, but this appears as a slight change accounted for by diffusion.

Is the hexagonal cellular convection an equilibrium solution of any of the decoupled systems referred to above? De Coninck (1979) answered this question in the affirmative but only when diffusion is neglected. He has also shown that this equilibrium solution is always unstable.

We do not know what happens when diffusion is taken into account. We think that it is not possible to make more precise statement on the basis of the present investigation, and we stress again that our objective was to derive the extended bifurcation equation, not to discuss its solution.

9.4 Some Facets of the RB and BM Problem

9.4.1 Rayleigh–Bénard Convective Instability

In Sect. 7.5 of Chap. 7, we first derived the RB problem (7.74). In the RB problem are three similarity parameters, Gr, δ^*, and We*, and three dimensionless parameters, Pr, Ma, and Bi.

In Dauby and Lebon (1996), precisely this RB model problem was recently considered. For the derivation from (7.74) of the starting problem considered in Dauby and Lebon (1996), it is sufficient to introduce the following new functions and new variables in RB model problem (7.74):

$$\Theta = \theta_0 + z - 1, \quad \Pi = \Pr\left[\operatorname{Gr} z\left(\frac{z}{2} - 1\right) + \pi_0\right],$$

$$(u, v, w) = \Pr(v_{10}, v_{20}, v_{30}),$$

$$\tau = \frac{t}{\Pr}, \quad (x, y, z) = (x_1, x_2, x_3). \tag{9.85}$$

In this case, in place of problem (7.74), we obtain the following dimensionless RB model problem for the velocity vector $\boldsymbol{v} = (u, v, w)$, the perturbation of pressure Π, and the perturbation of temperature Θ as functions of τ, x, y, z:

$$\nabla \cdot \boldsymbol{v} = 0,$$

$$\frac{1}{\Pr}\frac{d\boldsymbol{v}}{d\tau} = -\nabla\Pi + \operatorname{Ra}\Theta\boldsymbol{k} + \nabla^2\boldsymbol{v},$$

$$\frac{d\Theta}{d\tau} = w + \nabla^2\Theta, \tag{9.86a}$$

$$\boldsymbol{v} = \Theta = 0 \quad \text{at } z = 0, \tag{9.86b}$$

$$w = 0 \quad \text{at } z = 1,$$

$$\frac{\partial u}{\partial z} = -\operatorname{Ma}\frac{\partial \Theta}{\partial x} \quad \text{at } z = 1,$$

$$\frac{\partial v}{\partial z} = -\operatorname{Ma}\frac{\partial \Theta}{\partial y} \quad \text{at } z = 1,$$

$$\frac{\partial \Theta}{\partial z} + \operatorname{Bi}\Theta = 0 \quad \text{at } z = 1. \tag{9.86c}$$

In second of the equations (9.86a), $\boldsymbol{k}$ is the unit vector along the z axis, the material time derivative is denoted by $d/d\tau = \partial/\partial\tau + \boldsymbol{v} \cdot \nabla$, the gradient operator is $\nabla = (\partial/\partial x, \partial/\partial y, \partial/\partial z)$, and the Rayleigh number Ra is defined by $\operatorname{Ra} = \Pr\operatorname{Gr}$. The last boundary condition in (9.86c) for Θ is written in classical form when $\operatorname{Bi}_s \equiv \operatorname{Bi}^0 \equiv \operatorname{Bi}$ (see (7.68c)).

The RB classical thermal convection problem. From the formulated RB problem (9.86a–c), if we assume that the Marangoni and Biot effects are *both negligible*, we derive the classical RB problem for the thermal instability of a layer of fluid heated from below for the *"rigid-free"* case:

$$\nabla \cdot \boldsymbol{v} = 0,$$

$$\frac{1}{\Pr}\frac{D\boldsymbol{v}}{D\tau} = -\nabla\Pi + \Theta\boldsymbol{k} + \nabla^2\boldsymbol{v},$$

$$\frac{D\Theta}{D\tau} = \mathrm{Ra}\, w + \nabla^2 \Theta\,,$$

$\boldsymbol{v} = \Theta = 0 \quad \text{at } z = 0\,,$

$w = 0 \quad \text{at } z = 1\,,$

$$\frac{\partial u}{\partial z} = \frac{\partial v}{\partial z} = 0 \Rightarrow \frac{\partial^2 w}{\partial z^2} = 0 \quad \text{at } z = 1\,,$$

$$\frac{\partial \Theta}{\partial z} = 0 \quad \text{at } z = 1\,. \tag{9.87}$$

Usually, in place of the last boundary condition $\partial \Theta / \partial z = 0$, at $z = 1$, it is assumed that the surface $z = 1$ is maintained at constant temperature $\Theta = 0$.

For the *rigid-rigid* case, in place of the boundary condition,

$$\frac{\partial^2 w}{\partial z^2} = 0\,, \quad \text{at } z = 1\,, \text{ we write } \quad \frac{\partial w}{\partial z} = 0\,, \quad \text{at } z = 1\,. \tag{9.88a}$$

Finally, for the *free-free* problem, we impose

$$\Theta = 0\,,\ w = 0\,,\ \frac{\partial^2 w}{\partial z^2} = 0 \quad \text{at } z = 0 \text{ and } z = 1\,. \tag{9.88b}$$

Linear and nonlinear RB stability problems with the effects of Ma and Bi. The linear RB stability problem consists of determining the critical value of the Marangoni or Rayleigh number above which convection sets in. This is achieved by first linearizing (9.86a). The boundary conditions (9.86b,c), which are linear, keep the same form. Then, an exponential time dependence of the form $\exp(\sigma t)$ for all of the variables is introduced in the linear equations, which results in an eigenvalue problem for the Marangoni number, the Rayleigh number, or the growth rate σ:

$$\nabla \cdot \boldsymbol{v} = 0\,,$$

$$\nabla^2 \boldsymbol{v} - \nabla \Pi + \mathrm{Ra}\, \Theta \boldsymbol{k} = \left(\frac{\sigma}{\mathrm{Pr}}\right) \boldsymbol{v}\,,$$

$$\nabla^2 \Theta + w = \sigma \Theta\,. \tag{9.89}$$

To account for the evolution of the convective pattern above a threshold, it is convenient to follow a method first introduced by Eckhaus (1965). This method consists first of expanding the solution of the nonlinear equations in a series of eigenmodes of the linear problem. The time-dependent coefficients are the so-called "amplitudes" of the different modes of convection. The eigenmodes used here are the eigenmodes of (9.89) considered as an eigenvalue problem for the growth rate σ, when the Marangoni number Ma is fixed and equal to critical Marangoni number $\mathrm{Ma_c}$ – the smallest eigenvalue of this problem is zero and corresponds to the marginally stable convective mode.

Note that the adjoint linear eigenvalue problem is required to apply the Eckhaus method. The equations of the adjoint linear eigenvalue problem can be written as

$$\nabla \cdot \boldsymbol{v}^* = 0,$$

$$\nabla^2 \boldsymbol{v}^* - \nabla \Pi^* + \Theta^* \boldsymbol{k} = \left(\frac{\sigma}{\mathrm{Pr}}\right) \boldsymbol{v}^*,$$

$$\nabla^2 \Theta^2 + \mathrm{Ra}\, w^* = \sigma \Theta^*, \qquad (9.89^*)$$

where the starred quantities refer to the adjoint problem's unknown fields. The boundary conditions of the above adjoint equations (9.89*) are:

$$\boldsymbol{v}^* = \Theta^* = 0 \quad \text{at } z = 0,$$

$$\frac{\partial u^*}{\partial z} = \frac{\partial v^*}{\partial z} = w^* = 0 \quad \text{at } z = 1,$$

$$\frac{\partial \Theta}{\partial z} + \mathrm{Bi}\, \Theta + \mathrm{Ma}\, \frac{\partial w^*}{\partial z} = 0 \quad \text{at } z = 1. \qquad (9.86\mathrm{b,c}^*)$$

The direct system (9.89) and its adjoint (9.89*) have the same eigenvalues. Moreover, if both systems are considered as eigenvalue problems for the growth rate σ, when the Rayleigh and Marangoni numbers are fixed, the following bi-orthogonality relations between the eigenmodes of both problems are satisfied [see Dauby and Lebon (1996, Appendix B)]:

$$(\sigma_p - \sigma_q) \left\langle \Theta_p^* \Theta_q + \left(\frac{1}{\mathrm{Pr}}\right) \boldsymbol{v}_p^* \cdot \boldsymbol{v}_q \right\rangle = 0. \qquad (9.90)$$

In (9.90), $\langle \ldots \rangle$ denotes the integral over the fluid volume (a rigid rectangular container – see below) and indexes p and q characterize eigenmodes of the direct and adjoint problems, respectively. $(\boldsymbol{v}_p, \Theta_p)$ denotes the eigenmodes of the linear eigenvalue problem (9.89), (9.86b,c), for σ with $\mathrm{Ma} = \mathrm{Ma_c}$, and $(\boldsymbol{v}_p^*, \Theta_p^*)$ denotes an eigenmode, with eigenvalue σ_p, of the adjoint eigenvalue problem, (9.89*), (9.86b,c*) considered as an eigenvalue problem for the growth rate when $\mathrm{Ma} = \mathrm{Ma_c}$.

Dauby and Lebon (1996) consider the RB instability in rigid rectangular containers (if the thickness of the layer is equal to d, as in Sect. 9.5, then the length and width of the container considered are $a_1 d$ and $a_2 d$, respectively, and a_1 and a_2 are the aspect ratios) and in such a case, it is also necessary to impose the following boundary conditions for the direct problem, when the sidewalls are adiabatically insulated and rigid:

$$\boldsymbol{v} = \frac{\partial \Theta}{\partial x} = 0 \quad \text{at } x = 0 \text{ and } a_1, \qquad (9.91\mathrm{a})$$

$$\boldsymbol{v} = \frac{\partial \Theta}{\partial y} = 0 \quad \text{at } y = 0 \text{ and } a_2. \qquad (9.91\mathrm{b})$$

Correspondingly, for the adjoint problem, we write

$$v^* = \frac{\partial \Theta^*}{\partial x} = 0 \quad \text{at } x = 0 \text{ and } a_1 ; \tag{9.91a*}$$

$$v^* = \frac{\partial \Theta^*}{\partial y} = 0 \quad \text{at } y = 0 \text{ and } a_2 . \tag{9.91b*}$$

Now, if we assume that the eigenmodes form a complete set (it is assumed that the eigenmodes are ordered by deceasing growth rate σ_p, $\sigma_p \leq \sigma_q \leq 0$ if $p > q$), one may write

$$[v, \Theta] = \sum_p A_p(t) \left[v_p, \left(\frac{\text{Ma}_c}{\text{Ma}}\right) \Theta_p \right] , \tag{9.92}$$

and derive the following evolution equation for the amplitudes $A_p(t)$:

$$\frac{dA_p}{dt} = \sigma_p A_p + \varepsilon \sum_q T_0(p,q) A_q + \sum_{q,r} T_1(p,q,r) A_q A_r , \tag{9.93}$$

where

$$\varepsilon = \frac{\text{Ma}_c}{\text{Ma}} - 1 , \tag{9.94}$$

is the relative distance to the threshold.

The matrices in (9.93) are given by

$$T_0(p,q) = \frac{\langle \Theta_p^* w_q \rangle}{\tau_{pq}} , \tag{9.95a}$$

$$T_1(p,q,r) = \frac{\langle \Theta_p^* \alpha_{q,r} + (1/\Pr) v_p^* \cdot U_{q,r} \rangle}{\tau_{pq}} , \tag{9.95b}$$

with

$$\tau_{pq} = \langle \Theta_p^* \Theta_p \rangle + \left(\frac{1}{\Pr}\right) \langle v_p^* \cdot v_p \rangle , \tag{9.96a}$$

where

$$\alpha_{q,r} = (v_q \cdot \nabla \Theta_r) \quad \text{and} \quad U_{q,r} = (v_q \cdot \nabla) v_r . \tag{9.96b}$$

The set (9.93) forms an infinite-dimensional system of nonlinear ordinary differential equations for the amplitudes of the eigenmodes and is the basis of the nonlinear analysis. The use of a finite number of equations can be justified by noting that, in the weakly nonlinear regime, only a small number of eigenmodes are linearly unstable whereas all other modes have negative growth rates. Obviously, it is convenient to share out the eigenmodes into two sets. The first set contains the "unstable" eigenmodes which bifurcate in the vicinity of the threshold and which are thus actually unstable in the neighborhood of $\varepsilon = 0$. The ε value defining the bifurcation of the successive eigenmodes is easily obtained from (9.93):

$$\varepsilon_p = -\frac{\sigma_p}{T_0(p,q)}, \qquad (9.97)$$

where ε_p denotes the value of ε at which eigenmode p bifurcates.

The second set of eigenmodes is made up of "stable" eigenmodes whose linear growth rate remains negative in the weakly nonlinear regime. In the weakly nonlinear regime, this growth rate is given by

$$\sigma_p^* = \sigma_p + \varepsilon T_0(p,q). \qquad (9.98)$$

These stable eigenmodes do not really participate in the dynamics of the system, but they cannot be omitted because they represent the nonlinear response of the system to the growth of unstable modes. These modes are generated by quadratic interactions of the first unstable modes, and their amplitudes should be expressed as a quadratic expression in the amplitudes of unstable modes.

Equations (9.93) for stable modes will be simplified in the following way.

First, the term proportional to ε in (9.93) may be disregarded with respect to the first term, $\sigma_p A_p$, because ε remains small in the weakly nonlinear regime and σ_p is quite negative for a well-damped eigenmode.

Second, the time derivative of the amplitude is neglected because stable modes do not participate in the dynamics of the system.

Finally, the quadratic term may be restricted to a quadratic expression containing only the amplitudes of unstable modes because, in the neighborhood of the threshold, the amplitudes of unstable modes remain rather small and terms of order higher than two in these amplitudes may be neglected.

Such a procedure for reducing the dynamics of the system to a finite number of ordinary differential equations for the most unstable modes is the well-known *adiabatic elimination of slaved modes* [see, for instance, Manneville (1990)]. This procedure has been applied by Dauby and Lebon (1996, pp. 39–41) to the study of RB convection with Marangoni and Biot effects in no-slip rectangular vessels.

One of the main successes of the Dauby and Lebon (1996) analysis is the theoretical interpretation of Koschmieder and Prahl's experiments (1990). These authors have been able to recover all of the convective patterns observed by Koschmieder and Prahl (1990, Fig. 2) for aspect ratios smaller that eight, but for Ra and Bi, both $= 0$, and Pr $= 10^4$.

In Figs. 9.1a,d,e and 9.1h,k, [reproduction of Fig. 12 of Dauby and Lebon (1996)], the flow patterns (the iso-values of w at the mid-depth of the fluid layer) in the weakly nonlinear pure Marangoni convection regime (Ra $=$ Bi $= 0$) for quasi-square containers at Pr $= 10^4$ are represented. In these figures, ε in percent is the relative distance to the threshold.

9.4 Some Facets of the RB and BM Problem   343

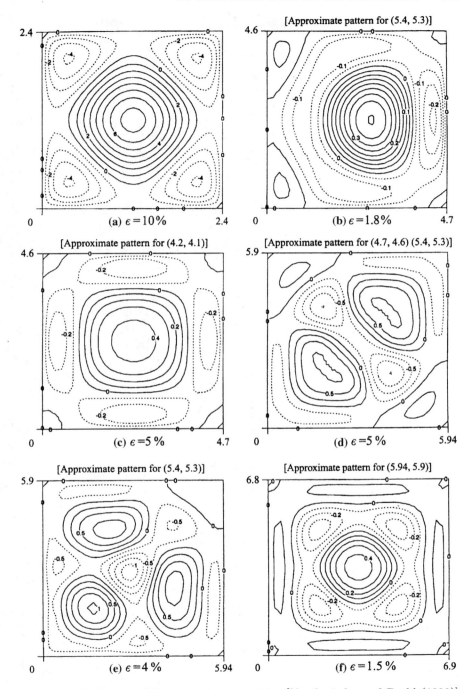

Fig. 9.1a–f. Corresponding experiment pattern [Koschmieder and Prahl (1990)]: (**a**) one-cell solution; (**d**) two-cell solution; (**e**) three-cell solution

344 9 Linear and Nonlinear Stability of Fluid Motion

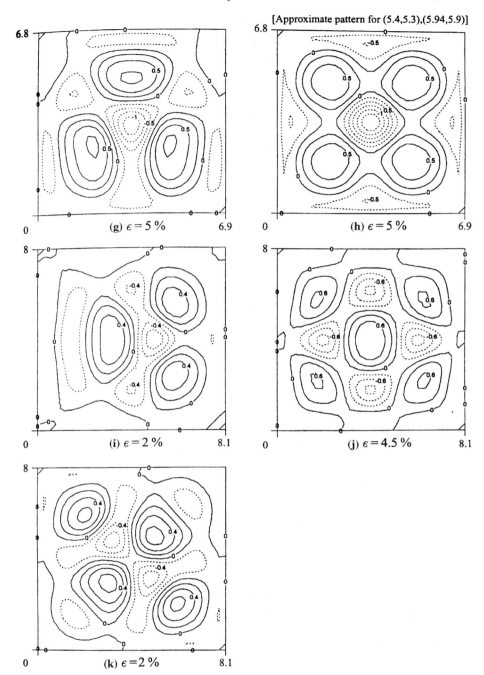

Fig. 9.1g–k. Corresponding experiment pattern [Koschmieder and Prahl (1990)]: (**h**) and (**k**) four-cell solution; (**j**) five-cell solution

Similarity between the observations of Koschmieder and Prahl and the theoretical prediction [according to the Eckhaus method via the evolution equation (9.93) for the amplitudes] does not usually occur very near the threshold – the patterns of Koschmieder and Prahl are thus probably not critical structures.

Asymptotic derivation of an amplitude equation for the RB free-free convection problem. Below we consider the Boussinesq equations:

$$\nabla \cdot \boldsymbol{v} = 0,$$

$$\frac{1}{\Pr} \frac{D\boldsymbol{v}}{D\tau} = -\nabla \Pi + \Theta \boldsymbol{k} + \nabla^2 \boldsymbol{v},$$

$$\frac{D\Theta}{D\tau} = \operatorname{Ra} w + \nabla^2 \Theta, \tag{9.99a}$$

with the boundary conditions (*free-free problem*),

$$\Theta = 0, \ w = 0, \ \frac{\partial^2 w}{\partial z^2} = 0 \quad \text{at} \quad z = 0 \ \text{and} \ z = 1. \tag{9.99b}$$

As usual in weakly nonlinear analyses, one is constrained to values of Ra close to the critical Rayleigh $\operatorname{Ra_c}$, and a "supercriticality" parameter r of $O(1)$ is introduced such that

$$\operatorname{Ra} = \operatorname{Ra_c} + \varepsilon^2 r. \tag{9.100}$$

In the free-free case, the system first becomes unstable when $\operatorname{Ra_c} = 27\pi^4/4$, where upon a periodic array of convection rolls sets in a critical wave number $k_c = \pi/\sqrt{2}$.

First, it is necessary to introduce the slow scales

$$\xi = \varepsilon x \quad \eta = \varepsilon^{1/2} y, \quad \tau = \varepsilon^2 t. \tag{9.101}$$

The choice (9.101) of slow scales is motivated by the behavior of the linear growth rate in the vicinity of $(k_c, \operatorname{Ra_c})$, and by the expected form of the leading-order nonlinearity in the final amplitude equation. All dependent variables are expanded according to

$$(u, v, w, \Pi, \Theta) \equiv \mathcal{U} = \varepsilon \mathcal{U}_1 + \varepsilon^{3/2} \mathcal{U}_{3/2} + \varepsilon^2 \mathcal{U}_2$$

$$+ \varepsilon^{5/2} \mathcal{U}_{5/2} + \varepsilon^3 \mathcal{U}_3 + \ldots \tag{9.102}$$

where

$$\mathcal{U}_n = \mathcal{U}_n^{(0)}(\xi, \eta, \tau, z)$$

$$+ \operatorname{Real} \left[\sum_{m=1}^{N} \mathcal{U}_n^{(m)}(\xi, \eta, \tau, z) \exp(imk_c x) \right], \tag{9.103}$$

with

$$n = 1 + \frac{p}{2}, \quad p = 0, 1, 2, 4, \ldots$$

In Zeytounian (1991b, pp. 378–392), the reader can find a detailed derivation of the amplitude evolution equation for the amplitude $A(\xi, \eta, \tau)$ of the $O(\varepsilon)$ problem. Below, we give only the main steps of this asymptotic derivation.

The first step is the substitution of expansion (9.103) in the governing equations (9.99a) and boundary conditions (9.99b) after introducing of (9.100) and the slow scales (9.101).

Because $\mathcal{U}_1^{(0)} = 0$, describing the periodic array of convection rolls, the solution of the $O(\varepsilon)$ problem ($n = 1$, $m = 1$) for $\mathcal{U}_1^{(1)}$ is obtained in the following form:

$$u_1^{(1)} = \pi A \cos(\pi z), \quad v_1^{(1)} = 0, \quad w_1^{(1)} = -i k_c A \sin(\pi z),$$

$$\Theta_1^{(1)} = -\frac{9i}{2\sqrt{2}} \pi^3 A \sin(\pi z),$$

$$\Pi_1^{(1)} = \frac{i}{k_c} \pi (\pi^2 + k_c^2) A \cos(\pi z), \tag{9.104}$$

where the complex amplitude $A(\xi, \eta, \tau)$ is, at this stage, an unknown function to be determined at higher order by applying of a suitable orthogonality condition [elimination of secular terms according to the multiple-scale method (MSM)].

For the $O(\varepsilon^{3/2})$ problem, one finds $\mathcal{U}_{3/2}^{(0)} = 0$ in a straightforward manner, and then,

$$u_{3/2}^{(1)} = 0, \quad v_{3/2}^{(1)} = -\frac{\pi}{k_c} i \cos(\pi z) \frac{\partial A}{\partial \eta},$$

$$w_{3/2}^{(1)} = 0, \quad \Theta_{3/2}^{(1)} = \Pi_{3/2}^{(1)} = 0. \tag{9.105}$$

At the $O(\varepsilon^2)$ order, the problem is more complicated because $\mathcal{U}_2^{(0)}$ is different from zero and $w_2^{(1)}$ and $\Theta_2^{(1)}$ are solutions of a nonhomogeneous system of two equations. We derive the following system of equations for the components of $\mathcal{U}_2^{(0)}$:

$$\frac{\partial u_2^{(0)}}{\partial z} = 0, \quad \frac{\partial v_2^{(0)}}{\partial z} = 0, \quad \frac{\partial w_2^{(0)}}{\partial z} = 0,$$

$$\frac{\partial \Pi_2^{(0)}}{\partial z} - \Theta_2^{(0)} = -\frac{\pi}{2 \Pr} k_c^2 |A|^2 \sin(2\pi z),$$

$$\frac{\partial^2 \Theta_2^{(0)}}{\partial z^2} = \frac{9}{4\sqrt{2}} \pi^4 k_c |A|^2 \sin(2\pi z) - \text{Ra}_c\, w_2^{(0)}. \tag{9.106}$$

9.4 Some Facets of the RB and BM Problem

The solution of (9.106) is simply

$$u_2^{(0)} = 0, \quad v_2^{(0)} = 0, \quad w_2^{(0)} = 0,$$

$$\Theta_2^{(0)} = -\frac{9}{32}\pi^3 |A|^2 \sin(2\pi z),\qquad(9.107)$$

$$\Pi_2^{(0)} = \frac{1}{8}\left[\frac{1}{\Pr} + \frac{9}{8}\right]\pi^2 |A|^2 \cos(2\pi z).$$

For the components of $\mathcal{U}_2^{(1)}$, we obtain first $v_2^{(1)} = 0$, and then for the two functions $w_2^{(1)}$ and $\Theta_2^{(1)}$, we derive a nonhomogeneous system of two equations:

$$\left(\frac{\partial^2}{\partial z^2} - k_c^2\right) w_2^{(1)} - k_c^2 \Theta_2^{(1)} = F_2,\qquad(9.108\text{a})$$

$$\left(\frac{\partial^2}{\partial z^2} - k_c^2\right) \Theta_2^{(1)} + \mathrm{Ra}_c\, w_2^{(1)} = G_2,\qquad(9.108\text{b})$$

where

$$F_2 = -\frac{3}{2}\pi^4 \left[\frac{\partial A}{\partial \xi} - \frac{i}{2k_c}\frac{\partial^2 A}{\partial \eta^2}\right] \sin(\pi z),\qquad(9.109\text{a})$$

$$G_2 = -\frac{9}{2}\pi^4 \left[\frac{\partial A}{\partial \xi} - \frac{i}{2k_c}\frac{\partial^2 A}{\partial \eta^2}\right] \sin(\pi z),\qquad(9.109\text{b})$$

with boundary conditions

$$w_2^{(1)} = \frac{\partial^2 w_2^{(1)}}{\partial z^2} = \Theta_2^{(1)} = 0 \quad \text{at} \quad z = 0 \text{ and } 1.\qquad(9.110)$$

For this above problem to admit a nontrivial solution, the forcing term (9.109a,b) must be orthogonal to the adjoint eigenfunctions of the homogeneous problem, i.e., to the adjoint eigenfunctions of the equation at $O(\varepsilon)$ with boundary conditions similar to the above (but written for $w_1^{(1)}$ and $\Theta_1^{(1)}$). One readily obtains the adjoint eigenfunctions as $w^* = -3\sin(\pi z)$, and $\Theta^* = \sin(\pi z)$, and the orthogonality condition is the following:

$$\int_0^1 [F_2 w^* + G_2 \Theta^*]\,\mathrm{d}z = 0,\qquad(9.111)$$

and is identically satisfied. Then, the solution for $w_2^{(1)}$ and $\Theta_2^{(1)}$ is

$$w_2^{(1)} = 0, \quad \Theta_2^{(1)} = 3\pi^2 \left[\frac{\partial A}{\partial \xi} - \frac{i}{2k_c}\frac{\partial^2 A}{\partial \eta^2}\right] \sin(\pi z),\qquad(9.112\text{a})$$

and for $u_2^{(1)}$ and $\Pi_2^{(1)}$, we obtain

$$u_2^{(1)} = -\frac{\pi}{ik_c}\left[\frac{\partial A}{\partial \xi} + \frac{1}{ik_c}\frac{\partial^2 A}{\partial \eta^2}\right]\cos(\pi z),\tag{9.112b}$$

$$\Pi_2^{(1)} = i\left(\frac{\pi}{k_c}\right)^3\left[2ik_c\frac{\partial A}{\partial \xi} + \frac{\partial^2 A}{\partial \eta^2}\right]\cos(\pi z).\tag{9.112c}$$

At the $O(\varepsilon^{5/2})$ order, all field variables admit a solution of the form (9.103) with $p = 3$ and $m = 1$, and in this case, we obtain easily the following solution for the components of $\mathcal{U}_{5/2}^{(0)}$:

$$u_{5/2}^{(0)} = 0, \quad \Theta_{5/2}^{(0)} = 0, \quad w_{5/2}^{(0)} = 0,\tag{9.113a}$$

$$v_{5/2}^{(0)} = -\frac{3}{32}\left[\frac{1}{\Pr} + \frac{3}{8}\right]\frac{\partial |A|^2}{\partial \eta}\cos(2\pi z).\tag{9.113b}$$

Then for the components of $\mathcal{U}_{5/2}^{(1)}$, we obtain

$$u_{5/2}^{(1)} = 0, \quad \Theta_{5/2}^{(1)} = 0, \quad w_{5/2}^{(1)} = 0,\tag{9.113c}$$

$$v_{5/2}^{(1)} = \frac{4}{\pi}\frac{\partial}{\partial \eta}\left[\frac{\partial A}{\partial \xi} - \frac{i}{2k_c}\frac{\partial^2 A}{\partial \eta^2}\right]\cos(\pi z).\tag{9.113d}$$

At the $O(\varepsilon^3)$ order, all field variables admit a solution of the form (9.103) with $p = 4$, $m = 1$ to 3. But only the components of $\mathcal{U}_3^{(0)}$ and $\mathcal{U}_3^{(1)}$ are of interest in determining the evolution equation for the leading-order amplitude $A(\xi, \eta, \tau)$. First, we obtain the following two equations for $u_3^{(0)}$ and $w_3^{(0)}$:

$$\frac{\partial^2 u_3^{(0)}}{\partial z^2} = S_3,\tag{9.114a}$$

$$\frac{\partial^2 w_3^{(0)}}{\partial z^2} = Q_3,\tag{9.114b}$$

where

$$S_3 = \frac{\pi^2}{8}\left[\frac{1}{\Pr} + \frac{9}{8}\right]\frac{\partial |A|^2}{\partial \eta}\cos(2\pi z),\tag{9.114c}$$

$$Q_3 = \frac{3}{32}\left[\frac{1}{\Pr} + \frac{3}{8}\right]\frac{\partial |A|^2}{\partial \eta}\cos(2\pi z).\tag{9.114d}$$

Consequently, to satisfy boundary conditions

$$\frac{\partial u_3^{(0)}}{\partial z} = 0, \quad w_3^{(0)} = 0 \quad \text{at } z = 0, 1,$$

we must enforce the following two compatibility conditions on the forcing terms in (9.114a,b):

$$\int_0^1 S_3 \, dz = 0, \quad \int_0^1 Q_3 \, dz = 0,$$

which are again identically satisfied.

Next, for $w_3^{(1)}$ and $\Theta_3^{(1)}$, we derive a system of two nonhomogeneous equations analogous to the above system for $w_2^{(1)}$ and $\Theta_2^{(1)}$, but with F_3 and G_3 on the right-hand side, such that

$$F_3 = 3i \frac{\pi^3}{2\sqrt{2}} \left\{ \frac{1}{\Pr} \frac{\partial A}{\partial \tau} + \frac{\partial^2 A}{\partial \xi^2} - \frac{16}{3} \left[\frac{\partial A}{\partial \xi} - \frac{i}{2k_c} \frac{\partial^2 A}{\partial \eta^2} \right]^2 \right\} \sin(\pi z), \tag{9.115a}$$

$$G_3 = -9i \frac{\pi^3}{2\sqrt{2}} \left\{ \frac{\partial A}{\partial \tau} - \frac{\partial^2 A}{\partial \xi^2} + \frac{4}{3} \left[\frac{\partial A}{\partial \xi} - \frac{i}{2k_c} \frac{\partial^2 A}{\partial \eta^2} \right] \right.$$
$$\left. - \frac{2}{9\pi^2} rA - \frac{\pi^2}{8} \cos(2\pi z) A|A|^2 \right\} \sin(\pi z). \tag{9.115b}$$

The boundary conditions are

$$w_3^{(1)} = \frac{\partial^2 w_3^{(1)}}{\partial z^2} = \Theta_3^{(1)} = 0 \quad \text{at } z = 0, 1.$$

Again, the orthogonality with adjoint eigenfunction requires that

$$\int_0^1 [F_3 w^* + G_3 \Theta^*] \, dz = 0,$$

thereby leading to the *amplitude evolution equation* for $A(\tau, \xi, \eta)$,

$$\left[1 + \frac{1}{\Pr}\right] \frac{\partial A}{\partial \tau} = 4 \left[\frac{\partial A}{\partial \xi} - \frac{i}{2k_c} \frac{\partial^2 A}{\partial \eta^2} \right]^2$$
$$+ \frac{2}{9\pi^2} rA - \frac{\pi^2}{16} A|A|^2. \tag{9.116}$$

If

$$X = 2k_c \xi, \quad Y = 2k_c \eta, \quad T = 16 \left[1 + \frac{1}{\Pr}\right] k_c^2 \tau, \tag{9.117a}$$

then for the new amplitude function $B(T, X, Y)$, such that:

$$B = \frac{\pi}{16k_c} A \left(\frac{X}{2k_c}, \frac{Y}{2k_c}, \frac{T}{16k_c^2} \left[1 + \frac{1}{\Pr}\right] \right), \tag{9.117b}$$

the evolution equation for $B(T,X,Y)$ takes the final form,

$$\frac{\partial B}{\partial T} = \left[\frac{\partial B}{\partial X} - i\frac{\partial^2 B}{\partial Y^2}\right]^2 + \mu B - B|B|^2 \,, \tag{9.118}$$

with

$$\mu = \frac{r}{36\pi^4} \,.$$

Equation (9.118) is the amplitude equation previously derived by Newell and Whitehead (1969). Our derivation of the amplitude equation (9.118) is directly suggested by the paper of Coullet and Huerre (1986) and see also Coullet and Spiegel (1983).

Phase dynamics analysis. For (9.118), we can obtain first a family of stationary periodic (in X) solutions:

$$B_s = Q\exp(iqX) \,, \tag{9.119}$$

where the amplitude Q is given by

$$(\mu - q^2)Q - Q^3 = 0 \Rightarrow Q = (\mu - q^2)^{1/2} \,. \tag{9.120a}$$

To study the stability of this pattern, we make a change of variables:

$$B(T,X,Y) = [Q + \rho(T,X,Y)]\exp[i(qX+\phi)] \,, \tag{9.120b}$$

with $\phi = \phi(T,X,Y)$. In this case, from (9.118) we obtain two equations:

$$\frac{\partial \rho}{\partial T} = -2Q^2\rho - 2\rho Q\frac{\partial \phi}{\partial X} + \frac{\partial^2 \rho}{\partial X^2} + 2q\frac{\partial^2 \rho}{\partial Y^2} \,; \tag{9.120c}$$

$$\frac{\partial \phi}{\partial T} = \frac{2q}{Q}\frac{\partial \rho}{\partial X} + \frac{\partial^2 \phi}{\partial X^2} + 2q\frac{\partial^2 \phi}{\partial Y^2} \,. \tag{9.120d}$$

Thus, the spatial pattern may be subject to two possible modes of perturbations. The amplitude mode associated with the variable ρ is governed by (9.120c), and in the *long-wavelength approximation*, $\partial/\partial X \ll 1$, $\partial/\partial Y \ll 1$, this amplitude mode is highly *damped*. By contrast, the remaining variable ϕ corresponds to a *marginal phase mode*; its dynamics is governed by (9.120d) and in the *long wavelength-limit*, $\partial\phi/\partial T = 0$ – this mode is *neutrally stable*. To describe the long-wavelength dynamics of the phase mode ϕ, it is legitimate to assume that the amplitude ρ is adiabatically slaved to the slowly-varying phase. To leading-order, the amplitude equation (9.120c) can then be approximated by

$$\rho \sim -\frac{q}{Q}\frac{\partial \phi}{\partial X} \,, \tag{9.120e}$$

and substituting in (9.120d) gives rise to the phase evolution equation

$$\frac{\partial \phi}{\partial T} = \left[1 - \frac{2q^2}{Q^2}\right]\frac{\partial^2 \phi}{\partial X^2} + 2q\frac{\partial^2 \phi}{\partial Y^2}, \tag{9.120f}$$

where, according to (9.120a),

$$1 - \frac{2q^2}{Q^2} = \frac{\mu - 3q^2}{\mu - q^2} = \beta. \tag{9.120g}$$

Finally, phase fluctuations are governed by the single diffusive equation

$$\frac{\partial \phi}{\partial T} = \beta\frac{\partial^2 \phi}{\partial X^2} + 2q\frac{\partial^2 \phi}{\partial Y^2}. \tag{9.121}$$

The signs of β and q control the so-called *Eckhaus* and *zig-zag instability*, respectively. We note that q can change sign *if* the basic pattern is zig-zag unstable, and if $q > 0$, the phase is *diffusive* in Y, and no zig-zag instability can occur. If $q < 0$, the medium is zig-zag unstable, and additional terms need to be brought into (9.121) to describe possible two-dimensional soliton lattices!

Instability and a route to chaos in RB (buoyancy force effect) convection. RB thermal convection in a fluid layer heated from below, with a nondeformable upper surface and without surface tension, represents the simplest example of hydrodynamic instability and transition to turbulence in a fluid system. In this case, the more important effect is the buoyancy effect in the fluid; the reader can find an excellent account of the various features of this buoyancy effect in Turner (1973). A qualitative description of the convection motions is given in Velarde and Normand (1980). In Normand et al. (1977), a physicist's approach to convective instability is presented. This review paper is a pertinent account of the theoretical and experimental results on convective instability up to 1977. For the classical theory of the RB instability problem, see the books by Chandrasekhar (1961/1981) and Drazin and Reid (1981). The only systematic method for analyzing the manifold of 3-D nonlinear steady-state solutions of the OB equations (see Sect. 2.3.2 in Chap. 2) is the perturbation approach based on the amplitude ε of convection as a small parameter. This approach is particularly appropriate for convection because instability occurs in the form of infinitesimal disturbances. Obviously, the perturbation expansion is of limited usefulness when the Rayleigh number Ra is increased much beyond its critical value. Thus, direct numerical methods have been used to solve the problem of fully nonlinear RB convection, and, for this, it is convenient to define five values of the Ra that distinguish various flow regimes. But in any given system, some or all of these Ra may be nonexistent! First, the linear critical Ra_c is defined so that the heat-conducting motionless state of the fluid is stable to infinitesimal disturbances for Ra $<$ Ra_c and is unstable for Ra $>$ Ra_c. As Ra increases beyond Ra_c steady-state convecting rolls appear, and these rolls are 2-D in character. Next, Ra_1 is defined as that Raleigh number at which these rolls undergo a bifurcation to a periodic, possibly 3-D oscillatory state – periodic convection ensues as Ra increases above Ra_1. At Ra_2, a second (normally incommensurate) frequency appears, so the flow is quasi-periodic, but if this second frequency is commensurable with the first, then a phase locking occurs, so the flow is still periodic but has a new frequency. Then at Ra_t, the flow undergoes transition to a chaotic state with a broadband frequency response. Of course, there may also be transitional numbers Ra_n for $n \geq 3$ in which n distinct incommensurate frequencies are observable. In fact, the Ruelle et al. scenario, considered in Chap. 10, Sect. 10.3, suggests that $Ra_t = Ra_3$. Finally, there is another critical Rayleigh number that is useful to define, although its existence is not anticipated by the generic mathematical analysis outlined in Curry et al. (1984). This Ra'_1 is defined as that value of Ra at which a reverse transition from quasi-periodic or chaotic flow to periodic flow occurs as Ra increases. Though the flow just below Ra'_1 has at least two incommensurate frequencies present, that just above Ra'_1 has but one significant frequency. Furthermore, there may even be bands of Rayleigh numbers between Ra_t and Ra'_1 in which the flow reverts to quasi-periodic behavior. Finally, we emphasize once again that some or all of these putative critical Ra values may not exist in any particular realization of a real thermal RB convection flow. Below, we present some numerical results obtained by Curry et al. (1984) with free-slip (no-stress) conditions and periodic conditions in x and y through a spectral method [Orszag (1971)]. The dependent flow variables are expanded in a Fourier series, and then the nonlinear terms are evaluated by fast-transform methods with aliasing terms usually removed; time-stepping is done by a leapfrog scheme for the nonlinear terms and by an implicit scheme for the viscous terms. The pressure term is computed in Fourier representation by local algebraic manipulation of the constraint $\nabla \cdot \boldsymbol{u} = 0$.

In Figs. 9.2 and 9.3, for the 3-D case, some results for the transition are presented for 16^3 and $32^2 \times 16$ runs, respectively. The curves show $u(\boldsymbol{p}, t)$ versus $w(\boldsymbol{p}, t)$, where $\boldsymbol{p}$ is a coordinate near the midpoint of the box, for $3 \leq t \leq 4$. The two plots, in the case of 16^3 runs, at $Ra = 60\,Ra_c$ are obtained for different initial conditions, that show dependence of the final quasi-periodic state on initial data. This difference may also suggest alternative routes to chaos in addition to the Ruelle et al. scenario! At $Ra = 65\,Ra_c$ (in the case of 16^3 runs), the phase portrait suggests chaotic flow, although the spectrum of the flow is still dominated by phase-locked lines. Only the velocity components have phase plots that project onto a torus; those that involve temperature appear much more random.

In contrast with the 16^3 results plotted in Fig. 9.2, the plot of (u, w) for the $32^2 \times 16$ runs at $50\,Ra_c$ is now a simple circle, corresponding to the presence of only a single frequency. The phase plot at $Ra = 60\,Ra_c$ has much the same appearance as with 16^3 resolution. At $Ra = 70\,Ra_c$, we again observe a chaotic

9.4 Some Facets of the RB and BM Problem   353

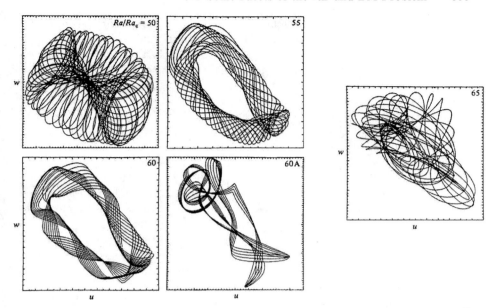

Fig. 9.2. Two-dimensional phase projections of the (u, w) fields for resolution 16^3 runs

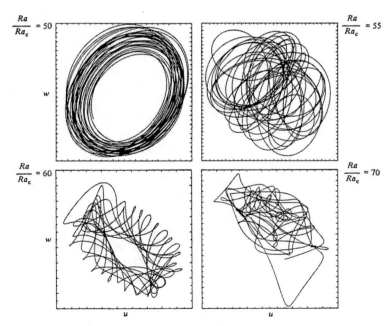

Fig. 9.3. Two-dimensional phase projections of the (u, w) fields for resolution $32^2 \times 16$ runs

regime. The transition scenario reported here for 3-D closely parallels route I described by Gollub and Benson (1980) – the qualitative differences are related to the existence or nonexistence of phase-locked regimes. Although such regimes may be present for some range of parameters, they have been not observed by the above authors because of the coarseness of the partition through parameter space.

In the survey by Busse (1981), the reader can find a comprehensive review of the basic physical properties of RB convection. The transition to chaos in 2-D double-diffusive (thermosolutal – when we also take into account the solute concentration) convection is analyzed in Knobloch et al. (1986). In Howard and Krishnamurti (1986), large-scale flow in (turbulent) convection is considered, and to this end, the three Fourier components that lead to the well-known Lorenz equations (see, Chap. 10, Sect. 10.1) were augmented with three additional components, leading to a sixth-order system.

The main results of the study of the bifurcations of this system are that (i) after the second bifurcation, steady-state tilted cells are the stable flow. This symmetry change is like that observed in the laboratory.

Though steady-state tilted cells are not observed in a convecting layer of fluid, they have been seen in Hele–Shaw cell convection. (ii) After the third bifurcation, stable limit cycles are found for a range of Ra $[= (g\alpha/\nu\kappa)\Delta T d^3$, where α is the thermal expansion coefficient, κ the thermal diffusivity, πd the layer depth, and ΔT the imposed temperature difference between the bottom and top boundaries of the layer and $\Pr = \nu/\kappa]$, with the same symmetry as in (i). The flow and thermal structure can be described as hot transient plumes that form periodically and tilt as they rise from below and cold ones that sink from above with the same angle of tilt. In this range, there is a net Lagrangian transport of mass in one horizontal direction near the top of the layer and in the opposite direction near the bottom. (iii) Within this range of Ra where stable limit cycles are found, there are narrow sub-ranges of aperiodic flows.

The occurrence of this chaotic behavior is related to the existence of heterocline orbits pairs. One example of the temperature field at times equally spaced within one period is shown in Fig. 9.4, for $\Pr = 1.0$, Ra $= 55$, and $\alpha = 1.2$. Similar orbits were found for $\Pr = 0.1$ and $\Pr = 10.0$ for Ra slightly in excess of the critical Ra for the onset of oscillatory convection.

As noted above, a hot plume or bubble forms in the lower part of the region, rises and tilts from lower left to upper right, and later, a cold plume forms in the upper part, sinks, and tilts from upper right to lower left. It also shows a leftward-propagating wave in the isotherms near the bottom of the layer and a rightward-propagating wave near the top of the layer.

For convective stability of incompressible fluids, see the book by Gershuni and Zhukhovitskii (1976). The book by Platen and Legros (1984) is pertinent to the studies of convection in liquids.

9.4 Some Facets of the RB and BM Problem 355

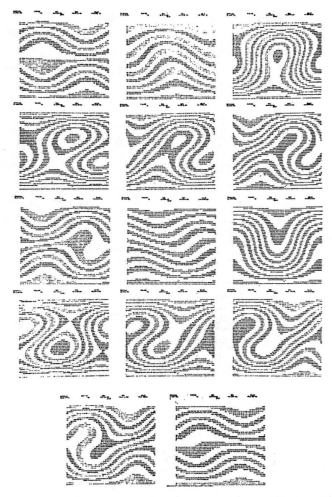

Fig. 9.4. Temperature field at successive time intervals within one oscillation period

In the Kelly (1994) review paper, the reader can find a detailed account of the onset and development of RB thermal convection in fully developed shear flow. For instance, the interesting consequences of the convective rather than absolute nature of initial instability have been elucidated recently, and our understanding of finite amplitude states following initial instability has been greatly enhanced. The onset of RB convection in most shear flow differs from zero shear in a fundamental way that affects how convection might be observed in an experiment. This difference has to do with the concepts of absolute and convective instabilities that rise from the analysis of the initial-value problem. These concepts have been discussed in detail for isothermal shear flow by Huerre and Monkewitz (1990). Absolute instability is pertinent to zero shear – if a localized disturbance of the conduction state is created impulsively for Ra > Ra$_c$, it spreads by diffusion and, as $t \to \infty$, occurs throughout the fluid. Although the disturbance itself is spatially dependent, clearly, its amplitude grows locally with time on a linear basis. Now, say that a mean flow with a small Reynolds number but nonzero net mass flux exists in the positive x direction. For simplicity, consider first a 2-D disturbance. Due to the initial disturbance, a wave packet is generated that propagates in the positive x direction at a group velocity c_g proportional to the average velocity.

At the same time, diffusion occurs in both the positive and the negative x directions and if the net effect is that the motion decays at a fixed value of x, as $t \to \infty$, the flow is said to be convectively unstable. The amplitude of the wave packet increases with time because Ra > Ra$_c$, but only the fluid close to the propagating wave packet is disturbed.

For the absolute instability of RB convection in a time-periodic shear flow, see the recent paper by Li et al. (1997). These authors conclude that absolute instability is characteristic of the development of thermal convection in this oscillatory flow, regardless of frequency and Reynolds number.

Convective instability can, of course, still occur if there is a nonzero mean flow component in addition to the oscillatory component.

For that case, the boundary between convective and absolute instability seems to be the same as if the flow were steady-state. The mean value of Re required for this to occur is exactly the same as would be obtained on the basis of the relation

$$\frac{\text{Ra}}{\text{Ra}_c} = 1 + 0.255 \, \text{Re}^{4/3} + O(\text{Re}^2) \, .$$

For still higher values of the mean Reynolds number, the disturbance would be swept away in the downstream direction.

9.4.2 Bénard–Marangoni (BM) Thermocapillary Instability Problem for a Thin Layer (Film) with a Deformable Free Surface

The full BM model problem, (7.77a), (7.77b,c), derived in Sect. 7.5.4, Chap. 7, is very complicated, but for thin film flow, we can apply the long-wave approximation, and then, we can derive a simplified BM, boundary-layer model, the BM$_{\text{BL}}$ problem for high Reynolds and Weber numbers.

BM$_{\text{BL}}$ Problem. According to long-wave approximation, we assume that the characteristic value for the horizontal wavelength $\lambda^0 \gg d$. Then, instead of dimensionless variables (t', x'_1, x'_2, x'_3), it is judicious to introduce the following new dimensionless variables:

$$x = \varepsilon x'_1 \, , \quad y = \varepsilon x'_2 \, , \quad z \equiv x'_3 \, , \quad \tau = \varepsilon \, \text{Re} \, t' \, ,$$

9.4 Some Facets of the RB and BM Problem

new functions (in place of dimensionless velocity components):

$$u = \frac{v_{10}}{\text{Re}}, \quad v = \frac{v_{20}}{\text{Re}}, \quad w = \frac{v_{30}}{\varepsilon \,\text{Re}},$$

and new pressure perturbation: $\Pi = \pi/(\text{Re})^2$, where

$$\varepsilon = \frac{d}{\lambda^0} \quad \text{and} \quad \text{Re} = U^0 \frac{d}{\nu_0}. \tag{9.122}$$

Finally, in place of dimensionless parameters Fr, We, and Ma, we introduce corresponding modified Froude, Weber, and Marangoni numbers based on the characteristic velocity U^0 [see (9.132)]:

$$F = \frac{U^0}{(gd)^{1/2}}, \quad \text{We} = \frac{\sigma_0}{\rho_0 d U^{02}}, \quad \text{Ma} = \frac{\gamma \Delta T^0}{\rho_0 d U^{02}}. \tag{9.123}$$

When [see, Zeytounian (1997) and (1998)]

$$\varepsilon \to 0, \quad \text{Re} \to \infty \quad \text{and} \quad \text{We} \to \infty, \tag{9.124a}$$

with the following similarity relations:

$$\varepsilon \,\text{Re} = \text{Re}^* = O(1) \quad \text{and} \quad \varepsilon^2 \,\text{We} = W^* = O(1), \tag{9.124b}$$

in place of the full BM model problem, (7.77a), (7.77b,c), for the functions $\boldsymbol{V} = [u(\tau,x,y,z), v(\tau,x,y,z)]$, $w(\tau,x,y,z)$, $\Pi(\tau,x,y,z)$, $\Theta(\tau,x,y,z)(=\vartheta_0)$, and $H(\tau,x,y)(=h')$, we derive the following approximate BM_{BL} model problem (9.125a–c), where $\text{Bi}_s \equiv \text{Bi}^0 = \text{Bi}$:

$$\boldsymbol{D} \cdot \boldsymbol{V} + \frac{\partial w}{\partial z} = 0,$$

$$\frac{d\boldsymbol{V}}{d\tau} + \boldsymbol{D}\Pi = \frac{1}{\text{Re}^*} \frac{\partial^2 \boldsymbol{V}}{\partial z^2},$$

$$\frac{\partial \Pi}{\partial z} = 0 \Rightarrow \Pi = \frac{H-1}{F^2} - W^* \boldsymbol{D}^2 H,$$

$$\text{Pr} \frac{d\Theta}{d\tau} = \frac{1}{\text{Re}^*} \frac{\partial^2 \Theta}{\partial z^2}, \tag{9.125a}$$

$$\text{at} \quad z = 0: \quad \boldsymbol{V} = w = 0, \quad \Theta = 1, \tag{9.125b}$$

$$\text{at} \quad z = H(\tau,x,y): \begin{cases} \dfrac{\partial \boldsymbol{V}}{\partial z} = -\text{Re}^* \text{Ma}\left[\boldsymbol{D}\Theta + (\boldsymbol{D}H)\dfrac{\partial \Theta}{\partial z}\right], \\ \dfrac{\partial \Theta}{\partial z} + 1 + \text{Bi}\,\Theta = 0, \\ w = \dfrac{\partial H}{\partial \tau} + \boldsymbol{V} \cdot \boldsymbol{D}H. \end{cases} \tag{9.125c}$$

358   9 Linear and Nonlinear Stability of Fluid Motion

In these BM$_{\text{BL}}$ model equations, we have the following operators:

$$\frac{\mathrm{d}}{\mathrm{d}\tau} = \frac{\partial}{\partial \tau} + \boldsymbol{V} \cdot \boldsymbol{D} + w\frac{\partial}{\partial z} \quad \text{and} \quad \boldsymbol{D} = \left(\frac{\partial}{\partial x}, \frac{\partial}{\partial y}\right),$$

$$\boldsymbol{D}^2 = \frac{\partial^2}{\partial x^2} + \frac{\partial^2}{\partial y^2}.$$

In (9.125a,c), the parameters F, Ma, Pr, Bi, Re* and W^* are $O(1)$. The BM$_{\text{BL}}$ problem is a *very significant approximate model for investigating thin, slightly viscous, film instability*, and this model deserves further consideration.

The limit case: Pr $\to$ 0. When Pr $\to$ 0, the solution for temperature perturbation Θ is very simple. According to the fourth equation of (9.125a), last condition of (9.125b) for Θ, and the second condition of (9.125c), we obtain

$$\Theta = 1 - \frac{1 + \text{Bi}}{1 + \text{Bi}\, H} z. \tag{9.126}$$

On the other hand (for all Pr and Re* fixed) from the first equation of (9.125a) with the conditions

$$w = 0, \text{ at } z = 0, \quad \text{and} \quad w = \frac{\partial H}{\partial \tau} + \boldsymbol{V} \cdot \boldsymbol{D} H, \text{ at } z = H,$$

we derive the following averaged (evolution) equation

$$\frac{\partial H}{\partial t} + \boldsymbol{D} \cdot \int_0^H \boldsymbol{V}\, \mathrm{d}z = 0. \tag{9.127a}$$

Consequently, for two functions, $\boldsymbol{V}(\tau, x, y, z)$, and $H(\tau, x, y)$, we derive the following system of two evolution equations, (9.127a) and

$$\frac{\mathrm{d}\boldsymbol{V}}{\mathrm{d}\tau} - \frac{1}{\text{Re}^*}\frac{\partial^2 \boldsymbol{V}}{\partial z^2} = -\frac{1}{F^2}\boldsymbol{D}H + W^*\boldsymbol{D}[\boldsymbol{D}^2 H], \tag{9.127b}$$

because

$$w = -\int_0^z (\boldsymbol{D} \cdot \boldsymbol{V})\, \mathrm{d}z. \tag{9.127c}$$

For (9.127a,b) with (9.127c), as boundary conditions (in z), we write:

$$\boldsymbol{V} = 0 \quad \text{at} \quad z = 0, \tag{9.128a}$$

$$\frac{\partial \boldsymbol{V}}{\partial z} = \text{Ma}\, \text{Re}^* \frac{1 + \text{Bi}}{(1 + \text{Bi}\, H)^2} \boldsymbol{D}H, \quad \text{at} \quad z = H. \tag{9.128b}$$

9.4 Some Facets of the RB and BM Problem

The case of $Re^* \ll 1$. If $Re^* \ll 1$, then in the limit $Re^* \to 0$, from (9.127b), with the boundary condition (9.128a,b) in $z = 0$ and $z = H$, we obtain the following limit solution for horizontal velocity:

$$\mathbf{V} = \mathbf{V}(H) = \frac{1}{3}\frac{d}{dH}\left\{H^3[\alpha^0 \mathbf{D}(\mathbf{D}^2 H) - \mathbf{D}H]\right.$$

$$\left. +\mu^0 \frac{1+\text{Bi}}{(1+\text{Bi}\,H)^2} H^2 \mathbf{D}H\right\}, \qquad (9.129)$$

and consequently, in place of (9.127a), we derive a single evolution equation for the thickness of the film $H(\tau, x, y)$:

$$\frac{\partial H}{\partial \tau} + \frac{1}{3}\mathbf{D}\cdot\left\{H^3[\alpha^0 \mathbf{D}(\mathbf{D}^2 H) - \mathbf{D}H]\right.$$

$$\left. +\mu^0 \frac{1+\text{Bi}}{(1+\text{Bi}\,H)^2} H^2 \mathbf{D}H\right\} = 0. \qquad (9.130)$$

In (9.129) and (9.130), we have the following significant coefficients

$$\alpha^0 = \frac{\sigma_0}{\rho_0 g \lambda^{02}}, \quad \text{and} \quad \mu^0 = \frac{1}{2}\gamma \frac{\Delta T^0}{g\rho_0 d^2}, \qquad (9.131)$$

when the following characteristic velocity, introduced above [see (9.122) and (9.123)], is:

$$U^0 = \frac{gd^3}{\lambda^0 \nu_0}. \qquad (9.132)$$

In (9.131) which is similar to that derived by Oron and Rosenau [1992, Eq. (2) and 1994, Eq. (15)] and also by Kopbosynov and Pukhnachev (1986), the term proportional to α^0 has a stabilizing effect whereas the term proportional to μ^0 has, on the contrary, a destabilizing impact. It should be noted that the gravity term $-(1/3)\mathbf{D}\cdot[H^3 \mathbf{D}H]$, in (9.130) stabilizes the evolution of the interface when the film is supported from below.

In the *one-dimensional case* ($\partial/\partial y = 0$), we obtain the following evolution equation for the thickness $H \equiv h(\tau, x)$ of the thin film:

$$\frac{\partial h}{\partial \tau} - \frac{\partial}{\partial x}\left\{\left[\frac{h^3}{3} - \frac{\beta^0}{3}\frac{h^2}{(1+\text{Bi}\,h)^2}\right]\frac{\partial h}{\partial x}\right\}$$

$$+\frac{\alpha^0}{3}\frac{\partial}{\partial x}\left[h^3 \frac{\partial^3 h}{\partial x^3}\right] = 0, \qquad (9.133)$$

with $\beta^0 = \mu^0(1+\text{Bi})$. When $\beta^0 = 0$ (the Marangoni effect is disregarded), and (9.133) then becomes

$$\frac{\partial h}{\partial \tau} + \frac{\partial}{\partial x}\left\{\frac{h^3}{3}\frac{\partial}{\partial x}\left[\alpha^0\frac{\partial^2 h}{\partial x^2} - h\right]\right\} = 0. \qquad (9.134)$$

Note that (9.134) preserves the total mass of the film (not subject to the Marangoni effect!)

$$\frac{\mathrm{d}}{\mathrm{d}\tau}\int_\Omega h\,\mathrm{d}\Omega = 0, \qquad (9.135)$$

where Ω is either unbounded or a periodic domain.

Another conservation law is available: multiply (9.134) by $(1/h^2)$ and integrate over Ω with periodic boundary conditions to obtain [by analogy with the conservation law derived in Oron and Rosenau (1992, Eq. (47)]:

$$\frac{\mathrm{d}}{\mathrm{d}\tau}\int_\Omega \left(\frac{1}{h}\right)\mathrm{d}\Omega = -\frac{2}{3}\int_\Omega \left[\left(\frac{\partial h}{\partial x}\right)^2 + \alpha^0\left(\frac{\partial^2 h}{\partial x^2}\right)^2\right]\mathrm{d}\Omega. \qquad (9.136)$$

The full one-dimensional evolution equation (9.133) can be obviously rewritten in the conservative form,

$$\frac{\partial h}{\partial \tau} = -\frac{\partial}{\partial x}\left\{h^3 \frac{\partial}{\partial x}\left[Q(h) + \frac{\alpha^0}{3}\frac{\partial^2 h}{\partial x^2}\right]\right\}, \qquad (9.137\mathrm{a})$$

with

$$Q(h) = -\frac{h}{3} + \frac{\beta^0}{3}\left\{\log\left[\frac{h}{(1+\mathrm{Bi}\,h)}\right] + \left[\frac{1}{(1+\mathrm{Bi}\,h)}\right] - \frac{A^0}{6}\right\}, \qquad (9.137\mathrm{b})$$

where A^0 is an integration constant.

Equation (9.137a) can also be rewritten as a so-called, Cahn–Hilliard type form,

$$\frac{\partial h}{\partial \tau} = \frac{\partial}{\partial x}\left\{h^3 \frac{\partial}{\partial x}\left(\frac{\delta L}{\delta h}\right)\right\}, \qquad (9.138\mathrm{a})$$

with

$$L = \int_\Omega \left[U(h) + \frac{\alpha^0}{6}\left(\frac{\partial h}{\partial x}\right)^2\right]\mathrm{d}\Omega, \qquad (9.138\mathrm{b})$$

playing the role of *free energy* and h^3 the role of *mobility* coefficient.

In (9.138b),

$$\frac{\mathrm{d}U(h)}{\mathrm{d}h} = -Q(h).$$

The right-hand side of the equation,

$$\frac{\mathrm{d}L}{\mathrm{d}\tau} = -\int_\Omega h^3\left\{\frac{\partial}{\partial x}\left[Q(h) + \frac{\alpha^0}{3}\frac{\partial^2 h}{\partial x^2}\right]\right\}^2\mathrm{d}\Omega, \qquad (9.139)$$

derived from (9.137a), is, on the one hand, always *nonnegative*:

$$\frac{dL}{d\tau} \leq 0, \tag{9.140}$$

and L is, on the other hand, a *decreasing function along any trajectory*. Therefore, L is a *Lyapunov function* for (9.138a), and, therefore, *if steady states of the system exist*, they are solutions of the equation,

$$Q(h) + \frac{\alpha^0}{3}\frac{\partial^2 h}{\partial x^2} = \text{const}. \tag{9.141}$$

Relations (9.137a)–(9.141) provide a start for any meaningful stability analysis of the one-dimensional equation (9.133).

We recall that

if a system has a Lyapunov function bounded from below (a free energy function decreasing along all trajectories) then, any initial data evolves into a steady state.

Free-surface deformation and instability in thin liquid films. Although Bénard was aware of the role of surface tension and surface tension gradients in his experiments, it took five decades to determine unambiguously, experimentally, and theoretically [see, for instance, the papers by Block (1956) and Pearson (1958)], that surface tension gradients rather than buoyancy was the cause of Bénard cells in thin liquid films. Only in 1997 was this almost evident physical fact rigorously proved through an asymptotic approach by Zeytounian (1997). It seems that the first author to explain the effect of surface tension gradients on Bénard convection was Pearson (1958). The review articles by Normand et al. (1977) and by Davis (1987) considered the role of both buoyancy and surface tension gradients in triggering convective instability. Recently, a review article by Cross and Hohenberg (1993), devoted to nonequilibrium pattern formation, had a sketchy section dealing with genuine Bénard cells, i.e., Bénard–Marangoni convection. Koschmieder (1993) who has been for decades a key figure in the experimental investigation of the Bénard problem, wrote a valuable monograph recently. The more recent review article by Bragard and Velarde (1997) provided salient findings, old and recent, about Bénard convection flow in a liquid layer heated from below and open to the ambient air. In Zeytounian (1998), the author considers three main situations for a thin film in relation to the magnitude of the characteristic Reynolds number and derives various model equations. These model equations are analyzed from various points of view, but the central intent of the Zeytounian review paper is to elucidate the role of the Marangoni number in the evolution of a free surface in space-time. Myers (1998) is a review of work on thin films when (high) surface tension is a driving mechanism. Its aim is to highlight the substantial amount of literature dealing with relevant physical models and also analytic work on the resultant equations. The recent paper of Ida and Miksis (1998a,b) considers the dynamics of a general 3-D thin film subject to van der Waals forces, surface tension, and surfactants. Using an asymptotic analysis based upon the thinness of the film with respect to the lateral extent, evolution equations for leading-order film thicknesses, tangential velocities, and surfactant concentrations are obtained. Scaling was chosen by these authors so that the surface tension effects occur at leading order in the dynamics model of the thin film. Note that the analysis applies to the breaking off of a thin liquid film from a stable center surface. Unfortunately, the model equations presented form a complicated set of evolution equations and cannot be solved until the center surface is prescribed. In Part II of their work, Ida and Miksis (1998b, pp. 474–500), consider a series of special center surfaces and in each case, consider the linear stability and solve the resulting nonlinear equations numerically. In particular, it was shown that increasing surface tension is stabilizing, whereas increasing the effects of van der Waals forces is destabilizing. The effects of surfactants, although irrelevant in determining the neutral stability curves, is stabilizing. The results obtained by solving the full evolution equations agreed numerically with the stability results obtained analytically. Time-dependent BM (oscillatory) instability and waves were considered recently by Velarde and Rednikov (1998). As the liquid layer is subjected to a thermal gradient orthogonal to its open surface, attention is focused on the role played by surface tension nonuniformity, which induces surface stresses and (Marangoni) convective motions (beyond an instability threshold).

Note that the presence of a free surface introduces additional interesting effects of surface tension and gravity, which change the character of the instability dramatically in a parallel flow [for classical instability in parallel flow (between two rigid walls), see Sect. 9.6, where some recent stability results for viscous (Navier) fluid flows are given].

The instability of parallel flow between two rigid walls takes the form of short shear waves, but the instability in a liquid film takes the form of long gravity-capillary waves at a relatively small Reynolds number. Another interesting feature of film instability is that no finite critical wavelength exists according to the linear theory, in contrast to the case of rigid boundaries. Linear theory predicts that the instability will take place as an infinitely long wave. On the other hand, surface waves of finite wavelengths were observed, as a consequence of instability, by Kapitza and Kapitza (1949/1965) and Binnie (1957). This led to the conjecture that the observed waves are the most amplified waves whose wavelength λ_m is predicted by linear theory. However, referring to a free-falling vertical film [see Sect. 10.4.2, in Chap. 10], Benjamin (1957) pointed out that "one can scarcely expect waves to appear with a strictly uniform and distinct periodicity, because under all conditions infinitesimal waves with a wide range of wavelengths are unstable, and the wave with length λ_m comes into prominence only through a rather uncritical selection process depending on differences in the rates of amplification of different wavelengths. The ultimate state of the amplified waves is, of

9.4 Some Facets of the RB and BM Problem

course, determined largely by nonlinear effects which remain unknown." This statement is consistent with the experiment of Kapitza and Kapitza (1949), who found that distinctively periodic waves could not be observed unless the disturbances were introduced at precisely controlled frequencies. Thus, Lin (1969, 1970) was led to investigate the nonlinear evolution of the so-called Benjamin–Yih (1963) wave of a given mode. In Lin (1974), the nonlinear instability to disturbances of a finite-frequency bandwidth is studied for a layer of an incompressible viscous fluid flowing down a plane inclined at an angle β to the horizontal. Lin considered weakly nonlinear wave motion that perturbs the free surface only slightly and derived an amplitude equation for the leading-wave envelope by a multiple-scale asymptotic expansion [similar to (9.116) but for 2-D]. The Lin [1974, Eq. (18)] equation is appropriate for describing the weak nonlinear evolution of relatively short waves near the upper branch of the neutral curve where the amplification rate c_i is $O(\varepsilon^2)$; see in Lin (1974) amplitude equation (18) and Fig. 1 [which represents the stability curves for water in the (Re, α) plane at $15\,°\mathrm{C}$ with $\beta = 90°$ and Weber number $\mathrm{We} = 463.3$].

Near the lower branch of the neutral curve, the wave number α (in distance $2\pi d$, d is the constant film thickness) is zero, even where $c_i < O(\alpha^2) = O(\varepsilon)$, the modal interaction is stronger, and the Lin (1974) equations (25) or (26) – which are similar to the KS–KdV equation (derived in Sect. 10.4.3 of Chap. 10), but with $\gamma = 0$ – without the diffusion term, is then the governing equation of nonlinear evolution. Farther away from the neutral curve where $c_i = O(\varepsilon)$, the diffusion term in Lin equations (25) or (26) becomes important. Unfortunately, the Lin (1974) film stability study does not take into account the Marangoni effect.

In a recent paper by Wilson and Thess (1997), explicit analytical expressions for the linear growth (and decay) rates of long-wave modes in Bénard–Marangoni convection are derived and discussed. These analytical predictions are in good agreement with experimental observations [of Van Hook et al. (1995)] and are used to estimate the minimum experimental time necessary to observe long-wave instability under microgravitational conditions. This work is a natural extension of previous linear stability studies by Pérez-Garcia and Carneiro (1991), which concentrated on determinating marginal stability curves; the nonlinear analyses of Marangoni convection in a thin layer of fluid by Kopbosynov and Pukhnachev (1986) and Davis (1987); and the recent investigation of linear growth rates of the Marangoni problem near the onset of convection by Regnier and Lebon (1995). It is interesting to note that long-wave instability (L-WI) occurs even in highly viscous fluids, but its growth rate becomes so small as to be experimentally undetectable! On the other hand, in the limit of a small Biot number (which is typical of experimental situations), neither the dimensional growth rate S nor the corresponding critical temperature difference ΔT_L (for the onset of the L-WI at which $S = 0$) depends on the thermal diffusivity of the fluid, but S does not depend on the absolute value of the surface tension. The results obtained by Regnier and Lebon (1995) indicate that the influence of surface deformation on relaxation time and correlation length is weak for the non-zero wave number instability mode; in contrast, the zero mode, which is inherent in surface deformation, exhibits high sensitivity to the so-called crispation number (Cr = 1/Pr We).

The dispersion relation for the zero mode was found analytically by using appropriate scaling. As the main result of the analysis of Regnier and Lebon (1995), it is shown that the presence of the zero mode can be detected only in very large aspect ratio boxes and for very thin fluid layers. The results for the zero mode provide a first step toward (weakly) nonlinear analysis. Just a weakly nonlinear analysis of coupled surface tension and gravitationally driven instability in a thin fluid layer (but, again, with a flat upper free surface!) is presented in Parmentier et al. (1996). In a weakly nonlinear analysis, it is sufficient to take into acount the modes that are critical at the linear threshold and as a consequence for the critical modes, in Parmentier et al. (1996), a system of three coupled Ginzburg–Landau type equations for the three amplitudes A_1, A_2, A_3, is derived:

$$\tau \frac{\partial A_i}{\partial t} = \varepsilon A_i + a A_j^* A_k^* - b A_i \left[|A_j|^2 + |A_k|^2 \right] - c A_i |A_i|^2 , \qquad (9.142)$$

with,

$$i = 1 \text{ and } j = 2, k = 3; \quad i = 2 \text{ and } j = 3, k = 1;$$
$$i = 3 \text{ and } j = 1, k = 2 ,$$

wherein the coefficients τ, a, b, and c depend generally on the Prandtl and the Biot numbers and also on the ratio α (the percentage of the buoyancy effect with regard to the thermocapillary effect; $\alpha = 0$ corresponds to pure thermocapillarity, and $\alpha = 1$ to pure buoyancy). The relative distance from the threshold is

$$\varepsilon = 1 - \frac{\lambda}{\lambda_c} \quad \text{with } \lambda = \frac{\text{Ra}}{\text{Ra}^0} + \frac{\text{Ma}}{\text{Ma}^0} , \qquad (9.143)$$

where the wave number k corresponding to λ_c is the critical wave number k_c, Ra^0 is the critical Rayleigh number for pure buoyancy, and Ma^0 is the critical Marangoni number for pure thermocapillarity. According to Parmentier et al. (1996), when buoyancy is singly responsible for convection, only rolls will be observed. As soon as capillary effects are observed, it appears that a hexagonal structure is preferred at the linear threshold. The more the thermocapillary forces are dominant with respect to buoyancy forces, the larger the size of the region where hexagons are stable. It is shown that the direction of motion inside the hexagons is directly linked to the value of the Prandtl number, and for Pr > 0.23, the fluid moves upward at the center of the hexagons, in accord with experiments [see, Koschmieder and Prahl (1990)].

9.4 Some Facets of the RB and BM Problem   365

A subcritical region where hexagons are stable has also been displayed by these authors. The region is the largest when buoyancy does not act and in this case, the value of the subcritical parameter is in excellent agrement with direct numerical simulation by Thess and Orszag (1995). But all of these results correspond to the (RBM) case when the upper free (!) surface is flat. A detailed analysis of system (9.142) can be found in Cross (1980, 1982) and in Cross and Hohenberg (1993).

Recently, the instability of a liquid hanging below a solid ceiling [the so-called Rayleigh–Taylor (RT) instability] was considered by Limat (1993) in a short Note, according to a lubrication equation derived by Kopbosynov and Pukhnachev (1986). Limat discussed the influence of the initial thickness on RT instability, and the results are summarized in a diagram giving the different possible regimes. This diagram allows one to predict two different thickness dependences that are based on the physical properties of the liquid.

For a vertical film, the recent review paper by Chang (1994) gives an excellent survey mostly of the various transition regimes on a free-falling vertical film; for an extension of this review, see Chang and Demekhin (1996). But in both review papers, a discussion of the Marangoni effect is absent.

The experimental investigation of three-dimensional instability of film flow is presented in Liu et al. (1995), and several distinct transverse instabilities that are found which deform the traveling waves: a synchronous mode (in which the deformations of adjacent wave fronts are in phase) and a subharmonic mode (in which the modulations of adjacent wave fronts are out of phase – in this case, herringbone patterns result). The 3-D subharmonic weakly nonlinear instability is due to the resonant excitation of a triad of waves consisting of a fundamental two-dimensional wave and two oblique waves. The evolution of wavy films after the onset of either of these 3-D instabilities is complex. However, sufficiently far downstream, large-amplitude solitary waves absorb the smaller waves and become dominant. In Liu et al. (1995), a detailed study of these instabilities is presented, along with a qualitative treatment of further evolution toward an asymptotic "turbulent" regime.

Note that in Bénard–Marangoni convection in a liquid layer with a deformable interface, as previously shown by Takashima (1981a,b) by linear stability analysis (see Chap. 3, Sect. 3.5), *two monotonic modes* of surface tension driven instability exist. One, a *short-scale* mode, is caused by surface tension gradients alone without surface deformation, and it leads to formation of stationary convection with a characteristic scale of the order of the liquid layer depth. The other, a *long-scale* mode, is also influenced by gravity and capillary (Laplace) forces, and surface deformation plays a crucial role in its development. This instability mode results in large-scale convection and in the growth of long surface deformations whose the characteristic scale is large in comparison with the thickness of the liquid layer. As shown in the recent paper by Golovin et al. (1994), these two types of Marangoni convection, having different scales, can interact with each other in the course of nonlinear evolution. There are two mechanisms for the coupling between them. On the one hand, surface deformations change the Marangoni number locally, which depends on the depth of the liquid layer. This leads to a space-dependent growth rate of the short-scale convection, and, hence, its intensity also becomes space-dependent. On the other hand, the short-scale convection generates an additional mean mass/heat flux from the bottom to the free surface, which is proportional to its intensity (square of the amplitude). When the intensity is not uniform, this leads to additional long-scale surface tension gradients that affect the evolution of the long-scale mode. The coupling effects are most pronounced when the long- and short-scale modes have instability thresholds close to each other. In the paper cited, Golovin et al. studied these effects analytically in the vicinity of the instability thresholds. Close to the bifurcation point, the mean long-scale flow generated by the short-scale convection is very weak, of the order of ε^3, and it will considerably affect the long-scale surface deformations only if the latter are also small, of the order of ε^2, where ε is the amplitude of the deformationless convective mode. This happens when the surface tension is sufficiently large. According to Golovin et al. (1994), near the instability threshold, the nonlinear evolution and interaction between the two modes can be described by a system of two coupled nonlinear equations:

$$\frac{\partial A}{\partial T} = (\pm A) + \frac{\partial^2 A}{\partial x^2} + A|A|^2 + \eta A \,, \tag{9.144a}$$

$$\frac{\partial \eta}{\partial T} = -(\pm m)\frac{\partial^2 \eta}{\partial x^2} - w\frac{\partial^4 \eta}{\partial x^4} + s\frac{\partial^2 |A|^2}{\partial x^2} \,, \tag{9.144b}$$

where the parameters m, w, s are all positive. The parameter m characterizes the effect of surface tension gradients and gravity, w corresponds to the Laplace pressure, and s is the interaction parameter that characterizes the coupling between the two modes of Marangoni convection. The complex amplitude of the short-scale convection A undergoes long-scale evolution, described by the Ginzburg–Landau equation, but the latter, however, contains an additional term, ηA, connected with the surface deformation η. This term adds to the linear growth rate for the amplitude A and describes a nonuniform space-dependent supercriticality. It plays a stabilizing role when the surface is elevated ($\eta > 0$) and supresses short-scale convection under surface deflections. The surface deformation η is governed by a nonlinear evolution equation of the fourth order. However, the only nonlinear term in this equation is the coupling term proportional to $\partial^2 |A|^2/\partial x^2$ that describes the effect of the mean flow generated by short-scale convection. This term always plays a stabilizing role. In (9.144a,b), the various signs of the terms correspond to the four cases described by Golovin et al. (1994, §V). (i) When in (9.144a) we have $+A$ and in (9.144b) $+m$, both the short-scale deformational mode and the long-scale deformational mode are unstable; (ii) if in (9.144a), we have $+A$ and in (9.144b) $-m$, only the short-scale mode is unstable; (iii) when in (9.144a), we have $-A$ and in (9.144b) $+m$, the deformational mode is unstable; (iv) if in (9.144a), we have $-A$ and in (9.144b) $-m$, both modes are linearly stable, but their nonlinear interaction may lead to instability. The typical neutral stability curve, $\mathrm{Ma}(k)$, represented in Fig. 9.5 below, has two minima: first Ma_l corresponding to $k = 0$, describes long-wave instability, and then Ma_s, related to $k_c \neq 0$, indicates the threshold of short-scale convection.

In Fig. 9.5, the dashed line corresponds to a layer with a undeformable interface and then only one minimum exists, Ma_s.

The surface deformation η is a real quantity, whereas the amplitude A of the small-scale convection is complex. If we assume, for simplicity, that A is also real, thus considering the evolution of the short-scale convective structure with a fixed wave number $k = k_c$, then, for the derivation of a three-mode truncated model, we write

$$A(T) = A_0(T) + A_1(T)\cos(k_m x) + \ldots, \quad (9.145a)$$
$$\eta = B_1(T)\cos(k_m x) + \ldots, \quad (9.145b)$$

with $k_m = (m/2w)^{1/2}$, corresponding to the maximum linear growth rate of the first harmonic. In this case, substituting (9.145a,b) in (9.144a,b) and considering the third case $(-A, +m)$ after appropriate rescaling, we obtain the following dynamic system for $A_0(T)$, $A_1(T)$, and $B_1(T)$:

$$\frac{dA_0}{dT} = -A_0\left[1 + A_0^2 + \frac{3}{2}A_1^2\right] + \frac{1}{2}A_1 B_1,$$
$$\frac{dA_1}{dT} = -A_1\left[1 + \frac{m}{2w} + 3A_0^2 + \frac{3}{4}A_1^2\right] + A_0 B_1,$$
$$\frac{dB_1}{dT} = -\mu B_1 - \sigma A_0 A_1, \quad (9.146)$$

where

$$\mu(k_m) = mk_m^2 - wk_m^4, \quad \sigma = 2k_m^2 s. \quad (9.147)$$

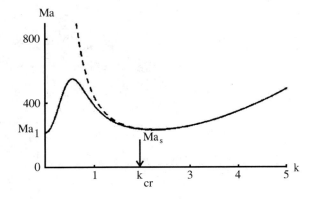

Fig. 9.5. Neutral stability curve for Marangoni convection

System (9.146) describes the temporal evolution of a periodic surface deformation (mode B_1) which can generate a periodic mode of short-scale convection (A_1), following the surface deformation, and also a uniform zero mode (A_0). Figure 9.6 shows a projection of one of the (strange) chaotic attractors on the planes: (a) (A_0, B_1) and (b) (A_1, B_1). The reader can find in Chap. 10, Sects. 10.3 and 10.4 a phenomenological theory of the strange attractors which appear in various routes to chaos. Thus, the coupling between short-scale convection and large-scale deformations of the interface can lead to stochastization of the system and can be one of the causes of interfacial turbulence of a thin film. The reader can find in Kazhdan et al. (1995) a numerical analysis of the system of two nonlinear coupled equations (9.144a,b), which confirms the predictions of weakly nonlinear analysis and shows the existence of either standing or traveling waves in the proper parametric regions at low supercriticality. With increasing supercriticality, waves undergo various transformations leading to the formation of pulsating traveling waves and nonharmonic standing waves as well as irregular wavy behavior resembling "interfacial turbulence."

When the coupling parameter s in (9.144b) increase, the stationary pattern becomes unstable and both long-scale surface deformations and the amplitude of short-scale convection undergo irregular oscillations, as shown in Fig. 9.7.

Although significant understanding has been achieved, yet surface tension gradient-driven (BM) convection flow still deserves further study. As a paradigmatic form of a spontaneous self-organizing system, the doctrine about the original Bénard problem has not reached the degree of sophistication in theory and experimentation attained in buoyancy-driven (RB) convection.

There are still challenging problems like the relative stability of patterns (hexagons, rolls, squares, ... , labyrinthine convection flows!), higher transitions and interfacial turbulence (at low Marangoni number), which is a space-

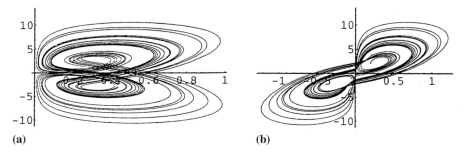

Fig. 9.6. Chaotic attractor of system (9.146) for $k_m = 1$, $\mu = 1$, $\sigma = 30$; time interval $T \in [0, 150]$

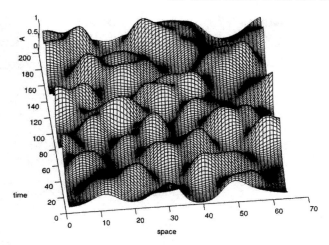

Fig. 9.7. Long-scale modulation of the amplitude of short-scale convection generated by surface deformation for $s = 70$, $m = 4.5$, $w = 1$ and $\text{Ma}_l < \text{Ma} < \text{Ma}_s$

time chaos with high dissipation. Finally, it should be pointed out that the ever increasing number of industrial applications of thin film flows and the richness of the behavior of the governing equations make this area particularly rewarding for mathematicians, engineers, and industrialists alike.

Actually, the number of works where the various aspects of the stability of a thin film are considered is very important. Here, obviously, we mention only a small part of these contributions. There are a number of excellent sources of information on various features of variable surface tension effects. Among them are the papers by Sterning and Scriven (1959), Norman et al. (1977), Velarde and Chu (1992), Sarpkaya (1996), and books by Levich (1962), Probstein (1994), Joseph and Renardy (1993), Colinet, Legros and Velarde (2001) and the book edited by Meyer (1983). For investigations where the Marangoni effect is taken into account, see the references in Davis (1987) and also, for more recent references, in Zeytounian (1998, Refs. [17] to [46]). Two pertinent recent contributions are Vince (1994) and Oron et al. (1997). For stability, we mention the following papers: Davis and Homsy (1980), Smith and Davis (1983), Cloot and Lebon (1984), Burelbach et al. (1988), Prokopiou et al. (1991), Joo et al. (1991), Joo (1995), Shipp et al. (1997), Sisoev and Shkadov (1997a,b). In 2002 the following two books have been edited: Velarde and Zeytounian (eds.), Nepomnyashchy, Velarde and Colinet.

9.5 Couette–Taylor Viscous Flow Between Two Rotating Cylinders

9.5.1 A Short Survey

In Sect. 3.3 we considered flow between two concentric circular (infinite) cylinders in the framework of linear theory. Below, we want to give some informations concerning nonlinear (the *weakly nonlinear*) theory, which is well investigated (in particular, thanks to the theoretical results of Gérard Iooss). Our presentation is mainly phenomenological; nevertheless, theoretical evidence concerned with this phenomena is also discussed.

For a rigorous investigation (via center-manifold reduction, bifurcations, and amplitude equations) of various facets of the Couette–Taylor problem, see the recent book by Chossat and Iooss (1994). In chapter II of the book by Iooss and Adelmeyer (1992), the reader can also find applications of center manifolds, normal forms, and bifurcations of vector fields near critical points to Couette–Taylor which is a paradigm for the transition from regular to turbulent behavior (chaos – see Chap. 10).

In the idealized classical problem, the cylinders are infinitely long and rotate about their common axis at constant angular velocities (see Fig. 9.8):

$$\Omega_1, \text{ at } r = R_1 = a \quad \text{and} \quad \Omega_2 \text{ at } r = R_2 = b, \quad \text{with } b > a.$$

The motion of the incompressible viscous fluid satisfies the Navier equations, written with polar cylindrical coordinates (r, θ, z), for flow between the cylinders and the adherence condition at the cylinder walls. There is a unique, steady-state solution, of the Navier equations (written for an incompressible

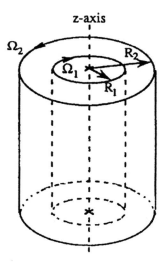

Fig. 9.8. Couette–Taylor Problem

9.5 Couette–Taylor Viscous Flow Between Two Rotating Cylinders

viscous homogeneous fluid), which depends on r alone and takes on prescribed values of the cylinders:

$$\boldsymbol{U} = \left[Ar + \frac{B}{r}\right]\boldsymbol{e}_\theta = V(r)\boldsymbol{e}_\theta \,, \tag{9.148a}$$

where

$$A = \frac{b^2\Omega_2 - a^2\Omega_1}{b^2 - a^2} \,, \tag{9.148b}$$

$$B = -a^2 b^2 \frac{\Omega_2 - \Omega_1}{b^2 - a^2} \,. \tag{9.148c}$$

The flow (9.148a) with (9.148b,c) is called Couette flow between rotating cylinders. This idealized steady-state Couette flow is a good representation of the flow which is observed between cylinders of finite length, away from the ends, when

$$R_B = \left|\frac{B}{\nu}\right| = a^2 b^2 \frac{|\Omega_2 - \Omega_1|}{\nu(b^2 - a^2)} \text{ is small.} \tag{9.149}$$

Couette (1890) experimented with two long concentric cylinders; the inner was fixed ($\Omega_1 = 0$), but the outer was rotating about the axis. He made the following crucial observations: If Ω_2, the angular velocity of the outer cylinder, was less than a critical value (Ω_{2c}), the torque, which was required to sustain the steady-state rotation, was linear in Ω_2. But if Ω_2 exceeded Ω_{2c}, the torque increased more rapidly than this, perhaps because of the development of some form of turbulent motion.

Six years later, Mallock (1896) published his observations, which confirm those of Couette. In addition, however, he observed the case where the outer cylinder was fixed ($\Omega_2 = 0$) and the inner cylinder was rotating about its axis. He found that the flow was unstable, in the sense that it was not circumferential laminar Couette flow at all of the speeds he tested. Thus, Mallock brought to light the important distinction between two cases [according to Stuart (1986)]: (i) the outer cylinder rotating and the inner one fixed; (ii) the inner cylinder rotating and the outer one fixed. This led Rayleigh (1916) to recognize the role of angular momentum in promoting the stability or instability of inviscid (revolving) fluid motion between two concentric rotating cylinders. He couched his argument in terms of the concept of an energy well. On the other hand, inviscid stability requires the square of the circulation to increase outward, and the instability requires the reverse behavior.

However, Synge (1938a,b) showed that the stability criterion,

$$\Omega_2 b^2 > \Omega_1 a^2 \,, \tag{9.150}$$

is a sufficient condition for stability when $\Omega_2 > \Omega_1$, even if the perturbation satisfies viscous flow equations. Couette flow is unstable when R_B is large,

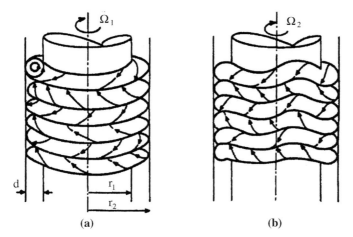

Fig. 9.9. (a) Taylor vortices and (b) wavy vortices-azimuthal waves

[see (9.149)], and then there can be many other stable flows which need not be uniquely determined by the given data. Two other stable flows are the so-called Taylor vortices. In this case, the laminar Couette flow between rotating cylinders has lost its stability and is replaced by the secondary motion. Next, this secondary flow in Taylor vortices has lost its stability and is replaced by a time-periodic motion in which undulations of the vortices propagate around the cylinders (see Fig. 9.9).

The Taylor vortices arise as a steady bifurcation of Couette flow, and the wavy vortices arise as a time-periodic bifurcation of Taylor vortices. It is necessary to mention that Taylor (1923) worked experimentally with very long cylinders and noted the appearance of toroidal vortices (now known as Taylor vortices) under some circumstances of rotation. In this situation, (see Fig. 9.9), plane surfaces separate neighboring toroidal vortices that rotate in opposite directions when viewed in a meridional plane.

We observe many different types of flow dependent on the velocity of rotation of (inner and outer) cylinders. There are many published experimental results since Taylor (1923). In particular, we mention Fenstermacher et al. (1979) and Andereck et al. (1986).

An overview of important experimental and some theoretical work is presented in Cognet (1985). In Fig. 9.10, an experimental stability diagram is presented according to the paper by Andereck et al. (1986); in this diagram, Re_1 and Re_2 are the Reynolds numbers of the inner and outer cylinder.

The theoretical curves in Fig. 9.11 come from Taylor's formula, which was derived from the eigenrelation [when both $[1 - (\Omega_2/\Omega_1)]$ and d/a, with $d = b - a$, are small], in the form:

$$T_c = \frac{\pi^4}{P}\left[\frac{(1+\mu)}{(1-\mu)}\right], \tag{9.151a}$$

9.5 Couette–Taylor Viscous Flow Between Two Rotating Cylinders

with

$$P \equiv 0.0571 \left\{ \frac{(1+\mu)}{(1-\mu)} - 0.652\frac{d}{a} \right\}$$
$$+ 0.00056 \left\{ \frac{(1+\mu)}{(1-\mu)} - 0.652\frac{d}{a} \right\}^{-1}, \qquad (9.151\text{b})$$

where $\mu = \Omega_2/\Omega_1$.

In comparison with later calculations, this formula gives remarkable accuracy for the critical condition (subscript c) above which toroidal vortices occur. In Fig. 9.11, the dotted line corresponds to the Rayleigh (1916) inviscid

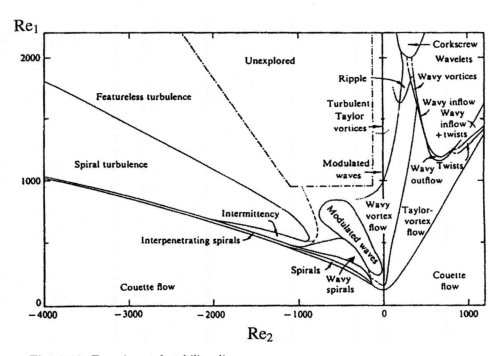

Fig. 9.10. Experimental stability diagram

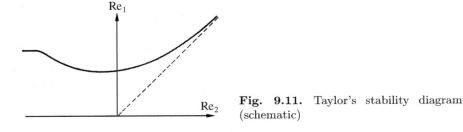

Fig. 9.11. Taylor's stability diagram (schematic)

result. Parameter T is now known as a Taylor number. For $\mu \equiv 0$ and $d/a \equiv 0$, (9.151a,b) give $T_c = 1689$, whereas 1695 is the correct value.

But Couette flow will not bifurcate into Taylor vortices under all circumstances. If the absolute value of μ is large, the loss of stability of Couette flow can be more complicated, and various types of unsteady-state and nonaxisymmetric motions may arise. Besides the various stable and unsteady-state motions that can be seen when Couette flow loses stability, there are still more motions that satisfy the Navier equations but are not observed! Toward the end of his paper, Taylor commented on a number of features: (i) nonlinearity acts to restrain the motion; (ii) there are often pulsating motions of the toroidal vortices, when $0 > \mu > -1$; (iii) spiral vortices can occur due to pre-existing circulation in the annulus, and then neighbouring vortices and are of different sizes. Subsequent improved theoretical and experimental work generally confirmed Taylor's results (Taylor considered only axisymmetric disturbances); however, it was shown by Krueger et al. (1966) that, for counter-rotating cylinders with μ sufficiently negative (beyond the range studied experimentally by Taylor), the mode that first becomes unstable is not axisymmetric but has an angular dependence of the form $\exp(im\theta)$; m increases through the values $1, 2, 3, \ldots$ as μ is made more and more negative. Hence, the left most part of Taylor's curve in Fig. 9.11 must be revised downward (see Fig. 9.10), but only by quite a small amount, because the stability of the mode depends only slightly on m, in the rather narrow gap $\eta = a/b = 0.880$ studied by Taylor. These predictions were confirmed experimentally by Snyder (1970), who also showed that the nonaxisymmetric modes, when they appear, are helical. We observe that linear theory cannot distinguish helical vortices from wavy ring vortices; the four normal modes containing $\exp(\pm inz)\exp(\pm im\theta)$ are all equally likely and can be combined to give real dependence either as $[\sin/\cos](nz)$, $[\sin/\cos](m\theta)$ or $[\sin/\cos](nz \pm m\theta)$, and only a nonlinear theory can say which of these is preferred. Nonlinear theory developed by Davey (1962), Davey et al. (1968), and Eagles (1971) led to an understanding of the structure and stability of finite-amplitude Taylor vortices, the second bifurcation to wavy vortices, and the structure and stability of wavy vortices. In a problem like this involving a sequence of bifurcations, the theory consists ideally of a sequence of alternately linear and nonlinear investigations. After each bifurcation, the structure and amplitude of the new flow is found by a nonlinear calculation, and its stability is then investigated by linearizing the equations about the new flow and studying the growth of infinitesimal disturbances to find the next bifurcation, and so on! In the book by Richtmyer (1981, Chap. 30), a method is described in general terms for calculating the *unstable manifold* developed by the cited four authors above on the basis of earlier work by Stuart and Watson. A significant equation is derived that represents a finite-dimensional dynamical system (see Chap. 10) in that unstable manifold and can be studied by any of the standard methods; analytical search for fixed

9.5 Couette–Taylor Viscous Flow Between Two Rotating Cylinders

points and cycles or calculations of orbits by numerical methods for ordinary differential equations.

In Iooss and Adelmeyer (1992, Chap. II), the authors from the basic equations, taken into account symmetries, and linearization about Couette (purely azimuthal) flow (9.148a–c) investigate the bifurcations from Couette flow (solution $u = 0$ of perturbation system) and its stability which is determined by the spectrum of the (unbounded) linear operator

$$L(\chi) = -\pi \left[(\boldsymbol{u} \cdot \boldsymbol{\nabla})\boldsymbol{U} + (\boldsymbol{U} \cdot \boldsymbol{\nabla})\boldsymbol{u} - \frac{1}{\mathrm{Re}} \boldsymbol{\Delta u} \right] , \qquad (9.152)$$

which emerges from the Navier equations [see (2.61) in Sect. 2.2.6, Chap. 2 – the solutions of this equation are assumed to be *periodic* in z), where the (bifurcation) parameter $\chi \in \boldsymbol{R}^3$ depends on η^{-1}, μ, and Re. Then a *center-manifold reduction* is made by these authors leading (see Sect. 9.5.2) to an ordinary differential equation in a *finite-dimensional space*.

We observe [see Iooss and Adelmeyer (1992, Chap. II, Sect. II.3.3)], first, that the *spectrum* of $L(\chi)$, given by (9.152), contains only *isolated eigenvalues* with *finite multiplicities*; then, for *small Reynolds* number Re, all eigenvalues of $L(\chi)$ lie on the left half of the complex plane, but, when Re is *increased*, a critical value $\mathrm{Re}_c = (\eta^{-1}, \mu)$ is reached where one (at least) eigenvalue crosses the imaginary axis. For Ω_2 positive or Ω_1 (in μ) slightly negative, instability occurs via a zero eigenvalue that corresponds to *axisymmetric eigenvectors* of the form

$$\zeta(r, \theta, z) = \zeta^*(r) \exp\left(\frac{2\mathrm{i}nz}{h}\right) , \qquad (9.153\mathrm{a})$$

for some $n \in \boldsymbol{Z}$, where h is the period in z. But for Ω_2 *sufficiently negative*, *instability occurs via a pair of purely imaginary eigenvalues* which correspond to non-axisymmetric *eigenvectors* of the form

$$\zeta(r, \theta, z) = \zeta^*(r) \exp\left(\mathrm{i}m\theta + \frac{2\mathrm{i}nz}{h}\right) , \qquad (9.153\mathrm{b})$$

for some $(n, m) \in \boldsymbol{Z}^2$, $m \neq 0$. Figure 9.12 shows a typical stability curve in the $(\mathrm{Re}_1, \mathrm{Re}_2)$-plane, according to Iooss and Adelmeyer (1992, p. 96).

Intersections of the bifurcations for Couette flow were studied by Kolesov and Yudovich (1994) and Kolesov et al. (1995), and various other types of intersections of bifurcations were investigated by Andreichikov (1977), Chossat et al. (1987), Chossat and Iooss (1985), and also by Frank Gerd and Meyer-Spashe Rita (1981). The detailed theory of the intersection of bifurcations in the Couette–Taylor problem is presented in the book by Chossat and Iooss (1994). Chossat, Demay, and Iooss consider bifurcations corresponding to the interaction of non-axisymmetric modes that have the same axial but different azimuthal wave numbers. In Andreichikov and Frank Gerd and Meyer-Spashe

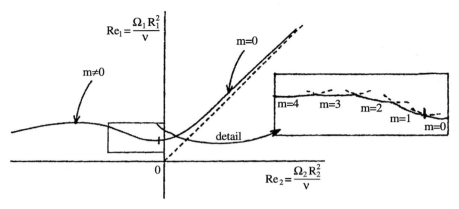

Fig. 9.12. Stability curve of Couette flow

Rita, the intersection of axisymmetric bifurcations with different axial wave numbers (one value is two or three times greater than the other) is studied. In a recent paper by Kolesov and Shapakidze (2000), the oscillatory modes in viscous incompressible liquid flow between two counterrotating *permeable* cylinders is considered. It was found that the first instability of the main stationary regime occurred via either axisymmetric or non-axisymmetric disturbances. In the first case, the main flow is replaced by the secondary stationary flow that has the form of toroidal vortices, and for non-axisymmetric disturbances, the secondary regime is a time-periodic flow with azimuthal waves; the objective of Kolesov and Shapakidze (2000) was the study of oscillatory modes arising in a vicinity of the intersection point of the neutral curves corresponding to two of the above-mentioned kinds of instability. In Shearer and Walton (1981), symmetries of a nonlinear boundary-value problem governing Taylor vortices and also Bénard convection cells are described. The effects of end walls on the onset of Taylor vortices is considered in Walton (1979). In Denath (1987, Sects. 8 and 12.2), the bifurcations and nonlinear instability phenomena involved in the classic Taylor problem are discussed.

9.5.2 Bifurcations

Bifurcation theory for Couette flow is very complicated (as a consequence of *multiple eigenvalues of the spectral problem*) when μ is negative. In this case, it is necessary to determine the number of solutions that bifurcate and to compute the different bifurcating branches. Some bifurcation results for the multiple eigenvalue problem with $\mu < 0$ have been given by DiPrima and Grannick (1971). One interesting result is that Taylor vortices bifurcate subcritically when $-0.73 < \mu < -0.70$ and $\eta = a/b = 0.95$. The bifurcation picture for $\mu < -0.73$ is very complicated. On the contrary, bifurcation theory for Couette flow when $\mu > 0$ is well developed and in good agreement

9.5 Couette–Taylor Viscous Flow Between Two Rotating Cylinders

with experiments. This bifurcation can be called a nonlinear Taylor problem because it leads to steady-state supercritical Taylor vortices studied first by Stuart (1958). Mathematically rigorous theories for bifurcating Taylor flow have been first given by Velte (1966), Yudovich (1966), Kirchgässner and Sorger (1969); in Joseph (1976, Chapter V), the reader can find information and results obtained up to 1975.

In a more recent paper by Cognet (1985), the laminar-turbulent transition of Couette–Taylor flow is presented: Taylor cells, periodic flow, then quasi-periodic and weak turbulence. A large part of this paper (in French) is concerned with a review of the literature (up to 1984). Analysis of the flow is based on visualization and local measurements, and the influence of geometry is discussed, as well as the dynamics of transition. Theoretical models are also presented which are developed to explain the observed phenomena. Computation of bifurcated solutions for the Couette–Taylor problem (both cylinders rotating) is considered in the paper by Demay and Iooss (1985). A method is described for computing the coefficients in the differential system on the center manifold for the Couette–Taylor problem with several symmetries. This system describes all of the dynamics of the solutions of the initial-value problem in the neighborhood of a bifurcation. The numerical results confirm previous results and moreover allow us to predict, for specified values of the parameters, a new type of periodic flow, horizontal cells that look like uniformly rotating ribbons (see Fig. 9.13).

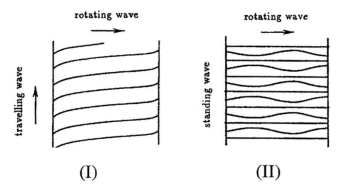

Fig. 9.13. (I) Spirals and (II) Ribbons

Bifurcations from Couette flow. The application of bifurcation theory gives, first, the possibility of describing rigorously *steady-state bifurcation* from Couette flow, when we assume that $\lambda = 0$ is a *double eigenvalue* of $L(0)$ given by (9.152), and that the axial wave number $m = 0$. In this case, Taylor *axisymmetric vortex flow*, periodic along the z axis of the period h/n, appears. Moreover, the fluid particles are confined to horizontal cells of height $h/2n$ (see Fig. 9.14).

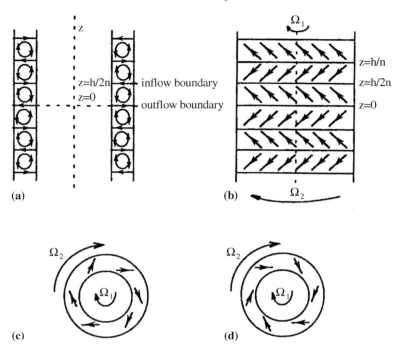

Fig. 9.14. Taylor vortex flow: (a) vertical cut view; (b) front view; (c) horizontal cut view at $z = 0$ and (d) horizontal cut view at $z = h/2n$

Now, when we assume that $\lambda = \pm i\omega$, $\omega > 0$, are double *eigenvalues* of $L(0)$ and that the axial wave number $m \neq 0$, we have *Hopf bifurcation* from Couette flow (for Hopf bifurcation, see Sect. 10.2, in the Chap. 10) and two types of periodic solutions which bifurcate from the (Couette) solution $\boldsymbol{u} = 0$.

Type I – Spirals: a rotating wave around the z axis and a traveling wave along the z axis, and this gives a *spiral structure* (see Fig. 9.13).

Type II – Ribbons: flow is confined to horizontal cells of height $h/2n$, and we have a standing wave in the axial direction (see Fig. 9.13). Unlike Taylor flow, this flow is not axisymmetric but azimuthally periodic (a rotating wave in the azimuthal direction) and all invariance properties propagate to the velocity vector field.

Steady bifurcations from Taylor vortex flow. First, it is necessary, via center-manifold reduction for group orbits of steady-state solutions, to derive a *reduced problem* so that all symmetries of the original problem propagate to the reduced problem. Then, the phase equation and amplitude equation reduces to

$$\frac{d\psi}{dt} = 0 \text{ and } \frac{dA}{dt} = f(A, \mu), \qquad (9.154a)$$

where $\psi \in \boldsymbol{R}/2\pi\boldsymbol{Z}$, $A \in \boldsymbol{R}$.

9.5 Couette–Taylor Viscous Flow Between Two Rotating Cylinders

If $L(0)$ has a simple eigenvalue 0 (which is not the general case for the instability of Taylor vortex flow), then we have a steady-state bifurcation from Taylor vortices. It is necessary to consider, on the one hand, case (I), when a *saddle-node bifurcation* occurs and the solutions are built up of horizontal cells and also when a *pitchfork bifurcation* occurs producing solutions that have *twice the period in the axial direction*.

On the other hand, in case (II), when the right-hand side in the phase equation is different from zero,

$$\frac{d\psi}{dt} = A \text{ and } \frac{dA}{dt} = f(A, \mu), \qquad (9.154b)$$

the bifurcation is pitchfork type, the bifurcating solutions are traveling waves in the axial direction under the circle group orbit of Taylor vortices, and this case of traveling waves appears (according to Iooss) even though the critical eigenvalue is zero. It seems that this situation has not been observed yet in the Couette–Taylor problem. Finally, when the bifurcation is again of the pitchfork type, bifurcating solutions look like case (I), with doubled axial periodicity; but we have *horizontal cells* of height h/n, *shifted by half* of the basic period of Taylor vortices with respect to case (I), when a saddle-node bifurcation occurs.

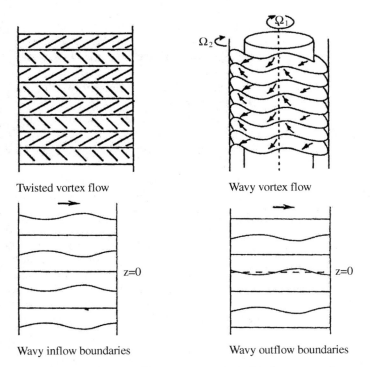

Fig. 9.15. Hopf bifurcating solutions from Taylor vortex flow

Hopf bifurcation from Taylor vortices: twisted vortices, wavy vortices, vortices with wavy inflow and outflow boundaries. Here, the phase equation and amplitude equation reduce to

$$\frac{\mathrm{d}\psi}{\mathrm{d}t} = 0 \text{ and } \frac{\mathrm{d}A}{\mathrm{d}t} = AQ(|A|, \mu), \tag{9.154c}$$

where Q is even in $|A|$; we have the following amplitude equation for A:

$$\frac{\mathrm{d}A}{\mathrm{d}t} = (\mathrm{i}\omega + a_{11}\mu)A + a_{30}A|A|^2 + o(|\mu| + |A|^2). \tag{9.155}$$

In this case, the solutions represent *rotating waves*, and, according to the symmetry assumptions, we distinguish *four type* of solutions represented schematically in Fig. 9.15.

9.6 Concluding Comments and Remarks

First we mention that, in a very recent survey paper, Bodenschatz et al. (2000) summarize results for RB convection that have been obtained during the past decade or so. It is now well known that convection occurs in a spatially extended sytem when a sufficiently steep temperature gradient is applied across a fluid layer, and a "pattern" appears that is generated by the spatial variation of the convection structure – the nature of such convection patterns is at the center of the survey cited above. In Fig. 9.16 (see pages 382 and 383), the reader can find some "spectacular" RB convection patterns selected just from the survey cited.

In the survey cited, the authors concentrate their attention on convection in compressed gases and gas mixtures with Prandtl numbers near one and smaller. In addition to the classical problem of a horizontal stationary fluid layer heated from below, it also briefly covers convection in such a layer rotating about a vertical axis with inclination and modulation of the vertical acceleration.

We emphasize again that in the case of RB thermal convection, the more important effect is the buoyancy effect in the fluid; the reader can find an excellent account of the various features of this buoyancy effect in Turner (1973) book.

For the stability of Navier flows in exterior domains, see Kozono and Ogawa (1994). In Brevdo (1995), the initial boundary-value stability problem for 2-D disturbances in the Blasius boundary layer [boundary layer on a flat plate (in fact, a half-plane) placed edgewise in a uniform steady-state stream] is treated formally by means of the Fourier–Laplace transform. The resulting nonhomogeneous boundary-value problem for the Orr–Sommerfeld equation is studied analytically.

There is extensive literature on the nonlinear stability of viscous shock waves; the reader can find various references in Szepessy and Xin (1993).

9.6 Concluding Comments and Remarks

In this last paper, the authors considered the nonlinear stability of viscous conservation laws and analyzed the large-time asymptotic stability of a shock wave in a strictly hyperbolic system. A shock wave in Szepessy and Xin (1993) is referred to as a smooth traveling wave solution that satisfies

$$\frac{\partial u}{\partial x} + \frac{\partial f(u)}{\partial x} = \frac{\partial^2 u}{\partial x^2}, \ x \in \mathbf{R}^1, \ t > 0, \ u \in \mathbf{R}^m \ , \tag{9.156a}$$

$$u(.,0) = u^0(.). \tag{9.156b}$$

We note an important observation from Liu's (1986a) works that a generic perturbation of a viscous shock profile produces not only translation and but also diffusion waves. We note also that shock waves for compressible NS equations (in one dimension) are stable [see Liu (1986b)]. For mathematically rigorous results, we mention also the books by Iooss and Joseph (1980), Joseph (1976), and Eckhaus (1965) where the reader can find some early studies in nonlinear stability theory.

The recent book *Hydrodynamics and Nonlinear Instabilities*, edited by Godreche and Manneville (1998), provides a review of current research in the areas of specific interest of the contributors, rather than a comprehensive and traditional textbook that covers every introductory aspect. The authors consistently relate the mathematics to the physical phenomena under consideration in an informal and intuitive way, resulting in a style that is very readable (see, for instance, the detailed review by R. Lingwood (1999) in *J. Fluid Mech.*). It is important to note that this book has, to its credit, many topics that are not easy to find elsewhere. For example, Chap. 2 is a very significantly updates Chap. 4 of Drazin and Reid's *Hydrodynamic Stability* (1981), which makes only brief mention of spatial, as opposed to temporal, instability of open flows, and Chap. 4 significantly adds to the current textbook by extending the discussion of pattern-forming instabilities to include sections on secondary (Eckhaus, zig-zag, and drift) instabilities. Finally, the topics in Chap. 5 (flames, shocks, and detonation in initially premixed homogeneous reactants) are largely new for advanced students and scientists not familiar with these fields and will make a rather intriguing impression; on the contrary, Chap. 3 is a fairly standard (and very incomplete) introduction to singular perturbation theory and omits many interesting and important research papers and books.

The book *Wave and Stability in Fluids* by Hsieh and Ho (1994), provides a modern introduction to two major fields of fluid dynamics. Hydrodynamic instabilities are introduced in a long chapter of nearly 150 pages and the major mechanisms, Kelvin–Helmholtz instability, and Orr–Sommerfeld, Faraday, RB, and Taylor–Couette problems are described; in a few cases, nonlinear aspects of the evolution of disturbances are considered. The book also includes a short chapter on chaos, based on a discussion of the Lorenz equations and of the logistic map.

The problems surveyed in the book *Explosive Instabilities in Mechanics* by Straughan (1998), include the open question of finite-time blowup for

Fig. 9.16. Various RB convection patterns

9.6 Concluding Comments and Remarks 383

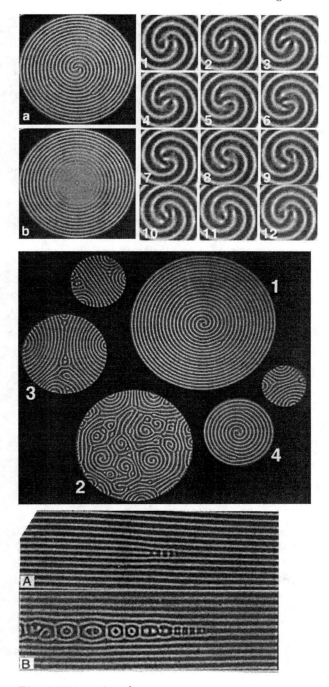

Fig. 9.16. continued

the Navier equations, the development of infinite temperature gradients in BM thermocapillary convection (in the limit of an infinite Prandtl number), blowup backward in time for nonlinear viscous fluids, flow with pressure-dependent viscosity, and the rapid growth of instabilities in parallel shear flow at high Reynolds numbers. The style of the book is similar to a review article, summarizing work that appeared in the literature, and for this, the book will be valuable as a survey and a reference. References to the literature are extensive and are a crucial component of the value of this book as a resource. Finally, we observe that the selection of topics for blowup in fluid dynamics, which have been included in this book, is quite broad and should be of interest to readers from many different subdisciplines of fluid dynamics.

In relation to the Couette–Taylor problem, we mention the book by Batchelor (1996), *The Life and Legacy of G.I. Taylor*, where the reader can find many accounts of work which became classic, like that on the motion of particles in rotating flows and centrifugal instabilities. The main success of Taylor was that the Rayleigh (inviscid) criterion for stability in the problem should be revised to account for viscosity!

The problems of hydrodynamic stability are numerous and have attracted fluid dynamicians, applied mathematicians, numericians, and experimentalists for all of this past century. Below, we mention some work on stability–instability for various fluid dynamics problems where viscosity plays a fundamental role.

The stability of three-dimensional boundary layers is discussed in the review paper by Reed and Saric (1989); the main driving force for the study of *boundary-layer stability* was understanding prediction and control of the transition to turbulence. We observe that all 3-D flows exhibit similar characteristics (streamwise vorticity) and all appear to depend heavily on initial conditions. In general, it is uncertain how to define transition in 3-D, and, clearly, the details of the transition process in 3-D are still missing. For example, it is important to know whether the formation of 3-D structures (the so-called K-type and H-type) and characteristic stages (e.g., one-spike stage) will occur as in 2-D flow transition and mechanisms for secondary instabilities remain to be determined! Finally, we mention the monograph (AGARd Report) of Mack (1984) on *boundary-layer linear stability theory*, which is the primary source of basic information on the subject, and also the Arnal (1986), AGARd Report, which is an extensive review of *transition in 3-D flow* and, as such, is complementary to the material covered in Reed and Saric (1989).

For the *secondary instability of boundary layers*, see the pertinent rewiew by Herbert (1988). We observe that, although Squire's (1933) transformation highlights the stronger instability of 2-D waves, oblique waves or streamwise vortices cannot be considered irrelevant and the natural background must be viewed as an irregular pattern of 3-D wave packets with nonuniform spectral content. A critical evaluation and interpretation of the 1983 state of the art was given by Morkovin (1983, 1985).

Instability mechanisms in shear-flow transition are considered in the review paper by Bayly et al. (1988), and the *onset of instability in unsteady-state boundary-layer separation* at high Reynolds number is investigated in Cassel et al. (1996) paper (their results appear to *alter considerably* current thinking about the physical picture of unsteady-state boundary-layer separation). Solution of the unsteady-state non-interactive, classic boundary-layer equations develop a *generic separation singularity* in regions where the pressure gradient is prescribed and adverse. As the boundary layer starts to separate from the surface, however, the external pressure distribution is *altered* through viscous–inviscid interaction just prior to the formation of the separation singularity (the *first interacting stage*), and the (numerical) solutions exhibit *high-frequency inviscid instability resulting in an immediate finite-time breakdown of this stage*, which is confirmed by linear analysis! The implications for the theoretical description of unsteady boundary-layer separation are discussed in Cassel et al. (1996).

Rayleigh-Taylor instability (RTI) occurs whenever fluids of *different density* are subjected to *acceleration in a direction opposite that of the density gradient* [Rayleigh (1883), Taylor (1950), Chandrasekhar (1955, 1961), Duff et al. (1962), Menikoff et al. (1977), Sharp (1984), and Iooss and Rossi (1989)], and is encountered in a variety of contexts, such as combustion, rotating machinery, inertial-confinement fusion, supernovae explosions, and geophysics). In a recent paper by Cook and Dimotakis (2001) devoted to transition stages of RTI between miscible fluids, the reader can find recent pertinent references. RTI flow represents an important simulation and modeling test case for hydrodynamic processes, in general, involving many aspects of turbulence, transport, and diffusion in non uniform-density flow subject to external body forces.

The survey paper by Kelly (1994) is devoted to the *onset and development of thermal convection in fully developed shear flow*. Although various investigations have already been made in these areas [see the references in Kelly (1994)], many of the phenomena discussed by Kelly in his survey of fully developed flows have not yet been studied. Kelly, first, considers the stability of unstably stratified parallel shear flow when the shear is established primarily by some mechanical means (at first, by conduction, and varies only in the vertical direction; the effects of a horizonal variation of temperature upon the instability are considered later). Any contribution to the basic velocity distribution due to the variation of temperature acting via the buoyancy term in the equations of motion is assumed to be of secondary importance. Then, Kelly considers the stability of mean flows that are thermally induced, i.e., they are of the "natural" or "free" convection type and occur due to horizontal temperature and density variations.

In the Hele–Shaw geometrical configuration, a viscous fluid in a narrow slot between vertical parallel plates is subjected to a constant gradient of temperature either in the same direction as $\boldsymbol{g}$ or opposite. In the recent paper by Aniss et al. (2000) an asymptotic study of convective parametric instability in a Hele–Shaw cell is investigated, and the reader can find various references [in particular, the stability of nonlinear convection in a Hele–Shaw cell is considered by Kvernvold (1979)].

Finally, in conclusion of this chapter devoted to stability of fluid motions we mention some recent papers and books; namely: Bowles (2000), Cassel (2001), Iooss et al. (eds.) (2000), Riahi (2000).

10 A Finite-Dimensional Dynamical System Approach to Turbulence

Today turbulent phenomena are not yet completely understood even from a physical point of view, so that there is no general agreement on what "turbulence" is! In reality, the term turbulence describes very different fluid behaviors, and it seems that many notions of turbulence/chaos exist!

There is no complete mathematical theory describing turbulent flow, and consequently, any exposition must be, at least partially, phenomenological.

It often happens in practice that real fluid flows for certain values of the characteristic (control) parameters behave chaotically, and we have only a statistical way to describe this flow. Consequently, turbulent motion is not only complex but also needs, for intrinsic reasons, a statistical description.

However, recent progress has been made in understanding why turbulence/chaos develops. We mean the so-called onset of turbulence or the transition from laminar to turbulent flow, for which the general theory of dynamical dissipative systems provides a convincing explanation of many features that accompany this transition. This is mainly the content of this chapter.

10.1 A Phenomenological Approach to Turbulence

In many cases, it is allowable to deal with a dynamical system of finite dimensions by using a model of a continuous fluid. This is particularly the case at the stage of generating turbulence from a laminar motion, at which only a limited number of freedoms of motion have been excited. In practice, so-called "chaos" is studied mostly via a dynamical system of finite (low) dimensions, whereas "turbulence" is generally dealt with as motion in a fluid of infinite degrees of freedom.

The approximation of a fluid in terms of a model system of finite dimensions provides us with a very powerful means of analysis. For this reason, recent progress in the theory of chaos has enabled us to look directly at the fundamental mechanism of generating turbulence.

On the other hand, it is generally recognized that turbulence in its fully developed stage has a singular structure in space and time. Such singular behavior of a fluid cannot be described correctly by a model system of finite dimensions. Thus, in this restricted sense, chaos in fluids covers only a part of turbulent phenomena.

If chaos is not exactly the study of fluid turbulence, nevertheless, the image of turbulent, erratic motion serves as a powerful icon to remind physicists of the sorts of problems they would ultimately like to comprehend.

Fluid turbulence presents us with highly erratic and only partially predictable phenomena. Much of chaos as a science is connected with the notion of "sensitive dependence on initial conditions." Technically, scientists term as "chaotic" those non - random complicated motions that exhibit very rapid growth of errors that, despite perfect determinism, inhibits any pragmatic ability to render accurate long-term prediction. Temporal chaos is known to appear in certain systems having only a few degrees of freedom.

Take, for example, the Lorenz model (1963) which has only three degrees of freedom. It is a crude truncation of the Rayleigh–Bénard (RB) classical shallow convection 2-D problem with only one Fourier component in the horizontal and vertical directions, so that the motion can in no way be considered spatially chaotic!

Nevertheless, there is strong numerical evidence that the temporal spectrum becomes continuous when the Rayleigh number crosses a certain threshold, an indication that temporal chaos has developed.

Very different is the problem of fully developed turbulence which is essentially spatio-temporal chaos. When the Reynolds number goes to infinity, all space and timescales, down to infinitesimal are excited. Such chaos may develop in a finite time and have universal scaling properties (e.g., a power-law energy spectrum). A universal property of fully developed turbulence is related to intermittency, or in other words spottiness of small scales. But transition to turbulence (chaos!) through intermittency is also possible (the Pomeau–Manneville route to chaos). In hydrodynamics, one often distinguishes between the behavior of closed systems and open systems, and from a purely kinematic point of view, one is tempted to say that a flow is closed when fluid particles are recycled within the physical domain of interest (Bénard convection within a horizontal fluid layer). If all particles ultimately leave the domain (as in plane Poiseuille flow), the flow is said to be open.

In transitional closed flows, the motion is absolutely unstable (Rayleigh–Bénard convection or Taylor–Couette flow) and can give rise, under carefully controlled conditions, to chaotic motion on a low-dimensional strange attractor. Such is not the case for open, convectively unstable flow! The onset of chaos in convectively unstable flows is discussed in Huerre (1987).

In 1944, Landau (1944) and then, in 1948, Hopf (1948) advanced a theory to explain the genesis of turbulence by suggesting that the physical parameters describing turbulence phenomena are *quasi-periodic functions* of time that take the following form:

$$f(\omega_1 t, \omega_2 t, ..., \omega_k t),$$

10.1 A Phenomenological Approach to Turbulence

where f is a periodic function, with a period of unity in relation to each of variables and where the ω_i, $1 < i < k$, are not connected by a linear combination with rational coefficients. *Quasi-periodic movement* at a high number of independent frequencies can give rise to very complex behavior which resembles that generated by a turbulent regime!

Nonetheless, a major objection to the Landau–Hopf (LH) theory (scenario/route to chaos) is that quasi-periodic movements do not exhibit the crucial phenomenon called *"sensitive dependence on initial conditions"* (SDCI), an essential characteristic of turbulence (chaos).

This is amply demonstrated by the fact that two initially quasi-identical turbulent flows differ strongly ulteriorly in relation to the "exponential divergence" phenomenon.

Nearly 20 years after Landau's theory came the work of Lorenz (1963), who attempted an approximate description of the RB convective instability phenomenon as it applies to the atmosphere.

If temperature is sufficiently intense, convection takes place in a disorderly turbulent fashion, and this turbulence in its temporal evolution is linked to the appearance in the phase space – from the associated Galerkin system – of a so-called Lorenz (strange) attractor.

An illustration of such an attractor can be found in Fig. 10.1, which is the work of O.E. Lanford (Berkeley University).

The Lorenz dynamic system of three equations is the following:

$$\frac{dX}{dt} = -10X + 10Y ,$$
$$\frac{dY}{dt} = -XZ + 28X - Y ,$$
$$\frac{dZ}{dt} = XY - \frac{8}{3}Z .$$
(10.1)

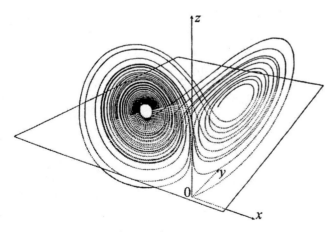

Fig. 10.1. Lorenz (strange) attractor

Lorenz expanded the Oberbeck–Boussinesq equations for RB thermal shallow convection of a horizontal layer of fluid heated from below in a triple Fourier series with respect to the space variable (Galerkin method), then truncated the resulting system of ODEs for the time dependence of the Fourier coefficients to three equations.

If the Fourier coefficients in those equations are denoted by $X(t)$, $Y(t)$, and $Z(t)$, the Lorenz equations are just (10.1), but in place of 10, in the first equation of (10.1), we have $\sigma > 0$ (the Prandtl number), in the second equation of (10.1), we have $r > 0$ [proportional to the Rayleigh number – the ratio of the Rayleigh number over the critical one – and is a measure (control parameter) of the intensity of the heating] in place of 28, and in the last equation of (10.1) in place of 8/3 we have $b > 0$ (a geometrical factor of the box).

With the Lorenz equations (10.1) as a starting point, Lanford chooses the solution which goes from the origin ($X = 0, Y = 0, Z = 0$) to time $t = 0$. During numerical simulation, this initial solution is not prescribed "exactly," but rather a "very close" initial solution. With this solution, one has a first loop on the right, then a few on the left, then another on the right, and so on and so forth haphazardly. The solution can be followed for up to fifty loops in Fig. 10.1, where the part below the level $Z = 27$ is represented by a dotted line. But, if, instead of $(0,0,0)$, adjoining initial conditions were chosen at $t = 0$, the new solution of (10.1) would quickly deviate from the former (exponential divergence), and the number of loops on the left or right would also change due to SDIC. On the basis of this phenomenon, Lorenz would aptly justify the inaccuracy of more or less long term weather forecasting; long term means more than a week! Unfortunately, fluid mechanics specialists of the time never even suspected the connection between Lorenz's research and their own in the field of turbulence. It was thus that Lorenz's (1963) article went unheeded in the community of mathematicians and physicists.

The first publication [8 years after Lorenz's (1963) paper] containing the expression "*strange attractor*" (Lorenz's attractor falls into this category!) was an article by Ruelle and Takens (1971a,b). The very notion of SDIC lies at the heart of the RT theory. It closely links turbulence to a strange attractor (the term "strange" is narrowly connected to SDIC) which appears in the phase space of the associated dynamical system and concerning the Lorenz attractor and the problem of turbulence see Ruelle (1976).

This attractor, which is an abstract mathematical object, is formed by an infinite number of thin sheets or rings stretched out and folded over on each other. It must be understood that for a dissipative system (like that characterized by Navier's viscous incompressible equations), the trajectories don't cover the entire space of phases but rather are attracted toward a part of it, called attractor, and there remain confined. Because of this, the structure of the attractor is closely tied to the nature of the established limiting regime in time.

10.1 A Phenomenological Approach to Turbulence

Therefore, contrary to the LH theory for which the attractor associated with a turbulent regime is a *torus* T^k, where k is large, Ruelle and Takens set forth the idea that turbulence could be described by a *strange attractor* that exhibits the phenomenon of SDIC; the motion on the strange attractor is also *Lyapunov unstable* (see Chap. 9) and hence has a purely continuous power spectrum.

This new RT theory has since engendered a great many articles on the theory of dynamical systems as well as numerous experiments and numerical simulations concerning transition toward "turbulence" for various physical systems – see, for example, the various strange attractors associated with the so-called Zeytounian *deep Bénard convection* problem (1989) presented in the paper by Errafiy and Zeytounian (1991b).

Another 8 years went by after Ruelle and Takens' work was published before a new direction opened up in the field of turbulence research. The name Feigenbaum (1978, 1980/1983) immediately comes to mind on this subject, but it must be emphasized that Tresser and Coullet (1978) obtained similar results. In Feigenbaum's (FTC!) theory, transition toward chaos is caused by a *periodic doubling* phenomenon and for this theory, the technique known as the *renormalization group* (RG) constitutes an important tool.

A great deal of research is presently devoted to studying the transition to chaos developing from a quasi-periodic regime (e.g., in convective flow) or by intermittency [according to Pomeau–Manneville (1980) – PM route to chaos]. In the latter case, turbulent "blasts" of great amplitude occur. Several *scenarios* have been observed for the passage from the periodic to the chaotic regime which are summed up in Swinney (1983).

Note also that chaos, a subject that has developed through the study of dynamical systems, has connections with *fractal dimensions* and fractal geometry. In the context of dynamics, attracting sets with fractal properties have been called strange attractors!

Mathematically, a fractral is a shape "by definition for which the *Hausdorff–Besicovitch dimension* strictly exceeds the *topological dimension*." Attractors are geometric forms that characterize long-term behavior in state (phase) space. Roughly speaking, an attractor is what the behavior of a system settles down to, or is attracted to. The simplest kind of attractor is a fixed point, which is simply an equilibrium or steady-state solution of a dynamical system.

Depending on the stability of these *fixed points*, they might attract (stable fixed point) or repel (unstable fixed point) nearby trajectories in phase space. Hence, in studying a dynamical system, it is useful to determine its fixed points and their stability, which in turn motivates because changes in the stability of the fixed points are associated with these bifurcations.

The next most complicated attractor (in comparison with the fixed points) is a *limit cycle*, which forms a closed loop in state space. A limit cycle describes stable oscillations (periodic solution). If a stable limit cyle existed in a 3-D phase space and a 2-D *Poincaré* surface (or *section*) were introduced, the trajectory or orbit would repeatedly penetrate the surface at the same point. Compound oscillations, or *quasi-periodic* behavior, correspond to a *torus* attractor.

All three attractors are *predictable*: their behavior can be forecast as accurately as desired. *Chaotic attractors*, on the other hand, correspond to *unpredictable motions* and have a more complicated geometric form. The key to understanding chaotic behavior lies in understanding a simple *stretching* and *folding* operation, which takes place in the state space. *Exponential divergence* is a *local* feature. Because attractors have *finite size*, two orbits on a chaotic attractor *cannot diverge exponentially* forever. Consequently, the attractor *must fold over onto itself*. Although orbits diverge and follow increasingly different paths, they eventually must pass close to one another again. The orbits on a chaotic attractor are *shuffled* by this process, much as a deck of cards is shuffled by a dealer! The *randomness* of chaotic orbits is the result of the *shuffling process*.

The process of *stretching and folding happens repeatedly, creating folds within folds ad infinitum*. A chaotic (strange) attractor is, in other words, a *fractal*, an object that reveals more detail as it is increasingly magnified – for a "different point of view" of strange attractors, see the article reviews by Guckenheimer (1986). Experiments designed to adress the strange (chaotic) attractor theory of turbulence have been conducted primarily on convecting layers (the RB system) and on flow between rotating cylinders (the Taylor–Couette experiment).

10.2 Bifurcations in Dissipative Dynamical Systems

We shall consider the nonstationary initial-boundary value problem for Navier equations,

$$\frac{\partial \boldsymbol{V}}{\partial t} - \nu^0 \Delta \boldsymbol{V} + \boldsymbol{V} \cdot \boldsymbol{\nabla} \boldsymbol{V} = -\boldsymbol{\nabla} p + \boldsymbol{f} ,$$
$$\boldsymbol{\nabla} \cdot \boldsymbol{V} = 0 ,$$
$$\boldsymbol{V}|_{\partial \Omega} = 0 , \boldsymbol{V}|_{t=0} = \boldsymbol{V}^0 , \qquad (10.2)$$

in the domain $\Omega_T = \Omega \times [0, T]$, where $\boldsymbol{f}(\boldsymbol{x}, t)$ is a given vector field of external forces and $(\boldsymbol{V}, p)$ are the unknowns (velocity vector and pressure). Moreover, we assume that the given vector (data) $\boldsymbol{V}^0(\boldsymbol{x})$ is solenoidal and that it satisfies also the boundary condition $\boldsymbol{V}^0|_{\partial \Omega} = 0$.

Let $\{\phi^k(\boldsymbol{x})\}$ be a set of functions complete in the Hilbert space $H(\Omega)$ and simultaneously orthonormal in $L_2(\Omega)$. Furthermore, suppose that the function ϕ^1 is identical with the given initial data of the sought for solution,

$\phi^1(x) = V^0(x)$. We now approximate the solution by means of the function V^n that has the form

$$V^n = \sum C_{ln}(t)\phi^l(x), \quad l = 1, ..., n . \tag{10.3}$$

In this case, the functions $C_{ln}(t)$ are determined by the relations:

$$C_{ln}(0) = \delta_{l1} , \tag{10.4}$$

and

$$\left\langle \frac{\partial V^n}{\partial t} + V^n \cdot \nabla V^n - f ; \phi^l \right\rangle + \nu^0 \langle \nabla V^n ; \nabla \phi^l \rangle = 0 , \tag{10.5}$$

with $l = 1, 2, ..., n$.

We see that this is the known Galerkin method (see Sect. 8.2 for a sligthly different approach), where the notation $\langle . ; . \rangle$ is the scalar product in $L_2(\Omega)$ space. The system of equations (10.5), by virtue of the assumed orthonormality of system $\phi^1(x)$ in $L_2(\Omega)$, takes the form of a system of ordinary differential equations for the unknown functions $C_{ln}(t)$. This system of ODEs is a finite-dimensional dynamical system:

$$\frac{dC_{ln}(t)}{dt} = \nu^0 \sum_{l=1 \text{ to } n} a_{ll} C_{ln} + \sum_{l,p=1 \text{ to } n} a_{llp} C_{ln} C_{pn} + f_l , \tag{10.6}$$

with $l = 1, ...n$.

The a_{ll} and a_{llp} are numerical coefficients, and $f_l = \langle f ; \phi^l \rangle$. The right-hand side of (10.6) are analytical functions of unknowns $C_{ln}(t)$. Therefore, the existence of a unique solution of system (10.6), with the initial conditions (10.4), in the interval where functions on the right-hand side are defined is ensured. Their boundedness being a consequence of restrictions [see, Ladyzhenskaya (1967)],

$$V^0 \in W_2^2(\Omega) \quad \text{and} \quad \int_0^{T^*} \left[\|f\| + \left\| \frac{\partial f}{\partial t} \right\| \right] dt < \infty , \tag{10.7}$$

with T^*, a positive constant time, will follow, provided that the boundedness of the solutions $C_{ln}(t)$ in this interval is established.

System (10.6) is a *finite n-dimensional dynamical system* for the amplitude functions of time $C_{ln}(t)$, $l = 1, ...n$, and plays a fundamental role in investigating routes to chaos (see Sect. 10.3).

We consider, now, a general finite-dimensional dynamical system with N degrees of freedom:

$$\frac{dX}{dt} = F(X; \lambda), \text{ with } X(t) \in R^N . \tag{10.8}$$

The term *dynamical system* (DS) is often reserved for sufficiently *small* values of N. For a *dissipative* DS, the volume in the phase space are contracting, and we then look at trajectories and their attractors in the phase space [for a Course related on generic instabilities and nonlinear dynamics, see Thual (1988/1993)]. When the *control parameter* λ varies, destabilization of these attractors may occur with topological changes in phase portraits – these are called bifurcations. The Lorenz model (10.1) is the first example of a dissipative DS.

Obviously, a fixed point (equilibrium) of a DS is a point $\boldsymbol{X}^*$ in the phase space such that

$$F(\boldsymbol{X}^*, \lambda) = 0 , \tag{10.9}$$

and to study its stability, we need to investigate the eigenvalues of the $N \times N$ operator obtained by *linearizing* the equations around $\boldsymbol{X}^*$:

$$\boldsymbol{X} = \boldsymbol{X}^* + \boldsymbol{U} , \quad \frac{d\boldsymbol{U}}{dt} = L^* \boldsymbol{U} + N(\boldsymbol{U}, \lambda) , \tag{10.10a}$$

with

$$L^* = DF(\boldsymbol{X}^*, \lambda) , \quad \text{the differential of } F(\boldsymbol{X}^*, \lambda) \text{ at } \boldsymbol{X}^* . \tag{10.10b}$$

A *fixed point* is *stable* when all eigenvalues of the (linear) operator L^* have a *negative real part*. *Instability* occurs when *one real eigenvalue* or *two complex conjugated eigenvalues* (for the generic case) *cross the imaginary axis*. The *topological singularity* associated with an instability is called a *bifurcation*. The generic bifurcation that occurs when a real value crosses the imaginary axis is called a *saddle-node bifurcation*. Two fixed points, one unstable and the studied stable one, collapse at the critical value of the parameter. There is a change in the number of solutions of the equation $F(\boldsymbol{X}^*, \lambda) = 0$ because the X differential of F is *noninvertible* (and this obviously violates the implicit function theorem).

When there is symmetry in the equations (nongeneric case), a so-called *pitchfork bifurcation* can occur. For the Lorenz model (written with σ, r, and b), the differential of $F(\boldsymbol{X}^*, \lambda)$ is an application which associates a 3×3 matrix to each point $\boldsymbol{X}(X, Y, Z)$; namely:

$$DF(\boldsymbol{X}^*, \lambda) = \begin{pmatrix} -\sigma & \sigma & 0 \\ r - Z & -1 & -X \\ Y & X & -b \end{pmatrix} \tag{10.11}$$

For $\boldsymbol{X} = 0$, the eigenvalues are found by solving the characteritic equations,

$$(s + b)[s^2 + (\sigma + 1)s + \sigma(1 - r)] = 0. \tag{10.12}$$

When $r = 1$, the null solution encounters a pitchfork bifurcation.

The equations are invariant under the symmetry $X \Rightarrow -X, Y \Rightarrow -Y$, and $Z \Rightarrow Z$. For $r > 1$, the 0 solution is unstable, and two symmetrical solutions, $\boldsymbol{X}^+(a, a, r-1)$ and $\boldsymbol{X}^-(-a, -a, r-1)$, with $a = [b(r-1)]^{1/2}$, are stable. These new fixed points correspond to convective rolls (rolling clockwise or counterclockwise). The generic situation for complex conjugated eigenvalues is a Hopf bifurcation, and in this case, L^* remains invertible. But at the bifurcation, a periodic trajectory centered around the fixed point appears, with a radius growing from zero.

10.2.1 Normal Form of the Pitchfork Bifurcation

A pitchfork bifurcation occurs when there is a real eigenvalue crossing the imaginary axis at $\lambda = \lambda_c$, plus a particular symmetry of the equations. Let $L = L_c$ be the critical operator, $L\phi = 0$; ϕ is the marginal mode, $L\phi_i = s_i \phi_i$, and ϕ_i are the damped modes. For any λ, we decompose $\boldsymbol{U}(t)$ on the base of L:

$$\boldsymbol{U}(t) = A(t)\phi + \sum B_i(t)\phi_i , \tag{10.13}$$

and from the equation,

$$\frac{d\boldsymbol{U}}{dt} = L\boldsymbol{U} + N(\boldsymbol{U}, \lambda) , \tag{10.14}$$

we derive

$$\frac{dA}{dt} = \mu A + g(A, B) , \tag{10.15a}$$

$$\frac{dB_i}{dt} = s_i B_i + g_i(A, B) , \tag{10.15b}$$

where μ is the value of the *destabilizing eigenvalue* and is equivalent to $(\lambda - \lambda_c)$ in the vicinity of λ_c. The characteristic evolution time of the damped modes is of order one, and the evolution time of the marginal mode is of order $(1/\mu)$.

For this reason, we suppose that B follows the evolution of A : $B = h(A)$ adiabatically, or more rigorously invokes the *central-manifold theorem*,

$$B = h(A) = aA^2 + bA^3 + O(A^4) . \tag{10.16a}$$

We then can eliminate B in the evolution equation for A:

$$\frac{dA}{dt} = \mu A + g(A, h(A)) =: \mu A + f(A) . \tag{10.16b}$$

Finally, because of the symmetry $A \Rightarrow -A$ that leads to a pitchfork bifurcation, the asymptotic expansion of $f(A)$ starts with a cubic term

$$f(A) = -\alpha A^3 + O(A^5) . \tag{10.17}$$

Consequently, we obtain the normal form of the pitchfork bifurcation at its lowest order:

$$\frac{dA}{dt} = \mu A - \alpha A^3 \;. \tag{10.18}$$

This is the well known *Landau equation*, and depending on the sign of α, the pitchfork bifurcation is supercritical ($\alpha > 0$) or subcritical ($\alpha < 0$). The Landau equation for the pitchfork bifurcation of the Lorenz model corresponds to

$$\mu = \frac{\sigma(r-1)}{(\sigma+1)} \quad \text{and} \quad \alpha = \frac{\sigma}{b(\sigma+1)} > 0 \;. \tag{10.19}$$

10.2.2 Normal Form of the Hopf Bifurcation

Two complex conjugated eigenvalues, $i\omega$ and $-i\omega$, cross the imaginary axis at $\lambda = \lambda_c$. We still use the notations [see Thual (1988/1993)]: $L = L_c$, the critical operator, $L\phi = i\omega\phi$, and $L\phi^* = -i\omega\phi^*$; ϕ and ϕ^* are the marginal modes, $L\phi_i = s_i\phi_i$, and ϕ_i are the damped modes. For any λ, we decompose $\boldsymbol{U}(t)$ on the base of L:

$$\boldsymbol{U}(t) = W(t)\phi + W^*(t)\phi^* + \sum B_i(t)\phi_i \;. \tag{10.20}$$

We recall that the solution of the linear problem $d\boldsymbol{U}/dt = L\boldsymbol{U}$ is a family of ellipses,

$$\boldsymbol{U}(t) = 2a\cos(wt - \gamma)\boldsymbol{V}_r - 2a\sin(wt - \gamma)\boldsymbol{V}_i \;, \tag{10.21}$$

with $\boldsymbol{V} = \boldsymbol{V}_r + i\boldsymbol{V}_i$, and in this case, the nonlinear problem $d\boldsymbol{U}/dt = L\boldsymbol{U} + N(\boldsymbol{U}, \lambda)$ leads to

$$\frac{dW}{dt} = (\mu + i\omega)W + g(W, W^*, B_i) \;, \tag{10.22a}$$

$$\frac{dW^*}{dt} = (\mu - i\omega)W^* + g^*(W, W^*, B_i) \;, \tag{10.22b}$$

$$\frac{dB_i}{dt} = s_i B_i + g_i(W, W^*, B_i) \;, \tag{10.22c}$$

where μ is the real value of the destabilizing eigenvalues and is proportional to $\lambda - \lambda_c$.

Now, the central-manifold expansion reads

$$\begin{aligned} B_i = h(W, W^*) &= aW^2 + bWW^* + cW^{*2} + dW^3 \\ &+ eW^2W^* + pWW^{*2} + qW^{*3} + O(|W|^4). \end{aligned} \tag{10.23}$$

We then can eliminate B_i in the evolution equation for W:

$$\begin{aligned} \frac{dW}{dt} &= (\mu + i\omega)W + g(W, W^*, h(W, W^*)) \\ &= (\mu + i\omega)W + f(W, W^*) \;. \end{aligned} \tag{10.24}$$

10.2 Bifurcations in Dissipative Dynamical Systems

The asymptotic expansion of function $f(W, W^*)$ starts with a quadratic term, but with an arbitrary nonlinear change of variable [see Thual (1988, p. 12)] we can eliminate all the coefficients of the asymptotic expansion of $f(W, W^*)$ except the resonant terms [see Thual (1988, p. 12)]. Finally, this leads to the normal form of the Hopf bifurcation, which reads at its lowest order,

$$\frac{dW}{dt} = (\mu + i\omega)W - \alpha|W|^2 W , \qquad (10.25)$$

with W and α complex.

In Fig. 10.2, the Hopf bifurcation from a fixed point to a limit cycle is represented with the behavior of the eigenvalues λ.

Depending on the sign of Real(α) the bifurcation is supercritical ($\alpha > 0$) or subcritical ($\alpha < 0$) at $\mu = 0$. It is not trivial to decide whether the Hopf bifurcation is supercritical or subcritical.

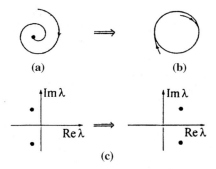

Fig. 10.2. Hopf bifurcation from a fixed point (**a**) to a limit cycle (**b**) and behavior of the eigenvalues λ (**c**)

The calculation of the normal form associated with this bifurcation will give the answer to this question. We note also that the calculation (derivation) of normal form can be performed either by asymptotic methods involving small parameters (Poincaré–Lindstedt, multiple scale, ...) or by a direct method called the Krylov, Bogolyubov and Mitropolsky (KBM) method [see, Thual (1988, p. 12–17)].

But, using a technique of the Lyapunov–Schmidt type, derived from bifurcation theory and a perturbation method coupled with a multiple-scale technique, we can also [as in Guiraud (1980)] derive the set of equations which rules the evolution of the amplitude of the most rapidly amplified modes of the linear theory. If we have only one amplified mode (and in this case, we have the so-called exchange of stabilities), then for the corresponding amplitude, we derive again the classical Landau equation (10.18). For a heuristic justification of the Landau equation, see Guiraud and Iooss (1968).

10.2.3 Bifurcation from a Periodic Orbit to an Invariant Torus

The next bifurcation, after one that results in a closed orbit, hence a periodic motion, can result in an invariant (stable) *2-D torus* T^2. But the appearance of an invariant 2-D torus at a bifurcation from a periodic stable orbit does not imply the appearance of orbits *dense* on that torus. It is in general unlikely that any single orbit covers the resulting 2-D torus densely, and hence yields a quasi-periodic function of time, owing to Peixoto's theorem [Richtmyer (1981, p. 306)]. Instead, the appearance of finitely many periodic orbits and fixed points is generic, and quasi-periodic motions are non-generic. The appearance of an invariant *3-D torus* at the next bifurcation depends on the existence of an orbit dense on the 2-D torus and hence, the bifurcation to an invariant 3-D torus seems unlikely! If a periodic orbit on the 2-D torus goes around the long way n times before closing, the bifurcation is *subharmonic* and has a sudden n-folding of the period at the bifurcation. For the Bénard experiment, after the appearance of two fundamental frequencies, the *power spectrum* becomes *continuous* (chaos). Even after two Hopf bifurcations, regular motion becomes *highly unstable* in favor of *chaotic motion at a strange attractor*. It is understandable that chaotic motion becomes possible only after two Hopf bifurcations, when the trajectory can explore additional dimensions, because *doubly periodic* motion corresponds to a trajectory on a 2-D torus (i.e., on a two-dimensional manifold), on which chaos is forbidden by the Poincaré–Bendixson theorem (which states that *there is no chaotic flow in a bounded region in two-dimensional space*). However, Newhouse et al. (1978) showed that a strange attractor is not only possible but generic (i.e., practically *unavoidable*, after two Hopf bifurcations). More precisely, Newhouse et al. showed that in a system with a phase-space flow consisting of *three incommensurate frequencies*, arbitrarily small changes in the system convert the flow from *quasi-periodic* three-frequency flow to *chaotic* flow. One might naively conclude [see, for instance, the book by Schuster (1984)], that three-frequency flow is improbable because it can be destroyed by small perturbations. But it has been shown numerically that the addition of smooth nonlinear perturbations does not typically destroy three-freqencies quasiperiodicity. The reason is that, in the proof by Newhouse et al. (1978), the small perturbations required to create chaotic attractors have small first and second derivatives but do not necessarily have small third- and higher order derivatives, as expected for physical applications!

10.3 Transition to Turbulence: Scenarios, Routes to Chaos

At first, bifurcation may be followed by further bifurcations, and we may ask what happens when a certain sequence of bifurcations has been encountered.

10.3 Transition to Turbulence: Scenarios, Routes to Chaos

In principle, there is an infinity of further possibilities, but, in some sense to be specified, not all of them are equally probable. The more likely will be called scenarios, and below we shall examine three prominent scenarios which have been theoretically and experimentally successful. In general, a scenario deals with the description of a few attractors. On the other hand, a given dynamical system may have many attractors. Therefore, several scenarios may evolve concurrently in different regions of phase space. A scenario does not describe its domain of applicability!

10.3.1 The Landau–Hopf "Inadequate" Scenario

After the first bifurcation, motion is generally periodic; after the second, it is generally quasi-periodic with two periods, and so on. It was shown that if the first bifurcation leads to a closed orbit, the second can lead to an attracting invariant 2-D torus in phase space.

If, furthermore, the motion is such that its orbit covers the 2-D torus densely, then a resulting function of time, such as one of coordinates in phase space, is quasi-periodic with two periods. Specifically, one can define two intrinsic angle-coordinates on the 2-D torus such that

$$\theta = \omega^* t + \text{const} , \quad \phi = \omega^{**} t + \text{const} , \tag{10.26}$$

and the orbit is dense on the 2-D torus if and only if ω^* and ω^{**} are incommensurable. After the next bifurcation, there may be motion on a 3-D torus, and so on. The idea behind the LH scenario was that, as soon as there are many independent frequencies, the motion is so irregular in appearance that it must be regarded for practical purposes as chaotic. There are various ways in which this scenario can be inappropriate: (i) One of the bifurcations may be subcritical; then, as soon as the corresponding critical value of the control parameter λ is exceeded, there is no nearby stable motion for the system to follow, and there is a so-called *explosive transition* to motion involving more or less remote parts of the phase space; (ii) Although an invariant 2-D torus generally appears at the second bifurcation, the orbit need not be dense on it, and it may return to its starting point after winding finitely many times around; then the orbit is closed and the motion is periodic. It is now believed, on the basis of Peixoto's theorem, that closed orbits on the 2-D torus are more likely than dense ones. This may lead to the Feigenbaum scenario: (iii) A possibility discussed by Ruelle and Takens is that, after a few bifurcations, an invariant point set in phase space appears, which is not a torus but a so-called strange attractor; then, as explained below, the motion is not quasi-periodic, but *aperiodic*.

10.3.2 The Ruelle–Takens–Newhouse Scenario

In the scenario of the early onset of turbulence proposed by Ruelle and Takens (1971), it is assumed that the first four bifurcations, as in the LH scenario, are supercritical and lead to invariant tori T^k, $k = 1, 2, 3, 4$, each is attracting between its appearance and the next bifurcation. For the existence of tori, see the discussion of the Feigenbaum scenario below. Ruelle and Takens prove that, on T^4, motion on a particular kind of strange attractor contained in T^4 is rather likely. The strange (*chaotic*) attractor is locally the Cartesian product of a two-dimensional *Cantor set* and a two-dimensional surface. The vector fields that yield the strange attractor cannot be dismissed as unlikely; however, their particular choice of strange attractor is somewhat arbitrary; one can imagine many variations of it; each has the property stated. Apparently (!), no one has found a specific vector field on a specific manifold that leads to a strange attractor precisely according to the RT scenario. The important idea in their paper is that motions on strange attractors are in some sense likely, or at least not unlikely, and are possibly even generic in certain circumstances. Their theorem does not say that the existence of a strange attractor is a generic property of the vector field on T^4. It says simply that,

> *once an invariant T^4 has been established, motion on a strange attractor is more likely, than quasi-periodic motion on T^4.*

A strange attractor is one on which the motions are unstable in the sense of Lyapunov and hence are characterized by a continuous power spectrum. In fact, the *strangeness of the chaotic attractor* is stable under small perturbations of the dynamical system; in other words, it is *not exceptional*.

> *According to Eckmann (1981), if a system, starting from a stationnary solution, undergoes three Hopf bifurcations, as a control parameter is varied, then it is likely that the system possesses a strange attractor with sensitivity to initial conditions after the third bifurcation.*

The power spectrum of such a system will exhibit one, then two, and possibly three independent basic frequencies. When the third frequency is about to appear, simultaneously, some *broad-band noise* will appear if there is a strange attractor. We interpret this as a chaotic, turbulent evolution of the system. The RTN [Newhouse, Ruelle and Takens (1978)] scenario is not destroyed by the addition of small external noise to the evolution equations. The nature of chaotic systems may be totally insensitive to small external noise. The systems most sensitive to noise seem to be deterministic systems near transition (bifurcation) points. In effect, the chaos of the scenario is so strong that order cannot be accidentally established by small noise terms, much like a very attracting fixed point is locally not much altered by noise, and globally there is, at most, a small probability of changing stochastically from one *basin of attraction* to another (if A is an attractor, its basin of attraction is defined as the set of points $\boldsymbol{X}$ such that, for the flow F^t, $F^t \boldsymbol{X}$ approaches A as $t \to \infty$).

Although the flow F^t (for dissipative systems) *contracts in volume*, it need not contract in length. In particular, even if all points in a finite volume V in state space $\boldsymbol{R}^n$ converge to a single attractor A, one still may find that

10.3 Transition to Turbulence: Scenarios, Routes to Chaos

points that are arbitrarily close initially may get *macroscopically* separated on the attractor after a sufficiently long time interval (*sensitive dependence on initial conditions – SDIC*).

Such an attractor (whose existence cannot be excluded even for *area-contracting flow*) is said to be strange. Solutions of dissipative dynamical systems separate exponentially (divergence) with time, have a positive Lyapunov characteristic exponent, and thus the movement is characterized as chaotic with the appearance of a strange attractor. The *Lyapunov characteristic exponent* σ of the flow F^t is defined as

$$\sigma = \lim \left\{ \frac{1}{t} \ln \left[\frac{D(t)}{D(0)} \right] \right\}, \quad \text{when } t \to \infty \text{ and } D(0) \to 0, \quad (10.27)$$

where the $D(t)$ are values of distances (a function of time) between initially neighboring solutions. This gives us a measure of the mean exponential rate of divergence of two initially neighboring solutions or of the chaoticity of the turbulence. The fact that σ is *positive* is important, indicating that the movement is chaotic. It would be expected that, as the control parameter (for instance, the Reynolds number) increases, the movement would become more chaotic, with a consequent increase in σ. The next property, after the Lyapunov exponent, which might be calculated to characterize the turbulence is the *dimension* of the strange attractor on which it resides. That dimension should be a measure of the number of *active degrees of freedom* or active modes of the turbulence. The problem of choosing which modes are active may prove to be formidable. The early discussion of hydrodynamic chaos was largely based on a geometric reconstruction of strange attractors, which is possible only in *low dimensions* (i.e., at the *onset of turbulence*).

For moderately excited systems, other tools are available from the *ergodic theory* of differential dynamical systems (the ergodic hypothesis implies that the trajectory uniformly covers the energetically allowed region of classical phase space such that *time averages can be replaced by the average over the corresponding phase space*). Various parameters are associated with *ergodic measure* ρ: (a) *Characteristic* exponents, $\lambda_1 > \lambda_2 > \lambda_3 > \ldots$ (also called *Lyapunov exponents*) from the multiplicative ergodic theorem and the λ_1 give the rate of exponential divergence of nearby orbits of the dynamical system; (b) *Entropy* $H(\rho)$ is the mean rate of creation of information by the system, or the *Kolmogorov–Sinaï invariant*; (c) *Information dimension*, $\dim_H \rho$, the smallest Hausdorff dimension of a set E such that $\rho(E) = 1$.

In Fig. 10.3 below, we represent the connection between the dimensions of simple attractors embedded in 3-D phase space with the signs of the corresponding three Lyapunov exponents (given in parentheses, where the zero value means that the corresponding Lyapunov exponent is zero).

On the other hand, when the system is ergodic, the entropy

$$H(\rho) = \sum \lambda_i, \quad i = 1, 2, 3, \ldots, N, \quad (10.28)$$

where the $\sum (i = 1, 2, 3, \ldots, N)$ is extended over all *positive* λ_i.

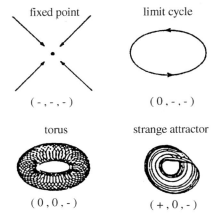

Fig. 10.3. Connection between dimensions of simple attractor with the sign of three Lyapunov exponents

A dimension is perhaps the most basic property of an attractor. The relevant definitions of the dimension are of two general types, those that depend only on metric properties and those that depend on the frequency with which a typical trajectory visits different regions of the attractor. We define here the dimension of chaotic attractor d by

$$d = N - \sum \frac{\lambda_i}{\lambda_{N+1}}, i = 1, 2, 3, \ldots, N , \qquad (10.29a)$$

where

$$\sum_{i=1,..,N+1} \lambda_i < 0 \quad \text{and} \quad \sum_{i=1,..,N} \lambda_i \equiv H(\rho) > 0 , \qquad (10.29b)$$

and

$$0 < \sum_{i=1,..,N} \frac{\lambda_i}{\lambda_{N+1}} < 1 . \qquad (10.29c)$$

For the Lorenz chaotic (strange) attractor, $d = 2.06$. The dimension of an attractor provides a way of quantifying the number of relevant degrees of freedom in dynamical motion. The dimension d of an attractor, if small and nonintegral, confirms that the dynamics admits a low-dimensional deterministic mathematical description characterized by a strange attractor.

Finally, for a finite-dimensional dynamical system

$$\frac{dX_i}{dt} = F_i, \quad i = 1, 2, 3, \ldots, N , \qquad (10.30a)$$

dissipativity means that

$$\text{div} F_i = \sum_{i=1,\ldots,N} \frac{\partial}{\partial X_i} \left(\frac{dX_i}{dt} \right) < 0 . \qquad (10.30b)$$

For the Lorenz system, we obtain

$$\mathrm{div} F_i = -(\sigma + 1 + b) < 0, \quad \text{because } \sigma > 0 \text{ and } b > 0 , \tag{10.31}$$

and, consequently, the volume element contracts exponentially in time t:

$$V(t) = V(0) \exp[-(\sigma + 1 + b)t] . \tag{10.32}$$

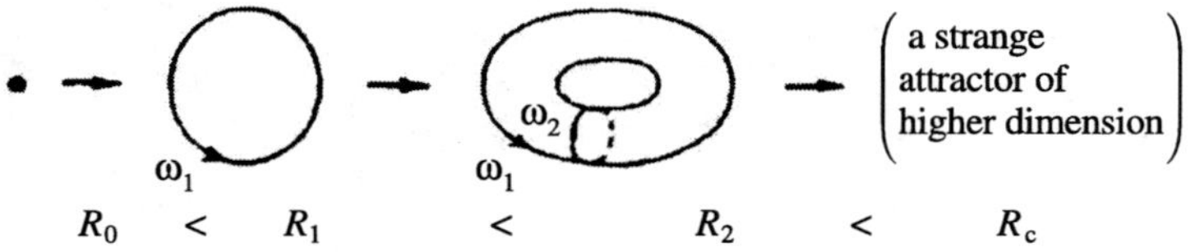

$$\bullet \rightarrow \underset{\underset{R_0}{\omega_1}}{\bigcirc} \; < \; \underset{\underset{R_1}{\omega_1}}{\bigodot}_{\omega_2} \; < \; \underset{R_2}{\text{(torus)}} \; < \; \begin{pmatrix} \text{a strange} \\ \text{attractor of} \\ \text{higher dimension} \end{pmatrix} \; R_c$$

Fig. 10.4. The RTN route to chaos (R is the control parameter and R_c the critical value of R)

The Ruelle–Takens–Newhouse (RTN) route to chaos is sketched in Fig. 10.4.

10.3.3 The Feigenbaum Scenario

The Lorenz attractor appears in connection with a subcritical Hopf bifurcation. The LH scenario and the RTN scenario both require a sequence of supercritical bifurcations that lead to invariant tori of successively higher dimension, arbitrarily high in the former scenario and of dimension at least four in the latter. However, such a sequence is unlikely according to Peixoto's theorem. Feigenbaum (1978, 1980/1983) developed a scenario based on a sequence of subharmonic bifurcations with period doubling (in this case the existence of an invariant torus breaks down, and the bifurcation may lead to one or more further periodic orbits). It turns out that such doubling occurs in many examples of iterated mapping and simple dynamical systems. Furthermore, as the number n of doubling increases, the behavior of the system is governed by certain asymptotic laws that involve universal constants and functions, independent of the system under study. In addition, the asymptotic laws appear to hold quite accurately for rather small values of n. In particular, the values μ_n of the dimensionless parameter μ at which the bifurcations (doubling) take place converge geometrically to a value μ_∞, with

$$\frac{(\mu_{n+1} - \mu_n)}{(\mu_n - \mu_{n-1})} \approx 0.21416938\ldots , \quad \text{for large } n . \tag{10.33}$$

As $n \to \infty$ (at least in the cases so far investigated), the power spectrum of the motion approaches a continuous spectrum with certain universal features. At $\mu = \mu_\infty$, the motion is presumably aperiodic on a (strange) chaotic attractor.

There is evidence for an example of this behavior in the Lorenz system at considerably higher values of the dimensionless parameter r than values studied by Lorenz. The Lorenz strange attractor that appears at $r = 24.74$ persists up to a value $r = r^*(\approx 250)$. For r considerably greater than r^*, there is a periodic orbit, and as r is decreased toward r^*, there is a sequence of doubling at values r_n of r that converge to r^* from above, with: $(r_{n+1}-r_n)/(r_n-r_{n-1}) \approx 0.21416938\ldots$. There is a universal (Feigenbaum) number

$$\frac{1}{(0.21416938\ldots)} = 4.669201545\ldots = \delta, \quad (10.34a)$$

such that

$$|\mu_j - \mu_\infty| \approx \text{const } \delta^{-j}, \quad \text{as } j \to \infty. \quad (10.34b)$$

After the cascade of period doubling, one expects an inverse cascade of noisy periods beyond the accumulation point μ_∞. In an experiment, if one observes subharmonic bifurcations at μ_1, μ_2, then, according to the Feigenbaum scenario, it is very probable that a further bifurcation will occur near $\mu_3 = \mu_2 - (\mu_1 - \mu_2)/\delta$. In addition, if one has seen three bifurcations, a fourth bifurcation becomes more probable than a third after only two, etc. At the accumulation point, one will observe aperiodic behavior, but not a broad-band spectrum. This Feigenbaum scenario is extremely well tested on numerical and physical grounds. Periodic doublings have been observed in most current, low-dimensional, dynamical systems and this phenomenon has also been investigated by *renormalization group analysis*. The functional renormalization group (RG), which is used in these cases, has been constructed analogously to the RG method for critical phenomena, see, for example, the papers by Argoul and Arneodo (pp. 241–288), and Coullet and Tresser (pp. 217–240), in Iooss and Peyret (1984). The power (energy) spectra and bifurcation diagrams for the Feigenbaum scenario are sketched in Fig. 10.5.

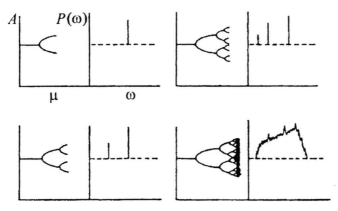

Fig. 10.5. Power spectra and bifurcation diagrams for the Feigenbaum scenario

10.3 Transition to Turbulence: Scenarios, Routes to Chaos 405

In Errafiy and Zeytounian (1991), the reader can find an example of the Feigenbaum route to chaos for the so-called Bénard problem of deep convection [see Sect. 2.3, Chap. 2, and especially (2.80) with the parameter δ defined by (2.81)]. In this case, we have a series of periodic regimes with period doublings (for the six values of the control parameter: $\mu = 200$, 217.5, 219.5, 220, 240, and 250). In Fig. 10.6 we have represented the phase diagrams for $\mu = 220$, 240, and 250.

Chaos appears at $\mu = 270$, and in Fig. 10.7, we have represented the corresponding Feigenbaum chaotic attractor for $\mu = 290$.

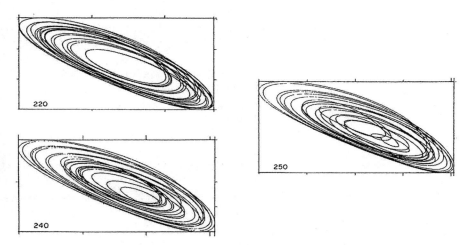

Fig. 10.6. Phase diagrams for $\mu = 220$, 240, and 250

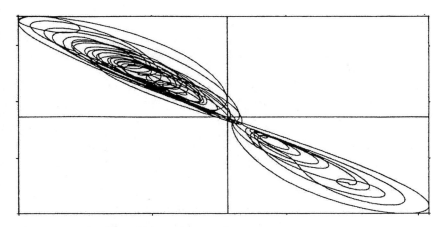

Fig. 10.7. The Feigenbaum attractor for $\mu = 290$

Strictly speaking, this Feigenbaum chaotic attractor is not a strange attractor [see Schuster (1984; Chap. 5) for a precise definition of this object)], but it is representative of chaos when the power spectrum is continuous. Hence, in this case, the route to chaos involves successive periodic doubling (subharmonic) bifurcations of a (simple) periodic flow (corresponding to $\mu = 200$).

10.3.4 The Pomeau–Manneville Scenario

This scenario is correctly named *"transition to turbulence through intermittency."* Its mathematical status is somewhat less satisfactory than that of the two other scenarios presented here because in the parameter region, the scenario to be described contains an infinity of (very long) stable periods and because there is no mention when the "turbulent regime" is reached or what the exact nature of this turbulence is! The two other scenarios have been associated with Hopf bifurcations (RTN case) and pitchfork bifurcations (Feigenbaum case), but this one is associated with *"saddle-node bifurcation,"* i.e., the *collision of a stable and unstable fixed point* which then both disappear (into complex fixed points).

By intermittency, we mean *the occurence of a signal which alternates randomly between long regular (laminar) phases (so called intermissions) and relatively short irregular bursts.*

It has been observed that the number of chaotic bursts increases with an external parameter, which means that intermittency offers a continuous route from regular to chaotic motion. The transition to chaos via intermittency has universal properties and represents one of the rare examples where the (linearized) RG equations can be solved exactly. The intermittency route to chaos has been investigated in a pioneering study by Manneville and Pomeau (1979). They solved numerically the differential equations of the Lorenz model (10.1), and they found that, for $r < r_c$, the $Y(t)$ component executes a stable periodic motion.

Above the threshold r_c, the oscillations are interrupted by chaotic bursts, which become more frequent as r is increased, until the motion becomes truly chaotic. In fact, the stable oscillations for $r < r_c$ correspond to a stable fixed point in the Poincaré map. Above r_c, this fixed point becomes unstable.

Because there are essentially three ways in which a fixed point can lose its stability, Pomeau and Manneville distinguished three types of intermittency: types I, II, and III; see Schuster (1984, Chap. 4). For example, the Lorenz model displays intermittency of type I, and in this case, the transition to chaos is characterized by an inverse tangent bifurcation. After the disappearance of the pair of fixed points, a stable and an unstable one, the iteration will spend a lot of time in a "channel region" (between the map and the bisector). This leads to long laminar regions for values of r just above r_c. After the trajectory has left the channel, the motion becomes chaotic until reinjection into the vicinity of the point ("ghost of the fixed point"), while attracts trajectories (on the left-hand side and repels them on the right-hand side) and starts a new regular phase. The mechanism for type I intermittency is sketched in Fig. 10.8.

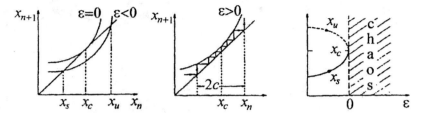

Fig. 10.8. Poincaré maps for $\varepsilon = r - r_c \leq 0$ and $\varepsilon > 0$; motion of the trajectory for $\varepsilon > 0$; inverse tangent bifurcation

The PM theory explains only laminar motion and gives no information about the mechanism that generates chaos! Note that inverse tangent bifurcations provide (in contrast to pitchfork bifurcations in which the number of fixed points is doubled) the only mechanism by which an uneven number of fixed points can be generated in a logistic map.

Intermittency type III appears simultaneously with period-doubling bifurcation. One observes the growth of a subharmonic amplitude together with a decrease in the fundamental amplitude. When the subharmonic amplitude reaches a high value, the signal loses its regularity, and turbulent bursts appear.

In type II intermittency, periodic (but unstable) trajectories still exist beyond the onset of instability.

Thus, contrary to type I intermittency, the maximum duration of seemingly periodic oscillations between two bursts is not bounded from above, and even in the fully turbulent regime, sequences of regular oscillations still occur from time to time. Finally, we note that, as the parameter is varied further to μ from the critical parameter value μ_c, one will see intermittent turbulent behavior of random duration, and laminar phases of mean duration

$$\sim |\mu - \mu_c|^{-1/2}$$

in between can be observed.

In Fig. 10.9, we have represented the power spectra and the attractors for the value of the control parameter $\varepsilon = 125$, for the Bénard deep convection problem, again according to Errafyi and Zeytounian (1991b).

For the value $\varepsilon = 120$, we have a pure periodic regime. In this case, instability occurs through the intermittent regime, and for $\varepsilon = 130$, the attractor is chaotic (see Fig. 10.10).

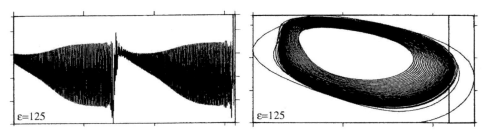

Fig. 10.9. Numerical evidence of intermittence for $\varepsilon = 125$

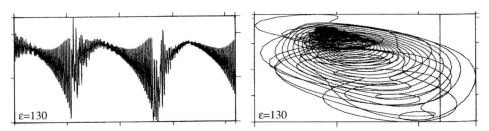

Fig. 10.10. Numerical evidence of intermittency for $\varepsilon = 130$. The bursts are relatively large, and the attractor is chaotic

This numerical result (by M. Errafiy) of the route to chaos via intermittency is very "fascinating" and confirms the Pomeau–Manneville scenario very well.

For $\varepsilon = 130$, temporal evolution occurs which alternates randomly between pseudo-regular (pseudo-laminar) relatively short phases (so-called intermissions) and "relatively large" irregular bursts.

The number of chaotic bursts increases with ε from 122 to 130, which means that intermittency offers a continuous route from regular to chaotic motion.

In Zeytounian (1991b), the reader can find the derivation of equations for the Bénard deep convection problem (see Sect. 17.3) and the analysis of the influence of the depth parameter δ defined by (2.81) in Sect. 23.2, where various strange attractors are also sketched (see pp. 603–611).

10.3.5 Complementary Remarks

We can say that the mechanism for the transition to chaos has been relatively well clarified from a logical point of view and an initial notion of turbulence (what we call here the onset of turbulence) has been achieved. However, many points must be investigated. For a rigorous proof of the existence of attractors for the Navier equations and the estimation of their (finite!) dimension, we refer the reader to Temam (1988a), where these arguments are treated in great detail. For example, we can prove that the Hausdorff dimension of the

10.3 Transition to Turbulence: Scenarios, Routes to Chaos

Navier attractor is finite (the compactness of the attractor suggests that it should have a finite Hausdorff dimension) and, usually, increases with the Reynolds number, going to infinity as $\text{Re} \to \infty$. This fact is interpreted as an increase in the complexity of the attractor (which has a very complicated geometric structure); an estimate of the "size" of this set and, in a sense, of its complexity, is a measure of its Hausdorff dimension (see below). However, it is not always true that the dimension of the attractor increases whenever $\text{Re} \to \infty$!

The nature of the Navier attractor and motion on it. First, we note that, in the space H of all divergence-free vector fields $\boldsymbol{u}$, with finite energy, that is, a Hilbert space equipped with norm

$$\|\boldsymbol{u}\| = \left(\int_{T^2} \boldsymbol{u}^2 \, d\boldsymbol{x} \right)^{1/2} \tag{10.35}$$

the initial-value problem associated with the Navier equations (for a 2-D viscous fluid moving in a flat torus $T^2 = [0, 2\pi] \times [0, 2\pi]$) possesses a unique solution belonging to $C([0, T]; H)$.

Then, we can consider the semigroup $S(t)$ defined by

$$S(t)\boldsymbol{u}^0(\boldsymbol{x}) = \boldsymbol{u}(\boldsymbol{x}, t), \quad \boldsymbol{u}^0(\boldsymbol{x}) \in H, \tag{10.36}$$

where $\boldsymbol{u}(\boldsymbol{x}, t)$ is the solution of the Navier problem associated with the initial value $\boldsymbol{u}^0(\boldsymbol{x})$. A global attractor $\mathcal{A} \subset H$ is an invariant set, such that

$$S(t)\mathcal{A} = \mathcal{A} \tag{10.37}$$

which has the property of attracting any orbit

$$\text{dist}(S(t)\boldsymbol{u}^0(\boldsymbol{x}); \mathcal{A}) \to 0 \quad \text{as } t \to \infty, \tag{10.38}$$

and it is also compact.

It is easy to show the existence of an absorbing ball in H, i.e., a ball which is invariant and attracts all of the orbits, as a consquence of the energy identity for the Navier equation

$$\frac{1}{2} \frac{d}{dt}(\|\boldsymbol{u}\|^2) = \nu^0(\boldsymbol{u}; \Delta\boldsymbol{u}) + (\boldsymbol{u}; \boldsymbol{f}),$$

where $\boldsymbol{f}$ is assumed independent of time.

But

$$-(\boldsymbol{u}; \Delta\boldsymbol{u}) > \|\boldsymbol{u}\|^2 \quad \text{and } (\boldsymbol{u}; \boldsymbol{f}) < \|\boldsymbol{u}\| \|\boldsymbol{f}\|,$$

and we arrive at

$$\frac{d}{dt}\left(\|S(t)\boldsymbol{u}^0\|\right) < \nu^0 \|S(t)\boldsymbol{u}^0\| + \|\boldsymbol{f}\|,$$

and finally
$$\|S(t)\boldsymbol{u}^0\| < \exp(-\nu^0 t)\|\boldsymbol{u}^0\| + (1/\nu^0)[1 + \exp(-\nu^0 t)]\|\boldsymbol{f}\| \ . \quad (10.39)$$
From estimate (10.39), we conclude that any ball in H of radius $R > \|\boldsymbol{f}\|/\nu^0$ is an invariant attracting set. The argument may be improved to get the existence of an attracting ball in Sobolev space V (which is the set of all divergence-free vector fields with the first square-summable derivative), and this is enough to prove the existence of a global attractor [see Temam (1988a)].

Unfortunately, it seems that very little is known about the nature of the Navier attractor and motion on it. The finite Hausdorff dimensionality of the attractor could induce us to believe that, for a fixed Re, everything goes as if the (nontransient) relevant motion were taking place on a manifold, a point of which is determined by few parameters, which are the relevant degrees of freedom of the system. However, this picture is too optimistic – no result concerning the smoothness of the attractor is yet known, and so the capture of the relevant degrees of freedom seems far from present knowledge.

There are attempts to study the Navier equation taking into account only a finite number of modes (as in this chapter); however, there is no reason to believe that a finite number of Fourier modes, evolved according to the truncated Navier equation, are enough to determine the long-time behavior of the solution of the Navier equation itself. In fact, when Re diverges, the dimension of the Navier attractor is also expected to diverge. In this case, the motion is chaotic and strongly unstable, as follows from experimental observations, so then a statistical approach seems more appropriate.

Inertial manifolds. In a series of papers, Constantin, Foias, Nikolaenko, Sell, Temam, Titi and others have shown that a fairly wide class of strongly dissipative PDEs not only possess attracting sets of finite dimension but also that these sets lie within smooth inertial manifolds: compact, finite-dimensional manifold $\mathcal{M}$, invariant under dynamical flow, and attracting all solutions at an exponential rate. However, observe that such manifolds are not unique. For example, one can always increase the dimension by adding additional "trivial" modes. Thus, if a suitable coordinate system can be found for $\mathcal{M}$, solutions are asymptotic to those of a finite (and maybe even computable) set of ODEs – the inertial form – on $\mathcal{M}$ and so, the long-term behavior is rigorously governed by ODEs on a flat, finite-dimensional model space $\boldsymbol{R}^n$. Essentially, inertial manifolds are global and attract center manifolds. Note that, in this global reduction to finite dimensions, one is not ignoring or truncating the high wave number modes; one is using the fact that they are "slaved" to low modes, as in local, center-manifold reduction. Higher modes are present in the inertial manifold, which is not flat. Its curvature describes the coupling ($\approx$ energy flow) to these higher modes.

Accounts of the theory can be found in Constantin et al. (1989), Foias et al. (1989), and Temam (1988b), and references therein. Inertial manifolds

10.3 Transition to Turbulence: Scenarios, Routes to Chaos

are known to exist for various equations, and even when they are not known to or probably do not exist in the strict case, such as the Navier equations, the notion of approximate inertial manifolds and the associated approximate inertial forms may be useful [Foias and Temam (1988)].

For the purpose of application and specific calculation, one must still bag the game by computing an approximate, reduced system and, in essentially all the work to date the inertial manifold $\mathcal{M}$ is sought as a graph over a subspace spanned by a finite set of eigenfunctions of a (trivially) linearized problem, and a procedure similar to the local one of Sect. 10.2 [(10.15a,b)] is followed. This choice of model space largely ignores nonlinear dynamics and often leads to poor estimates of a minimal dimension for $\mathcal{M}$. But, here a second idea may be useful, that of proper orthogonal decomposition and empirical eigenfunctions (Karhunen–Loève decomposition and collective coordinates). Its use was first suggested in a fluid dynamical context by Lumley (1970), who proposed the application of the decomposition theorem to turbulent flows, with a view to extracting "coherent structures." Sirovich (1997) provides a survey of these ideas with several applications to numerically generated data for fluid problems. The main idea is to derive, from a collection of data, a set of basic elements into which "typical" members of the collection can be decomposed "optimally." Let Φ be a basis for Hilbert space H – the finite dimensional subspace Φ_k, generated from an ensemble of solutions which are all (presumably) drawn toward the attracting set $\mathcal{A}$, is aligned in H to allow an "almost" (most of the time) one to one projection of typical solutions in $\mathcal{A}$. Now, if an inertial manifold exists, then it can be expressed as a graph over an appropriate finite-dimensional subspace of H. Clearly, the empirical subspace Φ_k, for some sufficiently large k, is a good choice because it represents a (finite) coordinate system adapted to nonlinear dynamics. The reduced system or inertial form can then be computed by projecting it onto Φ_k by (nonlinear) Galerkin's method and incorporating additional nonlinear "corrections," much as in the local analysis of Sect. 10.2 [(10.15a,b)], to account for the curvature of $\mathcal{A}$.

Therefore, the Karhunen–Loève or proper orthogonal decomposition gives a technique for generating "intelligent" sets of basic functions with which to represent field variables in strongly nonlinear situations.

Finally, what is turbulence? Although no definition of turbulence can be given at this time, there is widespread agreement concerning some of its attributes [Blackadar (1997)].

1. Turbulence is stochastic by nature – even though turbulent motions are subject to deterministic equations, we recognize that these equations are nonlinear, and as a result, the future characteristics of the motions are highly sensitive to small differences in the initial state. It is not possible to observe the initial state sufficiently accurately for us to treat (fully developed) turbulent motions deterministically!

2. Turbulence is 3-D, and any two marked particles that are free to move within a turbulent environment tend to become increasingly distant from each other as time goes on.
3. Turbulence is by nature rotational, and vorticity is an essential attribute.
4. Turbulence is dissipative – the energy of turbulence tends to shift from large, well-organized eddies (small wave numbers) toward smaller eddies and eventually into molecular motions. Vortices tend to be stretched by turbulence with a corresponding reduction in their diameter.
5. Turbulence is a phenomenon of large Reynolds numbers (large spatial dimensions but small viscosity). The space available for motions must be large compared to the dimensions of eddies that are quickly dissipated.

Roughly speaking, turbulence is inevitably an irreversible disorder, statistically "organized" according to several time and space scales, of mostly very sensitive, fast moving, dissipative rotational fluid flows at high Reynolds numbers. There are three main distinct but not mutually contradictory theoretical views of turbulence:

(a) The statistical view, in which the turbulence is considered as the observed behavior of the evolution of statistical distributions of flows instead of the evolution of one individual flow [see, for instance, the book by Marchioro and Pulvirenti (1994, §§ 7.4 and 7.5) for a short discussion of this statistical view].
(b) The viewpoint of regularity breakdown, in which it is considered that turbulence results from the blowup of vorticity in finite time, albeit necessarily on a set of small Hausdorff dimension (see, for instance, the discussion in Chap. 8). This might express the fact that the continuum mechanics model could become invalid, at least intermittently in time!
(c) The (finite-dimension) dynamical system view, in which turbulence is a phenomenological perception of the long-time complicated behavior of individual flows (as in this chapter).

Point of view (a) seems to be the view of most aeronautical engineers; on the contrary, mathematicians, theoretical physicists, and fluid dynamicians seems to hold the other views – especially (c)!

Note, however, that the phenomenon of turbulence cannot be equated with that of the *onset of turbulence*. We note also that the regularity breakdown viewpoint (b) is based on the assumption that weak solutions on $(0, \infty)$ exist for the 3-D Navier equations that are not regular.

It is premature to adopt one of the views (a), (b) or (c) exclusively, but it is reasonable to test them by trying to establish rigorous mathematical facts based on fluid dynamics equations (Navier equations!), whose interpretations are consistent with any of (a), (b), and (c). Foias and Temam (1987) present a pertinent contribution to this program from the dynamical systems point of view (c). But these authors also prove a regularity and backward uniqueness

10.3 Transition to Turbulence: Scenarios, Routes to Chaos

(new and the first of its kind) theorem on an open dense subset of a universal attractor of the 3-D Navier equations.

First, we now give a definition of the *Hausdorff dimension*: Let $\mathcal{M}$ be a metric space, and let $\mathcal{N} \subset \mathcal{M}$ be a subset of $\mathcal{M}$. Given two positive numbers, $d, \varepsilon > 0$, denote

$$\mu(\mathcal{N}, d, \varepsilon) = \inf \sum (r_i)^d, \quad i \in I, \tag{10.40}$$

where the infimum is taken over all of the covering of $\mathcal{N}$ by a family of balls $\{B_i\}$ of radii $r_i < \varepsilon$, $i \in I$. The number

$$\mu(\mathcal{N}, d) = \lim \mu(\mathcal{N}, d, \varepsilon), \quad \text{when } \varepsilon \downarrow 0, \tag{10.41}$$

is called the *d-Hausdorff measure of the set $\mathcal{N}$*.

But it is easy to realize that if $\mu(\mathcal{N}, d') < +\infty$ for some d', then the measure $\mu(\mathcal{N}, d) = 0$ for all $d > d'$.

Hence, a $d_H > 0$ exists such that

$$\mu(\mathcal{N}, d) = 0 \text{ for } d > d_H \quad \text{and} \quad \mu(\mathcal{N}, d) = +\infty \text{ for } d < d_H. \tag{10.42}$$

Such a d_H is called the *Hausdorff dimension of the set $\mathcal{N}$*.

Now, we give the definition of the *fractal dimension*:

Let Y be any compact subset of H, for instance, the closure of the space V of the divergence-free velocity vector that satisfies the respective boundary conditions in $L^2(\Omega)^n$, where Ω is an open connected bounded set $\boldsymbol{R}^n$, $n = 2, 3$.

On the other hand, the fractal dimension (capacity) $d_M(Y)$ of Y is defined by:

$$d_M(Y) = \lim \operatorname{Sup} \left[\frac{\log N_\varepsilon(Y)}{\log(1/\varepsilon)} \right], \quad \text{when } \varepsilon \downarrow 0, \tag{10.43}$$

where $N_\varepsilon(Y)$ is the smallest number of balls of radii $< \varepsilon$ covering Y.

This dimension, much emphasized by Mandelbrot (1982), is *larger* than the classical Hausdorff dimension d_H.

If X is the universal (chaotic-strange) attractor of Navier equations, then $d_M(X)$ can be viewed as a lower bound for the number N of asymptotic degrees of freedom of the phenomena described by the Navier equations. The number of real parameters necessary to describe, for $t \to \infty$, the asymptotic behavior of Navier equations must obviously exceed $d_M(X)$.

For 2-D, one has the following remarkable fact:

$$d_M(X) < C_2 \left[\frac{l^0}{l_d} \right]^2, \tag{10.44}$$

where $l_d = (\nu^3/\epsilon)^{1/4}$ is the Kolmogorov dissipation length, ϵ is the maximal mean dissipation of energy, and ν the kinematic viscosity; $l^0 = (\text{vol } \Omega)^{1/2}$.

Is the formula

$$d_M(X) < C_3 \left[\frac{l^0}{l_d}\right]^3 \qquad (10.45)$$

true in 3-D? For this lower bound on the dimension of the attractor for the Navier equations in space dimension three, see Ghidaglia and Temam (1990).

It is also interesting [again, according to Foias and Temam (1987)] to prove that for the universal attractor X of the Navier equations,

$$d_M(X) < C^* d_H(X)! \qquad (10.46)$$

10.4 Strange Attractors for Various Fluid Flows

Below, we present a collection of strange attractors for various viscous flow phenomena. In particular, the results of S. Godts, Christèle Bailly, R. Khiri, M. Errafiy, and S. Ginestet were obtained in the LML of the University of Lille I during the 1990s.

10.4.1 Viscous Isochoric Wave Motions

In Khiri (1992), the weakly nonlinear limit of 2-D unsteady-state isochoric wave motions in an incompressible, but viscous and stably stratified fluid is studied. In this case, we derive the following system of two dimensionless equations for the 2-D perturbed stream function $\psi(t, x, z)$ and the perturbation of the density $\rho(t, x, z)$:

$$\mathcal{L}\psi = \varepsilon\{\mathcal{M}(\psi, \rho) + a\frac{\partial^2}{\partial t^2}\left(\frac{\partial \psi}{\partial z}\right)$$
$$+ \frac{1}{R^*}\frac{\partial}{\partial t}\left[\mathbf{D}^2(\mathbf{D}\psi)\right]\} + O(\varepsilon^2) ; \qquad (10.47)$$

$$\frac{\partial \rho}{\partial t} - N^2 \frac{\partial \psi}{\partial x} = \varepsilon \mathcal{J}(\psi, \rho) + O(\varepsilon^2) , \qquad (10.48)$$

where $\varepsilon \ll 1$ is a small parameter characterizing the nonlinearity and

$$N^2 = a\left(\frac{1}{\mathrm{Fr}}\right)^2 > 0 , \qquad (10.49)$$

the dimensionless Brunt–Vâisälä frequency with Fr the Froude number. A hydrostatic background state with zero velocity is assumed, the background density is taken to have stably stratified form, and correspondingly, we also suppose that we have a background dynamic viscosity; in dimensionless form

$$\mu_b = \rho_b = \exp(-\varepsilon a z) \, . \tag{10.50}$$

If Re denotes the characteristic Reynolds number, then

$$R^* = \varepsilon \operatorname{Re} \, , \tag{10.51}$$

and

$$\mathcal{L} = \frac{\partial^2}{\partial t^2}(\mathbf{D}^2) - N^2 \frac{\partial^2}{\partial x^2} \, , \tag{10.52a}$$

$$\mathcal{M}(\psi, \rho) = \frac{\partial \mathcal{J}(\psi, \rho)}{\partial x} + \frac{\partial \mathcal{J}(\mathbf{D}^2 \psi, \psi)}{\partial t} \, , \tag{10.52b}$$

with $\mathcal{J}$ the Jacobian and $\mathbf{D} = (\partial/\partial x, \partial/\partial z)$.

For system (10.47), (10.48), we consider an initial-value problem, with only initial conditions at $t = 0$, for ψ and ρ. When $R^* = \infty$ in (10.47), we find again the equation used by Ibrahim (1987) for investigating the effects of weak nonlinearities on the behavior of vertically propagating waves in an inviscid, stably stratified atmosphere (but when the slow variation in $Z = \varepsilon z$ is not taken into account).

To derive amplitude equations for three-wave resonant interaction, the method of two scales is used to solve system (10.47), (10.48). The slow time $T = \varepsilon t$ is introduced to investigate the chaotic behavior of solutions of (10.47), (10.48), when time is very large.

With $\partial/\partial t = \varepsilon \partial/\partial T + \partial/\partial t$ and to obtain the relevant governing amplitude equations, it is assumed that an excited wave (of order ε) travels in a background atmosphere with waves of order ε or less. We let the leading-order solution for ψ be of the form,

$$\psi = \psi_0 + \varepsilon \psi_1 + O(\varepsilon^2) \, , \tag{10.53a}$$

$$\psi_0 = \sum A_m(T) \exp(i\theta_m) + C.C, \quad m = 1 \text{ to } 3 \, , \tag{10.53b}$$

where ($m = 1$ to 3)

$$\theta_m = k_m x + l_m z - \sigma_m t, \quad \text{with } \sum \theta_m = 0 \, , \tag{10.54a}$$

and

$$\sigma_m = +N\left(\frac{k_m}{\|\boldsymbol{K}_m\|}\right), \quad \boldsymbol{K}_m = k_m \boldsymbol{i} + l_m \boldsymbol{k} \, . \tag{10.54b}$$

From (10.53a,b) and (10.54a,b), as the leading solution for ρ, we write:

$$\rho = \rho_0 + \varepsilon \rho_1 + O(\varepsilon^2) \, , \tag{10.55a}$$

$$\rho_0 = -N^2 \sum \left(\frac{k_m}{\sigma_m}\right) A_m(T) \exp(i\theta_m) + C.C, \quad m = 1 \text{ to } 3 \, . \tag{10.55b}$$

Next, at order $O(\varepsilon)$, we derive an inhomogeneous equation for ψ_1 and the secular terms are removed from this equation if the reduced (and appropriately modified) amplitudes (assumed real), X, Y, Z, are the solution of the following dynamical system:

$$\frac{dX}{d\tau} = -YZ + \alpha X, \quad \frac{dY}{d\tau} = XZ - \beta X, \quad \frac{dZ}{d\tau} = XY - \gamma X. \quad (10.56)$$

Dynamical system (10.56), for $\gamma = 1$, was studied by Hughes and Proctor (1990) and is dissipative when

$$\alpha - \beta < 1, \quad \text{for } \gamma = 1. \quad (10.57)$$

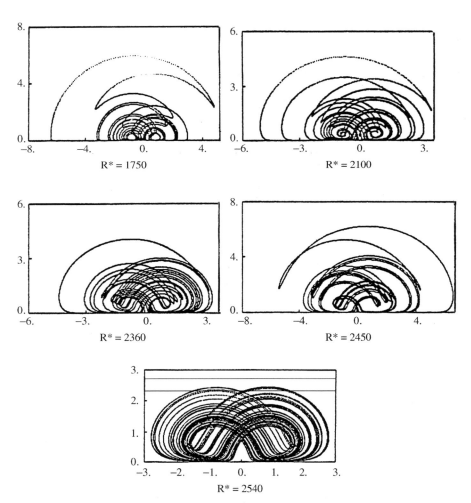

Fig. 10.11. Five strange attractors of the DS (10.56) corresponding to values $R^* = 1750, 2100, 2360, 2450,$ and 2540

For R^* (as a bifurcation parameter), we obtain a defined interval of variation (according to an analysis performed by S. Godts). For the numerical calculations in Khiri (1992), this interval is [1700, 2800] and for $R^* < 2540$, the behavior of the dynamical system (10.56) is chaotic, but with periodic windows. In above Fig. 10.11, we have shown five plots of X versus Y, for the values $R^* = 2540, 2450, 2360, 2100$ and 1750. In all of these, the solution has become chaotic. It can be seen that for the value $R^* = 2540$, the map is still remarkably one-dimensional and it is possible to investigate system (10.56) analytically [see, for instance, Sect. 4 of Hughes and Proctor (1990) and also Godts et al. (1993)].

10.4.2 The Bénard–Marangoni Problem for a Free-Falling Vertical Film: The Case of Re=$O(1)$ and the KS Equation

In what follows, we consider thermocapillary instabilities in a free-falling vertical (2-D) film. Most experiments and theories have focused on the latter, and wave dynamics on an inclined plane is quite analogous. Using dimensionless variables [see, below relations (10.61)], the governing equations and boundary conditions for (incompressible but thermally conducting) liquid motion down a vertical plane are as follows [see, for example, Zeytounian (1998, Sect. 3.4)]:

$$\begin{cases} \dfrac{\partial u}{\partial x} + \dfrac{\partial w}{\partial z} = 0 \, ; \\ \dfrac{du}{dt} + \dfrac{\partial p}{\partial x} - \dfrac{1}{\varepsilon F^2} = \dfrac{1}{\varepsilon \, \mathrm{Re}} \left[\dfrac{\partial^2 u}{\partial z^2} + \varepsilon^2 \dfrac{\partial^2 u}{\partial x^2} \right] \, ; \\ \varepsilon^2 \dfrac{dw}{dt} + \dfrac{\partial p}{\partial z} = \dfrac{\varepsilon}{\mathrm{Re}} \left[\dfrac{\partial^2 w}{\partial z^2} + \varepsilon^2 \dfrac{\partial^2 w}{\partial x^2} \right] \, ; \\ \mathrm{Pr} \dfrac{d\theta}{dt} = \dfrac{1}{\varepsilon \, \mathrm{Re}} \left[\dfrac{\partial^2 \theta}{\partial z^2} + \varepsilon^2 \dfrac{\partial^2 \theta}{\partial x^2} \right] \, . \end{cases} \quad (10.58)$$

In (10.58)

$$\frac{d}{dt} = \frac{\partial}{\partial t} + u \frac{\partial}{\partial x} + w \frac{\partial}{\partial z} \, .$$

Boundary conditions are:

$$\text{at } z = 0 : \quad u = w = 0 \quad \text{and } \theta = 1 \, , \quad (10.59)$$

and a free surface $z = h(t, x)$ we write:

$$\begin{cases}
\dfrac{\partial u}{\partial z} = -\varepsilon\,\mathrm{Ma}\left[\dfrac{\partial \theta}{\partial x} + \left(\dfrac{\partial h}{\partial x}\right)\dfrac{\partial \theta}{\partial z}\right] - \varepsilon^2 \left[\dfrac{\partial w}{\partial x} + 4\left(\dfrac{\partial h}{\partial x}\right)\dfrac{\partial w}{\partial z}\right] \\
\qquad - \dfrac{3}{2}\varepsilon^3\,\mathrm{Ma}\left(\dfrac{\partial h}{\partial x}\right)^2 \left[\dfrac{\partial \theta}{\partial x} + \left(\dfrac{\partial h}{\partial x}\right)\dfrac{\partial \theta}{\partial z}\right]; \\
p = p_a + 2\dfrac{\varepsilon}{\mathrm{Re}}\left[\dfrac{\partial w}{\partial z} - \left(\dfrac{\partial h}{\partial x}\right)\dfrac{\partial u}{\partial z}\right] - \varepsilon^2\,\mathrm{We}\,\dfrac{\partial^2 h}{\partial x^2} + \varepsilon^2\dfrac{\mathrm{Ma}}{\mathrm{Re}}\left(\dfrac{\partial^2 h}{\partial x^2}\right)\theta \\
\qquad + 2\dfrac{\varepsilon^3}{\mathrm{Re}}\left[\left(\dfrac{\partial h}{\partial x}\right)^3 \dfrac{\partial u}{\partial z} - 2\left(\dfrac{\partial h}{\partial x}\right)^2 \dfrac{\partial w}{\partial z} - \left(\dfrac{\partial h}{\partial x}\right)\dfrac{\partial w}{\partial x}\right]; \\
\dfrac{\partial \theta}{\partial z} = -(1+\mathrm{Bi}\,\theta) + \varepsilon^2\left[\left(\dfrac{\partial h}{\partial x}\right)\dfrac{\partial \theta}{\partial x} - \dfrac{1}{2}\left(\dfrac{\partial h}{\partial x}\right)^2 (1+\mathrm{Bi}\,\theta)\right]; \\
w = \dfrac{\partial h}{\partial t} + u\dfrac{\partial h}{\partial x}.
\end{cases} \qquad (10.60)$$

The free-surface boundary conditions (10.60) are written with an error of $O(\varepsilon^4)$. In (10.58) and (10.60), all functions and variables are dimensionless (the "*" marks dimensional quantities):

$$\begin{cases}
x = \dfrac{x^*}{\lambda^0}, \quad z = \dfrac{z^*}{h^0}, \quad t = \dfrac{t^*}{t^0}, \quad u = \dfrac{u^*}{U^0}, \\
w = \dfrac{w^*}{\varepsilon U^0}, \quad p = \dfrac{p^*}{\rho_0 U^{0\,2}}, \\
\theta = \dfrac{(T^* - T^0)}{\Delta T^0}, \quad t^0 = \dfrac{\lambda^0}{U^0}, \quad p_a = \dfrac{p_a^*}{\rho_0 U^{0\,2}},
\end{cases} \qquad (10.61)$$

and h^0, λ^0, U^0, ρ_0, T^0, $\Delta T^0 > 0$, are characteristic values for film thickness, wavelength, velocity, density, temperature, and temperature rate of increase at the lower horizontal boundary $z^* = 0$. In (10.58) and boundary conditions (10.60), we have the following dimensionless parameters:

$$\begin{cases}
\varepsilon = \dfrac{h^0}{\lambda^0} & \text{- long wave parameter,} \\
\mathrm{Re} = U\dfrac{h^0}{\nu_0} & \text{- Reynolds number,} \\
F = \dfrac{U^0}{\sqrt{gh^0}} & \text{- Froude number,} \\
\mathrm{Pr} = \dfrac{\nu_0}{\kappa_0} & \text{- Prandtl number,} \\
\mathrm{Ma} = \dfrac{\gamma \Delta T^0}{U^0 \rho_0 \nu_0} & \text{- Marangoni number,} \\
\mathrm{We} = \dfrac{\sigma_0}{h^0 \rho_0 U^{0\,2}} & \text{- Weber number,} \\
\mathrm{Bi} = \dfrac{q^0 h^0}{k_0} & \text{- Biot number.}
\end{cases} \qquad (10.62)$$

10.4 Strange Attractors for Various Fluid Flows

We note that x^* (along the vertical plate) and z^* (perpendicular to the vertical plate) are horizontal and vertical coordinates, respectively. u^* and w^* are the corresponding horizontal and vertical velocity components. For the surface tension $\sigma(T)$,

$$\sigma(T) = \sigma_0 - \gamma \Delta T^0 \theta, \quad \text{with } \gamma = \text{const.} \tag{10.63}$$

As characteristic velocity U^0, we can choose the interfacial velocity,

$$\frac{\text{Re}}{F^2} = 1 \Rightarrow U^0 = g\frac{h^{0\,2}}{\nu_0}. \tag{10.64}$$

The product

$$\text{Re We} = \frac{\sigma_0}{g\rho_0 h^{0\,2}} = \frac{1}{K}, \tag{10.65}$$

with K the so-called capillary number.

The KS equation. When $\text{Re} = O(1)$, ε is the main small parameter within (10.58) and conditions (10.60) for the full starting problem. We assume that the Weber number We is large, and we introduce a Weber similarity parameter,

$$W^* = \varepsilon^2 \text{We} = O(1). \tag{10.66}$$

For the Prandtl, Marangoni, and Biot numbers, we suppose that

$$\text{Pr} = O(1), \quad \text{Ma} = O(1) \quad \text{and} \quad \text{Bi} = O(1), \tag{10.67}$$

and we look for the solution of (10.58) with (10.59) and (10.60), with the above relations (10.63)–(10.67), for Re, F^2, We, Pr, Ma and Bi, in an expansion in the form proposed by Benney (1966)

$$U = [u, w, \pi, \theta]^T = U_0 + \varepsilon U_1 + O(\varepsilon^2), \quad \text{when } \varepsilon \to 0, \tag{10.68}$$

but we do not expand the film thickness $h(t, x)$ for the moment.

The solution U_0 is

$$u_0 = -z\left[\frac{z}{2} - h\right]; \quad w_0 = -\frac{1}{2}\frac{\partial h}{\partial x}z^2;$$
$$p_0 = p_a - W^* \frac{\partial^2 h}{\partial x^2};$$
$$\theta_0 = 1 - \frac{1+\text{Bi}}{1+\text{Bi}\,h}z, \tag{10.69}$$

and [see, below, (10.71) and (10.73)]

$$\frac{\partial h}{\partial t} + h^2 \frac{\partial h}{\partial x} = O(\varepsilon). \tag{10.70}$$

Now, writing out a set of equations and boundary conditions for U_1, and again, assuming that $h(t,x)$ is not expanded, it is easy to get an analytic expression for u_1 as a function of z. Using u_0 and u_1, we may compute q_0 and q_1 within the expansion of

$$q(t,x) = \int_0^{h(t,x)} u(t,x,z)\,dz = q_0 + \varepsilon q_1 + O(\varepsilon^2). \tag{10.71}$$

$q_0 = (1/3)h^3$ has already been taken into account in (10.70), and we get the following expression for q_1:

$$q_1(t,x) = \frac{1}{3}\operatorname{Re} W^* h^3 \frac{\partial^3 h}{\partial x^3}$$
$$+ \frac{1}{2}\operatorname{Ma} \frac{1+\operatorname{Bi}}{(1+\operatorname{Bi} h)^2} h^2 \frac{\partial h}{\partial x}$$
$$+ \frac{2}{15}\operatorname{Re} h^6 \frac{\partial h}{\partial x}. \tag{10.72}$$

From

$$\frac{\partial h}{\partial t} + \frac{\partial q}{\partial x} = 0, \tag{10.73}$$

as a consequence of the continuity equation in (10.58), the kinematic condition in (10.60), and condition $w = 0$ at $z = 0$ in (10.59), we may get, with (10.71), (10.72) and the value of $q_0 = (1/3)h^3$, an equation for the film thickness $h(t,x)$, involving the $O(\varepsilon)$ term in (10.71):

$$\frac{\partial h}{\partial t} + h^2 \frac{\partial h}{\partial x} + \varepsilon \frac{\partial}{\partial x}\left\{\frac{1}{3}\operatorname{Re} W^* h^3 \frac{\partial^3 h}{\partial x^3} + \frac{2}{15}\operatorname{Re} h^6 \frac{\partial h}{\partial x}\right.$$
$$\left. + \frac{1}{2}\operatorname{Ma}\frac{1+\operatorname{Bi}}{(1+\operatorname{Bi} h)^2} h^2 \frac{\partial h}{\partial x}\right\} = 0. \tag{10.74}$$

Evolution equation (10.74) of the "Benney type" contains the small paraemter ε because $h(t,x)$ has not been expanded by an expansion with respect to ε, as it should be in a fully consistent asymptotic approach. Of course, we may expand $h(t,x)$ in different ways, and we shall investigate, here, only the same kind of phenomenon that led to the so-called Kuramoto–Sivashinsky (KS) equation. To achieve this, we put in (10.74)

$$\tau = \varepsilon t, \quad \xi = x - t, \quad h(t,x) = 1 + \varepsilon \eta(\tau, \xi) + \ldots \tag{10.75}$$

and we assume that $\varepsilon = \delta$, where δ is the amplitude parameter for the free surface, $z^* = h^0[1 + \delta\eta]$. Because $\partial h/\partial t = -\varepsilon \partial \eta/\partial \xi + \varepsilon^2 \partial \eta/\partial \tau$ and

$\partial h/\partial x = \varepsilon \partial \eta/\partial \xi$, if we let $\varepsilon \to 0$ within the transformed version of (10.74), we derive the following KS equation:

$$\frac{\partial \eta}{\partial \tau} + 2\eta \frac{\partial \eta}{\partial \xi} + \alpha \frac{\partial^2 \eta}{\partial \xi^2} + \gamma \frac{\partial^4 \eta}{\partial \xi^4} = 0, \qquad (10.76)$$

where

$$\alpha = \frac{2}{15} \text{Re} + \frac{1}{2} \frac{\text{Ma}}{1 + \text{Bi}}; \quad \gamma = \frac{1}{3} \text{Re} W^* . \qquad (10.77)$$

The KS equation (10.76) is asymptotically consistent when $\varepsilon = \delta \to 0$ and is an approximate equation valid with an error of the order of $O(\varepsilon)$.

Hierarchy of bifurcations and strange attractors of the KS equation.
A more convenient reduced form of the KS equation (10.76) is derived if we introduce a new function $H(x,t)$ and new variables t and x by the relations

$$\eta = 2\alpha \left(\frac{\alpha}{\gamma}\right)^{1/2} H, \quad \tau = \left(\frac{\gamma}{\alpha^2}\right) t \quad \text{and} \quad \xi = \left(\frac{\gamma}{\alpha}\right) x .$$

Then, we obtain the following reduced KS equation for the amplitude $H(x,t)$:

$$\frac{\partial H}{\partial t} + 4H \frac{\partial H}{\partial x} + \frac{\partial^2 H}{\partial x^2} + \frac{\partial^4 H}{\partial x^4} = 0 . \qquad (10.78)$$

Now, we consider the nonstationary solution behavior and attractors of the KS equation in the form (10.78), and we assume that $(x,t) \in R^1 \times R^+$. We impose the initial condition, $H(x,0) = H^0(x)$, $x \in R^1$, and also a periodic boundary condition relative to x, $H(x,t) = H(x+2\pi/k,t)$, where k is a wave number. Below, we present mainly the results of the numerical investigations of Demekhin et al. (1991). A comprehensive review dedicated to researchs on the KS equation (up to 1986) is presented by Hyman et al. (1986). First, it is clear that for any (x,t) solutions $H(x,t)$ of (10.78) are invariant referring to transformations,

$$H(x,t) \to H(x+x^0, t+t^0); \quad H(x,t) \to -H(-x,t) ; \qquad (10.79a)$$

$$H(x,t) \to H(x-ct, t) + \frac{c}{4} . \qquad (10.79b)$$

Because $H = \text{const}$ is a trivial solution of the above problem for the KS equation (10.78), as a consequence of (10.79a,b), we can assume that $H = 0$ is a trivial solution of this problem. The linear characteristic equation is $c = ik(1-k^2) = c_i$, $c_r = 0$, where $c = c_r + ic_i$ is the phase velocity. When $k > 1$, the solution $H = 0$ is stable, and when $k < 1$, it is not. For $k = 1$, we have neutral stablity.

When $k = k_{\max} = \sqrt{2}/2$, the increment of growth (kc_i) has its maximum $(kc_i)_{\max} = 1/4$. For a periodic wave solution we write:

$$H(x,t) = \sum_{(n=1 \text{ to } \infty)} [A_n(t)\cos(knx) + B_n(t)\sin(knx)] \ . \tag{10.80}$$

By a simple linear transformation, $kx \to x$ and $kt \to t$, and by substituting (10.80) in KS equation (10.78), we will get an infinite system of ordinary differential equations for the amplitudes $A_n(t)$ and $B_n(t)$:

$$\frac{\mathrm{d}A_n(t)}{\mathrm{d}t} = k\left\{\gamma_n A_n(t) - \sum A_m(t)[B_{m+n}(t) - B_{m-n}(t)]\right.$$
$$\left. - B_m(t)[A_{m+n}(t) - A_{m-n}(t)]\right\} \ ; \tag{10.81a}$$

$$\frac{\mathrm{d}B_n(t)}{\mathrm{d}t} = k\left\{\gamma_n B_n(t) + \sum A_m(t)[A_{m+n}(t) + A_{m-n}(t)]\right.$$
$$\left. - B_m(t)[B_{m+n}(t) + B_{m-n}(t)]\right\} \ , \tag{10.81b}$$

where $m = 1$ to ∞, and

$$\gamma_n = nk[1 - (nk)^2], \quad n = 1, 2, 3, ...,$$
$$A_{-n} = A_n, \quad B_{-n} = -B_n, \quad A_0 = B_0 = 0.$$

But from (10.79a), the existence of a class of antisymmetric solutions of KS equation (10.78) follows. For such solutions, $A_n = 0$, for all n, and $B_n(t)$ are described by an infinite chain of ODEs:

$$\frac{\mathrm{d}B_n(t)}{\mathrm{d}t} = k\left\{\gamma_n B_n(t) + \sum B_m(t)[B_{m+n}(t) + B_{m-n}(t)]\right\} \ . \tag{10.82}$$
$n = 1, 2, 3, ..$ and $m = 1$ to ∞ .

In the process of numerical integration of the dynamical system, the series (10.80) seemed to be finite, containing N members. In the range of wave number: $0.15 < k < 1$, the number of harmonics usually satisfied the ratio $N = 2$ integer $(1/k)$ (that is, half of the modes was in the linear unstable zone, $nk < 1$, half of them in the linear stable one $nk > 1$, $n = 1, 2, .., N$).

In Demekhin et al. (1991), many of the calculations were corrected by the given $N = 4$ integer $(1/k)$. So, for the numbers $0.15 < k < 1$, the value varied from 3 to 20, and the dimension of the dynamical system (10.81a,b) varied correspondingly from 6 to 40. Bifurcations of this dynamical system appear with decreasing k from 1 to 0.15–0.2.

At $k > 1$, the origin 0 is the single global attractor in the phase space of the dynamical system (10.81a,b).

In Fig. 10.12, an evolution of the limit cycles with a decrease in k is shown. The cycle has the first period-doubling bifurcation at $k \approx 0.3482$, and it has the second at $k \approx 0.3481$.

It seems that this cascade of bifurcations is not infinite and is cut short during the third doubling. This is due to the fact that the cycle quickly "swells" with decreasing k, and at $k \approx 0.34799$, the phase trajectory comes

10.4 Strange Attractors for Various Fluid Flows   423

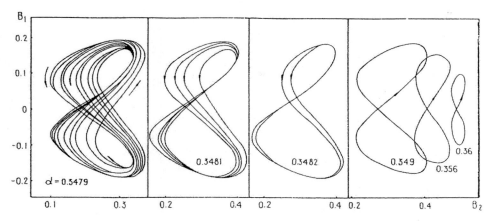

Fig. 10.12. The evolution of limit cycles with a decrease in k

onto the separatrix surface dividing the two centrosymmetric points from stationary solution branches of the KS equation. While doing this, the trajectory travels along the separatrix loop of the origin, and the cycle destructs. This develops into a situation which is in many respects similar to a Lorenz system!

The phase trajectory behavior in the B_2–B_4 plane projection at $k \approx 0.3$ is shown in Fig. 10.13. Let us point out that the value $k = 0.3$ is close to the critical one, $k^* = 0.2967$.

At $k \approx 0.2988$ the second of the tori families appeared, and the birth of this second family is connected with the family of stationary running waves.

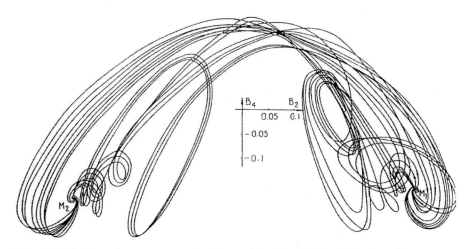

Fig. 10.13. Complex movement is observed in the vicinity of limit cycles

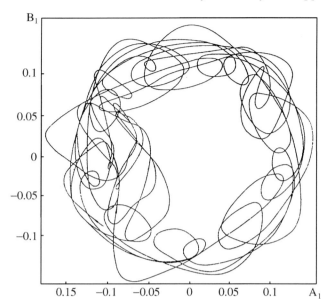

Fig. 10.14. The metastable chaos regime at $k = 0.273$

In the range of $0.283 < k < 0.295$, the tori are influenced by the cascade of period-doubling bifurcations of the Feigenbaum type.

The family of stationary running waves becomes stable at $k \approx 0.2831$ and loses its stablity due to Hopf bifurcation at $k \approx 0.2751$.

A nonstationary regime of the metastable chaos type (see above Fig. 10.14) can be observed at $k = 0.273$.

10.4.3 The Bénard–Marangoni Problem for a Free-Falling Vertical Film: The Case of $\mathrm{Re}/\varepsilon = O(1)$ and the KS–KdV Equation

The problem considered below is interesting when we look at (10.74). Equation (10.74) is a singular perturbation of the hyperbolic equation $\partial h/\partial t + h^2(\partial h/\partial x) = 0$. Curiously, we get again a singular perturbation of this same equation, but of an another type, if we assume (low Reynolds and Marangoni numbers)

$$\mathrm{Re} \ll 1, \ \mathrm{Ma} \ll 1, \quad \text{such that} \quad \frac{\mathrm{Re}}{\varepsilon} = R^0, \quad \text{and} \quad \frac{\mathrm{Ma}}{\varepsilon} = M^0 \,, \tag{10.83}$$

where R^0 and M^0 are both $O(1)$, and again we assume that

$$\varepsilon^2 \, \mathrm{We} = W^* = O(1) \quad \text{and} \quad \frac{\mathrm{Re}}{F^2} = 1 \,,$$

in the full starting problem (10.58), (10.59), (10.60). Then the formal Benney (1966) expansion in ε is modified by an expansion in ε^2; namely:

$$U = (u, w, p, \theta)^T = U_0 + \varepsilon^2 U_2 + \ldots, \quad \text{when } \varepsilon \to 0; \tag{10.84}$$

and, when $\varepsilon \to 0$, we obtain the following leading-order solution:

$$u_0 = -\frac{1}{2}z^2 + hz, \quad w_0 = -\frac{1}{2}\frac{\partial h}{\partial x}z^2,$$

$$p_0 = p_a - W^*\frac{\partial^2 h}{\partial x^2} - \frac{1}{R^0}\frac{\partial h}{\partial x}(h+z),$$

$$\theta_0 = 1 - \frac{1+\text{Bi}}{1+\text{Bi}\,h}z, \tag{10.85}$$

and, this time,

$$\frac{\partial h}{\partial t} + h^2\frac{\partial h}{\partial x} = O(\varepsilon^2), \tag{10.86}$$

because, again, $q_0 = (1/3)h^3$.

Writing out the set of equations and boundary conditions at the order ε^2 for U_2, and assuming, that $h(t, x)$ is not expanded, we may get an awkward expression for u_2 that may be integrated with respect to z to obtain an explicit expression for q_2 in

$$\frac{\partial h}{\partial t} + h^2\frac{\partial h}{\partial x} + \varepsilon^2\frac{\partial q_2}{\partial x} = O(\varepsilon^4). \tag{10.87}$$

The final result is analogous to (10.74), but with some additional terms:

$$\frac{\partial h}{\partial t} + h^2\frac{\partial h}{\partial x} + \varepsilon^2\frac{\partial}{\partial x}\left\{\frac{1}{3}h^3\left[R^0 W^*\frac{\partial^3 h}{\partial x^3} + 7\left(\frac{\partial h}{\partial x}\right)^2\right]\right.$$

$$+ h^4\frac{\partial^2 h}{\partial x^2} + \frac{2}{15}h^6\frac{\partial h}{\partial x}$$

$$\left. + \frac{1}{2}M^0\frac{1+\text{Bi}}{(1+\text{Bi}\,h)^2}h^2\frac{\partial h}{\partial x}\right\} = 0. \tag{10.88}$$

The evolution equation (10.88) for $h(t, x)$ is valid with an error of $O(\varepsilon^4)$.

The KS–KdV equation. Now, with this last evolution equation (10.88), we intend to play the same game as that considered for the reduction of (10.74) in KS equation (10.76). We use

$$\tau = \delta t, \quad \xi = x - t, \quad h = 1 + \left(\frac{1}{\phi}\right)\varepsilon^2\eta(\tau, \xi) + \ldots, \quad \delta = \left(\frac{1}{\phi}\right)\varepsilon^2, \tag{10.89}$$

where ϕ is the dispersive similarity parameter.

Carrying out, again, the limiting process $\varepsilon \to 0$, we find, in place of (10.88), an equation that cumulates the features of KdV, on the one hand, and KS, on the other hand:

$$\frac{\partial \eta}{\partial \tau} + 2\eta \frac{\partial \eta}{\partial \xi} + \alpha \frac{\partial^2 \eta}{\partial \xi^2} + \phi \frac{\partial^3 \eta}{\partial \xi^3} + \gamma \frac{\partial^4 \eta}{\partial \xi^4} = 0 , \qquad (10.90)$$

where

$$\alpha = \phi \left[\frac{2}{15} R^0 + \frac{1}{2} \frac{M^0}{(1 + \text{Bi})} \right], \quad \gamma = \frac{\phi}{3} R^0 W^* . \qquad (10.91)$$

The evolution KS–KdV equation (10.90) is again a significant model equation valid for large time with an error of $O(\delta)$. The coefficients α, γ, and ϕ are all positive constants characterizing instability, dissipation, and dispersion. As a consequence of deriving the KS–KdV equation (10.90) valid for low Reynolds numbers, we conclude that the features of a thin film of a strongly viscous liquid are quite different. The dispersive term, $\phi(\partial^3 \eta/\partial \xi^3)$, changes the behavior of the thickness $\eta(\tau, \xi)$ in space and time. According to Chang et al. (1993), the linear dispersion term, $\phi(\partial^3 \eta/\partial \xi^3)$, tends to arrest the irregular behavior (in the KS equation, which exhibits spatio-temporal chaos) in favor of a spatially periodic cellular structure, as is consistent with prior numerical and experimental observations.

For stationary waves in phase space, carrying out a moving coordinate transformation and stipulating that the waves are stationary in this moving frame, $x = \xi - c^0 \tau$, we obtain their governing equation from (10.90), integrating once:

$$\gamma \frac{d^3 H}{dx^3} + \phi \frac{d^2 H}{dx^2} + \alpha \frac{dH}{dx} - c^0 H + 2H^2 = Q , \qquad (10.92)$$

and

$$\int_0^{2\pi/k} H(x) \, dx = 0 , \qquad (10.93)$$

where

$$H(x) = \frac{1}{2} \eta(\xi - c^0 \tau) .$$

Relation (10.93) is a condition of zero mean, $2\pi/k$ is the wavelength of the stationary wave, and Q is an integration constant.

Balmforth (1995) presents a discussion of the theory of solitary-wave solution equilibria and dynamics, for the equation

$$\left[\frac{d^3}{d\xi^3} + \mu \frac{d^2}{d\xi^2} + \frac{d}{d\xi} \right] \Theta - c\Theta + \frac{1}{2} \Theta^2 = 0 , \qquad (10.94)$$

10.4 Strange Attractors for Various Fluid Flows   427

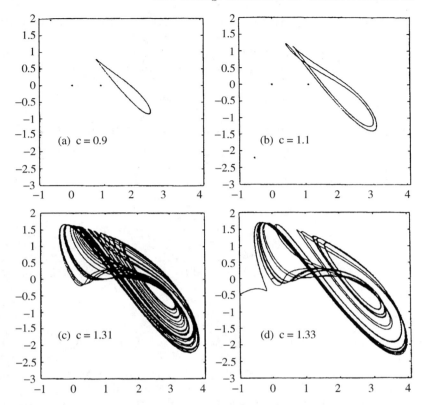

Fig. 10.15. Sequence of states of (10.94) for $\mu = 0.7$ and four values of c

analogous to (10.92), with $Q = 0$, within a framework of asymptotic analysis and dynamical system theory. Figure 10.15 according to Balmforth (1995), presents a succession of states that are related as c is varied ($= 0.9, 1.1, 1.31$, and 1.33) for $\mu = 0.7$ in (10.94).

10.4.4 Viscous and Thermal Effects in a Simple Stratified Fluid Model

Here, for the viscous layer $z \in [0, H^0]$, we consider the following model equations:

$$\frac{\partial w}{\partial t} + w\frac{\partial w}{\partial z} + RT_m\frac{\partial \pi}{\partial z} - g\theta = \nu^0\frac{\partial^2 w}{\partial z^2} \ ; \tag{10.95a}$$

$$\frac{\partial \theta}{\partial t} + w\frac{\partial \theta}{\partial z} + \mu_m w = \kappa^0\frac{\partial^2 \theta}{\partial z^2} \ ; \tag{10.95b}$$

$$\frac{\partial \pi}{\partial t} + \frac{\partial w}{\partial z} = \gamma^0 w \ , \tag{10.95c}$$

for the vertical velocity $w(t, z)$, perturbation of pressure $\pi(t, z)$ and temperature $\theta(t, z)$. The viscous and thermally conducting fluid considered is stratified, and $\mu_m = $ const characterizes this stratification in altitude and $g = |\boldsymbol{g}|$ is negative in the z direction, where $\boldsymbol{g}$ is the acceleration vector due to gravity. The constant parameter γ^0 takes into account the small effect of compressibility, T_m is a constant mean temperature, and R, ν^0, κ^0 are, respectively, the gas constant, the kinematic viscosity constant, and the thermal diffusivity constant. On the flat surface $z = 0$, we assume that

$$w = 0 \quad \text{and} \quad \theta = \theta^0 = \text{const}, \tag{10.96a}$$

and on $z = H^0$,

$$w = 0, \quad \theta = 0 \quad \text{and} \quad \frac{\partial \pi}{\partial z} = 0, \tag{10.96b}$$

because the terms $\partial^2 w/\partial z^2$ and $\partial^2 \theta/\partial z^2$ are very small in the vicinity of the upper plane $z = H^0$. Finally, as initial conditions at $t = 0$, the following conditions may be assumed:

$$w = \theta = \pi = 0. \tag{10.96c}$$

Below, we are concerned with the analysis of the chaotic behavior of model problem (10.95a–c), (10.96a–c) for *large values of time*. This model problem is a very good approximate model for studying the vertical structure of atmospheric motions when time increases [see, for instance, Martchuk (1969)], and gives the possibility of extending the classical Lorenz predictability atmospheric model, taking into account the *thermal effect*. First, in place of functions w, π, and θ, we can introduce new dimensionless functions,

$$\Psi = \frac{w}{(RT_m)^{1/2}}, \quad T = \left(\frac{gH^0}{RT_m}\right)\Theta, \quad \Omega = \left(\frac{H^0}{RT_m}\right)\Phi, \tag{10.97}$$

where Φ is such that

$$\frac{\partial \Phi}{\partial t} = -RT_m \left[\gamma^0 \frac{\partial w}{\partial z} - \frac{\partial^2 w}{\partial z^2}\right],$$

and

$$\Theta = \theta^0 \left(1 - \frac{z}{H^0}\right) - \theta.$$

In place of t and z, we have the following dimensionless variables:

$$\tau = \frac{(RT_m)^{1/2}}{H^0} t, \quad \zeta = \frac{z}{H^0}. \tag{10.98}$$

For new functions $\Psi(\tau, \zeta)$, $\Omega(\tau, \zeta)$ and $T(\tau, \zeta)$ we derive, in place of (10.95a–c), the following three dimensionless model equations:

10.4 Strange Attractors for Various Fluid Flows

$$\frac{\partial \Psi}{\partial \tau} = -\Psi \frac{\partial \Psi}{\partial \zeta} + T + \Omega + A^0 \frac{\partial^2 \Psi}{\partial \zeta^2} , \qquad (10.99\text{a})$$

$$\frac{\partial \Omega}{\partial \tau} = -H^0 \gamma^0 \frac{\partial \Psi}{\partial \zeta} + \frac{\partial^2 \Psi}{\partial \zeta^2} , \qquad (10.99\text{b})$$

$$\frac{\partial T}{\partial \tau} = -\Psi \frac{\partial T}{\partial \zeta} + B^0 \Psi + \sigma^0 A^0 \frac{\partial^2 T}{\partial \zeta^2} , \qquad (10.99\text{c})$$

where

$$A^0 = \frac{\nu^0}{H^0 (RT_m)^{1/2}} , \quad B^0 = \frac{gH^0}{RT_m} \left[\theta^0 - H^0 \mu_m \right] , \quad \sigma^0 = \frac{\kappa^0}{\nu^0} . \qquad (10.100)$$

Below, we assume that $\gamma^0 \equiv 0$ and, consequently, we do not take into account the small compressibility effect, although the competition between compressible and viscous/thermal effects is certainly of great interest. In this case, we can associate a homogeneous condition on $\zeta = 0$ *and* 1 for our unknown functions:

$$\Psi = \Omega = T = 0 . \qquad (10.101)$$

Finally, below we consider the system of three equations (10.99a,b,c) with the homogeneous boundary conditions (10.101) a $\zeta = 0$ and $\zeta = 1$. Now, if we use the Routh–Hurwitz classical criterion for the associated linear problem, we obtain the following critical value of bifurcation parameter B^0:

$$B_c^0 = n^4 \pi^4 \sigma A^{0\,2} + \frac{n^2 \pi^2}{1 + \sigma} . \qquad (10.102)$$

Naturally, for $B^0 > B_c^0$, the motion induced by viscous/thermal effects is unstable, and, on the contrary, for $B^0 < B_c^0$, the state of rest is a structurally stable focus – in this last case, trajectories tend exponentially to the origin in phase space. We have an oscillatory instability, and the principle of exchange of stabilities is not valid (see Chap. 9).

Evidently, linear theory gives very good results in the subcritical case, when $B^0 < B_c^0$, but for the supercritical case, when $B^0 > B_c^0$, it is necessary to analyze the full nonlinear problem (10.99a–c), (10.101). In the nonlinear case, according to the Galerkin technique, if we consider only *two modes* for each function Ψ, Ω, and T, in the associated Fourier series in $\sin(n\pi\zeta)$, with unknowns (which depend on time), $X_n(\tau)$, $Y_n(\tau)$, and $Z_n(\tau)$, $n = 1, 2$, we derive the following six-component dynamical system for the amplitudes $X_1(\tau)$, $X_2(\tau)$, $Y_1(\tau)$, $Y_2(\tau)$, and $Z_1(\tau)$, $Z_2(\tau)$:

$$\frac{dX_1}{d\tau} = Y_1 + Z_1 - r^0 X_1 + \frac{\pi}{2} X_1 X_2 ; \qquad (10.103\text{a})$$

$$\frac{dX_2}{d\tau} = Y_2 + Z_2 - 4r^0 X_1 - \frac{\pi}{2} X_1^2 , \qquad (10.103\text{b})$$

$$\frac{dY_1}{d\tau} = -\pi^2 X_1 \ ; \tag{10.104a}$$

$$\frac{dY_2}{d\tau} = -4\pi^2 X_2 \ , \tag{10.104b}$$

$$\frac{dZ_1}{d\tau} = B^0 X_1 - \sigma r^0 Z_1 - r^0 X_1 - \frac{\pi}{2}[X_2 Z_1 - 2X_1 Z_2] \ ; \tag{10.105a}$$

$$\frac{dZ_2}{d\tau} = B^0 X_2 - 4\sigma r^0 Z_2 - 4r^0 X_1 - \frac{\pi}{2} X_1 Z_1 \ , \tag{10.105b}$$

with $r^0 = \pi^2 A$. The NLDS (10.103a,b)–(10.105a,b) is invariant under the following transformation:

$$(X_1, X_2, Y_1, Y_2, Z_1, Z_2) \Rightarrow (-X_1, X_2, -Y_1, Y_2, -Z_1, Z_2) \ .$$

Nonadiabaticity (thermal effect) and viscosity in the NLDS rise to finite dimensionality of the phase space and also to dissipation of phase flow, i.e., an average compression of the phase volume "downstream," as $\tau \to \infty$:

Then, div $[\text{NLDS}(10.103a,b)-(10.105a,b)] = -5(1+\sigma)r^0 < 0$, (10.106)

and we note that the *minimum* critical value of the bifurcation parameter B^0 is

$$B_c^{0*} = \sigma r^{0\,2} + \frac{\pi^2}{1+\sigma}, \quad n = 1 \ . \tag{10.107}$$

Below, we present mainly various numerical results when $\sigma = 1$ and

$$r^0 = \pi^2 (A^0 = 1), \quad B_c^{0*} = \pi^4 + \frac{\pi^2}{2} \ . \tag{10.108}$$

Then we have a classical Hopf bifurcation and above the threshold B_c^{0*} or for

$$E = \frac{(B^0 - B_c^{0*})}{B_c^{0*}} > 0 \ , \tag{10.109}$$

the oscillations are interrupted by chaotic bursts, which become more frequent when B^0 increases (or E increases above unity) until the motion becomes truly chaotic, and a strange attractor appears in the phase space (see Figs. 10.16–10.20). This intermittency route to chaos is of type I, according to Pomeau and Manneville (1980), and is generated by an inverse tangent bifurcation. Some numerical results are also presented below for the values $\sigma \ll 1$ (when the viscous effect is dominant) and $\sigma \gg 1$ (when the thermal effect is dominant). Unfortunately, we do not have numerical results for the influence of a small compressibility (parameter $\alpha^0 = H^0 \gamma^0$), because in this case $\partial \Psi / \partial \zeta = 0$ at $\zeta = 0$ and 1, and as a consequence it is necessary to derive an other NLDS in place of (10.103a,b)–(10.105a,b).

10.4 Strange Attractors for Various Fluid Flows   431

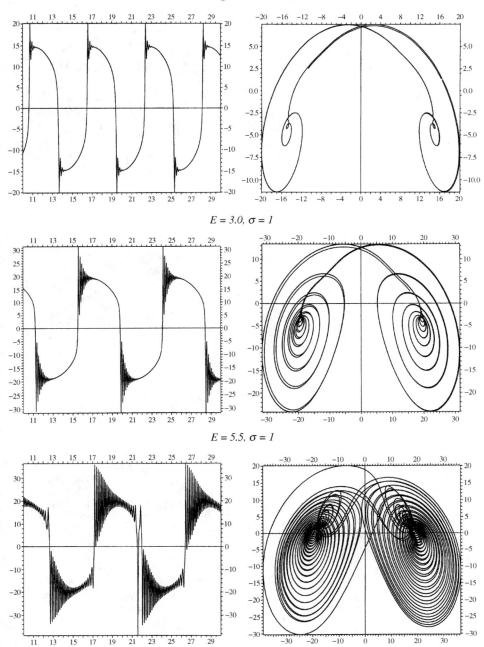

Fig. 10.16. The plots of X_1 as a function of time and X_1 versus X_2 for $E = 3.0$, 5.5, 6.5, and $\sigma = 1$

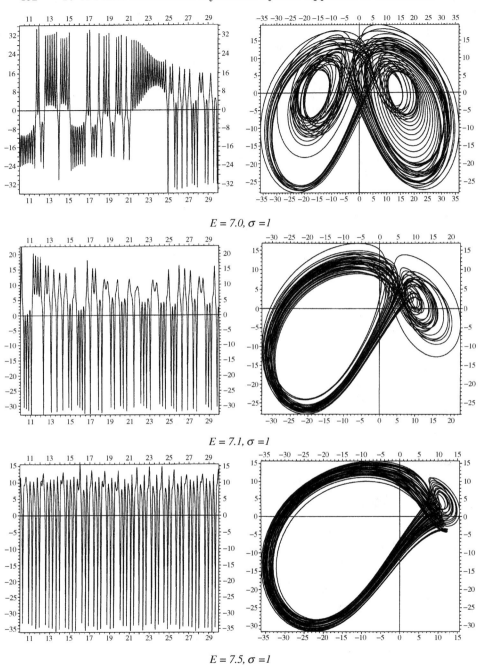

Fig. 10.17. Plots of X_1 as a function of time and X_1 versus X_2 for $E = 7.0$, 7.1, 7.5

10.4 Strange Attractors for Various Fluid Flows

I am very grateful to M. Errafiy and S. Ginestet for providing (in the framework of the LML of the University of Lille I) us with a computer code, for the numerical results presented (for the first times) in the various figures below, and for subsequent numerical advice. The plots of X_1 as a function of time and X_1 versus X_2 for various values of E and $\sigma = 1$ are shown in Figs. 10.16, 10.17 and 10.18, respectively, for $E = 3.0$, 5.5, 6.5; $E = 7.0$, 7.1, 7.5; and $E = 10.0$, 12.0. For $E = 12$, the strange attractor is very similar to the strange attractor obtained by Lorenz (see Fig. 10.1).

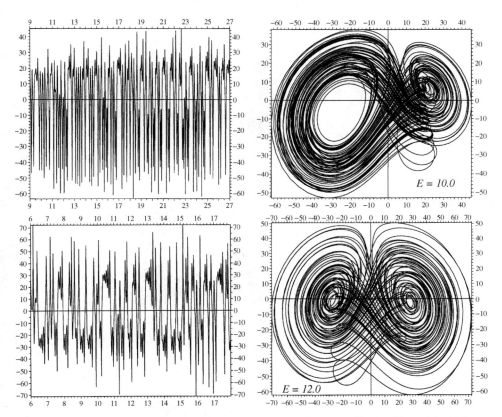

Fig. 10.18. Plots of X_1 as a function of time and X_1 versus X_2 for $E = 10.0$, 12.0, and $\sigma = 1$

In Figs. 10.19a,b and 10.20, we present some results for the case when the *viscous effect is dominant*, $\sigma \ll 1$, and for $\sigma \gg 1$, when the *thermal effect is dominant*.

When $\sigma \ll 1$, the evolution in time is periodic with in power spectra only, at first, one fundamental frequency appears, then a second fundamental frequency (Fig. 10.19a), next a third frequency, and, finally the temporal evo-

434 10 A Finite-Dimensional Dynamical System Approach to Turbulence

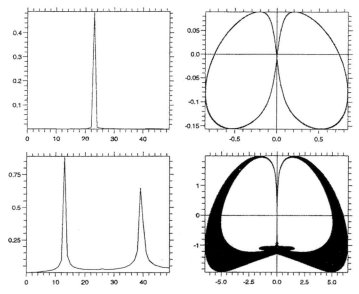

(a) one and second fundamental frequencies

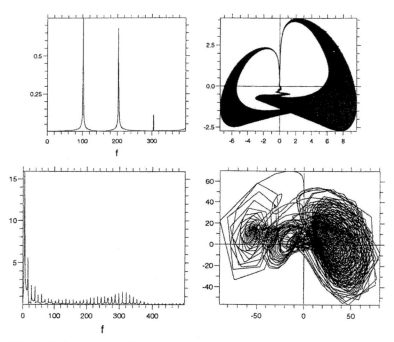

(b) third frequency and power spectrum with a continuous part

Fig. 10.19. Power spectra and attractors for the case $\sigma \ll 1$

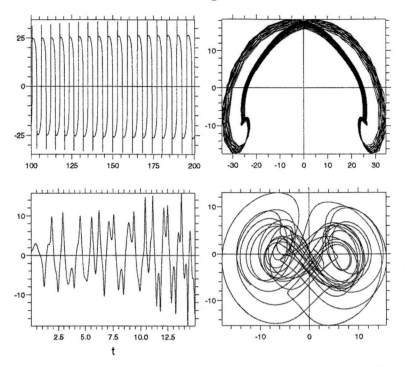

Fig. 10.20. Evolution of amplitude X_1 with time and attractors when $\sigma \gg 1$

lution is chaotic (Fig. 10.19b), and the power spectrum admits a continuous part.

When $\sigma \gg 1$, the periodic regime loses its stability abruptly! The attractor corresponding to the periodic regime is similar to a Smale horseshoe which is a hyperbolic limit set and arises whenever one has transverse homoclinic orbits. This Smale horseshoe is robust with respect to small changes in the equations (structural stability).

10.4.5 Obukhov Discrete Cascade Systems for Developed Turbulence

The discrete cascade system proposed by Obukhov (1981) to describe cascade processes in developed turbulence [see also Gledzer et al. (1988)] was derived formally by Gledzer (1986) by reducing the Navier equations to a system of ODEs relative to time with the help of a Fourier/Galerkin approximation.

At first, the simplest discrete model of a nonlinear cascade was constructed from triplets by the "superposition principle." The multilevel cascade mode is shown schematically in Fig. 10.21 where each triplet is represented by a triangle. The model is multilevel; each level consists of triplets of

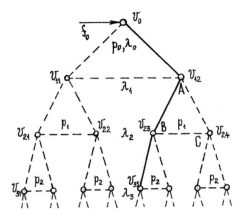

Fig. 10.21. Multilevel cascade model

the same scale $l_i = 1/p_i$ (in skew presentation) whose number doubles as we descend to the next level, and scales become smaller. The first level is formed by a single triplet, whose unstable mode v_0 is acted upon by an external force f_0. At the same time, stable modes of this triplet v_{11} and v_{12} are unstable modes of two second-level triplets, whose stable modes v_{21}, v_{22}, v_{23}, v_{24} are, in turn, unstable modes of four third-level triplets, and so forth. In system (10.110), the interaction parameters p_i and dissipative coefficients λ_i are the same for all modes of any particular level, and one may set

$$\lambda_i = \alpha^2 \nu p_i^2$$

where ν is the kinematic viscosity of the medium and α is a factor of the order of unity.

If $\lambda_i \equiv 0$ and $f_0 = 0$, then the energy $\mathcal{E} = (1/2) \sum v_{ij}^2$ in system (10.110) is conserved, whereas, generally it satisfies the balance equation

$$\frac{d\mathcal{E}}{dt} = \mathcal{W}_0 - \Phi,$$

where $\mathcal{W}_0 = f_0 v_0$ is the power input and $\Phi = \sum \lambda_i v_{ij}^2$ is the energy dissipation. The equation of motions are as follows:

$$\frac{dv_0}{dt} = p_0(v_{11}^2 - v_{12}^2) - \lambda_0 v_0 + f_0,$$

$$\frac{dv_{11}}{dt} = -p_0 v_0 v_{11} + p_1(v_{21} - v_{22}) - \lambda_1 v_{11},$$

$$\frac{dv_{12}}{dt} = p_0 v_0 v_{12} + p_1(v_{23}^2 - v_{24}^2) - \lambda_1 v_{12}, \tag{10.110a}$$

....
....
....

$$\frac{dv_{i,2k-1}}{dt} = -p_{i-1}v_{i-1,k}v_{i,2k-1} + p_i\left(v_{i+1,4k-3}^2 - v_{i+1,4k-2}^2\right) - \lambda_i v_{i,2k-1},$$

$$\frac{dv_{i,2k}}{dt} = p_{i-1}v_{i-1,k}v_{i,2k} + p_i\left(v_{i+1,4k-1}^2 - v_{i+1,4k}^2\right) - \lambda_i v_{i,2k}. \quad (10.110\text{b})$$

The ratio

$$q = \frac{p_{i+1}}{p_i} > 1,$$

is a basic invariant characteristic of the discrete cascade – it can be called the coefficient of refinement of turbulent disturbances.

When the external force f_0 in the first equation of (10.110a) is "switched on" very slowly, the excitation process can be represented as follows. First, v_0 increases, and the sign of v_0 is the same as that of f_0. Then, when $|f_0|$ exceeds the value $\lambda_0\lambda_1/p_0$, one of the components of the first level is excited: v_{11}, if $v_0(0) < 0$, or v_{12}, if $v_0(0) > 0$.

The sign of any excited mode (except v_0) depends entirely on small initial fluctuations, that is, it is random, and, at the same time, determines the directions of subsequent evolution of the excitation process. One such excited chain is shown in Fig. 10.21 above as a solid line. The development of excitations on a certain level is accompanied by growth of the corresponding quadratic terms on the preceding level, which simulates Reynolds stress.

It follows from (10.110b) that the component of the $(i+1)$st level is excited as soon as the intensity at the ith level exceeds the critical value λ_{i+1}/p_i.

Following is another cascade model, in dimensionless form (in the form of coupled gyrostats), considered in Gledzer et al. (1988):

$$\frac{du_1}{dt} = -u_2 u_3 - \mu_1 u_1 + 1,$$

$$\frac{du_2}{dt} = -u_3(u_1 - qu_4) - \sigma u_3 - \mu_2 u_2,$$

$$\frac{du_3}{dt} = qu_4(u_2 - qu_5) + \sigma(u_2 - q^{2/3}u_4) - \mu_3 u_3,$$

$$\frac{du_4}{dt} = q^2 u_5 u_3 + \sigma q^{2/3}(u_3 - q^{2/3}u_5) - \mu_4 u_4,$$

$$\frac{du_5}{dt} = \sigma q^{4/3} u_4 - \mu_5 u_5, \quad (10.111)$$

written here only for five components u_i, $i = 1, 2, 3, 4, 5$.

At $\sigma = 0.8$ and $\sigma = 1.26$, the steady states lose stability and within the above range $(0.8, 1.26)$ there is an interval (in the vicinity of $\sigma = 1.085$) where system (10.111) admits two stable periodic solutions (case (d) in Fig. 10.22). If σ increases from values in this interval, then period-doubling bifurcations of the periodic solution drawn by the solid line in (d) lead to the stochastic regime which arises when the steady state loses stability [see (d) → (f)]. If σ decreases from values in the interval, then a sequence of period-doubling

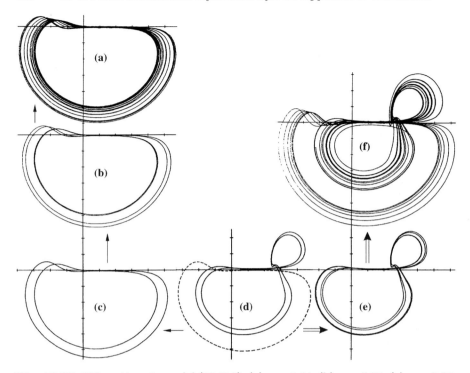

Fig. 10.22. Bifurcations in model (10.111): (a) $\sigma = 0.94$, (b) $\sigma = 0.95$, (c) $\sigma = 0.96$, (d) $\sigma = 1.085$, (e) $\sigma = 1.09$, (f) $\sigma = 1.1$

bifurcations of the periodic solution drawn by the dashed line in (d) leads to another stochastic regime [see (a) ← (d)]. Sonechkin (1984, § 5.4) considers a simplified Obukhov cascade system, when $f_0 > 0$, written in dimensional form:

$$\frac{du_0}{dt} = -p_0 u_{11}^2 - u_0 + f_0 ,$$

$$\frac{du_{11}}{dt} = u_0 u_{11} + q(u_{21}^2 - u_{22}^2) - q^2 u_{11} ,$$

$$\cdots\cdots\cdots\cdots\cdots\cdots\cdots\cdots\cdots\cdots\cdots\cdots\cdots\cdots\cdots\cdots\cdots\cdots\cdots$$

$$\cdots\cdots\cdots\cdots\cdots\cdots\cdots\cdots\cdots\cdots\cdots\cdots\cdots\cdots\cdots\cdots\cdots\cdots\cdots$$

$$\frac{du_{i,2k-1}}{dt} = -q^{i-1} u_{i-1,k} u_{i,2k-1} + q^i(u_{i+1,4k-3}^2 - u_{i+1,4k-2}^2) - q^{2i} u_{i,2k-1} ,$$

$$\frac{du_{i,2k}}{dt} = q^{i-1} u_{i-1,k} u_{i,2k} + q^i(u_{i+1,4k-1}^2 - u_{i+1,4k}^2) - q^{2i} u_{i,2k} . \quad (10.112)$$

In Fig. 10.23, we present the Obukhov strange attractor (Sonechkin, 1979), corresponding to system (10.112) in 3-D phase space (u_{11}, u_{21}, u_{22}), when

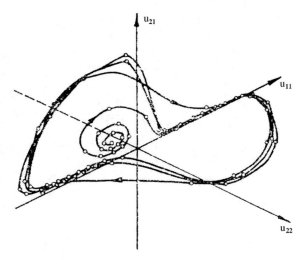

Fig. 10.23. The Obukhov strange attractor in phase space (u_{11}, u_{21}, u_{22})

$f_0 > q^6 + q^2 + 10$. This Obukhov attractor is obtained numerically after four bifurcations from NLDS (10.112), but it has only four excited modes, and its structure is hyperbolic.

10.4.6 Unpredictability in Viscous Fluid Flow Between a Stationary and a Rotating Disk

By definition, we know that, if A is an attractor for a dynamic system: $d\boldsymbol{X}/dt = \boldsymbol{F}(\boldsymbol{X}, \lambda)$, then the basin of attraction of A is the set of initial states $\boldsymbol{X}^0 = \boldsymbol{X}(0)$ such that $F^t \boldsymbol{X}^0$ approaches A, as $t \to \infty$. Here, we write $F^t \boldsymbol{X}^0$ for $\boldsymbol{X}(t)$ – the solution curve in state space, and F^t denotes the solution flow for the dynamical system considered.

For various (nonautonomous) dynamical systems the basin attraction geometry is fractal, and in particular, the fractal basin boundary implies *unpredictability* for the system in a certain range of initial conditions [see, for instance, Cheng et al. (1993)].

In Bailly (1995), the long-time behavior of viscous fluid flow between a stationary and a rotating disk is investigated in the framework of a finite-dimensional dynamical system. From the Navier equations written for this problem, Bailly derives a three-amplitudes, nonautonomous, dynamical system [see, below, problem (10.113a,b) and NLDS (10.116)]. The stationary solutions associated with this system (in the autonomous case) approximate well, similar classical solutions known in the literature for various values of the Reynolds number [for instance, Batchelor (1951) and Stewartson (1953) solutions] and also those obtained by Mellor et al. (1968) and Brady and Durlofsky (1987).

As in Bailly (1995), we assume that the upper disk is rotating at dimensionless angular velocity $\omega(t)$ and the lower disk is fixed. If the disks are considered "infinite" (with the Von Karman similarity hypothesis) then, after some tedious but straightforward algebra, we obtain the following problem for the two functions $H(t,z)$ and $Q(t,z)$:

$$\frac{\partial}{\partial t}\frac{\partial^2 H}{\partial z^2} + H\frac{\partial^3 H}{\partial z^3} + 4[Q + \omega(t)z]\left[\frac{\partial Q}{\partial z} + \omega(t)\right] = \frac{1}{\text{Re}}\frac{\partial^4 H}{\partial z^4},$$

$$\frac{\partial Q}{\partial t} + z\frac{d\omega}{dt} - [Q + \omega(t)z]\frac{\partial H}{\partial z} + H\left(\frac{\partial Q}{\partial z} + \omega(t)\right) = \frac{1}{\text{Re}}\frac{\partial^2 Q}{\partial z^2}, \quad (10.113a)$$

$$z = 0: \quad H = Q = \frac{\partial H}{\partial z} = 0,$$

$$z = 1: \quad H = Q = \frac{\partial H}{\partial z} = 0,$$

$$t = 0: \quad H = H^0, Q = Q^0. \quad (10.113b)$$

Bailly derives by a Galerkin technique, with

$$Q = \sum Q_n(t)\sin(n\pi z), \quad n = 1 \text{ to } \infty, \quad (10.114a)$$

$$H = \sum_\nu H_\nu(t)\psi_\nu(z), \quad (10.114b)$$

where

$$\psi_\nu(z) = (\sin\nu - \text{sh}\nu)(\text{ch}\nu z - \cos\nu z)$$
$$+ (\text{ch}\nu - \cos\nu)(\text{sh}\nu z - \sin\nu z), \quad (10.115)$$

a three-amplitude NLDS for $X = H_{\nu 1}$, $Y = Q_1$, and $Z = Q_2$, where ν_1 is the first nonzero solution of the equation $\text{ch}\nu\cos\nu = 1$:

$$\frac{dX}{dt} = pX + aX^2 + bY^2 + cYZ + dZ^2 + eY + fZ + g;$$

$$\frac{dY}{dt} = qY + a'XY + b'XZ + +e'X - \frac{1}{\pi}\frac{d\omega}{dt};$$

$$\frac{dZ}{Dt} = rZ + a''XY + b''XZ + e''X + \frac{1}{2\pi}\frac{d\omega}{dt}. \quad (10.116)$$

In the autonomous case, when the forcing terms [$\omega(t)$, dependent on the time] are absent, from (10.116), the associated stationary solutions tend to a fixed point (for moderate values of Re) and to a limit cycle (for high values of Re) in phase space.

In the non-autonomous case, Bailly considers first a sinusoidal small forcing

$$\omega(t) = 1 + \Delta\omega\sin t, \quad (10.117a)$$

and for this case a typical numerical result is shown in Fig. 10.24.

10.4 Strange Attractors for Various Fluid Flows

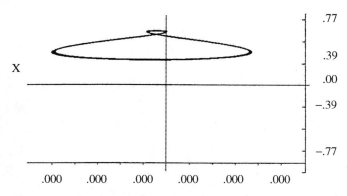

Fig. 10.24. Limit cycle (with a small buckle) in phase space (X, Y) for Re = 10000, $\Delta\omega = 0.4$ and initial conditions $X^0 = -1$, $Y^0 = Z^0 = 0$

To obtain chaotic behavior, it is necessary to consider stronger forcing:

$$\omega(t) = \cos(\varpi t) , \qquad (10.117b)$$

which simulates "turbulent feedback" and where ϖ is the forcing frequency. In this case, the chaotic dynamic is a function of the Re, ϖ, and also of the choice of initial conditions. For the influence of the Reynolds number, we note that, when Re increases, we obtain a cascade of attractors that is very similar to a series of periodic regimes with period doublings. In Figs. 10.25 and 10.26, the reader can find phase plots for Re = 100 and 1000 and also for Re = 2000, 3000, and 10000, when $\varpi = 1$. For Re = 10000 in phase space (X, Y), we have a (chaotic) strange attractor, and the corresponding 2-D Poincaré section (in Fig. 10.26) illustrates this chaotic structure very well.

Curiously, when $\varpi = 0.1$ or $\varpi = 10$ for the same value of Re = 10000 and the same initial conditions, we obtain a complicated cycle limit or a classical simple cycle limit in place of the chaotic structure; see Fig. 10.27.

Figure 10.28 exhibits the crucial phenomenon called "sensitive dependence on the initial conditions," for Re = 10000 and $\varpi = 1$, when the difference in the initial conditions X^0 is only in the fifth decimal place!

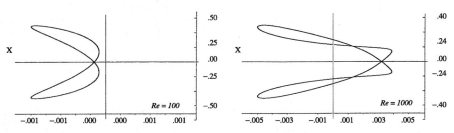

Fig. 10.25. Attractors in phase space (X, Y) for Re = 100 and Re = 1000 for $\varpi = 1$

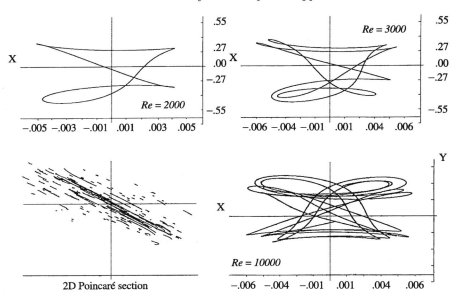

Fig. 10.26. Attractors in phase space (X, Y) for Re = 2000, 3000, and 10000 for $\varpi = 1$

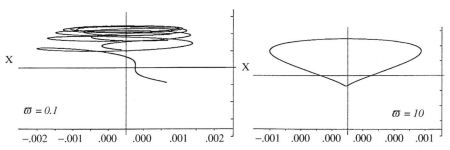

Fig. 10.27. Attractors in phase space (X, Y) for Re = 10000 and $\varpi = 0.1$ or $\varpi = 10$

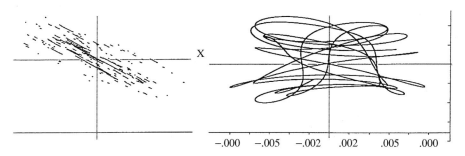

Fig. 10.28. Attractors in phase space (X, Y) for Re = 10000 and $\varpi = 1$, but the initial conditions are a little different

10.4 Strange Attractors for Various Fluid Flows

Finally, in Fig. 10.29, we show the basin of attraction (for $-10 \leq X^0 \leq +10$, $-10 \leq Y^0 \leq +10$, $Z^0 = 0$) and also (enlarged) the fractal structure of the left-hand side of the boundary (for $-8.2 \leq X^0 \leq -6.8$, $-10 \leq Y^0 \leq +10$, $Z^0 = 0$) of this basin of attraction for the bounded solutions, when Re = 3700 and $\varpi = 1$.

The existence of a fractal boundary of the basin of attraction (even if the limit motion for the large time is regular) causes sensitive dependence on initial conditions, which leads to system unpredictability.

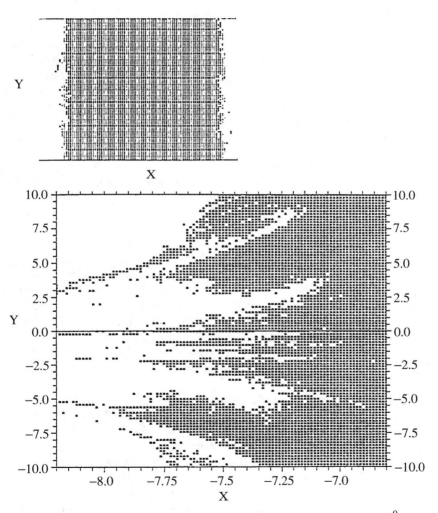

Fig. 10.29. Basin of attraction: Re = 3700, $\varpi = 1$, for $-10 \leq X^0 \leq +10$, $-10 \leq Y^0 \leq +10$, $Z^0 = 0$

10.5 Some Comments and References

In its most general form, bifurcation theory is a theory of equilibrium solutions of nonlinear equations. By equilibrium solutions, we mean, for example, steady-state, time-periodic, and quasi-periodic solutions. The purpose of the book by Iooss and Joseph (1980) is to teach the theory of bifurcation of equilibrium solutions of evolution problems governed by nonlinear differential equations. Curiously, the general theory is simpler than the detailed theory required for particular applications! It is generally believed that the mathematical theory of bifurcation requires some functional analysis and some of the methods of topology and dynamics. In Iooss and Joseph (1980) and also in Iooss (1984), the main application of the functional analysis of problems of bifurcation is justification of reducing problems posed in spaces of high or infinite dimensions to one and two dimensions. These low-dimensional problems are associated with eigenfunction projections, and in some special cases, like those arising in degenerate problems involving symmetry-breaking, steady-state bifurcations, analysis of problems of low dimension greater than two is required.

To derive low-dimensional problems, some authors use the "Lyapunov–Schmidt method" to decompose the space of solutions and equations into finite-dimensional and infinite-dimensional parts [see, for example, Guiraud (1980)]. The infinite part can be solved, and the resulting finite-dimensional problem has all of the information about bifurcation. For a rigorous discussion of this method, the reader is invited to consult the book by Vainberg and Trenogin (1962). Others authors use the "center manifold" to reduce problems to finite dimensions. This method uses the fact that, in various problems, the solutions are attracted to the center manifold, which is finite-dimensional. For a complete description of this method and the derivation of amplitude equations, see the book by Guckenheimer and Holmes (1983) and the paper by Coullet and Speigel (1983). Both methods are good for proving existence theorems. The last (third) method is the "implicit function theorem" to justify the direct, sequential computation of power series solutions in an (small) amplitude ε, using the Fredholm alternative. For a systematic use of this last method, see Iooss and Joseph (1980).

The reader can find a pertinent discussion of nonlinear dynamics and turbulence in the volume edited by Barenblatt, Iooss, and Joseph (1983). Many papers in this volume have as their aim the elucidation of the implications of the generalized concept of turbulence and the study of its utility in applications to observed turbulence in fluids. To achieve such an aim, it is necessary to combine analytical, computer, and experimental studies. It is also interesting to note that, if bifurcations occur in a parametrized dynamical system when a change in a (control) parameter causes an equilibrium to split into two, a catastrophe occurs when the stability of an equilibrium breaks down, causing the system to jump into another state. Zeeman (1982) describes some strange attractors and strange bifurcations, illustrates the difference between bifurcation and catastrophe in "non-elementary" theory, and also discusses some of their properties that resemble turbulence, such as stability, with sensitive dependence on initial conditions, and broadband frequency spectra similar to those observed by Swinney and Gollub (1978). For a short, but concise, critical and non-mathematical review of the less controversial results in catastrophe theory which will provide a useful introduction to the subject, see the book by Arnold (1986). The purpose of books by Abraham and Shaw (1984/1992) is to encourage the diffusion of mathematical ideas by presenting them visually. These books are based on the ideas that mathematical concepts (for example, critical points in 3-D, limit cycles, characteristic exponents, power spectra, chaotic limit sets, and attributes of chaos) may be communicated easily in a format that combines visual, verbal, and symbolic representations in tight coordination to attack math ignorance with an abundance of visual representations.

On the other hand, the study of the long-time behavior of solutions of dissipative partial differential equations is a major problem in mathematical physics, directly related to the understanding of turbulence. In general, the long-time behavior of systems can be characterized by the existence of global attractors. Associated with the global attractor, approximate inertial manifolds have been introduced recently in Foias et al. (1987). We recall that an approximate inertial manifold (according to Temam) is a finite-dimensional smooth manifold such that every solution enters its thin neighborhood in a finite time. In particular, the global attractor is contained in this neighborhood. The existence of approximate inertial manifolds has been obtained for many dissipative partial differential equations [see, for example, Temam (1989)]. More precisely, the long-time behavior of solutions of dissipative evolution equations is described by a compact attractor that attracts all bounded sets. At our present level of understanding of dynamical systems, little is known about the geometry of these attractors, which may be fractal sets [according to Mandelbrot (1982) and Guckenheimer and Holmes (1983)]. We know, thanks to general theorems that apply to diverse dissipative systems, that the attractor has a finite dimension in the sense of the Hausdorff dimension or of capacity (fractal dimension) – see, for instance, Temam (1988a). Foias with Temam introduce a method for approximating the global attractor of dissipative differential equations (including the 2-D Navier equations) based on the time analyticity (in a fixed infinite band) of all solutions lying on the attractor. In particular, these families of polynomial maps are constructed with properties that, under some suitable conditions, their zeros approximate the attractor. Other approximations of attractors by finite-dimensional algebraic or analytic sets were constructed elsewere [see, Temam (1989)], but the corresponding sets were graphs above a finite-dimensional space! It is worth mentioning that the Lorenz system (10.1), with the usual parameters $\sigma = 10$, $b = 8/3$, and $r = 28$, does not satisfy the assumptions mentioned by Foias and Temam!

If we restrict ourselves to a Galerkin approximation [for a general exposition, see, for instance, Kantorovich and Krilov (1958) and also Foias and Temam (1989) for the *nonlinear* Galerkin methods] of 2-D Navier equations based on the spaces V_m, i.e., the family w_j of eigenfunctions of the associated Stokes (linear) problem, then if m is sufficiently large, the behavior (as $t \to \infty$) of the Galerkin approximation u_m is completely determined by the behavior, as $t \to \infty$, of a certain number m^* of its mode [see the book edited by Barenblatt, Iooss, and Joseph (1983), pp. 139–155]. It is interesting to note that, as illustrated in the above book (pp. 334–342) by Treve, important properties of the Galerkin approximations can be established with great ease when they satisfy the energy balance equations. This is one more reason for looking for such approximations in fluid flow problems, and they can be constructed in many cases. For the classical Bénard convection problem, Treve proves that there is a unique way of constructing Galerkin approximations of any order such that the energy balance equations are satisfied exactly and gives a simple proof of the boundedness of the approximations so constructed, indicates how bounds can be obtained for quantities of physical interest such as the Nusselt number, and demonstrates the global asymptotic stability of a purely conductive solution for subcritical Rayleigh numbers. On the other hand, Foias and Saut (1986) show that, for the evolution Navier equations, the ratio of enstrophy to energy has a limit, as $t \to \infty$, which is an eigenvalue of the associated Stokes operator. This fact has direct consequences on the global struture of the Navier equations. This allows us to construct a flag of nonlinear spectral manifolds of the Navier operator in the space R of initial data. These are analytic manifolds of finite dimension that are invariant for the Navier equations and completely determine the energy decay of the solution. The paper by Ghidaglia (1986) was motivated by the study of attractors and of the long-time behavior of infinite-dimensional dynamical systems arising from fluid mechanics (Navier equations, magnetohydrodynamic equations, and thermohydraulic equations), from combustion (Kuramoto–Sivashinsky equations), and from optics (the nonlinear Schrödinger equation).

For the laminar-turbulent transition, the reader can find recent results in the Proceedings of the Fourth IUTAM Symposium, edited by Kobayashi (1995), held in Sendai, September 5–9, 1994. For an "asymptotic approaches to transition modeling," see the review paper by Cowley and Xuesong Wu (1994). For the structure and statistics of turbulence, see the Proceedings of a symposium held at the Research Institute for Mathematical Sciences, Kyoto University [January 17–18, 1994; abstract in Math. Rev. 96h: 76002]. On the other hand, the reader can find in the AMS short Course Lecture Notes, edited by Daveney and Keen (1989), fascinating and beautiful color plates that illustrate the appearance of fractals as a chaotic set for a dynamical system. For an introduction to chaotic dynamics, see also the book by Baker and Gollub (1996). This book is not a research review, but rather a focused introduction to the basic phenomena. The authors emphasize accessibility

10.5 Some Comments and References

rather than completeness. The book by Aref and Naschie (1995) presents a comprehensive picture of the state of development of chaotic advection, covering all areas, including the mathematical tools used to describe chaotic advection, detailed exploration of lobe dynamics, aperiodic systems, stretching and alignment, and chaotic advection and diffusion. The book by Beck and Schlögl (1995) deals with the various thermodynamic concepts used for analyzing nonlinear dynamical systems. The recent book by Marek and Schreiber (1995) is a graduate text surveying both the theoretical and experimental aspects of chaotic behavior. The book by Van Groesen and Jager (1994) provides an excellent review of the role of infinite-dimensional Hamiltonian and Poisson dynamical structure in a range of PDE problems. A very nice feature is that acknowledging that the Hamiltonian case is only an idealized limit in fluid mechanics. The aim of the book by Nicolis (1995) is to develop a unified approach to nonlinear science, which does justice to its multiple facets and to the diversity and richness of the concepts and tools developed in this field over the years, and Cercignani and Sattinger (1998) discuss the various aspects of scaling limits and models in physical processes.

In the recent paper by Shilnikov (1997), the reader can find a review of the theory of nonlinear systems, especially that of strange attractors; special attention being devoted to recent results concerning hyperbolic attractors and features of high-dimensional systems in the Newhouse region. An example of a "wild" strange attractor of topological dimension three is presented.

The paper by Ladyzhenskaya (1987, in Russian) is very interesting and is devoted to problems of the global stability theory for Navier equations via a dynamical systems approach. In the Introduction (§1) of that paper, the reader can find pertinent information relative to the theory of dynamical systems with some historical facts about the development of this theory in Russia and its relation to the results obtained by Ruelle, Foias, and Temam.

In the recent book edited by Malek, Necas, and Rokyta (2000), the reader can find six survey contributions that focus on several open problems of theoretical (mathematical!) fluid mechanics both for incompressible (Navier) and compressible (NS) fluids. In particular, the papers by Feireisl (pp. 35–66) and Masmoudi (pp. 119–158) are devoted to the global (in time) properties of solutions of the NS (isentropic) equations for compressible fluids. More precisely, in the Feireisl paper, a dynamical systems approach is used, and, once the dynamic is given, it is the task of dynamical systems theory to study the patterns of how states change in the long run. In an attempt to predict the long-time behavior of dynamical systems, we encounter several difficulties related to chaos, bifurcation, and sensitivity to initial data. We mention also the recent (and curious!) paper by Robert (2001), where the author concludes that the famous "butterfly (Lorenz) effect does not exist", and for this, Robert investigates the behavior of the motion governed by Euler (nonviscous) equations for the long time? In conclusion, it seems (!)

that the exponential instability of a microscopic system is not related to the concept of unpredictability!

Finally, as a conclusion, we mention the recent book edited in Cambridge University Press (2000), by Batchelor, Moffat and Worster concerning the "Perspective in Fluid Dynamics".

References[1]

1. Abraham R and Shaw CD. *Dynamics. The geometry of behavior.* Part 2: *Chaotic behavior.* Aerial Press, Santa Cruz, 1984.
2. Abraham R and Shaw CD. Dynamics. *The geometry of behavior.* Part 1: *Periodic behavior*, Part 2: *Chaotic behavior*, Part 3: *Global behavior*, Part 4: *Bifurcation behavior.* Addison-Wesley, Redwood City, 1992.
3. Afendikov AL and Babenko KI. In: *Mechanics and Scientific-Technical Progress*, Vol 2: *Mekhanika zhidkosti i gaza*, 49–66. Naouka, Moskva, 1987, (in Russian).
4. Afendikov A and Mielke A. ARMA **129** (1995), 101–127.
5. Alassar RS and Badr HM. J Engineering Math **36** (1999), 277–287.
6. Alekseev VA. Russian J Numer Anal Math Modelling **9** (1994), 315–336.
7. Allen T and Riley N. J Engng Maths **28** (1994), 345–364.
8. Amann H Arch Rat Mech Anal **126** (1994), 231–242.
9. Amick CJ. On steady Navier-Stokes flow past a body in the plane. In: Proceeding of Symposia in Pure Maths, 45, Part I, (1986), 37–50.
10. Amick CJ. Acta Math **161** (1988), 71–130.
11. Andereck CD, Liu SS and Swinney HL. *Flow regimes in a circular Couette system with indepentently rotating cylinders.* Preprint, 1986.
12. Anderson Jr JD. *Hypersonic and High Temperature Gas Dynamics.* McGraw-Hill Book Co, New York, 1989.
13. Andreichikov I. Bifurcation of secondary modes of fluid flow between the rotating cylinders, (in Russian). Isv AN SSSR, MJG, **1**, 47–53, (1977).
14. Aniss S, Souhar M and Belhaq M. Phys Fluids **12**(2) (2000), 262–268.
15. Antontsev SN, Kazhikhov AV and Monakhov VN. Boundary value problems in mechanics of nonhomogeneous fluids, Nauka, Novosibirsk, 1983, (in Russian) English Transl., North-Holland, Amsterdam, 1990.
16. Aref H and Naschie MS (eds). *Chaos applied to fluid mixing.* Pergamon, Oxford 1995. Reprinted from: *Chaos, Solitons and Fractals*, Vol 4, 1994.
17. Arnal D. Three-dimensional boundary layers: Laminar-turbulent transition. AGARD Rep number 741 – Special Course on calculation of three-dimensional boundary layers with separation. Von Karman Inst, Rhode-St-Genese, Belg, (1986).
18. Arnold VI. *Catastrophe theory.* 2nd ed., Nauka, Moskva 1990. English transl, Springer, Berlin 1986.

---

[1] In the list of references the reader can find not only the all references mentioned in Introduction and the Chapters 1 to 10, but also some complementary references which are pertinent for the discussed topics.

19. Arnold VI and Khesin BA. *Topological methods in hydrodynamics*. Springer, 1996.
20. Asano K. *Zero viscosity limit of incompressible Navier-Stokes equations*. Preprint, Kyoto University, 1989.
21. Asmolov ES. J Fluid Mech **305** (1995), 29–46.
22. Asmolov ES and Manuilovich SV. J Fluid Mech **365** (1998), 137–170.
23. Babenko KI. Mat Sbornik **91** (1973), (133).
24. Babenko KI. Doklady AN SSSR (Russian), **227**(3) (1976), 592–595.
25. Babenko KI. Doklady AN SSSR (Russian), **263**(3) (1982), 521–524.
26. Babenko KI and Vasiliev MM. PMM, **37**(4) (1973), 691.
27. Babin AV and Vishik MI. Russian Math Survey **38** (1983), 151–213.
28. Badr HM and Dennis SCR. J Fluid Mech **158** (1985), 447–488.
29. Badr HM, Dennis SCR and Kocabiyik S. J Engng Maths **29** (1995), 255–269.
30. Badr HM, Dennis SCR, Kocabiyik S and NGuyen P. J Fluid Mech **303** (1995), 215–232.
31. Bailly Ch. *Modelisation asymptotique et mumérique de l'écoulement dû à des disques en rotation*. Thése presentée à l'Université des Sciences et Technologies de Lille, Villeneuve d'Ascq Cedex. No. d'ordre **1512**, 1995
32. Baker GL and Gollub JP. *Chaotic dynamics. An introduction*. Cambridge University Press, Cambridge 1996.
33. Bakker PG. *Bifurcations in Flow Patterns*. Kluwer, Dordrecht, 1991.
34. Ball J. Quart J Math **28**(2) (1997), 473–486.
35. Balmforth NJ. Ann Rev Fluid Mech **27** (1995), 335.
36. Bardos C. From molecules to turbulence-an overview of multiscale analysis in fluid dynamics. In: *Advanced topics in theoretical fluid mechanics*, Malek J, Necas J and Rokyta M (eds), 1–88. Addison Wesley Longman Limited 1998.
37. Bardos C, Golse F, and Levermore D. J Statistic Phys **63** (1991), 323–344.
38. Bardos C, Golse F, and Levermore D. Commun Pure Appl Math **46** (1993), 667–753.
39. Barenblatt GI, Iooss G and Joseph DD. *Nonlinear Dynamics and Turbulence*. Pitman Adv Publ Program, **16**. Pitman, London, 1983.
40. Barnes DR and Kerswell RR. J Fluid Mech **417** (2000), 103–126.
41. Bartucelli M, Doering CR, and Gibbon JD. Nonlinearity, **4**, (1991), 531–542.
42. Bartucelli M, Doering CR, Gibbon JD, and Malham SJA. Nonlinearity, **6**, (1993), 549–568.
43. Batchelor GK. Q J Math Appl Math **4** (1951), 29–41.
44. Batchelor GK. J Fluid Mech **1** (1956), 177–190.
45. Batchelor G. *The Life and Legacy of G.I. Taylor*. Cambridge Univ Press, 1996
46. Batchelor GK. *An Introduction to Fluid Dynamics*. Cambridge Univ Press, 2000.
47. Batchelor GK, Moffat K and Worster G (eds). *Perpspective in Fluid Dynamics*. Cambridge Univ Press, 2000.
48. Bayly BJ, Levermore CD and Passot T. Phys Fluids **A4** (1992), 945–954.
49. Bayly BJ, Orszag SA and Herbert Th. Ann Rev Fluid Mech **20** (1988), 359–391.
50. Beale JT. Comm Pure Appl Math **34** (1981), 359–392.
51. Beck C and Schlögl F. *Thermodynamics of chaotic systems. An introduction*. Cambridge University Press, Cambridge, 1995.

52. Beirào da Veiga H. J Math Pures Appl **64** (1985), 77–86.
53. Beirào da Veiga H. Indiana Univ Math J **36**(1) (1987a), 149–166.
54. Beirào da Veiga H. Comm Math Phys **109** (1987b), 229–248.
55. Beirào da Veiga H. Ann Inst Henri Poincaré **11**(3) (1994a), 297–311.
56. Beirào da Veiga H. Arch Rat Mech Anal **128** (1994b), 313–327.
57. Beirào da Veiga H. CR Acad Sci, Paris, Série I **321** (1995), 405–408.
58. Beirào da Veiga H. Diff Integr Eqs **10**(6) (1997a), 1149–1156.
59. Beirào da Veiga H. Portugaliae Math **54**(3) (1997b), 271–286.
60. Beirào da Veiga H. Portugaliae Math **54**(4) (1997c), 381–391.
61. Beirào da Veiga H. Concerning the regularity of the solutions to the Navier-Stokes equations via the truncation method, Par II. In: *Equations aux dérivées partielles et applications* (Articles dédiés à Jacques-Louis Lions), 127. Gauthier-Villars, Paris 1998.
62. Beirào da Veiga H and Secchi P. ARMA **98**(1) (1987), 65–69.
63. Bellomo N, Le Tallec P and Pertham B. Appl Mech Rev **48**(12) (1995), 777–794.
64. Bellomo N and Pulvirenti M. *Modeling in Applied Sciences: A Kinetic Theory Approach.* Birkhäuser, 2000.
65. Belov YuYa and Yanenko NN. Math Notes **10** (1971), 480–483.
66. Bénard H. Rev Gén Sci Pures Appl **11** (1900), 1261.
67. Bénard H. Ann Chim (Phys) **23** (1901), 62–144.
68. Ben-Artzi M. Arch Rat Mech Anal **128** (1994), 329–358.
69. Benjamin TB. J Fluid Mech **2** (1957), 554–574.
70. Benjamin TB. Math Proc Camb Phil Soc **79** (1976), 373–392.
71. Benjamin TB. Math Proc Cambridge Phil Soc **79** (1978a), 373–392.
72. Benjamin TB. Proc Roy Soc, London A**359** (1978b), 1.
73. Benney DJ. J Math Phys **45** (1966), 150.
74. Benney DJ and Gustavsson LH. Studies Appl Maths **64** (1981), 185–209.
75. Bentwich M. J Eng Math **19** 21–26 and **20** (1985), 97–111.
76. Bentwich M and Miloh T. J Fluid Mech **88** (1978), 17–32.
77. Bentwich M and Miloh T. J Eng Math **16** (1982), 1–21.
78. Bentwich M and Miloh T. J Eng Math **18** (1984), 1–6.
79. Bernardi C, Metivet B and Pernaud-Thomas B. Math Modelling Num Analysis **29**(7) (1995), 871–921.
80. Bertelsen AF. J Fluid Mech **64** (1974), 589.
81. Bertelsen AF, Svardal A and Tjotta S. J Fluid Mech **59** (1973), 493.
82. Bestman AR. J Appl Maths Phys (ZAMP) **34** (1983), 867–885.
83. Biagioni H and Gramchev T. Mat Contemp **10** (1996), 1–20.
84. Bickley WG. Phil Mag **23** (1937), 727.
85. Binnie AM. J Fluid Mech **2** (1957), 551–553.
86. Birikh RV, Briskman VA, Velarde MG and Legros JC. *Liquid Interfacial Systems: Oscillations and Instability.* Marcel Dekker, New York 2002.
87. Birkhoff G. *Hydrodynamics. A Study in logic, fact and similitude.* Academic Press, New York, 1960.
88. Blasius H. Z Math Phys **56** (1908), 1–37.
89. Blackadar AK. *Turbulence and diffusion in the atmosphare.* Springer, Berlin, 1997.
90. Block MJ. Nature **178** (1956), 650–651.
91. Blokhintsev. *Acoustics of a non homogeneous moving medium.* Naouka, Moscow, Russian ed, 1981.

92. Bodenschatz E, Pesch W and Ahlers G. Ann Rev Fluid Mech **32** (2000), 709–778.
93. Boeck T and Thess A. J Fluid Mech **399** (1999), 251–275.
94. Bois PA, Dériat E, Gatignol R and Rigolot A (eds). *Asymptotic Modelling in Fluid Mechanics*. Lecture Notes in Physics, **442**, 1995.
95. Borchers W and Miyakawa T. Arch Rat Mech Anal **118** (1992), 273–295.
96. Boris AYu and Fridlender OG. Slow motions of a gas near a strongly heated or cooled sphere. Izv AN SSSR, Mekh Zhidk i Gaza, **6**, 170, (1981), in Russian.
97. Bouard R and Contanceau M. J Fluid Mech **101** (1980), 583–607.
98. Boussinesq J. *Théorie Analytique de la Chaleur*, Vol 2. Gauthier-Villars, Paris, 1903.
99. Bouthier M. J Mécanique **11** (1972), 599.
100. Bouthier M. J Mécanique **12** (1973), 76–95.
101. Bowles RI. Phil Trans Roy Soc, London A**358** (2000), 245–260.
102. Brady JF and Durlofsky L. J Fluid Mech **175** (1987), 363–394.
103. Bragard J and Velarde MG. J Non-Equilib Thermodyn **22** (1997), 1–19.
104. Brazier JPh, Aupoix B and Cousteix J. Second-order effects in hypersonic laminar boundary layers. ONERA-CERT, Note DERAT, Toulouse, 1991.
105. Brazier JPh, Aupoix B and Cousteix J. AIAA J **30**(5) (1992), 1252–1259.
106. Breach DR. J Fluid Mech **10** (1961), 306–314.
107. Breuer KS et al. J Fluid Mech **340** (1997), 395–411.
108. Brevdo L. ZAMM **75**(5) (1995), 371–378.
109. Brezis H. Arch Rat Mech Anal **128** (1994), 359–360.
110. Brons M. Phys Fluids **6**(8) (1994), 2730–2737.
111. Brooker AM et al. J Fluid Mech **352** (1997), 265–281.
112. Brown SN and Stewartson K. Ann Rev Fluid Mech **1** (1969), 45–72.
113. Brown SN and Stewartson K. J Fluid Mech **42**(3) (1970), 561–564.
114. Bruneau C-H and Fabrie P. Math Anal Numér **30**(7) (1996), 815–840.
115. Buldakov EV, Chernyshenko SL and Ruban AI. J Fluid Mech **411** (2000), 213–232.
116. Burelbach JP, Bankoff SG and Davis SH. J Fluid Mech **195** (1988), 463.
117. Bunyakin AV, Chernyshenko SI and Stepanov GYu. J Fluid Mech **358** (1998), 283–297.
118. Bush WB. J Fluid Mech **25** (1966), 51–64.
119. Busse FH. In: *Hydrodynamic Instabilities and the Transition to Turbulence*. Swinney HL and Gollub JP (eds). Springer, Berlin, 1981.
120. Caffarelli L, Kohn R and Nirenberg L. Com Pure Appl Math **35** (1982), 771–831.
121. Caflisch R and Sammartino M. CR Acad Paris, Série I **324** (1997), 861–866.
122. Caflisch R, Ercolani N and Steele G. Selecta Math (N.S.) **2**(3) (1996), 369–414.
123. Caltagirone J-P and Vincent S. CR Acad Sci, Paris, Série IIb **329** (2001), 607–613.
124. Cannone M and Planchon F. Comm Partial Diff Eqs **21** (1996), 179–193.
125. Cannone M. In: *Advances in Mathematical Fluid Mechanics*, 1–34. Springer, Berlin, 2000.
126. Cassel KW. J Fluid Mech **428** (2001), 107–131.
127. Cassel KW, Ruban AI and Walker JDA. J Fluid Mech **321** (1996), 189–216.
128. Cassel KW, Smith FT and Walker JDA. J Fluid Mech **315** (1996), 233–256.

129. Cauchy A. Ex. de Math. 3 = Oeuvres (2) **8**, 195–226, 227–252 and 253–277, (1828).
130. Cercignani C. *Theory and Application of the Boltzmann equation,* Scottish Academic Press, 1975.
131. Cercignani C. *Rarefied Gas Dynamics: From basic concepts to actual calculations.* Cambridge texts in applied mathematics, Cambridge Univ Press, Cambridge, 2000.
132. Cercignani C and Sattinger DH. *Scaling Limits and Model in Physical Processes.* DMV Seminar Band 28. Birkhäuser, Basel, 1998.
133. Chacon Rebollon T and Guillén Gonzalez F. CR Acad Scis, Paris, Série I **330** (2000), 841–846.
134. Chae Dongo and Nam Hee-Seok. Proc Roy Soc, Edinburgh Sect A**127**(5) (1997), 935–946.
135. Chandra K. Proc Roy Soc, London A**161** (1938), 231–242.
136. Chandrasekhar S. Proc Camb Phil Soc **51** (1955), 162–178.
137. Chandrasekhar S. *Hydrodynamic and Hydromagnetic Stability.* Oxford at the Clarendon Press, 1961, and see also Dover publications, New York, 1981.
138. Chang EJ and Maxey MR. J Fluid Mech **303** (1995), 133–153.
139. Chang H-Ch. Ann. Rev. Fluid Mech **26** (1994), 103–136.
140. Chang H-Ch and Demekhin EA. Advances in Appl Mech **32** (1996), 1–58.
141. Chang H-Ch, Demekhin EA and Kopelevich DI. J Fluid Mech **250** (1993), 433–480.
142. Chapman S and Cowling TG. *The Mathematical Theory of Nonuniform Gases,* Cambridge University Press, Cambridge, 1952.
143. Charki Z. J Math Sci Univ Tokyo **1** (1994), 435–459.
144. Charki Z. ZAMM **75**(12) (1995), 909–915.
145. Charki Z. Math Models Methods Appl Sci **6**(2) (1996), 269–277.
146. Charki Z and Zeytounian RKh. Intern J Engng Sci **32**(10) (1994), 1561–1566.
147. Charki Z and Zeytounian RKh. Intern J Engng Sci **33**(12) (1995), 1839–1847.
148. Charney JG. In: *Proc Int Sympos on Numerical Wearther Prediction in Tokyo.* Meteor Soc Japan, 1962, 82–111.
149. Cheatham S and Matalon M. J Fluid Mech **414** (2000), 105–144.
150. Chemin J-Y. Asymptotic Analysis **3** (1990), 215–220.
151. Chemin J-Y. SIAM J Math Anal **23**(1) (1992), 20–28.
152. Chemin J-Y. About Navier-Stokes system. Publ Lab Num, Univ Paris 6, **R 96023** 1–43, (1996).
153. Chemin J-Y. Journ d'Anal Math **77** (1997), 27–50.
154. Chemin JY. Jean Leray et Navier-Stokes. In: *Jean Leray 1906–1998,* 71–82, No. spécial de la Gazette des Mathématiciens, SMF, Paris, 2000.
155. Cheng He. J Math Anal Appl **210**(2) (1997), 512–430.
156. Cheng HK and Lee CJ. In: *Numerical and Physical Asoects of Aerodynamics Flows III,* (ed) Cebeci T, 102–125, Springer, New York, 1986.
157. Cheng HK and Smith FT. ZAMP **33** (1982), 151–180.
158. Cheng AH-D, Yang CY, Hackl K and Chajes MJ. Intern J Non-Linear Mech **28**(5) (1993), 549–565.
159. Chhabra RP and de Kee D (eds). *Transport processes in bulbes, drops and particles.* Hemisphere, New York, 1992.
160. Cho Chung Ki and Choe Hi Jun. Lecture Notes Ser **38** 73–81, Seoul Nat Univ. Seoul, (1997).

161. Choi Y-H and Merkle CL. In: *Proc 10th Intern Conf on Numer Methods in Fluid Dynamics*, 169–173. Beijing (1986).
162. Chossat P and Iooss G. Japan J Appl Math **2** (1985), 37–68.
163. Chossat P and Iooss G. *The Couette-Taylor problem*, AMS 102. Springer, 1994.
164. Chossat P, Demay Y and Iooss G. ARMA **99** (1987), 213–248.
165. Chu CK. Adv Appl Mech **18** (1978), 285–331.
166. Cloot A and Lebon G. J Fluid Mech **145** (1984), 447–469.
167. Cognet G. J Th Appl Mech, **Special No. 1984** (1985), 7–44.
168. Coifman R, Lions P-L, Meyer Y and Semmes S. J Math Pures Appl **72** (1993), 247-286.
169. Coimbra CFM and Rangel RH. J Fluid Mech **370** (1998), 53–72.
170. Coles D. J Fluid Mech **21** (1965), 385–425.
171. Colinet P, Legros JC and Velard MG. *Nonlinear Dynamics of Surface-Tension-Driven Instabilities*. Wiley-VCH, Berlin, 2001.
172. Collins WM and Dennis SCR. Q J Mech Appl Math **26** (1973a), 53.
173. Collins WM and Dennis SCR. J Fluid Mech **60** (1973b), 105.
174. Constantin P. Comm Math Phys **104** (1986), 311–326.
175. Constantin P. Notices Amer Math Soc **42**(6) (1995a), 658–663.
176. Constantin P. Some mathematical problems of fluid mechanics. In: *Proc of the Intern Congr of Mathematicians*. (Zürich, 1994) 1086–1095. Birkhäuser, Basel, (1995b).
177. Constantin P and Foias C. Comm Pure Appl Math **38** (1985), 1–27.
178. Constantin P and Foias C. *Navier-Stokes equations*, University of Chicago Press, Chicago, 1988.
179. Constantin P and Wu J. Inviscid limit for vortex patches. Institut Mittag Leffler **7** (1994).
180. Constantin P and Wu J. Indiana Univ Math J **45** (1996), 67–81.
181. Constantin P, Foias C and Temam R. *Attractors representing turbulent flows.* AMS **53** (1985) p. 67.
182. Constantin P, Foias C and Temam R. Physica D **30** (1988), 284–296.
183. Constantin P, Foias C, Manley O and Temam R. J Fluid Mech **150** (1985), 427–440.
184. Constantin P, Foias C, Nicolaenko B and Temam R. *Integral manifolds and inertial manifolds for dissipative partial differential equations.* AMS **70**. Springer, New York, Berlin, 1989.
185. Contantinescu VN. Rev Roumaine Sci Techn, Méc Appl **31**(1) (1986), 109–110.
186. Cook AW and Dimotakis PE. J Fluid Mech **443** (2001), 69–99.
187. Cottet GH. Two Dimensional Incompressible Fluid Flow with Singular Initial Data. In: *Mathematical Topics in Fluid Mechanics,* Rodrigues JF and Sequeira (eds), 32–49. Longman Scientific and Technical, New York, 1992.
188. Couette M. Ann Chim Phys **21** (1890), 433–510.
189. Coullet PH and Huerre P. Physica D **23** (1986), 27–44 and "errata", (1986).
190. Coullet PH and Spiegel EA. SIAM J Appl Math **43** (1983), 776–821.
191. Cousteix J. Ann Rev Fluid Mech **18** (1986), 173–196.
192. Cousteix J, Roget C, Brazier J-Ph and Mauss J. *Variations around the triple deck theory.* California State University, Long Beach, Aerospace Engineering Department. Paper dedicated to Cebeci T, 2000.

193. Cousteix J, Brazier J-Ph and Mauss J. CR Acad Sci, Paris **329**(IIb) (2001), 213–219.
194. Coutanceau M and Ménard Ch. J Fluid Mech **158** (1985), 399–446.
195. Cowley SJ. J Fluid Mech **135** (1983), 389–405.
196. Cowley SJ and Davis SH. J Fluid Mech **135** (1983), 175–188.
197. Cowley SJ and Van Dommelen LL and Lam ST. Phil Trans Roy Soc London A**333** (1990), 343–378.
198. Cowley SJ and Wu X. AGARD Rep **793**(3) (1994), 1–38.
199. Crane LJ. J Engin Maths **10**(2) (1976), 115–124.
200. Cross MC Phys Fluids **23**(9) (1980), 1727–1731.
201. Cross MC Phys Rev A**25** (1982), 1065.
202. Cross MC and Hohenberg PC. Rev Mod Phys **65** (1993), 851.
203. Crow SC. Stud Appl Math **49** (1970), 21–44.
204. Curle N and Davies HJ. *Modern Fluid Dynamics;* Volume 1: *Incompressible Flow.* Van Nostrand Reinhold Company, London, 1968.
205. Curry JH et al. J Fluid Mech **147** (1984), 1–38.
206. Curtiss CF. In: *High Speed Aerodynamics and Jet Propulsion,* Vol 1: Rossini F (ed). 1955.
207. Dauby PC and Lebon G. J Fluid Mech **329** (1996), 25–64.
208. Daveney RL and Keen L (eds). Proc Sympos Appl Math, 39, 1989.
209. Davey A. The growth of Taylor vortices in flow between rotating cylinders. J Fluid Mech **14** (1962), 336–368.
210. Davey A and Stewartson K. Proc Roy Soc, London A**338** (1974), 101–110.
211. Davey A, DiPrima RC and Stuart JT. J Fluid Mech **31** (1968), 17–52.
212. Davey A, Hocking LM and Stewartson K. J Fluid Mech **63** (1974), 529–536.
213. Davidson BJ and Riley N. J Fluid Mech **53** (1972), 287.
214. Davis H T (ed). *Statistical mechanics of phases, interfaces, and thin films.* Wiley-VCH New York, 1996.
215. Davis SH. Ann Rev Fluid Mech **8** (1976), 57–74.
216. Davis SH. Ann Rev Fluid Mech **19** (1987), 403–435.
217. Davis SH and Homsy G. J Fluid Mech **98** (1980), 527.
218. Davis SH and Lumley JL (eds). *Frontiers in Fluid Mechanics.* Springer, Berlin, 1985.
219. Dawes JHP. J Fluid Mech **428** (2001), 61–80.
220. Debnath L. Bifurcation and nonlinear instability in applied mathematics. In: *Nonlinear Analysis,* 161–285, Rassias ThM (ed). World Scientific Publ Co, Singapore, 1987.
221. De Coninck F. CR Acad Sci, Paris A**288** (1979), 665.
222. De Coninck F, Guiraud JP and Zeytounian RKh. Q J Mech Appl Math **36**(1) (1983), 1–18.
223. Degani AT, Li Q and Walker JDA. Phys Fluids **8**(3) (1998), 704–714.
224. Degani AT, Walker JDA and Smith FT. J Fluid Mech **375** (1998), 1–38.
225. De Gennes P-G, Brochard-Wyart F and Quéré D. *Gouttes, Bulles, Perles et Ondes.* Editions Belin, Paris, 2002.
226. Degond P and Lemou M. Eur J Mech, B **20** (2001), 303–327.
227. De Giorgi E. Mem Acad Sc Torino, cl Sci Fis Mat **3**(3) (1957), 25–43.
228. De Groot SR. *Thermodynamics of Irreversible Processes.* New York, 1951.
229. Delgado-Buscalioni R, Crespo del Arco E and Bontoux P. Eur J Mech, B **20** (2001), 657–672.

230. Delort M. J Amer Math Soc **4** (1991), 553–586.
231. Demay Y and Iooss G. Journal de Mécanique Théorique et Appliquée, **Special No. 1984** /1985), 193–216.
232. Demekhin EA and Shkadov VYa. Izv Akad Naouk USSR, Mekhanika Zhidkosti i Gaza **4** (1981), 9.
233. Demekhin EA, Tokarev GYu and Shkadov VYa. Physica D**52** (1991), 338–361.
234. Denbigh KG. *The Thermodynamics of the Steady State*. London, 1951.
235. Deriat E and Guiraud J-P. On the asymptotic description of turbulent boundary layers. Journal de Mécanique Théorique et Appliquée, **Special No. 1986** (1986), 109–140.
236. Desjardins B. Arch Rat Math Anal **137** (1997a), 135–158.
237. Desjardins B. Diff Integr Equations, 587–598, (1997b).
238. Desjardins B, Grenier E, Lions PL and Masmoudi N. J Maths Pures Appl **78**(5) (1999), 461–471.
239. Dijkstra D and Zandbergen PJ. J Engng Math **11** (1977), 167–188.
240. Dimitrakopoulos P and Higdon JJL. J Fluid Mech **336** (1997), 351–378.
241. Dimitrakopoulos P and Higdon JJL. J Fluid Mech **377** (1998), 189–222.
242. DiPerna R. (a) Comm Math Phys **91** (1983), 1–30, and (b) Arch Rat Math Anal **82** (1983), 27–70.
243. DiPerna R and Lions PL. In: *Séminaire EDP 1988–1989*, Ecole Polytechnique, Palaiseau, 1989a.
244. DiPerna R and Lions PL. Invent Math **98** (1989b), 511–547.
245. DiPerna R and Majda A. Comm Pure Appl Math **40** (1987), 301–345.
246. DiPrima RC, Eckhaus W and Segel A. J Fluid Mech **49** (1971), 505–744.
247. DiPrima RC and Grannick RC. A nonlinear investigation of the stability of flow between counter-rotating cylinders. IUTAM Symp on instability of continuous systems. Leipholtz (ct.), Springer, 1971.
248. DiPrima RC and Habetler GJ. ARMA **34** (1969), 218.
249. Dobrokhotov SYu and Shafarevich AI. Fluid Dynamics **31**(4) (1996), 511–514.
250. Doering CR and Gibbon JD. *Applied analysis of the Navier-Stokes equations*. Cambridge University Press, Cambridge, 1995.
251. Dopazo C, Lozano A and Barreras F. Phys Fluids **12**(8) (2000), 1928–1931.
252. Drazin PG. Tellus **13** (1961), 239–251.
253. Drazin PG and Reid WH. *Hydrodynamic Stability*. Cambridge University Press, Cambridge, 1981.
254. Dubois S. CR Acad Sc, Paris, **335**(I) (2002), 27–32.
255. Duchon J and Robert R. CR Acad Sci, Paris **329**(I) (1999), 243–248.
256. Duck PW and Dry SL. J Fluid Mech **441** (2001), 31–65.
257. Duck PW and Smith FT. J Fluid Mech **91** (1979), 93–110.
258. Duck PW, Stow SR and Dhanak MR. J Fluid Mech **400** (1999), 125.
259. Duff RE, Harlow FH and Hurt CW. Phys Fluids **5** (1962) 417–425.
260. Duhem PMM. In: Ann Toulouse **2** (1901/1902).
261. Duhem PMM. *Traité d'énergétique ou de thermodynamique générale*. t. 2, Gauthier-Villars, Paris, 1911.
262. Durban D and Pearson JRA (eds). *IUTAM Symposium on Non-Linear Singularities in Deformation and Flow*. Kluwer, 1999.
263. Dütsch et al. J Fluid Mech **360** (1998), 249–271.

264. Dwyer HA and Yam C. In: *Proced Intern Symp on Comput Fluid Dynamics*, 46–51. Nagoya, 1989.
265. Eagles PM. J Fluid Mech **49** (1971), 529–550.
266. Ebin DG. Comm Pure and Appl Math **35** (1982), 451.
267. Ece MC, Walker JDA and Doligalski TL. Phys Fluids **27** (1984) 1077–1089.
268. Eckhaus W. *Studies in Non-linear Stability Theory*. Springer, Berlin, 1965.
269. Eckmann JP. Rev Modern Phys **53**(4) (1981), 643–654.
270. Eichelbrenner EA. La couche limite laminaire à trois dimensions. Publications Scientifiques et Techniques du Ministère de l'Air. Note Technique **85**(Nov), 55–84, (1959).
271. Eichelbrenner EA (ed). *Recent research on unsteady boundary layers*. IUTAM Symposium 1971, Part 1 and 2. Les Presses de l'Université Laval, Québec, 1972.
272. Eichelbrenner EA. Ann Rev Fluid Mech **5** (1973), 339–360.
273. Eichelbrenner EA and Askovic. Journal de Mécanique **8**(3) (1967), 353–370.
274. Ekman VW. Ark Mat Astron Fys **2** (1905), 1–52.
275. El Hafi M. *Analyse asymptotique et raccordements, étude d'une couche limite de convection naturelle*. PhD thesis, Université Paul Sabatier, Toulouse, France, 1994.
276. Elizarova TG and Chetverushkin BN. In: *Mathematical Modelling Processes in Non-Linear Media*, Nauka, Moscow, 261–278, 1986.
277. Ellenrieder KD von and Cantwell BJ. J Fluid Mech **423** (2000), 293–315.
278. Elliott JW and Smith FT. J Fluid Mech **179** (1987), 489.
279. Emanuel G. Phys Fluids A**5**(2) (1993) 294–298.
280. Embid P. Comm PDE **12** (1987), 1227–1284.
281. Emisle AG, Bonner FJ and Peck LG. J Appl Phys **29** (1958), 858.
282. Eppler R and Fasel H (eds). *Laminar-Turbulent Transition*. Springer, Berlin 1980.
283. Ericksen JL. J Washington Acad Sci **44** (1954), 33.
284. Ericksen JL and Rivlin RS. J Rational Mech Anal **4** (1955), 323–425.
285. Erneux T and Davis SH. Phys Fluids A**5**(5) (1993), 1117–1122.
286. Errafyi M, and Zeytounian RKh. Intern J Engng Sci **29**(5) (1991a), 625–635.
287. Errafyi M, and Zeytounian RKh. Intern J Engng Sci **29**(11) (1991b), 1363–1373.
288. Esposito R, Marra R, Pulvirenti M and Sciarretta C. Commun Partial Differential Equations, **13**(12) (1988), 1601.
289. Exner A and Kluwick A. Phys Fluids **13**(6) (2001), 1691–1703.
290. Fabre A, Guitton H, Guitton J, Lichnerowicz A and Wolff E. *Chaos and Determinism*. Johns Hopkins Press, 1995.
291. Feigenbaum MJ. J Statist Phys **19** (1978), 25–52.
292. Feigenbaum MJ. Los Alamos Science **1**(2) (1980), 4–27. Reprinted in Phys D**7** (1983), 16–39.
293. Feireisl E. CR Acad Sci, Paris **331**(I) (2000a), 35–39.
294. Feireisl E. In: *Advances in Mathematical Fluid Mechanics*, 35–66. Springer, Berlin 2000b.
295. Feireisl E and Petzeltova H. Namuscripta Math **97** (1998), 109–116.
296. Feireisl E and Petzeltova H. ARMA **150** (1990), 77–96.
297. Feireisl E et al. ARMA **149** (1990), 69–96.

298. Feistauer M. *Mathematical Methods in Fluid Dynamics*. Longman Scientific and Technical, New York, 1993.
299. Fenstermacher P, Swinney H and Gollub J. J Fluid Mech **94** (1979), 103–128.
300. Feynman RP and Lagerstrom PA. In: *Proc IX International Congress on Applied Mechanics*, Vol 3, 342–343, Brussels, 1956.
301. Finn R. Arch Rat Mech Anal **19**(5) (1965), 363–406.
302. Finn R. and Smith DR. Arch Rat Mech Anal **25** (1967), 26–39.
303. Flori F and Orenga P. In: *Trends in Applications of Maths to Mechanics*, 293–305. Chapman and Hall/CRC, Boca Raton, London, 2000.
304. Foias C. In: *Approximation Methods for Navier-Stokes Problems*, Lecture Notes in Maths, Vol **771**, p. 196, Dodd A and Eckmann B (eds). Springer, Berlin, 1979.
305. Foias C and Saut JC. In: *Proceeding of Symposia in Pure Maths*, Vol **45**, Part I, 439–448, (1986).
306. Foias C and Temam R. J Math Pures Appl **58** (1979), 339–368.
307. Foias C and Temam R. The connection between the Navier-Stokes equations, dynamical systems, and turbulence theory. Direction in partial differential equations (Madison, WI, 1985), 55–73. Publ Math Res Center Univ Wisconsin, **54**, Academic Press, Boston, MA, 1987.
308. Foias C and Temam R. CR Acad Sci, Paris **307**(I) (1988), 5–8 and 67–70.
309. Foias C and Temam R. SIAM J Num Anal **26** (1989), 1139–1157.
310. Foias C , Manley O and Temam R. CR Acad Sci, Paris **305**(I) (1987), 497–500.
311. Foias C, Sell G and Temam R. J Diff Equations **73** (1988), 309–353.
312. Foias C, Sell GR and Titi ES. J Dynamics Equations **1**(2) (1989), 199–244.
313. Foias C , Manley O, Rosa R and Temam R. *Navier-Stokes Equations and Turbulence*. Cambridge Univ Press, Cambridge, 2001a.
314. Foias C , Manley O, Rosa R MS and Temam R. CR Acad Sc, Paris **333**(I) (2001b), 499–504.
315. Foias C, Manley O, Temam R and Treve YM. Physica D**9** (1983), 157–188.
316. Fourier J. Mémoire d'analyse sur le mouvement de la chaleur dans les fluides, *Mémoires de l'Académie Royale des Sciences de l'Institut de France (2)*, **12**, 507–530, (1833) = (with editorial changes), pp. 595–616 of *Oeuvres de Fourier*, Tome **2**, (1890).
317. Franchi F and Straughan B. Int J Engng Sci **30** (1992), 739–745.
318. Frank Gerd and Meyer-Spashe Rita. Computation of transitions in Taylor vortex flows Z angew Math und Phys **32**(6) (1981), 710–720.
319. Frehse J and Rùzicka M. Arch Rat Mech Anal **128** (1994), 361–380 and Ann Sci Norm Sup Pisa **21** (1994), 63–95.
320. Frehse J and Rùzicka M. In Acta Appl Math (1995).
321. Frisch U. *Turbulence: The Legacy of A.N. Kolmogorov*. Cambridge Univ Press, 1995.
322. Frisch U and Orszag S. Physics Today **24** (1990).
323. Fröhlich J and Peyret R. Intern J Num Meth Heat Fluid Flow **2** (1992), 195–213.
324. Fujita H and Kato T. ARMA **16** (1964), 269–315.
325. Fujita H and Morimoto H. A remark on the existence of the Navier-Stokes flow with non-vanishing outflow condition. In: *Nonlinear waves*, Agemi R et al. (eds) Proc 4th MSJ Int Res Institute, Sapporo, Japan, July 10–21,

1995. Tokyo, Gakkōtosho. GAKUTO Int Ser Math Sci Appl, **10**, 53–61, (1997).
326. Gadd GE. J Aeronaut Sci **24** (1957), 754.
327. Galdi GP. Ricerche Mat **27** (1978), 387–404.
328. Galdi GP. In: *Recent Developments in Theoretical Fluid Mechanics*, Galdi GP, and Necas J, (eds) Longman, 1993.
329. Galdi GP. *An Introduction to the Mathematical Theory of the Navier-Stokes equations: Linearized Steady Problems*, Springer, Berlin 1994a (Revised edition, 1998a).
330. Galdi GP. *An Introduction to the Mathematical Theory of the Navier-Stokes equations : Nonlinear Steady Problems*, Springer, Berlin 1994b (Revised edition, 1998b).
331. Galdi GP, and Necas J (eds). *Recent Developments in Theoretical Fluid Mechanics*, Longman, 1993.
332. Galdi GP and Padula M. ARMA **110** (1990), 187–286.
333. Galdi GP and Rionero S. Weighted energy methods in fluid dynamics and elasticity. In: Lecture Notes in Mathematics, Vol **1134**, Springer, New York, 1985.
334. Galdi GP, and Sohr H. Arch Rat Mech Anal **131** (1995), 101–119.
335. Galdi GP, Heywood JG and Shibata Y. On the global existence and convergence to steady state of Navier-Stokes flow past an obstacle that is started from rest, Arch Rat Mech Anal **138** (1997), 307–318.
336. Galdi GP, Màlek J, and Necàs J (eds). *Progress in theoretical and computational fluid mechanics,* Longman Scientific and Technical, New York, 1994.
337. Galdi GP, Màlek J, and Necàs J (eds). *Mathematical Theory in Fluid Mechanics.* Longman, 1997.
338. Galdi GP and Novotny A and Padula M. Pacific J Math **179**(65) (1997), 100.
339. Galdi GP, Padula M and Solonnikov VA. Indiana Univ Math J **45**(4) (1996), 961–995.
340. Gallagher I, Ibrahim S and Majdoub M. CR Acad Sci, Paris **330**(I) (2000), 791–794.
341. Gallagher I and Planchon F. On infinite energy solutions to the Navier-Stokes equations: global 2D existence and 3D weak-strong uniqueness. ARMA, (2001).
342. Gans RF. J Appl Mech **50** (1983), 251–254.
343. Gaponov-Grekhov AV and Rabinovich MI. *Nonlinearities in Action.* Springer, Berlin, 1992.
344. Gaster M. J Fluid Mech **66** (1974), 465–480.
345. Gatignol R and Prud'homme R. *Mechanical and Thermodynamical Modeling of Fluid Interfaces.* World Scientific, Singapore, 2001.
346. Geratz KJ et al. Multiple Pressure Variable (MPV) Approach for Low Mach Number Flows Based on Asymptotic Analysis. In: *Flow Simulation with High-Performance Computers II,* Hirschel EH (ed), 340–354. Vieweg, Braunschweig, 1996.
347. Germain P. Adv Appl Mech **12** (1972), 131–194.
348. Germain P. Entropie **55** (1974), 7–14.
349. Germain P. The "New" Mechanics of Fluids of Ludwig Prandtl. In: *Ludwig Prandtl, ein Führer in der Strömungslehre,* G Meier (ed), 31–40. Vieweg, Braunschweig, 2000.

350. Germain P and Guiraud JP. J Math Pures Appl **45** (1966), 311–358.
351. Germond J-L. CR Acad Sci, Paris **325**(I) (1997), 1239–1332.
352. Gershuni GZ and Zhukhovitskii EM. *Convective Stability of Incompressible fluids*. Tranlated from the Russian by Louvish D. Keter Publ, Jerusalem, 1976.
353. Gersten K. ZAMM **53** (1973), T99–T101.
354. Gersten K and Gross JF. Int J Heat Mass Trans **16** (1973), 2241–2260.
355. Gersten K and Herwig H. *Strömungsmechanik*. Vieweg, Wiesbaden, 1992.
356. Getling AV. *Rayleigh-Bénard Convection: Structures and Dynamics*. World Scientific, Singapore 1998.
357. Geymonat G et Lions J-L. Rôle des instruments mathématiques et numériques dans la modélisation. Matapli, SMAI' **65** 51–62, Paris, avril 2002.
358. Ghidaglia JM. Nonlinear Analysis: TMA **10**(8) (1986), 777–790.
359. Ghidaglia JM and Temam R. Lower bound on the dimension of the attractor for the Navier-Stokes equations in space dimension 3. In: *Mechanics, Analysis and Geometry:* Francaviglia M et al. (eds). North-Holland, (1990).
360. Ghosh S and Matthaeus WH. Phys Fluids **A4** (1992), 148–164.
361. Gibbs J.W. Trans Conn Acad **3** (1875), 108–248 and 343–524; see also Scientific Papers, Vol 1, Dover Publications, New York, 1961.
362. Giga Y. Regularity criteria for weak solutions of the Navier-Stokes system. In: *Proceeding of Symposia in Pure Mathematics,* Vol **45**, Part I, 449, 1986.
363. Giga Y and Miyakawa T. Comm Partial Diff Eqs **14** (1989), 577–618.
364. Giga Y, Miyakawa T and Osada H. ARMA **104** (1988), 223–250.
365. Girault V and Sequeira A. ARMA **114** (1991), 313–333.
366. Girault V and Sequeira A. In: *Mathematical Topics in Fluid Mechanics,* 50–63, Rodrigues JF and Sequeira A (eds), Longman, New York, 1992.
367. Giroire J. *Etude de quelques problèmes aux limites extérieures et résolution par équations intégrales*. Thèse UPMC, Paris, 1987.
368. Glauert WB. Proc R Soc, London **A1242** (1957), 108–115.
369. Gledzer EB. Sov Phys JETP **64** (1986), 483–486.
370. Gledzer EB et al. J Theoretic Appl Mech **Special No. 1988**(2nd supplement) (1988), 111–130.
371. Glowinski R. *Numerical Methods for Non-linear Variational Problems*. Springer, New York, 1984.
372. Glushko AV. Computationl Maths Math Phys **38**(1) (1998), 137–145.
373. Glushko AV and Glushko YeG. Comp Maths Math Phys **35**(8) (1995), 1019–1024.
374. Godrèche C and Manneville P (eds). *Hydrodynamics and Nonlinear Instabilities*. Cambridge University Press, Cambridge, 1998.
375. Godts S and Zeytounian KKh. ZAMM **70**(1) (1990), 67–69.
376. Godts S and Zeytounian KKh. J Engineering Math **28** (1991), 93–98.
377. Godts S, Khiri R and Zeytounian RKh. *Isochoric wave motions: bifurcations and chaotic behavior*. Unpublished paper, University of Lille I, 1993.
378. Goldshtik MA. Ann Rev Fluid Mech **22** (1990), 441–472.
379. Goldstein S. Proc Camb Phil Soc **25** (1930), 1–30.
380. Goldstein S. Q J Mech Appl Math **1** (1948), 43–69.
381. Goldstein S. Tech Rep Eng Res Inst Univ Calif **HE** (1956), 150–154.
382. Goldstein S. *Lectures on Fluid Mechanics*. Wiley, New York, 1960.
383. Goldstein S. Ann Rev Fluid Mech **1** (1969), 1–28.

384. Gollub JP and Benson SV. J Fluid Mech **100** (1980), 449.
385. Gollub JP and Swinney HL. Phys Rev Letters **35** (1975), 927–930.
386. Golovin AA, Nepomnyaschy AA and Pismen LM. Phys Fluids **6**(1) (1994), 35–48.
387. Golse F and Levermore CD. CR Acad Sci, Paris **333**(I) (2001), 897–902.
388. Golse F and Saint-Raymond L. CR Acad Sci, Paris **333**(I) (2001), 145–150.
389. Graffi D. J Rat Mech Anal **2** (1953), 99.
390. Greenspan HP. *The Theory of rotating fluids*. Breukelen Press, Brookline, 1990.
391. Grenier E and Gues O. J Diff Equations **143**(1) (1998), 110–146.
392. Gresho PM. Advances Appl Mechanics **28** (1992), 45–140.
393. Guckenheimer J. Ann Rev Fluid Mech **18** (1986), 15–31.
394. Guckenheimer J and Holmes P. *Nonlinear Oscillations, Dynamical Systems, and Bifurcations of Vector Fields*. Springer, Berlin, 1983.
395. Gues O. Ann Inst Fourier **45** (1995), 973–1006.
396. Guiraud JP. J Mécanique **13**(3) (1974), 409–432.
397. Guiraud JP. Ann Phys France **5** (1980), 33–59.
398. Guiraud JP. In: *Asymptotic Modelling in Fluid Mechanics,* Lecture Notes in Physics, Vol 442, 257–307, Bois PA, Dériat E, Gatignol R, and Rigolot A, (eds), 1995.
399. Guiraud JP and Iooss G. CR Acad Sci, Paris A **266** (1968), 1283–1286.
400. Guiraud JP and Zeytounian RKh. J Mécanique **17**(3) (1978), 337–402.
401. Guiraud JP and Zeytounian RKh. J Mécanique **18**(3) (1979a), 423–431.
402. Guiraud JP and Zeytounian RKh. J Fluid Mech **90**(1) (1979b), 197–201.
403. Guiraud JP and Zeytounian RKh. Geophys Astr Fluid Dyn **15**(3/4) (1980), 283–295.
404. Guiraud JP and Zeytounian RKh. Tellus **34**(1) (1982), 50–54.
405. Guiraud JP (Giro ZH) and Zeytounian RKh. USSR Comput Maths Math Phys **24**(4) (1984a), 92–95.
406. Guiraud JP (Giro ZH) and Zeytounian RKh. USSR Comput Maths Math Phys **24**(4) (1984b), 191–193.
407. Guiraud JP and Zeytounian RKh (Ed). *Asymptotic modelling of fluid flows*. J. Th. Appl. Mech., **Special No. 1986** (1986a), 1.
408. Guiraud JP and Zeytounian RKh. CR Acad Sc, Paris **302**(II)(7) (1986b), 383–386.
409. Guo BL and Yuan GW. Acta Math Sinica (N.S) **12**(2) (1996), 205–216.
410. Guo LiHui. Neimenggu Daxue Xuebao Ziran Kexue **29**(2) (1997), 147–151.
411. Gustafsson B and Kreiss HO. J Comput Phys **30** (1979), 333–351.
412. Gustafsson B and Sundström A. SIAM J Appl Math **35** (1978), 343–357.
413. Guyon E, Hulin J-P, Petit L and Mitescu CD. *Physical Hydrodynamics*. Oxford University Press, Oxford, 2001.
414. Hagstrom T and Lorenz J. Siam Math Anal **29**(3) (1998), 652–672.
415. Haddon EW, and Riley N. Q J Mech Appl Maths **32** (1979), 265.
416. Hall P. J Fluid Mech **304** (1995), 185–212.
417. Hanouzet B. Rend Sem Univ Padova **XLVI** (1971), 222–272.
418. Happel J and Brenner H. *Low Reynolds number hydrodynamics*. Prentice Hall, Englewood Cliffs, 1965.
419. Harris SD, Ingham DB and Pop I. European J of Mech, B **20** (2001), 187–204.
420. He Cheng. J Math Anal Appl **209**(1) (1997), 228–242.

421. Helmholtz H. Monatsberichte der Königlich Preussischen Akademie der Wissenschaften zu Berlin, Berlin, 215–228, (1868).
422. Henkes and Veldman. J Fluid Mech **179** (1987), 513–529.
423. Herbert Th. Ann Rev Fluid Mech **20** (1988), 487–526.
424. Herbert Th. Ann Rev Fluid Mech **29** (1997), 245–283.
425. Hewit RE, Duck PW and Foster MR. J Fluid Mech **384** (1999), 339–374.
426. Heywood JG. Acta Math **136** (1976), 61–102.
427. Heywood JG. Annali Sc Norm Pisa **6**(IV) (1979), 427–444.
428. Heywood JG. Lecture Notes Maths **771** (1980a), 235–248.
429. Heywood JG. Indiana University Maths J **29**(5) (1980b), 639–680.
430. Heywood JG. Fluid Dynamics Transactions **10** (1980c), 177–203.
431. Heywood JG. Open problems in the theory of the Navier-Stokes equations for viscous incompressible flow. In: *Lecture Notes in Mathematics,* Vol **1431**, Springer, 1–22, 1989.
432. Heywood JG. Remarks on the Possible Global Regularity of Solutions of the Three-Dimensional Navier-Stokes Equations, In: *Progress in theoretical and computational fluid mechanics,* Galdi GP, Màlek J, and Necàs J, (eds), 1–32. Longman, New York, 1994.
433. Hide R and Titman CW. J Fluid Mech **29** (1967), 39–60.
434. Higgins BG. Phys Fluids **29**(11) (1986), 3522–3529.
435. Higuera FJ. Phys Fluids **9**(10) (1997), 2841–50.
436. Hishida T. ARMA **150** (1999), 307–348.
437. Hoff D. Ach Rat Math Anal **114** (1991), 15–46.
438. Hoff D. J Diff Eqs **120** (1995), 215–254.
439. Hoff D. Discontinuous solutions of the Navier-Stokes equations for multidimensional, isentropic flow. *Hyperbolic problems: theory, numerics, applications* (Stony Brook, NY, 1994), 148–157. World Sci Publishing, River Edge, NJ, 1996.
440. Hoff D. Ach Rat Math Anal **139** (1997), 303–354.
441. Hoffman GH. J de Mécanique **13** (1974), 433.
442. Hofman M. J Fluid Mech **252** (1993), 399–418.
443. Holtsmark J, Johnsen I, Sikkeland T and Skavlem S. J Acoust Soc Am **26** (1954), 26.
444. Hopf E. Comm Pure Appl Math **1** (1948), 303–322.
445. Hopf E. Math Nachr **4** (1951), 213–231.
446. Horibata Ya. Computers Fluids **21**(2) (1992), 185–200.
447. Howarth LN. Q J Mech Appl Math **4**(2) (1951), 157–165.
448. Howarth LN. Laminar Boundary Layer. In: *Handbuch der Physik,* Vol VIII/1, Függe, S (ed), 264–350. Springer, Berlin,1959.
449. Howarth LN. Convection at high Rayleigh numbers. In: *Proc 11th Int Cong Appl Mech,* Görtler H (ed), 1109–1115, Munich, 1964.
450. Howard LN and Krishnamurti R. J Fluid Mech **170** (1986), 385–410.
451. Hsieh DY and Ho SP. *Wave and Stability in fluids.* World Scientific, Singapore 1994.
452. Huerre P. Nonlinear instability of free shear layers. In: *Special Course on Stability and Transition of Laminar Flow.* AGARD Report **No 709**, 5–12; see also Comments on papers 1–5, Session I, by Guiraud J-P, C1-1–C1-2, 1984.

453. Huerre P. Spatio-temporal instabilities in closed and open flows. In: *Instabilities and Nonequilibrium Structures*, Tirapegui E and Villarroel (eds), 141–177. Reidel, Dordrecht, 1987.
454. Huerre P and Monkewitz PA. Ann Rev Fluid Mech **22** (1990), 473–537.
455. Hughes DW and Proctor MRE. Nonlinearity **3** (1990), 127–153.
456. Hupper HE. George Keith Batchelor. J Fluid Mech **421** (2001), 1–14.
457. Hunt JCR. J Fluid Mech **49** (1971), 159.
458. Hunt JCR et al. J Fluid Mech **436** (2001), 353–391.
459. Hyman JM, Nicolaenko B and Zaleski S. Physica D**23** (1986), 265.
460. Ibrahim MM. J Atmospheric Sci **44**(4) (1987), 706–720.
461. Ida MP and Miksis MJ. J Appl Math **58**(2) (1998a), 456–473.
462. Ida MP and Miksis MJ. J Appl Math **58**(2) (1998b), 474–500.
463. Iguchi T. Math Methods Appl Sci **20**(11) (1997), 945–958.
464. Imai I. Application of the M^2-expansion method to the subsonic flow of a compressible fluid past a parabolic cylinder. In: *Proc 1st Japan Nat Congr Appl Mech* 349–352, (1952).
465. Imai I. J Aeronaut Sci **24** (1957a), 155–156.
466. Imai I. Tech Note Inst Fluid Dynamics and Appl Math. Univ Maryland, No. **BN-104**, 1957b.
467. Ingham DB. Acta Mech **42** (1984), 111–122.
468. Iooss G. *Bifurcation of maps and applications.* North-Holland, Amsterdam, 1979.
469. Iooss G. Bifurcation and transition to turbulence in hydrodynamics. In: *CIME Session on Bifurcation Theory and Applications,* Salvadori ed, Lecture Notes in Math, Vol **1057**, 152–201, Springer, New York 1984.
470. Iooss G and Adelmeyer M. *Topics in Bifurcation Theory and applications.* World Scientific, Singapore, 1992.
471. Iooss G and Joseph DD. *Elementary Stability and Bifurcation Theory.* Springer, New York, 1980.
472. Iooss G and Peyret R (eds). *Bifurcations and chaotic behaviors.* J Th Appl Mech, **Special No. 1984**, Paris 1985.
473. Iooss G and Rossi M. Eur J Mech, B **8**(1) (1989).
474. Iooss G, Helleman RHG and Stora R (eds). *Comportements chaotiques des systémes déterministes. Les Houches, Session XXXVI, 1981.* North-Holland, 1983.
475. Iooss G, Guès O and Nouri A (eds). *Trends in Applications of Mathematics to Mechanics.* Chapman and Hall, London, 2000.
476. Itaya N. J Math Kyoto Univ **16** (1976), 413–427.
477. Itoh N. J Fluid Mech **317** (1996), 129–154.
478. Janzen O. Phys Z **14** (1913), 639.
479. Jarvis GT and McKenzie DP. J Fluid Mech **96**(3) (1980), 515–583.
480. Jeffery GB. Proc Roy Soc A**101** (1922), 169–174.
481. Jeffreys H. Proc Camb Phil Soc **26** (1930), 170–172.
482. Jenkins AD and Jacobs SJ. Phys Fluids **9**(5) (1997), 1256–1264.
483. Joo SW. J Fluid Mech **293** (1995), 127–145.
484. Joo SW, Davis SH and Bankoff SG. J Fluid Mech **230** (1991), 117–146.
485. Joseph DD. ARMA **20** (1965), 59.
486. Joseph DD. ARMA **22** (1966), 163.
487. Joseph DD. *Stability of fluid motions, Vol II* Springer, Heidelberg, 1976.

488. Joseph DD. Stability and Bifurcation Theory. In: [473]
489. Joseph DD and Renardy YuR. *Fundamentals of Two-Fluid Dynamics. Part I: Mathematical Theory and Applications.* Springer, New York, 1993.
490. Kachanov YS. Ann Rev Fluid Mech **26** (1994), 411–482.
491. Kadanoff LP. Roads to chaos. Phys Today. December, 46–53, (1983).
492. Kaniel S and Shinbrot M. Arch Rational Mech Anal **24** (1967), 303–324.
493. Kantorovich LV and Krylov VI. *Approximate Methods of Higher Analysis.* English transl, Noordhoff, Groningen, 1958.
494. Kapitza PL and Kapitza SP. J Exp Theor Phys **19**(2) (1949), 105–120, in Russian. See also: *Collected Papers of Kapitza PL,* Vol 2. Ter Haar D. (ed). Pergamon, London, 690–709, 1965.
495. Kaplun S. ZAMP **5** (1954), 111–135.
496. Kaplun S. J Math Mech **6** (1957), 595–603.
497. Kaplun S, and Lagerstrom PA. J Rat Mech Anal **6** (1957), 585-593.
498. Kato T. J Funct Anal **9** (1972), 296–309.
499. Kato T. *Quasi-Linear Equations of Evolution with Application to Partial Differential Equation,* Lecture Notes in Mathematics, Vol **448**, Springer, New York, 25–70, 1975.
500. Kato T. Math Z **187** (1984a), 471–480.
501. Kato T. Remarks on the zero viscosity limit for nonstationary Navier-Stokes flows with boundary. In: *Seminar on PDE* (Edited by Chern SS), Springer, New York, 1984b.
502. Kato T. Remarks on the Euler and Navier-Stokes equations in $\mathbf{R}^2$. In: *Proceeding of Symposia in Pure Mathematics,* Vol **45**, AMS, Providence, Rhode Island, 1986.
503. Kato H. J Math Anal Appl **208**(1) (1997), 141–157.
504. Kato T and Fujita H. Rend Sem Math Univ Padova **32** (1962), 243–260.
505. Kato T and Ponce G. Commun Pure Appl Math **41** (1988), 893–907.
506. Kashdan D et al. Phys Fluids **7**(11) (1995), 2679–2685.
507. Kazhikov AV. Acta Math Appl **37**(1) (1994), 77–81.
508. Kazhikov AV. In: *Progress in theoretical and computational fluid mechanics,* Galdi GP, Màlek J, and Necàs J (eds), Longman Scientific and Technical, New York, 33–72, 1994.
509. Kazhikov AV and Shelukhin V. J Appl Math Mech (PMM) **41** (1977), 273–283.
510. Keller JB and Ward MJ. J Engng Math **30** (1996), 253–265.
511. Kelly RE. Quart J Mech Applied Math **4** (1966), 473–484.
512. Kelly RE. Adv in Appl Mech **31** (1994), 35–112.
513. Kerr R. Phys Fluids A**5**(7) (1993), 1725–1746.
514. Kevorkian J, and Cole JD. *Multi Scale and Singular Perturbation Methods.* AMS Vol **114**, Springer, New York (1996).
515. Kharab R. Int J Num Methods Fluids **24**(11) (1997), 1211–1223.
516. Khatskevich VL. Doklady Maths **53** (1996), 191–193.
517. Khiri R. *Waves, vortices and chaos in the isochoric motions.* Doctoral thesis in mechanics, Univesity of Lille I, LML, 1992. In French.
518. Kibel IA. Izv Akad Nauk SSSR, Ser Geogr Geofiz **5** (1940), 627–638.
519. Kibel IA. *An Introduction to the Hydrodynamical Methods of short Period Weather Forecasting.* Macmillan, London, 1963; English Translation from Russian edition, Nauka, Moskva, 1957.

520. Kim S-C. SIAM Appl Math **58**(5) (1998), 1394–1413.
521. Kirchgässner K and Sorger P. Quart J Mech Appl Math **22** (1969), 183.
522. Kirchhoff G. Ann Phys **134** (1868), 177–193.
523. Kirchhoff G. J Reine angew Math **70** (1869), 289–298.
524. Kiselev AA and Ladyzhenskaya OA. Izv Akad Nauk SSSR, ser Math **21** (1957), 655–680.
525. Klainerman S and Majda A. Comm Pure Appl Math **35**(5) (1982), 629–651.
526. Klein R. J Computational Physics **121** (1995), 213–237.
527. Klein R et al. J Engineering Maths **39** (2001), 261–343.
528. Kleiser L and Zang TA. Ann Rev Fluid Mech **23** (1991), 495–537.
529. Kliakhandler IL. J Fluid Mech **423** (2000), 381–390.
530. Knightly GH and Sather D. ARMA **97** (1987), 271–297.
531. Knobloch E, Moore DR, Toomre J and Weiss NO. J Fluid Mech **166** (1986), 409–448.
532. Kobayaschi T (ed). *Laminar-turbulent transition.* Springer, Berlin 1995.
533. Kobayaschi Takayuki. Proc Japan Math Sci **73**(7) (1997), 126–129.
534. Koch H and Tataru D. Adv Math **157** (2001), 22–35.
535. Kolesov V and Shapakidze L. On oscillatory modes in viscous incompressible liquid flows between two counter-rotating permeable cylinders. In: *Trends in Applications of Mathematics to Mechanics*, 221–227. Chapman and Hall/CRC, Boca Raton, London, 2000.
536. Kolesov V and Yudovich V. Transitions near the interactions of bifurcations producing Taylor vortices and azimutal waves. Cited in references of the above paper by Kolesov V and Shapakidze L, 2000, p. 227.
537. Kolesov V, Ovchinnikov S, Petrovskaya N and Yudovich V. Onset of chaos through intersections of bifurcations in Couette-Taylor flow. The Third ICIAM Congress, Book of abstracts, 201. Hamburg, 1995.
538. Kolmogorov AN. J Fluid Mech **13** (1962), 82–85.
539. Kopbosynov BK and Pukhnachev VV. Fluid Mech Sov Res **13** (1986), 95.
540. Korolev GL. Zh Vychisl Mat Mat Fiz **27** (1987), 1224.
541. Koschmieder EL. In: *Advances in Chemical Physics*, Vol 26, Progogine I and Rice SA (eds), 177–212. New York, 1974.
542. Koschmieder EL and Prahl S. J Fluid Mech **215** (1990), 571.
543. Koschmieder EL. *Bénard Cells and Taylor Vortices.* Cambridge Univ Press, Cambridge, (1993).
544. Kozono H and Ogawa T. Arch Rational Mech Anal **122** (1993), 1–17.
545. Kozono H and Ogawa T. Arch Rational Mech Anal **128** (1994), 1–31.
546. Kreiss HO and Lorenz J. *Initial-Boundary Value Problems and the Navier-Stokes Equations*, Academic Press, Boston, 1989.
547. Kreiss HO and Lorenz J, and Naughton MJ. Adv Appl Math **12** (1991), 187–214.
548. Kropinski MC, Ward MJ and Keller JB. SIAM J Appl Math **55** (1995), 1484–1510.
549. Krueger ER, Gross A and DiPrima RC. J Fluid Mech **24** (1996), 521–538.
550. Kuiken HK. J Engineering Maths **2** (1968), 355–371.
551. Kuiken HK (ed). *The Centenary of a Paper on Slow Viscous Flow by the Physicist H. A. Lorentz.* Kluwer, 1996.
552. KuoYH. J Math Phys **32** (1953), 83–101.
553. Kuramoto Y and Tsuzuki T. Prog theor Phys **55** (1976), 356.

554. Kvernvold O. Int J Heat Mass Transf **22** (1979), 395.
555. Ladhyzhenskaya OA. Regularity of the generalized solutions of the general nonlinear and nonstationary Navier-Stokes equations. In: *The Mathematical Problems in Fluid Mechanics,* 61–86. Institute of Basic Technical Problem, Polish Acad of Sci, Warszawa, 1967.
556. Ladhyzhenskaya OA. *The mathematical theory of viscous incompressible flow,* Gos Izdat Fiz-Mat Lit, Moskva, 1961 (in Russian) English trans: Gordon and Breach, New York, 1963.
557. Ladhyzhenskaya OA. A dynamical system generated by the Navier-Stokes equations. Translated from Zap Nauch Seminar Leningrad Otd Mat Inst Steklov **27** (1972), 91–115.
558. Ladhyzhenskaya OA. Ann Rev Fluid Mech **7** (1975), 249–272.
559. Ladhyzhenskaya OA. Uspekhi Mat Naouk **42**(6)(258) (1987), 25–60. In Russian.
560. Ladyzhenskaya OA and Solonnikov VA. Zap Nauchn Semin LOMI **96** (1980), 117–160.
561. Ladyzhenskaya OA, Solonnikov VA and Ural'ceva NN. *Linear and quasilinear equations of parabolic type.* Translations of Math Monographs, Vol 23, AMS, Providence, 1968.
562. Lagerstrom PA. Laminar Flow Theory, In: *Theory of laminar flows,* Moore FK (ed), Princeton University Press, 20–285, 1964.
563. Lagerstrom PA. SIAM J Appl Math **28** (1975), 202–214.
564. Lagerstrom PA. *Matched Asymptotic Expansions (Ideas and Techniques),* Applied Mathematics Sciences, Vol **76**, Springer, New York, 1988.
565. Lagrée P-Y. Int J Heat Mass Transfer **42** (1999), 2509–2524.
566. Lagrée P-Y. Removing the marching breakdown of the boundary-layer equations for mixed convection above a horizontal plate. In: Int J Heat and Mass Transfer, (2001).
567. Lamb H. *Hydrodynamics.* Cambridge University Press, 6th ed, 1932; Reprint, Dover, New York, 1945.
568. Lanchester FW. *Aerodynamics,* Constabble, London, 1907.
569. Landau LD. Dokl Akad Nauk SSSR **44** (1944), 339–342, in Russian.
570. Landau LD and Lifshitz EM. *Fluid Mechanics. Pergamon,* London, 1959.
571. Lauvstad VR. J Soun Vib **7** (1968), 90–105.
572. LeBalleur J-C. Couplage visqueux-non visqueux: méthode numérique et applications aux écoumements bidimensionnels transsoniques et supersoniques. La Recherche Aérospatiale, **1977-6**, 349-358, (1977).
573. LeBalleur J-C. La Recherche Aérospatiale **183** (1978), 65–76.
574. Lee SS. J Fluid Mech **347** (1997), 71–103.
575. Legendre R. La Recherche Aérospatiale **54** (1956), 3.
576. Lemarie-Rieusset PG. *Recent Developments in the Navier-Stokes Problem.* Research Notes in Mathematics, Chapman and Hall/CRC, 2002.
577. Leonard A. Ann Rec Fluid Mech **17** (1985), 523.
578. Leray J. J Math Pures Appl **12** (1933), 1–82.
579. Leray J. Acta Math **63** (1934), 193–248.
580. Leray J. La Vie des Sciences, Comptes rendus, série générale **11** (1994), 287–290.
581. Lesueur M. *Turbulence in Fluids.* Third Revised and Enlargedd edition, Kluwer, Dordrecht, 1997.

582. Levich VG. *Physicochemical Hydrodynamics*. Englewood Cliffs, New Jersey: Prentice Hall, 1962.
583. Levich VG and Krylov VS. Ann Rev Fluid Mech **1** (1969), 293.
584. Li H-Sh, Kelly RE and Hall Ph. Phys Fluids **9**(5) (1997), 1273–1276.
585. Lighthill MJ. Proc Roy Soc A**211** (1952), 564–587.
586. Lighthill MJ. Proc Roy Soc A**217** (1953), 478–507.
587. Lighthill MJ. Proc Roy Soc A**222** (1954) 1.
588. Lighthill MJ. Attachment and separation in three-dimensional flow. In: *Laminar Boundary Layers*, Oxford, Rosenhead L (ed), 72–82, Oxford University Press, 1963a.
589. Lighthill MJ. Introduction: real and ideal fluids. In: *Laminar Boundary Layers*, Oxford, Rosenhead L (ed), 1–45, Oxford University Press, 1963b.
590. Lighthill MJ. J Sound Vib **61** (1978), 391.
591. Lim TT and Chew YT. Phys Fluids **10**(12) (1998), 3233–35.
592. Limat L. CR Acad Sci **217**, Série II, (1993), 563–568.
593. Lin CC. *The Theory of Hydrodynamical Stability*. Cambridge Univ. Press, New York, 1955.
594. Lin SP. J Fluid Mech **36** (1969), 113–126.
595. Lin, SP. J. Fluid Mech **40** (1970), 307.
596. Lin SP. J Fluid Mech **63**(3) (1974), 417–429.
597. Lin SP and Chen JN. Phys Fluids **10**(8) (1998), 1787–1792.
598. Lingwood R. J Fluid Mech **380** (1999), 377.
599. Lions J-L and Prodi G. CR Acad Sci **248**, (1959), 3519–3521.
600. Lions J-L. *Perturbations singulières dans les problèmes aux limites et en controle optimal*. Lecture Notes Math, Vol **323**. Springer, New York, 1973.
601. Lions J-L, Temam R and Wang S. Nonlinearity **5** (1992), 237–288.
602. Lions J-L, Temam R and Wang S. CR Acad Sci **316** (1993), 113–119, 211–215, (1993) and **318** (1994), 1165–1171.
603. Lions PL. CR Acad Sci **316**(1) (1993a), 1335–1340.
604. Lions PL. CR Acad Sci **317**(1) (1993b), 115–120.
605. Lions PL. CR Acad Sci **317**(1) (1993c), 1197–1202.
606. Lions PL. *Mathematical Topics in Fluid Mechanics:* Vol 1: *Incompressible Models*, Oxford University Press, 1996.
607. Lions PL. *Mathematical Topics in Fluid Mechanics*. Vol 2: *Compressible Models*, Oxford University Press, 1998.
608. Lions PL and Masmoudi N. J Math Pures Appl, **77** (1998), 585–627.
609. Lions PL, Pertham B, Tadmor E. C R Acad Sci **312**(1) (1991), 97–102.
610. Liu T-P. Memoirs of Amer Math Soc **328** (1986a).
611. Liu T-P. Comm Pure Appl Math **39** (1986b), 565–594.
612. Liu J., Schneider J.B. and Gollub J.P. Phys Fluids **7**(1) (1995), 55–67.
613. Long LN. J Engineering Math **21** (1987), 167–178.
614. Long RR. Tellus **5**(1) (1953), 42–58.
615. Lorenz EN. J Atmopheric Sci **20** (1963), 130–141.
616. Lorenz EN. *The Essence of Chaos*. Taylor & Francis, London, 1995.
617. Lumley JL. *Stochastic Tools in Turbulence*. Academic Press, 1970.
618. Lundgren TS. Phys Fluids **14**(2) (2002), 638–642.
619. Mack LM. AGARD Report **709** (1984), 31.
620. Majda A. *Compressible Fluid Flow and Systems of Conservation Laws in Several Space Variables*, AMS Vol **53**, Springer, New York, 1984.

621. Majda A. *Mathematical foundations of incompressible fluid flow.* Princeton University, Dep. of Mathematics, Lecture Notes, 1985.
622. Majda A. Comm Pure Appl Math **39** (1986), S187–S220.
623. Majda A and Sethian J. Combustion Sci Tech **42** (1985), 185–205.
624. Malek J, Necas J and Rokyta M (eds). *Advanced Topics in Theoretical Fluid Mechanics.* Longman, 1999.
625. Malek J, Necas J and Rokyta M (eds). *Advances in Mathematical Fluid Mechanics.* Springer, Berlin, 2000.
626. Mallock A. Proc Roy Soc A**45** (1896), 126–132.
627. Man C-S. Archive Rational Mech Analysis **106**(1) (1989), 1–61.
628. Manceau R, Wang M and Laurence D. J Fluid Mech **438** (2001), 307–338.
629. Mandelbrot B. *The Fractal geometry of nature.* Freeman, San Francisco, 1982
630. Manneville P. *Dissipative Structures and Weak Turbulence.* Academic Press, London, 1990.
631. Manneville P and Pomeau Y. Phys Lett A**75** (1979), 1–2.
632. Marangoni CGM. Ann Phys (Poggendorff) **143** (1871), 337.
633. Marangoni CGM. Nuovo Cimento, Ser 2 **5/6** (1872), 239.
634. Marchioro C and M. Pulvirenti M. *Vortex Methods in Two-Dimensional Fluid Dynamics,* Lecture Notes in Physics, Vol **203**, Springer, Berlin, 1984.
635. Marchioro C and Pulvirenti M. *Mathematical Theory of Incompressible Nonviscous Fluids,* AMS Vol **96**, Springer, New York, 1994.
636. Marek M and Schreiber I. *Chaotic behaviour of deterministic dissipative systems.* Cambridge University Press, Cambridge, 1995.
637. Maremonti P. Ann Mat Pura appl **97** (1985), 57–75.
638. Marsden JE and McCracken M. *The Hopf bifurcation and its applications.* Springer, Berlin, 1976.
639. Martchuk GI. Numerical methods for the solution of weather prediction and climate theory problems. In: *Lectures on Numerical Short-Range Weather Prediction,* 92–187, WMO Regional Training Seminar. World Meteorological Organization, Hydrometeoizdat, Leningrad, 1969.
640. Masmoudi N. ARMA **142** (1998), 375–394.
641. Masmoudi N. Asymptotic problems and compressible-incompressible limits. In: *Advances in Mathematical Fluid Mechanics,* 119–158. Springer, Berlin, 2000.
642. Masuda K. Thoku Math J **36** (1984), 623–646.
643. Masuda K. In: *Proceeding of Symposia in Pure Mathematics,* Vol **45**, part 2, 179, AMS, Providence, Rhode Island, (1986).
644. Mathieu J and Scott J. *An Introduction to Turbulent Flow.* Cambridge Univ Press, Cambridge, 2000.
645. Matsumura A. *An energy method for the equation of motion of compressible viscous and heat-conductive fluids.* University of Wisconsin, Madison, *MRC Technical Summary Report,* Vol **2194**, 1981.
646. Matsumura A and Nishida T. *The initial boundary value problem for the equations of motion of compressible viscous and heat-conducting fluid.* University of Wisconsin, Madison, MRC Technical Summary Report, Vol **2237**, 1981.
647. Matsumura A and Nishida T. Initial boundary value problems for the equations of motion of general fluids, In: *Computing methods in applied sciences and engineering.* V, Glowinski R and Lions JL, (eds), North-Holland, Amsterdam, 389–406, 1982.

648. Mauss J. Asymptotic modelling for separating boundary layers, Lecture Notes in Physics, Vol **442**, 239–254, Springer, 1994.
649. Mauss J an Cousteix J. *Uniformly valid asymptotic expansions for singular perturbations problems*. Institut de Mécanique des Fluides (UMR-CNRS) and ONERA (DMAE), Toulouse, 2002.
650. McCroskey WJ. Ann Rev Fluid Mech **14** (1982), 285–311.
651. McGrath FJ. Arch Rat Math Anal **27** (1968), 329.
652. McNown JS. Fluid Dynamics Research **7** (1991), 203–214.
653. Mei R and Lawrence CJ. J Fluid Mech **325** (1996), 79–111.
654. Meier GEA and Obermeier F (eds). *Flow of Real Fluids*. Lecture Notes in Physics, Vol **235**. Springer, Heidelberg 1985.
655. Méléard S. Matapli SMAI' **65** (2001), 35–49.
656. Mellor GL. Int J Eng Sc **10** (1972), 857–873.
657. Mellor GL, Chapple PJ and Stokes VK. J Fluid Mech **31** (1968), 95–112.
658. Menikoff R et al. Phys Fluids **20** (1977), 2000–2004.
659. Messister AF. SIAM J of Applied Math **18** (1970), 241–247.
660. Messister AF. Boundary layer separation. In: *Proc 8th US National Congress on Appl Mech*. Western Periodicals North Hollywood, 157–179, 1978.
661. Messister AF. J Appl Mech **50** (1983), 1104.
662. Meyer RE. *Introduction to Mathematical Fluid Dynamics*. Wiley-Interscience, New York, 1971.
663. Meyer RE. SIAM J Applied Math **43**(4) (1983), 639–663.
664. Meyer RE (ed). *Waves on fluid interfaces*. Academic Press, New York, 1983.
665. Mezic I. J Fluid Mech **431** (2001), 347–370.
666. Mihaljan JM. Astrophys J **136** (1962), 1126.
667. Miles J. Ann Rev Fluid Mech **16** (1984), 198.
668. Miller J. Phys Rev Lett **65** (1990), 2137–2140.
669. Misra JC, Pal B, Pal A and Gupta AS. J Non-linear Mechanics **36** (2001), 731–741.
670. Miyagi T and Kamei T. J Fluid Mech **134** (1983), 221–230.
671. Miyakawa T. J Math Sci Univ Tokyo **4**(1) (1997), 67–119.
672. Moffatt HK and Tsinober A (eds). *Topological Fluid Dynamics*. Cambridge University Press, Cambridge, 1989.
673. Monin AS. Izv AN SSSR, Ser Geophys **4** (1961), 602.
674. Monin AS. *Weather Forecasting as a Problem in Physics*. MIT Press, Cambridge, 1972; English Transl.
675. Monin AS. Uspekhi Fiz Naouk **150**(1) (1986), 61–105 (in Russian).
676. Monin AS. *Fundamentals of Geophysical Fluid Dynamics*. Gidrometeoizdat, Leningrad, 1988 – Russian Edition.
677. Moore DW and Saffman PG. Phil Trans Roy Soc A**2640** (1969), 597–634.
678. Moore FK. NACA TN **2471** (1951).
679. Moore FK and Ostrach S. J Aero Sciences **124** (1957), 77–85.
680. Morkovin MV. *Understanding transition to turbulence in shear layers*. AFOSR Final Rep, **AD-A134796**, 1983.
681. Morkovin MV. AIAA Prof Study Ser: *Instabilities and Transition to Turbulence*. Cincinnati, Ohio, Courses Notes, 1985.
682. Müller B. *Computation of Compressible Low Mach Number Flow*. Habilitation thesis, Institut für Fluiddynamik, ETH Zürich, 1996.
683. Müller W. ZAMM **16** (1936), 227–238.

684. Myers TG. SIAM Rev **40**(3) (1998), 441–446.
685. Nakabayashi K and Kitoh O. J Fluid Mech **315** (1996), 1–29.
686. Nakanishi K, Kida O and Nakajima T. J Fluid Mech **334** (1997), 31–59.
687. Nam S. J Fluid Mech **214** (1990), 89–110.
688. Napolitano M, Werle MJ and Davis RT. AIAA Journal **17**(7) (1979), 699–705.
689. Navier CL MH. Ann. de Chimie **19** (1821), 244–260.
690. Navier CL MH. Mém Acad Sci, Institut de France **6** (1820), 375–394.
691. Nayfeh AH. *Perturbation Methods.* Wiley, New York, (1973).
692. Nayfeh AH. *Introduction to perturbation techniques.* Wiley, New York (1981).
693. Nayfeh AH. Stability of compressible boundary layers. In: *Transonic Symposium: Theory, Application, and Experiment.* Nasa Conf Publ, **3020**, Vol 1, 629, 1989.
694. Nayfeh AH. Comput Fluids **20**(3) (1991), 269–292.
695. Nazarov VI and Pileckas K. Rend Sem Mat Univ Padova **99** (1998), 30–43.
696. Necas J, Ruzicka M and Sverak Vl. CR Acad Sci, Paris, Série I, Math **323**(3) (1996), 245–249.
697. Needham DJ and Merkin JH. J Fluid Mech **184** (1987), 357–379.
698. Neiland VYa. Isv Akad Nauk SSSR, Mekh Zhidk i Gaza **4** (1969), 53–57.
699. Nepomnyashchy A, Velarde MG and Colinet P. *Interfacial Phenomena and Convection.* Chapman and Hall/CRC, Boca Raton, 2002.
700. Nettel S. *Waves Physics (Oscillations-Solitons-Chaos),* Chapter 8. Springer, Berlin, 1992
701. Neumann C. Ber Verh Ges Wiss **46** (1894), 1–24.
702. Newell AC and Whitehead J'. J Fluid Mech **38**(2) (1969), 279–303.
703. Newhouse S, Ruelle D and Takens F. Comm Math Phys **64** (1978), 35–40.
704. Nickel K. Ann Rev Fluid Mech **5** (1973), 405–428.
705. Nickel K. Math Meth Appl Sci **1** (1979), 445–452.
706. Nicolis G. *Introduction to nonlinear science.* Cambrige Univ Press, Cambridge 1995.
707. Njamkepo S. Math Models Methods Appl Sci **6**(1) (1996), 59–75.
708. Noe JM. *Sur une théorie asymptotique de la convection naturelle.* Thèse de Doctorat de 3e Cycle (Mécanique). Université de Lille I, 1981.
709. Noll W. J Rational Mech Anal **4** (1955), 1.
710. Noll W. Arch Rational Mech Anal **2** (1958), 197.
711. Nordström J. Comput Fluids **24**(5) (1995), 609–623.
712. Normand C, Pomeau Y and Velarde MG. Rev Mod Phys **49**(3) (1977), 581–624.
713. Novotny A and Padula M. Arch Rat Mech Anal **126** (1994), 243–297.
714. Novotny A and Padula M. Math Ann **308**(3) (1997), 439–489.
715. Oberbeck A. Annalen der Physik und Chemie **7** (1879), 271–292.
716. Obermeier F. The application of singular perturbation methods to aerodynamic sound generation. In: *Lecture Notes in Mathematics,* Vol **594**, 400–421. Springer, Heidelberg, 1977.
717. Obukhov AM (ed). *Sistemy Guidrodinamicheskovo Tipa i ikh Primenenie.* Nauka, Moskva, 1981, (in Russian).
718. Ogura Y and Phillips NA. J Atmosph Sci **19** (1962), 173–179.
719. Oleinik OA. Zhurn Vychislit Mat Mat Fiz **3** (1963), 489–507.
720. Oleinik OA. PMM (Russian) **33**(3) (1969), 441.

721. Oleinik OA and Samokhin VN. *Mathematical models in boundary layer theory*. Chapman & Hall/CRC, Boca, Raton 1999.
722. Oliger J and Sundström A. SIAM J Appl Math **35** (1978), 419–446.
723. O'Neill ME and Chorlton F. *Viscous and compressible fluid dynamics*. Ellis Horwood Limited, Chichester, 1989.
724. Onsager L. Phys Rev **37** 405, and **38** (1931), 2265.
725. Oron A and Rosenau Ph. Journal Physique II France **2** (1992), 131–136.
726. Oron A and Rosenau Ph. J Fluid Mech **273** (1994), 361–374.
727. Oron A, Davis SH and Bankoff SG. Rev Modern Phys **69**(3) (1997), 931–960.
728. Orr WMF. Proc R Ir Acad **A27** (1907), 9–27, 69–138.
729. Orszag S. Studies Appl Math L4 (1971), 293–327.
730. Oseen CW. Ark Math Astronom Fys **6**(29) (1910).
731. Oseledec VI. Trans Moscow Math Soc **19** 197-231, (1968).
732. Ostrach S. NACA TN **3569**, 1955.
733. Ostrach S. Ann Rev Fluid Mech **14** (1982), 313.
734. Oswatitsch K. Die Ablösungsbedingung von Grenzschichten. *IUTAM Symposium on Boundary Layer Research,* 357, Görtler H (ed), Springer, Berlin, 1958.
735. Öztekin A, Seymour BR and Varley E. Pump Studies Appl Maths **107** (2001), 1–41.
736. Padula M. Boll UMI B **5**(6) (1986), 720.
737. Padula M. Trans Theory and Stat Phys **21** (1993), 593–613.
738. Padula M. Mathematical Properties of Motion of Viscous Compressible Fluids. In: *Progress in theoretical and computational fluid mechanics,* Galdi GP, Màlek J, and Necàs J, (eds). Longman, New York, 1994.
739. Padula M. On the stability of a steady state of a viscous compressible fluid. In: *Trends in Applications of Mathematics to Mechanics,* 317–326. Chapman and Hall/CRC, Boca Raton, London, 2000.
740. Padula M and Petunin I. Eur. J. Mech. B **13**(6) (1994), 701–730.
741. Palm E. J Fluid Mech **8** (1960), 183–192.
742. Palm E. Ann Rev Fluid Mech **7** (1975), 39.
743. Papanastasiou TC, Georgiou GC and Alexandrou AN. *Viscous Fluid Flow*. Springer, Heidelberg, 2000.
744. Papenfuß H-D. Achives Mech **26** 981-994, and Mech Res Comm **1** (1974), 285–290.
745. Papenfuß H-D. *Die Grenzschichteffekte 2. Ordnung bei der kompressiblen dreidimensionlen Staupunktströmung*. Dissertation, Ruhr-Universität Bochum, 1975.
746. Papenfuß H-D. J Mécanique **16**(5) (1977), 705–732.
747. Parmentier PM, Regnier VC and Lebon G. Physical Review E**54**(1) (1996), 411–423.
748. Pavithran S and Redeekopp LG. Studies Appl Maths **93** (1994), 209.
749. Payne LE. Int J Engng Sci **10** (1992), 1341–1347.
750. Payne LE and Pell WH. J Fluid Mech **7** (1960), 529–549.
751. Pearson JRA. J Fluid Mech **4** (1058), 489–500.
752. Pedley TJ. J Fluid Mech **55**(2) (1072), 359–383.
753. Pedlosky J. *Geophysical Fluid Dynamics*. Springer, Berlin, 1987.
754. Peetre J. *New thoughts on Besov Spaces*. Duke Univ Math Series. Duke University, 1976.

755. Peitgen H-O and Richter PH. *The beauty of fractals.* Springer, Berlin, 1992
756. Pellew A and Southwell RV. Proc Roy Soc **A176** (1940), 312–343.
757. Pérez-Garcia C and Carneiro G. Phys Fluids **A3** (1991), 292.
758. Perry AE and Chong MS. J Fluid Mech **173** (1986), 207–223.
759. Perry AE and Fairlie BD. Ad Geophys **B18** (1974), 299–315.
760. Perry AE and Hornung HG. Z Flugwiss Weltraumforschung **8** (1984), 155–160.
761. Peyret R. *Handbook of Computational Fluid Mechanics,* Academic Press, 1996.
762. Pfister G. *Deterministic Chaos in Rotational Taylor-Couette Flow.* Lecture Notes in Physics, 199–210. Springer, Heidelberg 1985.
763. Phillips NA. Ann Rev Fluid Mech **2** (1970), 251–292.
764. Pileckas K. Recent advances in the theory of Stokes and Navier-Stokes equations in domains with non-compact boundaries. Math Th in Fluid Mech (Paseky, 1995), 30–85. Pitman Res Notes Math Ser **354**, Longman, Harlow 1996.
765. Piquet J. *Turbulent Flows:* Models and Physics. Springer, 1999.
766. Planchon F. Ann Inst H Poincaré Anal Non Linéaire **13** (1996), 319–336.
767. Planchon F. Revista Matematica Iberoamericana **14**(1) (1998), 71–93.
768. Platten JK and Legros JC. *Convection in Liquids.* Springer, Berlin, 1984.
769. Pletcher RH and Chen K-H. *On solving the compressible Navier-Stokes equations for unsteady flows at very low Mach numbers.* AIAA Paper 93-**3368**, (1993).
770. Pogu M and Tournemine G. J de Mécanique Théorique et Appliquée **1**(4) (1982), 645–670.
771. Poincaré H. *Thermodynamique,* Gauthier-Villars, 1892.
772. Poisson SD. J Ecole Polytechnique **13**(20) (1831), 1–174.
773. Pomeau Y and Manneville P. Comm Math Phys **77** (1980), 189–197.
774. Pomeau Y. Turbulence: Determinism and chaos. In: *Problems of Stellar Convection,* Spiegel EA and Zhan JP (eds), 337–348. Lecture Notes in Physics, Vol **71**, Springer, Heidelberg 1977.
775. Pop I. In: Annali di Matematica pura ed applicata **XC**(IV) (1971), 87–98.
776. Pope SB. *Turbulent Flow.* Cambridge Univ Press, Cambridge, 2001.
777. Prandtl L. Über Flüssigkeitsbewegung bei sehr kleiner Reibung, In: *Internat. Math Congresses Heidelberg,* Teubner, Leipzig, 484–491, (1904).
778. Prandtl L. *Essentials of Fluid Dynamics.* Blackie and Sons Ltd, London, 1952.
779. Prandtl L. *Gesammelte Abhandlungen zur angewandten Mechanik, Hydro- und Aerodynamik,* Vol 2, Springer, Berlin, 575–584, (1961).
780. Probstein RF. *Physicochemical Hydrodynamics: An Introduction.* Wiley, New York, 1994.
781. Prodi G. Ann di Mat **48** (1959), 173–182.
782. Prodi G. Rem Sem Mat Padova **30** (1960), 1–15.
783. Prokopiou Th, Cheng M and Chang H-C. J Fluid Mech **222** (1991), 665–691.
784. Proudman I and Pearson JRA. J Fluid Mech **2** (1957), 237–262.
785. Pukhnachev VV. *Non-classical problems of the boundary-layer theory.* Novossibirsk State University, 1979, (in Russian).
786. Pumir A and Blumenfeld L. Phys Rev E**54** (1996), R4528–R4531.
787. Ragab SA and Nayfeh AH. AIAA (1980a), 1–12.

788. Ragab SA and Nayfeh AH. Phys Fluids **23** (1980b), 1091.
789. Rajagopal KR et al. Math Models Methods Appl Sci **6**(8) (1996), 1157.
790. Ranger KB. Studies Appl Math **91** (1994), 27–37.
791. Rath HJ (ed). *Microgravity Fluid Mechanics.* Springer, Berlin, 1992.
792. Rayleigh Lord. Proc London Math Soc **14** (1883), 170–175.
793. Rayleigh Lord. Pros Roy Soc **A93** (1916a), 148-155.
794. Rayleigh Lord. Philos Mag **32**(6) (1916b), 1–6.
795. Rayleigh Lord. Phil Mag VI **32** (1916c), 529–546.
796. Reed M and Simon B. *Methods of Modern Mathematical Analysis I: Functional Analysis.* Academic Press, New York, 1980.
797. Reed HL and Saric W. Ann Rev Fluid Mech **21** (1989), 235–284.
798. Rednikov AYe, Colinet P, MG Velarde and Legros JC. J Fluid Mech **405** (2000), 57–77.
799. Regnier VC and Lebon G. Q Jl Mech Appl Math **48**(1) (1995), 57–75.
800. Reiner M. Amer J Math **67** (1945), 350.
801. Reynolds O. Phil Trans Roy Soc **A186** (1894), 123–164.
802. Reynolds WC and Potter MC. J Fluid Mech **24** (1967), 467–492.
803. Riahi DN. *Flow Instability.* WIT Press, 2000.
804. Richtmyer RD. *Principles of Advanced Mathematical Physics,* Vol II, Springer, Berlin, 1981.
805. Riley N. Mathematika **12**(2) (1965), 161–175.
806. Riley N. Quart J Mech Appl Math **19** (1966), 461.
807. Riley N. J Inst Math Appl **3** (1967), 419–454.
808. Riley N. SIAM Review **17**(2) (1975), 274–297.
809. Riley N. J of Engng Maths **13**(1) (1979), 75–91.
810. Riley N. J Fluid Mech **180** (1978), 319–326.
811. Riley N and Stewartson K. J Fluid Mech **39** (1969), 193–207.
812. Riley N and Vasantha R. J Fluid Mech **205** (1989), 243–262.
813. Rionero S. Ann Mat Pura Appl **LXXVI** (1965), 75; Ann Mat Pura Appl **LXXVIII** 339, and Ricerche di Mat **XVII** (1968), 64; Bol UMI **4** 364, and Ricerche di Mat **XX** (1971), 285.
814. Rionero S. On the choice of the Lyapunov function in the stability of fluid motions. In: *Energy stability and convection* (Capri, 1986), 392–412). Pitman Res Notes in Math Ser, Vol **168**, Longman, Harlow, 1988.
815. Rionero S and Galdi GP. ARMA **62** (1976), 295–301.
816. Rivlin RS. Nature **160** (1947), 611.
817. Rivlin RS. Proc Roy Soc **A193** (1948), 260.
818. Robert R. J Stat Phys **65** (1991), 531–543.
819. Robert R. Gazette des Mathématiciens, SMF **90** (2001), 11-25.
820. Rodrigues JF and Sequeira A (eds). *Mathematical Topics in Fluid Mechanics.* Longman, New York, 1992.
821. Roget C, Brazier JPh, Mauss J and Coustiex J. Eur J Mech B **3** (1998), 307–329.
822. Rosenhead L (ed). *Laminar boundary layers an account of the development, structure, and stability of laminar boundary layers in incompressible fluids, together with a description of associated experimental techniques.* Clarendon Press, Oxford 1963.
823. Ross N. Nonlinear Anal **30**(1) (1997), 459–464.
824. Ruban AI. Izv Akad Nauk SSSR Mech Zhid Gaza **1** (1982), 42.

825. Rudin W. *Functional Analysis*. McGraw-Hill, New York, 1973.
826. Rudin W. *Principles of Mathematical Analysis*. McGraw-Hill, New York, 1976.
827. Ruelle D and Takens F. Comm Math Phys **20** (1971a), 167–192.
828. Ruelle D and Takens F. Comm Math Phys **23** (1971b), 343–344.
829. Ruelle D. The Lorenz attractor and the problem of turbulence. In: *Turbulence and Navier-Stokes equations*, Springer, Berlin 1976.
830. Ruyer-Quil C and Manneville P. Eur Phys J B**6** (1998), 277–292.
831. Sagaut P. *Large Eddy Simulation for Incompressible Flows – An Introduction*. Springer, Berlin, 2001.
832. Saint-Venant AJB (de). C.R. Acad. Sci **17** (1843), 1240–1243.
833. Saintlos S and Mauss J. Int J Engng Sci **34**(2) (1996), 201–211.
834. Salmon R. *Lectures on Geophysical Fluid Dynamics*. Oxford University Press, New York, 1998.
835. Salvi Rodolfo (ed). *The Navier-Stokes Equations: Theory and numerical methods*. Marcel Dekker, New York, 2001.
836. Salwen H and Grosch CE. J Fluid Mech **104** (1981), 445.
837. Sammartino M. Comm Partial Diff Eqs **22**(5–6) (1997), 749–771.
838. Sammartino M and Caflisch RE. Rend Circ Mat, Palermo, Suppl **45** Part II, (1996), 595–605.
839. Sano T. J Fluid Mech **112** (1981), 433–441.
840. Sarma GN. Proc of the Cambridge Phil Soc **61** (1965), 975–807.
841. Sarpkaya T. Ann Rev Fluid Mech **28** (1996), 83–128.
842. Sattinger DH. *Topics in Stability and Bifurcation Theory*. Springer, Berlin, 1973.
843. Scanlon JW and Segel LA. J Fluid Mech **30** (1967), 149.
844. Scardovelli R and Zaleski S. Ann Rev Fluid Mech **31** (1999), 567–603.
845. Scheffer V. Turbulence and Hausdorff dimension. *Turbulence and Navier-Stokes equations*, Lecture Notes in Math, Vol **565**, 94–112, Springer, 1976.
846. Scheffer V. Comm Math Physics **73** (1980), 1–42.
847. Schiehlen W and van Wijngaarden L. *Mechanics at the Turn of the Century*. IUTAM 2000, Shaker Verlag, Aachen, 2000.
848. Schiller S, Heisig U and Panzer S. *Electron Beam Technology*. Wiley, 1982.
849. Schlichting H. Phys Z **33** (1932), 327–335.
850. Schlichting H. *Boundary Layer Theory*. McGraw-Hill, New York, 1992.
851. Schlichting H and Gersten K. *Boundary-Layer Theory*. Springer, Berlin, 2000.
852. Schlüter A, Lortz D and Busse FH. J Fluid Mech **23** (1965), 129–144.
853. Schneider W. J Fluid Mech **108** (1981), 55–65.
854. Schonbek ME. Arch Rat Mech Anal **88** (1985), 209–222.
855. Schuster HG. *Deterministic chaos. An introduction*. Physik-Verlag, Weinheim, 1984.
856. Scriven LE and Sterling CV. Nature **187** (1960), 186.
857. Scriven LE and Sterling CV. J Fluid Mech **19** (1964), 321–340.
858. Sears WR and Telionis DP. SIAM J Appl Math **28** (1975), 215–235.
859. Secchi P. Rend Sem Mat Univ Padova **70** (1983), 73–102.
860. Sell GR. J Dynam Diff Eqs **8**(1) (1996), 1–33.
861. Sequeira A (ed). *Navier-Stokes Equations and Related Nonlinear Problems*. Plenum Press, 1995.
862. Séror S et al. J Fluid Mech **339** (1997), 213–238.

863. Serre D. C R Acas Sc **303**(I) (1986), 703–706.
864. Serre D. Physics D**48** (1991), 113–128.
865. Serre E, Crespo del Arco E and Bontoux P. J Fluid Mech **434** (2001), 65–100.
866. Serrin J. J Math Mech **8** (1955), 459–469.
867. Serrin J. Mathematical principles of classical fluid mechanics. *Handbuch der Physik* VIII/1, 125–263, Flügge S (ed), Springer, Berlin, 1959a.
868. Serrin J. Arch Rational Mech Anal **3** (1959b), 271–288.
869. Serrin J. Arch Rat Mech Anal **3** (1959c), 120–122.
870. Serrin J. The initial value problem for the Navier-Stokes equations. In: *Nonlinear Problems,* Langer RE (ed), 69–98, Univ of Wisconsin Press, Madison, (1963).
871. Serrin J. Proc Roy Soc A**299** (1967), 491–507.
872. Serrin J (ed). *New Perpectives in Thermodynamics.* Springer, Berlin, 1986.
873. Severin J and Herwig H. Z angew Math Phys **50** (1999), 375–386.
874. Sha W and Nakabayashi K. J Fluid Mech **431** (2001), 323–345.
875. Shah RK and London AL. Adv in Heat Transfer, Suppl **1** (1978).
876. Shajii A and Freidberg JP. J Fluid Mech **313** (1996), 131–145.
877. Shananin NA. Mat Zametki **59**(5) (1996), 791–796, in Russian.
878. Shankar V and Kumaran V. J Fluid Mech **434** (2001), 337–354.
879. Shapakidze L. On the stability of flows between two rotating cylinders. In: *Proc of the Intern Conf of Appl Mech,* 1, 450–454. Beijing, China, 1989.
880. Shapiro VL. Resonance and the stationary Navier-Stokes equations. In: *Proceeding of Symposia in Pure Mathematics,* **45**, Part 2, 359, AMS, Providence, Rhode Island, (1986).
881. Sharp DH. Physics D**12** (1984), 3–18.
882. Shearer M and Walton JC. Studies in Appl Maths **65** (1981), 85–93.
883. Shen SF. Adv in Appl Mech **18** (1978), 177–220.
884. Sheretov YuV. Comput Maths Phys **34**(3) (1994), 409–415.
885. Sherman FS. *Viscous Flow.* McGraw-Hill, New York, 1990.
886. Shibata Y. An exterior initial boundary value problem for Navier-Stokes equation. Math Research Note, Institute of Mathematics, University of Tsukuba, (1995).
887. Shilnikov LP. J Franklin Inst B**334**(5–6) (1997), 793–864.
888. Shinbrot M. *Lectures on fluid mechanics,* Gordon and Breach, New York, 1973.
889. Shipp DJ, Riley DS and Wheeler AA. Quart J Mech Appl Math **50**(3) (1997), 379–405.
890. Shkadov VYa. Izv Akad Nauk USSR, Mekhanika Zhidkosti i Gaza **1** (1967), 43.
891. Shkadov VYa. *Hydrodynamics of slopped falling films.* CISM Notes to Advanced School (CISM). Coordinated by M.G. Velarde and R.Kh. Zeytounian, Udine, 2000.
892. Shugai GA and Yakubenko PA. Eur J Mech B **17**(3) (1998), 371–384.
893. Shyy W and Narayanan R. *Fluid Dynamics at interfaces.* Cambridge Univ Press, Cambridge, 1999.
894. Simanovskii JB and Nepomnyaschy AA. *Convective Instabilities in Systems with Interface.* Gordon and Breach, New York 1993.
895. Simon J. *Non-homogeneous viscous incompressible fluids: existence of velocity, density and pressure.* Publ. Lab. d'Analyse Numérique, Univ. P et M. Curie, Paris, R **89004**, 1989, see also: SIAM J Math Anal **21** (1990).

896. Singh SN. J Engng Sci **13** (1975), 1085–1089.
897. Sirovich L. Quart Appl Math **45**(3) (1997), 95–97, 573–582, 583–590.
898. Sisoev GM and Shkadov VYa. Physics-Doklady **42**(12) (1997a), 683–686.
899. Sisoev GM and Shkadov V.Ya. Fluid Dynamics **32**(6) (1997b), 784–792.
900. Skeldon AC, Riley DS and Cliffe KA. J Crystal Growth **162** (1996), 96–106.
901. Skinner LA. Q Jl Mech Appl Math **XXVIII**(3) (1975), 333–340.
902. Smith JHB. RAE Tech Rept **77058** (1977).
903. Smith FT. J Fluid Mech **57** (1973), 803–824.
904. Smith FT. Proc Roy Soc A**356** (1977), 443–463.
905. Smith FT. RAE Tech Report **78095** (1978).
906. Smith FT. J Fluid Mech **92** (1979), 171–205.
907. Smith FT. Proc Roy Soc A**375** (1981), 65.
908. Smith FT. IMA J Appl Math **28** (1982), 207–281.
909. Smith FT. Ann Rev Fluid Mech **18** (1986), 197–220.
910. Smith FT. Mathematika **35** (1988), 256–273.
911. Smith FT. J Engineering Math **30** (1996), 307–320.
912. Smith FT. Philos Trans R Soc A (2000)-in press.
913. Smith FT and Merkin JH. Int J Comput FLuids **10** (1982), 7.
914. Smith FT, Sykes RI and Brighton PWM. J Fluid Mech **83**(1) (1977), 163–176.
915. Smith MK and Davis SH. J Fluid Mech **132** (1983), 119–144.
916. Smith SH. Q Appl Math **LV**(1) (1997), 1–22.
917. Snyder HA. Intl Jour Nonlinear Mech **5** (1970), 659–685.
918. Sobey IJ. *Introduction to Interactive Boundary Layer Theory*. Oxford University Press, 2000.
919. Socolescu D. CR Acad Sci **330**(I) (2000), 427–432.
920. Soehngen EE. Progress Heat Mass Transfer **2** (1969), 125–150.
921. Solonnikov VA. J Soviet Math **80** (1977), 467–528.
922. Solonnikov VA. Algebra and Analysis **5**(4) (1993a), 252–270.
923. Solonnikov VA. SAACM **3**(4) (1993b), 257–275.
924. Solonnikov VA. Zap Nauchn Semin POMI **213** (1994), 179–205.
925. Solonnikov VA. Pitman Research Notes Math Series **323** (1995a), 213–214.
926. Solonnikov VA. Math Ann **302** (1995b), 743–772.
927. Solonnikov VA. Bollettino UMI B**8**(1) (1998), 283–342.
928. Solonnikov VA and Kazhikov A.V. Ann Rev Fluid Mech **13** (1981), 79–95.
929. Sone Y. Asymptotic theory of a steady flow of a rarefied gas past bodies for small knudsen numbers. In: *Advances in Kinetic Theory and Continuum Mechanics,* 19–31. Springer, Berlin, 1991.
930. Sone Y, Bardos C, Golse F and Sugimoto H. Eur J Mech B **19** (2000), 325–360.
931. Sonechkin DM. J Statistical Phys **21**(1) (1979), 51–54.
932. Sonechkin DM. *Stochasticity in atmospheric general circulation models.* Guidrometeoizdat, Leningrad, 1984, (in Russian).
933. Speziale CG. J Comput Phys **73** (1987), 476.
934. Squire HB. Proc R Soc A**142** (1933), 621–628.
935. Stansby PK and Smith PA. J fluid Mech **229** (1991), 159–171.
936. Sterning CV and Scriven LE. AIChE Journal **5** (1959), 514.
937. Stewartson K. Proc Cambridge Phil Soc **49** (1953), 333–341.
938. Stewartson K. J Fluid Mech **3** (1957), 17–26.

939. Stewartson K. Q J Mech Appl Math **11** (1958), 399–410.
940. Stewartson K. *The Theory of Laminar Boundary Layers in Compressible Flow.* Oxford Univ Press, London, 1964.
941. Stewartson K. Mathematika **16** (1969), 106.
942. Stewartson K. J Fluid Mech **44** (1970), 347–364.
943. Stewartson K. Adv Appl Mech **14** (1974), 145–239.
944. Stewartson K. SIAM J Appl Math **28** (1975), 501–518.
945. Stewartson K. *Some aspects of non linear stability theory.* Fluid Dyn Trans, Polish Acad of Sci, Institute of Fundamental Technological Research. Vol **7**, part 1, 1976.
946. Stewartson K. D'Alembert's Paradox., SIAM Review **23** (1981), 308–345.
947. Stewartson K and Jones LT. J Aeronautical Sciences **24** (1957), 379–380.
948. Stewartson K and Stuart JT. J Fluid Mech **48**(3) (1971), 529–545.
949. Stewartson K and Williams PG. Proc Roy Soc A**312** (1969), 181–206.
950. Stewartson K and Williams PG. Mathematika **20** (1973), 98–108.
951. Stewartson K, Smith FT and Kaups. Stud Appl Math **67** (1982), 45.
952. Stokes G.G. Trans Cambridge Phil Soc **viii** (1845), 287–319.
953. Stokes G G. Trans Cambridge Phil Soc **9**(II) (1851), 8.
954. Straughan B. *The Energy Method, Stability, and Nonlinear Convection.* Springer, New York, 1992.
955. Straughan B. *Mathematical aspects of penetrative convection.* Longman, 1993.
956. Straughan B. *Explosive Instabilities in Mechanics.* Springer, 1998.
957. Strikwerda JC. Comm Pure Appl Math **30** (1977), 797–822.
958. Struwe M. Regular solutions of the stationary Navier-Stokes equations on $\mathbf{R}^5$. Preprint, (1995).
959. Stuart JT. J Fluid Mech **4** (1958), 1–21.
960. Stuart JT. J Fluid Mech **9** (1960a). 353–370.
961. Stuart JT. Nonlinear effects in hydrodynamic stability. In: *Proc 10th Intern Congress Appl Mech (Stresa, Italy)*, Rolla F and Koiter WT (eds). Amsterdam 1960b.
962. Stuart JT. In: *Laminar boundary Layers,* Rosenhead L (ed), Oxford University Press, London and New York, 1963.
963. Stuart JT. J Fluid Mech **24** (1966), 673–684.
964. Stuart JT. Ann Rev Fluid Mech **3** (1971a), 347–370.
965. Stuart JT. In: *Recent Research of Unsteady Boundary Layers,* Vol 1, 1–46, 1971(b).
966. Stuart JT. Z Flugwiss Weltraumforsch **19** (1986), 379–392.
967. Stuart JT. Ann Rev Fluid Mech **18** (1986), 1–14.
968. Swann H. Trans Amer Math Soc **157** (1971), 373–397.
969. Swinney HL. Phys D**7** (1083), 3–15.
970. Swinney HL and Gollub JP. Physics Today **31**(8) (1978), 41–49.
971. Sychev VV. Isv Akad Nauk SSSR, Mekh Zhidk i Gaza **3** (1972), 47–59.
972. Sychev VV, Ruban AI, Sychev Vik V and Korolev GL. *Asymptotic Theory of Separating Flows.* Cambridge Univ Press, Cambridge, 1998.
973. Synge JL. Semicentenn Publ Am Math Soc **2** (1938a), 227–269.
974. Synge JL. Proc Roy Soc A**167** (1938b), 250–256.
975. Szepessy A and Xin ZP. Arch Rat Mech Anal **122** (1993), 53–103.
976. Szepessy A and Zumbrun K. Arch Rat Mech Anal **133** (1996), 249–298.

977. Tai-Peng Tsai. ARMA **143** (1998), 29–51; Erratum: ARMA **147** (1999), 363.
978. Takashima MJ. Physical Soc Japan **50**(8) (1981a), 2745–2750.
979. Takashima MJ. Physical Soc Japan **50**(8) (1981b), 2751–2756.
980. Tamada K, Miura H and Miyagi T. J Fluid Mech **132** (1983), 445–455.
981. Tani A. Arch Rat Mech Anal **133** (1996), 299–331.
982. Tani A and Tanaka N. Arch Rat Mech Anal **130** (1995), 303–314.
983. Ta Phuoc Loc and Bouard R. J Fluid Mech **160** (1985), 93–117.
984. Tartar L. Compensated compactness and applications to partial differential equations. In: *Nonlinear Anal and Mech, Heriot-Watt Sympos*, 136–211. RJ Knops (ed). Pitman, Boston, 1975.
985. Taylor GI. Phil Trans Roy Soc **A223** (1923), 289–343.
986. Taylor GI. Proc Roy Soc London **A201** (1950), 192–196.
987. Telionis DP. Calculations of Time Dependent Boundary Layers. In: *Unsteady Aerodynamics*, Vol 1, 155–190. Kinney RB (ed), 1975.
988. Telionis DP. *Unsteady Viscous Flows*. Springer, New York, 1981.
989. Telionis DP and Gupta TR. AIAA Journal **15**(7) (1977), 974–983.
990. Temam R. *The Navier-Stokes Equations and nonlinear functional analysis*, CBMS-NSF Regional Conference Series in Applied Mathematics. SIAM, Philadelphia, 1983 (second edition 1995).
991. Temam R. *Navier-Stokes equations, Theory and Numerical Analysis*. North-Holland, Amsterdam, 1977, (1984, third ed).
992. Temam R. In: *Nonlinear Functional Analysis and its Applications*, Browder F (ed). AMS, Proceedings of Symposia in Pure Mathematics, Vol **45**, Part 2, 431–445, 1986.
993. Temam R. *Infinite-Dimensional Dynamical Systems in Mechanics and Physics*, Springer, New York, 1988a.
994. Temam R. CR Acad Sci, Série II **306** (1988b), 399-402.
995. Temam R. J Fac Sci Univ Tokyo, Sect IA, Math **36** (1989), 629–647.
996. Temam R. Navier-Stokes Equations, a talk at the Special Session: *On The Diverse Mathematical Legacy of Jean Leray*. Amer Math Soc, Austin, Texas, October 8, 1999.
997. Temam R. Some Development on Navier-Stokes Equations in the Second Half of the 20th Century. In: *Development of Mathematics 1950–2000*, Pier J-P (ed), 1049–1106. Birkhäuser, Basel 2000.
998. Temam R and Miranville A. *Mathematical Modelling in Continuum Mechanics*. Cambridge University Press, Cambridge, 2001.
999. Temam R. and Wang X. Indiana Univ Math Journal **45** (1996a), 863–916.
1000. Temam R. and Wang X. Asymptotic Analysis **9** (1996b), 1–30.
1001. Temam R. and Wang X. Annali della Scuola Normale Superiore di Pisa, **XXV** Serie IV (1998), 807–828.
1002. Temam R. and Wang X. ZAMM **80** (2000), 835–843.
1003. Temam R. and Wang X. Applied Math Letters **14** (2001), 87–91.
1004. Tennekes H and Lunley JL. *A first Course on Turbulence*. MIT Press, 1972.
1005. Teramoto Y. J Math Kyoto Univ **32** (1992), 593–619.
1006. Thess A and Orszag SA. (1995). J Fluid Mech **283** (1995), 201–230.
1007. Thual O. J Fluid Mech **240** (1992), 229–258.
1008. Thual O. Generic instabilities and nonlinear dynamics. In: *Astrophysical Fluid Dynamics*, Zhan JP and Zinn-Justin J (eds), Elsevier, Amsterdam 1988; See also: *Dynamique des Fluides Astrophysiques*, Zhan JP and Zinn-Justin J (eds), Comptes Rendus de la Session 47–Les Houches, 29 juin–31 juillet 1987, North-Holland, Amsterdam 1993.

1009. Timoshin SN. J Fluid Mech **308** (1996), 171–194.
1010. Ting L and Klein R. *Viscous vortical flows.* Springer, Berlin, 1991.
1011. Tobak M and Peake DJ. Ann Rev Fluid Mech **14** (1982), 61–85.
1012. Tokuda N. Fluid Mech **33**(4) (1968), 657–672.
1013. Tollmien W. Nachr Ges Wiss Göttingen Math-Phys Kl (1929), 21–44.
1014. Tourrette L. J of Comput Phys **137** (1997), 1–37.
1015. Townsend AA. *The Structure of Turbulent Shear Flow.* Cambridge Univ Press, Cambridge, 1976.
1016. Travis B, Olson P and Schubert G. J Fluid Mech **216** (1990), 71–91.
1017. Tresser C and Coullet PH. CR Acad Sci **287** Série I, (1978), 577–580.
1018. Treve YM. In: *Nonlinear Dynamic and Turbulence*, Barenblatt GI, Iooss G and Joseph DD (eds), 334–342. Pitman Adv Publ Program, London, 1983.
1019. Triebel H. *Theory of Function Spaces.* Birkhäuser, Basel, 1983.
1020. Truesdell C. J Math Pures Appl **29**(9) (1950), 215.
1021. Truesdell C. J Math Pures Appl **30**(9) (1951), 111.
1022. Truesdell C. *Vorticity and the thermodynamic state in a gas flow.* Mémorial des Sciences Mathématiques, Fascicule **119**, Paris, 1952(a).
1023. Truesdell C. J Rational Mech Anal **1** (1952b), 125.
1024. Truesdell C. Z Physik **131** (1952c), 273.
1025. Truesdell C. J Rational Mech Anal **2** (1953a), 593.
1026. Truesdell C. Amer Math Monthly **60** (1953b), 445.
1027. Truesdell C. *The Mechanical Foundations of Elastcity and Fluid Dynamics,* Gordon and Breach, New York, 1966.
1028. Truesdell C. *Rational Thermodynamics.* McGraw-Hill, New York, 1969.
1029. Truesdell C. *A first course in rational continuum mechanics,* Vol 1, Academic Press, New York, 1977.
1030. Truesdell C. *The Tragicomical history of thermodynamics 1822–1854.* Springer, New York, 1980.
1031. Truesdell C and Noll W. *The Nonlinear Field Theories of Mechanics.* 2nd ed Springer, Berlin, 1992.
1032. Truesdell C and Rajagopal KR. *Introduction to the Mechanics of Fluids.* Birkhäuser, Boston, 2000.
1033. Tsutsumi M and Hatano H. TMA **22**(2) (1994), 155–171.
1034. Turner JS. *Buoyancy Effects in Fluids.* Cambridge University Press, UK, 1973.
1035. Vaigant VA and Kazhikov AV. Sibirski Mat Z **36**(6) (1995), 1283–1316.
1036. Vaigant VA and Kazhikov AV. Doklady Maths **56**(3) (1997), 897–900.
1037. Vainberg MM and Trenogin VA. Uspekhi Mat Nauk **17**(2) (1962), 13–75, (1962). English trans in: Russian Math Surveys **17**(2) (1962).
1038. Valli A. Boll Un Mat It, Anal Funz Appl **18-C**(5) (1981), 317–325.
1039. Valli A. Ann Scuola Norm Sup Pisa **10**(4) (1983), 607–647.
1040. Valli A. In: Proc Symposia Pure Maths (AMS) **45**(2) (1986), 467–476.
1041. Valli A. Ann Inst Henri Poicaré Anal Non Linéaire **4**(1) (1987), 99-113.
1042. Valli A. Mathematical results for compressible flows, In: *Mathematical Topics in Fluid Mechanics,* 193–229, Rodrigues JF and Sequeira A (eds), Longman, New York, 1992.
1043. Valli A and Zajaczkowski WM. Comm Math Phys **103** (1086), 259–296.
1044. Van de Vooren AI. J Engineering Math **26** (1992), 131–152.
1045. Van de Vooren AI and Dijkstra D. J Engng Math **4** (1970), 9–27.

1046. Van Dommelen LL. *Lagrangian Description of Unsteady Separation.* 701–718, Springer, Berlin, 1991.
1047. Van Dommelen LL and Cowley SJ. J Fluid Mech **210** (1990), 593–626.
1048. Van Dommelen LL and Shen SF. J Comput Phys **38** (1980), 125–140.
1049. Van Dyke M. ZAMP **III**(5) (1952), 343–353.
1050. Van Dyke M. J Fluid Mech **14** (1962a,b), 161–177 and 481–495.
1051. Van Dyke M. Second-order compressible boundary-layer theory with application to blunt bodies in hypersonic flow. *Hypersonic Flow Research,* Vol **7**, 37–76, Riddell FR (ed). Academic Press, New York, (1962c).
1052. Van Dyke M. Ann Rev Fluid Mech **1** (1969), 265–292.
1053. Van Dyke M. AGARDOgraph **147** (1970), 1–12.
1054. Van Dyke M. *Perturbation Methods in Fluid Mechanics,* Parabolic Press. Stanford, 1975.
1055. Van Dyke M. *An Album of Fluid Motion.* Parabolic Press, Stanford 1982.
1056. Van Dyke M. SIAM Review **36** (1994), 415–424.
1057. Van Dyke M. Growing up with Asymptotics. In: *Asymptotic Modelling in Fluid Mechanics,* Bois P-A et al (eds), 3–10. Springer, Berlin 1995.
1058. Van Groesen E and Jager EM. *Mathematical structures in continuous dynamical systems. Poisson systems and complete integrability with applications from fluid dynamics.* North-Holland, Amsterdam 1994.
1059. VanHook SJ et al. Phys Rev Lett **75** (1995), 4397.
1060. Van Oudheusden BW. J Fluid Mech **353** (1997), 313–330.
1061. Van Vaerenberg S, Colinet P. and Legros J.C. *The Role of the Soret Effect on Marangoni-Bénard Stability.* Springer, Berlin, 1990.
1062. Vasiliev MM. PMM (Russian ed) **41** (1977), 1072–1078.
1063. Vasilieva AB, and Boutouzov BF. Acoustics oscillations in a media with a small viscosity. In: *Asymptotic methods in singular perturbations theory,* 191–200. Vischaya Chkola, Moskva (in Russian), 1990.
1064. Vassilicos JC (ed). *Intermittency in Turbulent Flows.* Cambridge Univ Press, UK, 2000.
1065. Velarde MG (ed). *Physicochemical Hydrodynamics; Interfacial Phenomena.* Plenum Press, New York 1987.
1066. Velarde MG (ed). *Physicochemical Hydrodynamics.* NATO ASI Series, Vol **174**, Plenum Press, 1988.
1067. Velarde MG. Phil Trans R Soc A**356** (1998), 829–844.
1068. Velarde MG and Chu X-L. Dissipative thermohydrodynamic oscillators. In: *Flow, Diffusion, and Rate Processes,* Sienutycz S and Salamon P (eds), 110–145. Taylor & Francis, New York 1992.
1069. Velarde MG and Normand C. Sci Amer **243**(1) (1980), 92.
1070. Velarde MG and Rednikov A Ye. Time-dependent Bénard-Marangoni instability and waves. In: *Time-Dependent Nonlinear Convection,* 177–218, Tyvand PA (ed). Computational Mech. Publ, Southampton, and Boston (1998).
1071. Velarde MG and Zeytounian RKh. *Interfacial Phenomena and the Marangoni Effect.* CISM Courses and Lectures, Vol **428**. Springer, Wien, New York, 2002.
1072. Veldman AEP. AIAA J **19**(1) (1981), 79–85.
1073. Veldman AEP. J Engng Math **39** (2001), 189–206.
1074. Veldman AEP and van de Vooren AI. In: *Drag of a finite flat plate.* 422–430, Springer, Berlin 1974.

1075. Velte W. ARMA **22** (1966), 1–14.
1076. Vimala CS and Nath G. J Fluid Mech **70** (1975), 561–572.
1077. Vince Jean-Marc. *Ondes propagatives dans des systèmes convectifs soumis à des effets de tension superficielle.* Doctoral thesis, University Paris VII, number 345, 1994.
1078. Vischik MI and Lyusternik LA. Uspekhi Mat Nauk **12** (1957), 3–122.
1079. Viviand H. J Mécanique **9**(4) (1970), 573–599.
1080. Voizat C. *Calcul d'écoulements stationnaires et instationnaires à petit nombre de Mach, et en maillages étirés.* Thèse de Doctorat en Sciences (de l'ingénieur). Université de Nice-Sophia Antipolis, 1998.
1081. Volpe G. AIAA J **31** (1993), 49–56.
1082. Wahl W von. *The equations of Navier-Stokes and abstract parabollic equations.* Vieweg, Braunschweig, 1985.
1083. Wahl W von. Regularity of weak solutions of the Navier-Stokes equations. In: Proceeding of Symposia in Pure Mathematics, Vol **45**, Part 2, 497–503, 1986.
1084. Wahl W von. Arch Rat Mech Anal **126** (1994), 103–129.
1085. Wall DP and Wilson SK. Phys Fluids **9**(10) (1997), 2885.
1086. Walter W. Arch Rat Mech Anal **39** (1970), 169–188.
1087. Walton IC. The effect of end walls on the onset of Taylor vortices. Abstract of lecture given at inaugural meeting of Taylor Vortex Working Party, Leeds, 1979.
1088. Wang C-Y. J Sound Vib **2** (1965), 25.
1089. Wang C-Y. In: *Recent Research on Unsteady Boundary Layers.* IUTAM Symposium (1971), Quebec, 1653. Laval University Press, Quebec 1972.
1090. Wang C-Y. J Acoust Soc Am **71** (1982), 580.
1091. Wang C-Y. J Acoust Soc Am **75** (1984), 108.
1092. Wang C-Y. Rev Fluid Mech **23** (1991), 159–177.
1093. Wang C-Y. Eur J Mech B **20** (2001), 651–656.
1094. Wang M, Lele SK and Moin P. J Fluid Mech **318** (1996), 197–218.
1095. Watson J. Proc Roy Soc A**254** (1960a), 562–569.
1096. Watson J. J Fluid Mech **9** (1960b), 371–389.
1097. Watson EJ. Mathematika **42**(1) (1995), 105–126.
1098. Weidman PD. Physics Fluids **9**(5) (1997), 1470–1472.
1099. Weinan E. Acta Math Sin **16**(2) (2000), 207–218.
1100. Weyl H. Ann Math **43** (1942), 381–407.
1101. Wilcox CH. *Scattering theory for the d'Alembert equation in exterior domain.* Springer, Berlin 1975.
1102. Williams JC III. Ann Rev Fluid Mech **9** (1977), 113–144.
1103. Williams RF. *The structure of Lorenz attractors.* Turbulence Seminar, Univ of California, 94–112, Berkeley, 1976–77.
1104. Wilson S.K. and Thess A. Phys Fluids **9**(8) (1997), 2455–2457.
1105. Wong H, Rumschitzki D and Maldarelli C. J Fluid Mech **356** (1998), 93–124.
1106. Wood WW. J Fluid Mech **2** (1957), 77–87.
1107. Wu JZ. Phys Fluids **7** (1995), 2375.
1108. Wu JZ and Wu JM. Advances Appl Mech **32** (1996), 119–275.
1109. Wu JZ, Zhou Ye and Fan M. Phys Fluids **11**(2) (1999), 503–505.
1110. Xin ZP. J Partial Diff Equations **11** (1998), 97–124.
1111. Yan B and Riley N. J Fluid Mech **316** (1996), 241–257.

1112. Yih C.-S. Phys Fluids **6**(3) (1963), 321–330.
1113. Yih C.-S. ARMA **46** (1972), 218–240, and 47, 288–300.
1114. Yudovich VI. J Appl Math Mech **30** (1966), 1193–1199.
1115. Zajaczkowski WM. SIAM J Math Anal **25**(1) (1994), 1–84.
1116. Zakharenkov MN. Comput Maths Maths Phys **38**(5) (1998), 810–823.
1117. Zandbergen PJ and Dijkstra D. Ann Rev Fluid Mech **19** (1987), 465–491.
1118. Zank GP and Matthaeus WH. Phys Rev Lett **64** (1990), 1243.
1119. Zank GP and Matthaeus WH. Phys Fluids A**3** (1991), 69–82.
1120. ZankGP and Matthaeus WH. Phys Fluids A**5** (1992), 357.
1121. Zeeman EC. In: *New directions in applied mathematics,* Hilton PJ and Young GS (eds), 109–153. Springer, New York 1982.
1122. Zeytounian RKh. Contribution to the study of the three-dimensional incompressible laminar boundary layer in the nonstationary regime. NASA Technical Translation (Note Technique number **131**, ONERA, 1968), NASA TTF- **12.250**, 1968a,
1123. Zeytounian RKh. Journal de Mécanique **7**(3) (1968b), 231–247.
1124. Zeytounian RKh. Quelques aspects de la théorie des couches limites laminaires compressibles instationnaires. Note Technique number **162**, ONERA, 1970a.
1125. Zeytounian RKh. Quelques aspects de l'écoulement de couche limite laminaire compressible instationnaire au voisinage d'une plaque plane. Recherche Aérospatiale 1, 16–21, (1970b).
1126. Zeytounian RKh. Arch Mech Stosow **26**(3) (1974), 499–509.
1127. Zeytounian RKh. Zh Vychisl Mat Mat Fiz **17**(1) (1977), 175–182, and **17**(5) 1256–1266, (Russian).
1128. Zeytounian RKh. Sur une forme limite des équations de Navier-Stokes, à grand nombre de Reynolds, au voisinage d'une paroi et de l'instant initial. CR Acad Sci, Paris, **290**, Série A, 567–569, (1980).
1129. Zeytounian RKh. Izv Akad Nauk SSSR, Atmospheric and Oceanic Physics, **18**(6) (1982), 593–601, Russian.
1130. Zeytounian RKh. J de Mécanique Appl et de Physique Tech, (PMTF, Russian ed) **2** (1983), 53–61.
1131. Zeytounian RKh. *Les Modèles Asymptotiques de la Mécanique des Fluides,* Vol I, Springer, Heidelberg, 1986.
1132. Zeytounian RKh. *Les Modèles Asymptotiques de la Mécanique des Fluides,* Vol II, Springer, Heidelberg, 1987.
1133. Zeytounian RKh. Int J Engng Sci **27**(11) (1989a), 1361–1366.
1134. Zeytounian RKh. Arch Mech Stosow **41**(2–3) (1989b), 383–418.
1135. Zeytounian RKh and Mahdjoub A. ZAMP **40** (1989c), 931–939.
1136. Zeytounian RKh. *Asymptotic Modeling of Atmospheric Flows.* Springer, Heidelberg, 1990.
1137. Zeytounian RKh. *Meteorological Fluid Dynamics.* Springer, Heidelberg, 1991a.
1138. Zeytounian RKh. *Mécanique des Fluides Fondamentale.* Springer, Heidelberg, 1991b.
1139. Zeytounian RKh. *Modélisation asymptotique en mécanique des fluides newtoniens,* Springer, Heidelberg, 1994.
1140. Zeytounian RKh. Long-Waves on Thin Viscous Liquid Film: Derivation of Model Equations. In: *Asymptotic Modelling in Fluid Mechanics,* Bois P-A et al (eds), 153–162. Springer, Heidelberg, 1995.

1141. Zeytounian RKh. Int J Engng Sci **35**(5) (1997), 455–466.
1142. Zeytounian RKh. Physics-Uspekhi **41**(3) (1998), 241–267.
1143. Zeytounian RKh. Russian Math Surveys **54**(3) (1999), 479–564.
1144. Zeytounian RKh. Intern J of Engineering Sc **38**(17) (2000), 1983–1992.
1145. Zeytounian RKh. Appl Mech Rev **54**(6) (2001), 525–562.
1146. Zeytounian RKh. *Theory and Applications of Nonviscous Fluid Flows.* Springer, Berlin, 2002a.
1147. Zeytounian RKh. *Asymptotic Modelling of Fluid Flow Phenomena.* FMiA **64**. Kluwer, Dordrecht, 2002b.
1148. Zeytounian RKh and Guiraud JP. CR Acd Sci B**290** (1980a), 75–77.
1149. Zeytounian RKh and Guiraud JP. *Evolution d'ondes acoustiques dans une enceinte et concept d'incompressibilité.* Preprint for the AAAF, 7ème Colloque d'Acoustique Aérodynamique, Lyon (Novembre 1980), 1980b.
1150. Zeytounian RKh and Guiraud JP. Asymptotic features of low Mach number flows in aerodynamics and in the atmosphere. *Adv in Comput Methods for Boundary and Interior Layers,* Miller JJH (ed), 95–100. Boole Press, Dublin, 1984.
1151. Zeytounian RKh and Guiraud JP. CR Acad Sci, Paris **300** Série II, No. 20 (1985), 981–984.
1152. Zeytounian RKh and Guiraud JP. Classe I di Scienze Fis Mat e Nat, **LXV** (1988), 103–122.
1153. Zhi-Min Chen and Price WG. Comm Math Phys **179** (1996), 577–597.
1154. Zhouping Xin. Comm Pure Appl Math **51**(3) (1998), 229–240.
1155. Zierep J and Oertel H (eds). *Convective transport and instability phenomena.* Braun-Verlag, Karlsruhe 1982.
1156. Zimmermann G. Reduction to finite dimensions of continuous systems having only a few amplified modes. 181–187. Springer, Heidelberg 1985.

Index

3D BL equations 132, 134

Ackerblom problem 245
Acoustics equations 172, 173, 177
Acoustics streaming 191–196
Adjustment problems 114, 117, 217, 241
Admissible solutions 220–222
Amplitude equation for RB waves 349, 350
Approach to turbulence 389–392
Asymptotic structure at Re $\gg 1$ 103–105

Basic state 184
Basin of attraction 443
Batchelor–Wood formula 246, 247
Bénard thermal problem 337
Bénard–Marangoni problem 54, 71–75, 224, 231, 232, 356
Bénard thermal problem 68–70, 224, 335, 336
Bernoulli pressure 39, 40
Bifurcation from a periodic orbit to an invariant torus 398
Bifurcations 376, 392–394
Bifurcations from Couette flow 376, 377
Bifurcations from Taylor vortex flow 378
Biot number 30, 72
Blasius canonical problem 97–102
Blasius problem with an interface 268
Boundary conditions 28–31, 108, 170, 171
Boundary layer equations 95, 96, 106, 258–260
Boundary layer separation 255

Bounded deformable cavity 215–218
Boussinesq limiting process 229

Cauchy equation of motion 13
Cauchy formula for vorticity 48
Cauchy–Poisson law 1, 15
Chaotic attractors 368, 389, 441, 442
Components d_{ij} 2
Composite approximation 194
Compressible flow at Re $\ll 1$ 153–157
Conditions for temperature 29, 30, 32, 108, 171
Confined hydrodynamic configuration 326, 328
Conservation of energy 3, 17
Couette–Taylor problem 370–376
Coupling effect 366, 367

Deep thermal convection 9, 55
Dimensionless parameters 22–25, 53
Dimensions 402, 413, 414
Discrete cascade system 435–439
Displacement thickness 102
Dissipation function 5, 15
Dissipativity 403
Dominant NSF equations 131–132, 234–235
Double-scale technique 60
Droplet formation 268

Eckhaus method 340
Ekman layer 208, 211, 243, 244
Entropy 401
Equation for RB waves 336, 337
Equation for specific entropy 4
Equation of continuity 12
Equation of Neuman for the total energy 3
Equation of state 2, 4, 5, 19, 35

Ergodicity 401
Euler equations 106
Euler–Prandtl coupling 110–112, 135
Exchange of stability 320, 321
Expansible liquid 22, 52, 224

Feigenbaum formula 403
Feigenbaum number 404
Feigenbaum scenario 403–406
Fictious pressure 38
Flow in bounded cavity 181, 182
Four significant degeneracies 105–107
Fourier equation 39, 175
Fourier model 175–178
Fourier's law 4, 18
Fractal dimension 413
Free-surface deformation 361–368

Galerkin approximation 283, 284
Ginzburg–Landau equation 364
Global attractor 409
Grashof number 6
Guiraud and Zeytounian formula 250
Guiraud–Zeytounian approach to nonlinear stability 324–337

Hagen–Poiseuille flow 270, 271
Hausdorff dimensions 413
Hele–Shaw instability 386
Hopf bifurcation 321, 380, 396, 397
HV atmospheric model equations 237
Hyposonic equations 51, 181, 182

Ideal gas 3, 19, 35
Impulsively started circular cylinder at $Re \ll 1$ 149–152
Incompressibility constraint 37, 43–45
Incompressible limit 165–168, 305, 306
Inertial manifolds 410–411
Initial conditions 5, 26, 108, 109, 169, 170
Initialization problem 171–174
Instability in thin liquid films 361–364, 366–368
Interacting BL 127, 263–266
Interface condition 30
Interfacial flows 268, 368

Intermediate matching equations 114–116
Intermittency 406–408
Internal aerodynamics 187
Inviscid limit 140, 142, 143
Isentropic NS equations 6
Isentropicity 35

KS equation 419, 421
KS–KdV equations 224, 426

L-SSHV atmospheric equations 236
Landau and Stuart equations 330
Landau–Hopf scenario 399
Large Prandtl number flow 272–274
Large Reynolds number flow 89–92, 101, 109–112
Limit processes 110
Lorenz three equations 389
Low Mach number asymptotic expansions 81, 84, 87, 169
Low Mach number compressible flow 183
Low Prandtl number flow 273, 274
Low Reynolds number flow 145–147, 215–218
Lower deck problem 125–126
Lyapunov characteristic exponent 401

Mach number 22
Main deck 121, 123
Marangoni effect 54, 71, 74, 367
Marangoni number 72
Marginal and massive separation 128
Matching 96, 102, 110, 111, 113, 135–137, 174

Navier equations 6, 36, 38, 93, 94, 168
Navier evolutionary equation 48, 49
Navier–Fourier model 168, 179
Navier–Stokes (NS) equations 35, 36
Navier–Stokes–Fourier (NSF) equations 4, 20
Neuman problem 174
Newtonian fluid 1, 15
No-slip condition 29, 31
Nonadiabatic viscous atmospheric flow 233–235
Nonlinear stability 322–324

NS barotropic equation   36
NS equation   2, 19
NSF one-dimensional equations   79

Oberbeck–Boussinesq (OB) equations   6, 54, 231
One-dimensional unsteady gas dynamics equations   106, 107
Orr–Sommerfeld equation   58–60
Oscillatory disturbances   220
Outer flow   192, 201

Parameter Re_S 198
Phase dynamics analysis   350–351
Phase projections   353
Pitchfork bifurcation   395
Poincaré maps   407
Poisseuille flow   57
Poisseuille law   63
Pomeau–Manneville scenario   406–408
Power spectra   404, 432, 434
Prandtl–Batchelor condition   32, 246–247
Prandtl number   6, 19, 22
Prandtl theorem   246
Pressure Poisson equation   41

Quasi-geostrophic model   240

Rate-of-deformation tensor   1
Rayleigh–Bénard convection   335, 337–339
Rayleigh–Bénard problem   56, 229–231, 335
Rayleigh compressible problem   79–87, 113, 114, 180
Rayleigh–Taylor instability   385
Rayleigh viscous compressible equations   107, 112–114
RB convection patterns   382, 383
RB problem   230–231
RB thermal convection   338–342
Reynolds number   22
Rigorous results   28, 277, 278, 289, 295, 298, 300
Rotating and translating cylinder   203
Rotating disc   75, 77, 78, 158
Route to chaos   398–408

Route to chaos in RB convection   351–356
Ruelle–Takens scenario   399–403

Separation   128–130, 256
Similarity relations   22, 23
Singular incompressible limits   167, 168
Specific entropy   3, 17, 19
Specific internal energy   3
Specifying relation   5
Stability concepts   311–316, 319
Stability curve   376
Static pressure   1
Steady-state streaming effect   194, 196, 199
Stewartson layer   208, 211–213, 215
Stokes postulates   14
Stokes relations   1, 20
Stokes–Oseen flow   147–149
Strange attractor   400, 402, 405, 408, 416, 434
Stream surfaces   271
Stress vector   13
Sufficient stability criterion   316
Sychev proposal   128–130

Taylor number   67
Taylor problem   65, 66
Taylor vortices   372
Temperature   3
Temporal limit   143
Thermodynamic pressure   3, 17
Thermodynamic restrictions   2
Thermodynamics laws   17
THV atmospheric model equations   238, 239
Tollmien–Schlichting waves   334, 335
Topological fluid dynamics   266, 267
Trace $[(\mathbf{D}(\boldsymbol{u}))^2]$ 5
Trailing edge (asymptotic structure)   92
Transport theorem   11
Triple-deck concept   118–120, 124, 260–262

Unconditionally stability   319
Unconfined perturbations   327, 331, 335
Uniqueness   221, 222

Unpredictability 443
Unsteady-state adjustment 27, 28, 114–116, 118, 215, 217, 241
Unsteady-state at $Re \ll 1$ 148
Unsteady-state boundary layers 137, 138, 253–255
Upper deck 124

Viscous Boussinesq equations 184–187
Viscous–inviscid interaction problem 262–265

Vortex surfaces 271
Vorticity 39
Vorticity transport equation 40, 42

Wake 218, 219
Weak and strong solutions 278–282
Weak compressibility 178, 179
Weakly nonlinear stability 312–315
Weber number 72
Well-posed problem 46
What is turbulence 411–414